Briana
Andromeda
Purcell

Galaxy Girl's
Inquisitive Pioneer Adventures

The book of At-Home Basic-Materials Science Activities solving with a Slide Rule

This book is dedicated to my Cosmic Binary, Linda

Briana Andromeda Purcell

Introduction

Introductions to myself and the Galaxy Girl's Inquisitive Pioneer Adventures :

Let me introduce myself – I am Briana Andromeda Purcell – aka Galaxy Girl aka The Inquisitive Pioneer. I am an educator who has been in education for quite some time and has taught from the 6th grade to the 1st year of college and in both the sciences and maths. My main area of interests are in the physical sciences – such as Physics, Astronomy, Chemistry, along with Geology, Earth Science, Physical Science, and Meteorology. I describe myself as a Teacher, Writer, Artist, and Scientist in terms of what I do, how I think – feel – and live. Presently I am a college instructor of astronomy and physics in addition to being an author of this series of Science & Math Activity books.

My goal in this adventure in science and math is interesting imaginative and fun. Imagine you look about the house and go to some stores, dollar store variety and some hardware stores and have mostly basic items on hand – the question is this posed and answered here in my book – can I explore the basic ideas in science and math, which in reality are rules of thumb, with these basic materials? The answer is a resounding Yes! Use a super ball, a measuring tape and a stopwatch to determine the acceleration due to gravity for the Earth, explore topics of density, friction, throw darts to determine pi, examine Archimedes Principle using marbles – measuring cups – water – and a spring weight scale, find the number of calories in a peanut, make an electric motor out of paper clips – some wire, magnets and a battery, measure the speed of light with a microwave, a ruler and a slice of cheese or better still use insulated cups, water, a measuring cup, a thermometer, and a watch to measure the energy given off by the Sun itself. And I will accompany you here on this journey.

Galaxy Girl's Inquisitive Pioneer Adventures book series are for an inquisitive and thinking person wanting to take a personal journey into fundamental science principles through investigation. The primary person it is for is one who is inquisitive, one who not only likes science and math but prefers that science is explored in a hands-on method using basic, even everyday materials. This science-minded person views the world from the scientific method approach. In life, there is a phenomenon or question, which sparks one's curiosity to do some research and reading. This then leads to posing a measurable hypothesis. From there the science-minded person gathers materials that can readily be at hand or are easy to obtain for use and be used to reach the goals set forth in the hypothesis. Experiments are ran, measurements taken, and numbers analyzed via mathematics to reach conclusions. All along it is the mind of the student of life that assembles these thoughts, these ideas, these numbers measured and mentally constructs a model of what is.

Science itself is the story of what is and involves us – so in reality we are looking for where we come from, where we were, what things are, and where we are going when it comes to the bigger questions in the realm of science, and this story is the fabric of our curiosity woven through the whole of the universe and ourselves. We as scientists, uncover the story of science through qualitative descriptions utilizing science nomenclature and the quantitative depictions from the language of mathematics.

Science in this book looks at the story of what is when it comes to basic fundamental ideas that are part of this weave of cosmic fabric that are the foundation of understanding – such as the laws of conservation (energy, momentum, angular momentum and others), Newton's Laws, Archimedes Principle, among many topics.

We as teachers, parents, and even as students undertaking the journey of science need to emphasize it is our mind's ability to take in what is, organize the information, and decipher the story of the fabric of the whole of the cosmos from the subatomic to the galactic. Every journey in life begins in the mind. Here we each need to be curious, open, creative, and retain that spark of child-like wonder and joy in all things.

In essence, one can say that the best way science is learned is by doing. That is part of the goal of Galaxy Girl's Inquisitive Pioneer Adventures – science is not just to be read, seen in pictures, but explored through actions of doing the experiment, taking measurements and reflecting on results. Mentally creating the pictures of ideas in action.

Also in many of the Activities there are historical asides to illustrate the changing ideas and conclusions reached through time and the famous people behind such ideas (some tales of Galileo, Newton, Archimedes and the like are here). It is useful to have good historical stories of the story of science and the history of those who endeavored before us and see what is, but it is also useful to plant seeds to allow a platform to speak from, a place to explore and experiment and refine our knowledge as necessary. We make their story our own story of exploration and add to the ongoing adventure.

The Science activities in this book are called hands-on, but many books say this. Here the science explorations go from observing to actually doing the science. All activities have as their core the need to have controlled experiments with measurements and the analysis of the data. This is the heart of true science. Observational and qualitative science are the beginning positions for many things, but this book goes the extra step to all things measured, all measurements used for reasoning and reaching conclusions. We become the researcher, the experimenter, and the mathematician. Beyond this however is the creative mind to see and use materials to find the science in them and test these ideas.

The philosophy of the book in these Science activities is to encourage an independent thinker and learner. Find the pace you want from the activities. Many activities have a prelude that gives some of the historical and/or mathematical foundations of the exploration. When encountering new ideas, read and reflect on them as well as doing further research. When approaching the activity – read it through and envision it and what should happen first. In setting it up, test it and watch what happens even before measurements. Be willing to redo experiments as needed. I have had to redo many of these experiments and try different materials when and where needed that work better. With measurements, watch the data, its trends, make mental predictions, estimate the answers sought and the conclusions your experiment is pointing to. Develop a mental picture of the idea, how it works. Your imagination is a tool to understanding and connecting all of the puzzle pieces to make the picture.

There are many activities that make math a part of everyday life and the science activities use basic mostly everyday items so that a student of science, which is really a student of life – we are all born curious and inquisitive and we see that science is not just in a book or a given classroom but is found everywhere.

The materials in the lists are mostly low in cost – they can mostly be found at home right from the start – such as rulers, measuring tapes, tape, measuring cups, string, rubber bands, paper clips and others perhaps. If not, then they are readily found in inexpensive resource places such as dollar and dollar & more sort of stores. In some cases there are a few recommended items that can readily be found online with Amazon and the like, and are often low in cost as well, such as a plastic graduated cylinder or a good lab-grade thermometer. One of the goals of this book was to keep most items low in cost and readily reusable for another experiment in most cases. When we see basic materials as tools, it encourages our imaginations to consider other ideas and other questions. These basic materials are part of the philosophy of learning explored here – where common items become the templates to build ideas upon in our minds, which in turn not only help understanding but are then easily adapted to new situations as needed. It then allows us to explore other frontiers.

Think, too, to the history of science and math – most of it was forged not only in times of only basic materials but more importantly in the minds of those who gleaned it from nature itself by creating analogies between basic materials used in the lab to explore ideas and phenomena on all scales right outside the lab doors. Here too we as readers of this book and the doers of the science become the inquisitive pioneers like the scientists and mathematicians of yesteryear.

There is a basic reasoning to this book. Besides making science and math easy to do, it brings it to the everyday and brings it home. It allows science to jump off the page of a text book, allows it to leave the classroom, and takes it home and has it spring forth from the mind of the scientifically-minded student. The use of basic materials is to help to create tangible models to think about, recall and learn from in order to grasp key ideas in science. This is not only true for the use of marbles and such to act as masses in one activity or as objects in motion in another activity – these marbles can then be reasoned to be any other needed item of a different scale when needed. In one Activity, for example, you use a super ball, an item we almost all have had at one time or another and in the drawers left in childhood – here it is re-examined and seen as a tool to help us find in one case its Elasticity, in another situation we calculate the Earth's Acceleration due to Gravity with it, with a measuring tape and a timer only. In another lab we use a slice of cheese, a ruler and a microwave and find the Speed of Light itself! There is even a lab where you use insulated cups, water, a measuring cup, a thermometer, a watch and not much else and determine how much energy the Sun is giving off! Imagine that!

This reasoning of simplicity also to applies to the use the slide rule as the math tool to prompt the mind to take in the data, the numbers, and use them in the relations between the variables as found in the equations and uncover the relations that are indeed found in nature.

The following are topics we have touched upon and here are concluded from the introduction with some description from this book and the whole of the series, Galaxy Girl's Inquisitive Pioneer Adventures. The basic goals are these that are summarized :

1) Science Activities are essential in helping to build conceptual models of science concepts and to foster a growth in Scientific Literacy
2) Science Activities allow for the full expression of the Scientific Method
3) The Scientific Method's core is Data Analysis stemming from Data Measurement
4) Science Activities with basic materials are the best choice for building models of the science concepts under investigation
5) Ordinary Materials brings science and math to everyone – allows the imagination to soar since the Ordinary becomes Extraordinary
6) The Slide Rule is a visual tool that forces the user to employ and sharpen Numeracy skills
7) Between the Slide Rule and the Data Measurement the nearly lost art of Estimation in Numeracy is regained
8) Galaxy Girl's Inquisitive Pioneer Adventure philosophy is to move from the Conceptual, Descriptive, Qualitative to the Mathematical, Relational, Quantitative
9) To be the Inquisitive Pioneer, one must Resolve, Solve, and Evolve! :)

A good and personal definition of Science is as follows : Science is the ongoing investigative story of the universe as told by human translation of the descriptions, the interconnections, and the relations of all things in the Universe. We discover this story and make it our own by being involved in science directly as here in Galaxy Girl's Inquisitive Pioneer Adventures.

Why then, the Slide Rule as a tool in this book ?

First let me say that any and all tools when it comes to performing the calculations are recommended – such as a scientific multi-function calculator and in some of Activities since graphing is oftentimes recommended, using a spreadsheet is a great tool too. But where do the numbers begin once we have measured them? All in our minds, just like visualizing the key ideas in science using simple materials to do so, I am recommending the use of a slide rule in calculations if you wish to do so and the following is the rationale for it.

This opening chapter is my original essay examining the idea of using a slide rule in the classroom of today (see ch. I) – despite all of the available technology. The primary conclusions and goals of the book in using an antique item are quite simple :

1) The Slide Rule though seen by most as an antique is still a valuable tool today in math and science education because it is a tangible tool that strengthens math skills by acting as a bridge between reality and the math that measures it. It enables one to think about and visualize problems; forcing the user to come up with an estimated answer, and the slide rule, itself, yields a reasonable answer that is acceptable in a real-world context.
2) To present a set of basic Science and Math Activities that range in time from middle school to high school using very basic materials, utilizing a recipe of the scientific method as a procedural method to ask a measurable question, use basic materials, engage in a logical procedure, take measurements, and use these measures to do one or more necessary calculations to reach scientifically valid conclusions.
3) Present a brief essay on slide rule history takes a snapshot view of the last 3 centuries (1600s-1900s) where the inclusion of the Slide Rule was a necessary catalyst in the realm of the two major Industrial Revolution periods that affected all nations on Earth and ushered in new sciences, mathematics, innovations and inventions, along with scales of production.

Interestingly I love to use my slide rule, the Pickett B1, from about the time I was born, in the 1960s and is a classic piece and a cool math tool I employ regularly. In fact one of my other nicknames other than Galaxy Girl is the B1 Girl too.

The slide rule, (templates to make one are in the book) the tangible bridge for the student and here the Inquisitive Pioneer, can connect all the core classes and is connecting math to the world of science, carries with it history, famous people and places, and promotes active math engagement by the Inquisitive Pioneer or student of life and science. This is used in conjunction with the rationale of the opening of the book where science is tangible and can readily be extended from basic everyday materials which promotes models in the mind to understand concepts and which are explored through carefully crafted experiments by being the Inquisitive Pioneer!

Our journey begins in our mind with the basic tools and even toys we experienced as children and now can allow us to explore the realms of science and math through many ideas, basic foundations and concepts that encompass us all and apply to the Universe as a whole. Here we experience and explore the realm of science which is the story of the Universe and us. Here we become the Inquisitive Pioneer! This Galaxy Girl is certainly one.

A side note too about all these wonderful adventures we are to embark upon - With any and all Activities read through, then set up the process for the Activity and run it without measurement and watch where you may have bumps in the road so as to anticipate these. Be patient – Repeat processes. Once all is well when measuring see to the values – do you notice patterns or trends – what do you predict will happen next as you change one thing or another from one trial in an experiment to the next. I have had to redo and find new solutions to many obstacles here and everywhere and that is part of the adventure. Always Explore & Enjoy. Be Brilliant!

Also let your creativity flow too – Science and Math are Creative endeavors – be sure to read, research, color, draw, paint, and represent your experiences with these new adventures in different ways. Make for yourself a Badge for each Activity accomplished – to become the Inquisitive Pioneer. Galaxy Girl, Ms Briana Andromeda Purcell here has her own collection of artistic emblems of these adventures and the badge is really cool too and have that as well. We use our imagination to create the Universe in all ways in our selves and then express it in our art and communication.

Wow! All things have a story to tell us – we have to be creative in thought, curious in nature, and passionate about our need to explore to learn and be Brilliant!
So come along on a scientific adventure with Galaxy Girl and become the Inquisitive Pioneer – make learning fun and adaptive – think creatively, be curious and learn about different ideas. Find my book and many other ideas plus some of these Activities and such at : http://galaxygirl-brianaandromedapurcell.com/index.html. Some of my mantras are

Resolve, Solve, Evolve!
Explore and Enjoy! –
Be Brilliant – I am

Briana Andromeda Purcell
The one and only Galaxy Girl

Briana Andromeda Purcell
Galaxy Girl
Teacher, Writer, Artist, Scientist

Table of Contents

Important Note of Responsibility :

In the case of all of these Activities the following is to be adhered to – All children and students – you need the support, permission, and help from parents, guardians, and/or teachers to do these Activities. Parents and children alike – read ahead through the whole of the Activity so as to anticipate where there may be areas of concern and important levels of awareness. Always employ safe practices when using any items, such as wearing goggles. Be intelligent, mature, and cautious in the use of items – these are tools and you are using them for specific ends in order to learn. Be aware of all concerns – such as the use of hot or cold items, avoid foods that may have allergies associated with them (nuts, et al), sharp objects (knives, scissors, and other edges) – and act appropriately, make smart decisions, and be safe. When done with Activities always put away materials, clean up the area, and then address the calculations portion of the Activity.

Master List of Items

This is a nearly comprehensive list of all the items needed for all of the Activities. Recognize that any one Activity will require only a small handful of these and in most cases you might find alternatives that can act as good substitutions. Most items can be found at Dollar and Dollar and More type stores along with hardware stores. A few items left off the list are either very common or in the other case very exclusive to that particular Activity and will probably only be used there.

Slide Rule	Stopwatch	Paper
Ruler	Marbles	Graph Paper
Meter or Yard Stick(s)	(Slinky) Spring Toy	Pencil / Pen / Markers
Measuring Tape	Desk Lamp	String (kite)
Sewing Measuring Tape	Light Bulbs (various)	Styrofoam Cups
Measuring Cup Set	Penlight Laser	Plastic Cups
Kitchen Mass Scale	Diffraction Grating	Nuts & Bolts (various sizes)
Graduated Cylinder	Ceramic Magnet	Plastic Wrap
Multimeter	Magnet Wire	Aluminum Foil
Fondue Pot	Scissors	Paper Clips
Thermometers (lab quality)	Solar Cell(s)	Styrofoam Peanuts
Travel Coffee Pot	Stack of Books	Food : Sugar, Salt, Nuts
Rubber Bands	Pressure Gauge for Tires	Plastic sandwich bags
Stairway	Super ball	Water
Tension Scale	Wires with alligator clips	Suction Cup with hook
Protractor	Balloon(s)	Skateboard or Roller Skates
Goggles	Tape (regular & duct)	Gallon container
Hole Punch	Iron Nails	Dowel Rods
Hooks	Pop Bottle	Chairs
Use of Indoor Items : Refrigerator, Freezer, Stove, Table, Picture	Outdoor Items : Car or Bike, Tree with leaves, Moon, Sun	Batteries

Ch.I

USING OLD ANTIQUE TOOLS TO SOLVE A MODERN PROBLEM

The Introduction of the Slide Rule to the Classroom

Plus a Return to Basic and Essential Multi-Modal Learning
By Using Everyday Items

Briana Andromeda Purcell

Interestingly our childhood and nature herself answers all the questions explored here in this presentation about what are simple and effective solutions to increase communication so that compassionate logic becomes the standard, so that each of us rises to a good and effective level of Scientific Literacy which is a necessity and empower us to exercise and learn and grow through increased Numeracy (ie basic and essential math skills for everyday life). The solutions are already in our possession I feel.

That statement seems so enormous, larger than an Ocean, larger than a Planet, even larger than a Galaxy – how is this possible Ms Bri?

In our childhood we are born curious, wondering and exploring and seeking answers and the patterns of things about us so that we can understand what they are, how they work, and how to work with them as needed. This curiosity is the basis of our imagination, our hopes, our dreams, our very nature.

We are the children of the stars, that is made of star stuff – the expression of the Cosmos becoming conscious of itself and the fact we are here and we are students of life itself and on a journey. We need to go from ego to soul and develop the understanding that unconditional love is the basis of all true endeavors and that compassionate rationalism as I call it is the best way to think through feeling in life. We need to have understanding and empathy and see all of us as unique but equal and at the same round table of life and the best tools to work with this cosmic conversation is thus : first we all have the freedom of being and the freedom of expression as long as we do not affect others negatively. Second we should employ a simple set of guidelines on any and all relations in life – the 4 C's as I refer to them : Communication – where each of us not only has a voice but we listen and we ask questions since we care and we see each other as equals. This is the fundamental goal of all relations of all sorts and is an absolute need since this is at the heart of this whole paper not just in these introductory comments and considerations but the whole idea of the paper, the story of the need to have Numeracy and Scientific Literacy as essential architectural components of the world table so that we can live, work, grow, and prosper together. This leads to the next C, Commitment – committed to common goals having essentially the same values in one or more fashions – ideals such as respect, truth, hope, honor, integrity, sincerity, humility, honesty, curiosity, and a fundamental belief we are all equal. Next is Compromise – to not see any effort on any of our parts as a sacrifice and expect something in return, but instead to give to the whole since that relation in question here (be it two people, family members, community members, work members, ethnicities, cultures, religions, nation states and such – regardless of the relation in question since the 4 C's applies to any and all relations in the defined capacity of the parties concerned) but not a cost to ourselves but a compromise since this assistance helps the whole team and we all win in our own ways both collectively and individually and there is no expectation of a return on this since success is the reward. And finally the fourth C but just as important as all the other pieces – Compassion – without a love of oneself, of life, of one another and our success and supporting each other then we have nothing at all in the long run of things I feel. All of this is combined and achieved individually through having a passion in life – a purpose in life – and though this paper is about Numeracy and Scientific Literacy at its heart, realize that it is the fact we are all artists – that is a person who interprets our own unique version of ourselves, the universe and our place in it – we are the story tellers, the painters, the dancer in life, we are the creative imaginative thoughts behind all things. Every journey begins in the mind hence the heart – facts are merely points in the space of our mind much life planets and stars in the Cosmos, but it is our pathways we craft through our feelings that connects all these dots into the beautiful image we have of reality and our lives. And remember always our journey begins with our passions and thoughts in our mind and a resolve (that is we should always resolve, solve, and evolve – my annual New Years' if not daily resolution) that is half the journey – the rest is the outer expression of it and how we navigate it.

This Cosmic Piece now explores to the two most essential skills that enable understanding of ourselves, the world at large and all of its components from the atomic to the galactic namely those needed skills of Scientific Literacy and Numeracy – I will argue – in order to allow each of us to remain the curious child we always were, to be students of life, to be able to measure, estimate, calculate, and see the relations of one variable (the Independent Variable) and its effects on the other variable in question (the Dependent Variable).

Here in my argument I present the notion that using common everyday objects, such as String, Marbles, Measuring Cups, Water, Thermometers, Plastic Cups, Skateboards, Rubber Bands, Clocks, Magnets, Batteries, Wires, Hangers, Paper Clips and such and explore any and all the key critical concepts of Science that we need to know as the basis of all of Science and can apply these ideas to larger scale and smaller scale things from Galaxies to Atoms so as to conceptualize using everyday items the models of ideas that are embedded in the nature of nature herself. We turn the Ordinary into the Extraordinary in my awesome Science and Math Activities in Galaxy Girl's Inquisitive Pioneer Adventures.

But there is more to this –

Numeracy along with Scientific Literacy are the two most essential skills needed in everyday life – we cannot farm and cook without measuring, without experimentation, without connecting cause to effect. If I double this amount how does this affect the outcome? Imagine you can and I do make the claim that any and all calculations involve only the numbers 1 to 10 and we can employ Scientific Notation to handle the magnitude of the questions and outcomes at hand. This is all that is needed in all aspects of our daily lives both in the home and in the job or workplace.

In education there is a continual need to look for 'new' ideas to affect or solve problem areas in the access to success for learning. The problem to explore today is rooted in mathematics and moves through science and technology courses as well.

First a side note on the importance of math and science education so as to assert its necessity. Few would argue that education beyond high school today is not required and that the primary access to the employment market is through math and science related fields. In fact one could argue that one's math skills will determine the primary economic layer that one ascends to in society. The greater one's math skills are, often they have greater mastery of related science concepts if needed and between the first and with the second here it grants one the adaptability in the world today.

In such a complex world we need to be Scientifically Literate minimally so as to understand data, see the concepts in action – understand the importance of enacted ideas – to save the bees, the bears, the butterflies, the tigers, the trees, to grasp our impact on each and every biome since we can go everywhere and tap into all of the Earth's resources, to find ways to clean the air, land, and water not only for ourselves but all inhabitants – we are the only ones who have the farthest reach and capacity to be in places and utilize resources unlike all other beings on this world and our actions affect not only ourselves, not only each other, but all life on this planet. We can see that through sound understanding and reasoning and having compassion and emoting empathy and seeking solutions. There is no science, science reasoning, compassionate rationalism in considerations or debates without fundamental Scientific Literacy which necessitates Numeracy skills as well – otherwise we are nothing more than people shouting at each other with senseless opinions and hyperbole. Sound familiar?

But all of it requires fundamentals in science education which also therefore invokes mathematics for proper understanding. It was Galileo and da Vinci as well as Archimedes that Nature has in it these harmonic tendencies, these formulaic approaches to solution and life – it is in the very fabric of reality and there for us to grasp and understand. Less we forget this is the heart of the Renaissance times and Thinkers such as Newton and Kepler as well – To question, to measure, to see the patterns and interpret them so that we collectively uncover the language of math relations present in nature at all levels from the atoms to the galaxies. – To succeed, to survive, to understand ourselves, the world, its beings, its issues, and the cosmos at large it is essential to see science and math as skills to be a part of life in a multitude of ways so that solutions can be rationally discussed and ideas enacted to benefit all life.

The primary problem seems to be the barriers for learners to grasping, conceptualizing, and finally mastering some level of proficiency in various math skills which are found throughout most math courses and are necessities in science classrooms as well.

The first solution area has been simply to have more technology in the classroom. In the 1970s students were introduced to calculators. One of the stated reasons by the manufacturers as a selling point was that it could engage those who were not inclined for math, though no studies were done to reveal this claim. As time went on, there has been smaller yet more powerful calculators and there are now graphing ones that can store more information than a NASA computer system in the 1960s. Despite the technology, the problems persist.

I personally am not opposed to interesting, useful and captivating technology as well as great graphing tools such as programmable calculators in the classroom, but often there is not enough, the technology falls behind often and there are insufficient funds to keep up and upgrade as needed, plus this does not look at the real problems that exist – do the students know how to estimate, understand the application of the formula which in itself is a definition of a concept that is directly found in nature, can the students see the relation of one variable to another instead of reaching for the calculator since it has the answer as they say themselves – we as teachers have even said often 'plug and chug' when it comes to these problems – not asking them to understand the ideas themselves. Note I feel that a calculator should be used as it originally was – to check your answer – to see if your estimation and reasoning are sound and on track.

To explore options we need to look at the problems though. The problem I identify is a growing trend in society's students to have little to no what I call computing skills. These skills are the foundation for all of the mathematical and logical reasoning needed in the fields of math and science as a student progresses through education.

The following details these skills, the rationale as to why they are important, why the calculator may not be able to meet this need, and where looking at a historic device in the form of the slide rule could actually hold the solution to this puzzle.

What are these computing skills?

1) Basic Operation Calculations – These are the basics : adding, subtracting, multiplying, and dividing. – How many teachers can note that when asked a basic calculation task, a student does not know the answer, gets it wrong, or needs a calculator. Even those with some skills seem slower than students of yesteryear. Many teachers can note the number of times that students have difficulty in multiplying two numbers together in class. When asked, the students do not know the answer, where the answer should fall nor have any strategies for solving such a problem.
2) Little to No ability to Estimate an Answer (noted in number 1) but is a skill unto itself. Even the best calculator user is known to make mistakes. How often do we hear of the wrong values punched into machines by workers? We let the machines do our calculations and not us. Calculators, in fact, do not promote estimation as a skill. Simply type in the values and the answer magically appears in the window. Without estimation there is an overreliance on the magic machine. We have shifted the definition of computer of yesteryear, which was when we did the actual computing to that of the machine today. To illustrate how far this has gone, consider today's work environment, where pictures are used on cash registers to make the task easier so little to no math skill is needed for that layer of society's workforce. Thus the place to begin in Estimation with

Basic Math Computation Skills in place. These are found in early Elementary School and extend all the way unto college.

3) Inadequate skills at Simple Formula Manipulation along with Computation- Simple Formulae here are ones of the form A = B * C (or B/C). This is noted not only for Math but Science as well. Note that most ideas conveyed in school involve just such a formula. Take for example the formulae for speed (v = d/t), force (F=m*a), pressure (P=F/A), density (p = m/V), Ohm's Law (V=I*R), Area of a Rectangle or Square (A=L*W), Work (W=F*d) , and even everyday calculations like miles per gallon (mpg=mi/gal), and computed costs C=R*t (cost=rate*time). Why let someone or a machine tell you the answer – why not check your answer versus your estimation of the needed answer?
4) Often unable to realize the effects of one variable in a simple equation on another as it is changed by some factor in a given situation. (direct and inverse relations). To be able to see the world in terms of Independent Variable (the X often) – the one we control and change – and the Dependent Variable (the Y often) – the one that is affected by our actions in this case and we look at the outcomes and relations of the two (noting the Slope and what does this tell us from a simple presentation of the data). This problem is due to not being able to visualize the problem. – How does one understand the application of an idea in nature if we as teachers cannot ask – what then happens if I double the value of this variable – how does it affect the outcome?
5) Poor skills at using Scientific Notation computations (tied to both basic computational skills and estimation skills (here magnitude)). – Consider for a moment how critical this really is – Recognize that all of computational math really involves only the values of 0 through 10 and nothing more since the power of 10 is the needed concept here – which by the way, is the world of the Slide Rule – it has only 1 to 10 and expects you to keep track of the formula, keep track of known basic math tables to estimate outcomes, and understand what powers of 10 are involved in your problem – we become the problem solver and the device is checking our abilities – viola.
6) Poor skills in using fractions and converting back and forth to decimal equivalents. The trouble with fractions has been noted in numerous studies. This is also connected to Converting values from one measurement form to another (metric to English units, time units, and so on), as well as the critical skill of Solving Proportions which appears in many formulae and can be employed with all 3-variable functions noted above and others as well as with Similar Triangles, Geometric Parallax Measurements, and so on.
7) Poor Significant Figures skills. This can also be attributed to the calculator since it provides us with a flood of decimal places, virtually all of which are unnecessary.

The above list is not to say that all students lack these skills. Instead there is a very broad spectrum of students and hence skill levels. However, this is important to note. What we do with this spectrum of students is to create natural segregations. The top percentages, those with the sharpest of these skills often are enrolled in what is

traditionally referred to as the 'college-tract' classes and the others are labeled as 'regular education'. I use these too in my discussion.
Clearly, it is the regular education students who have the most troubles in these areas noted above, and are now being places in the college-tract system since that is the wave of the future.

Plus note the connection of Math and Science in this list – the Cause and Effect relations so that ideas are understood and a part of truth in nature and not an opinion or something not to care about.

Are the problems serious? Certainly. *It is primarily Math and Science which are the main determinants of where a person's occupation and income levels will fall.* Realize the necessity of Numeracy in math in the workplace today at jobs of higher needed skills, since going to college is not a luxury anymore and most if not all degrees necessitate some level of math competency. There are a number of studies that note the need for math in the workplace and for success. Lacking or being weak in these skills will result in many doors closed. Society is reflecting the problem. Today in many fast food places they use pictures instead of numbers. And when was the last time a teen counted back your change to you? Clearly too, the problems noted do exist. All the data proves this (teacher's grades, teacher's anecdotes, national and state test scores, and the like). So what are the solution possibilities?

This is also especially true for Science Literacy as well as Numeracy. All of technology and business today is wrapped and shaped by Science – nearly every single news story of consequence is about Science in one avenue or another – hence there is a need to have a compassionate reasoning or logic involving these ideas – to be able to ask what is going on and why, how do we work with this, change the outcome and such? All of engineering, all of the Health Services, all of our everyday life technology are directly tied to both Scientific Literacy and Numeracy.

To find a solution we need a common thread in all of these elements. That thread is the fact that we have transferred the skill of computation from ourselves to our tools. We rely too much on the calculator and not on ourselves for determining the outcome of a problem. Also recognize that many other solutions have been explored, from repackaged books, motivational exercises in the class, and all sorts of other materials. The following explores and explains my hypothesis for helping to solve the computation problem by using the slide rule as the tool. The slide rule is a logarithmic scale system that can be called a visual analog computer which promotes mental computation and is analogous to the number line which is still in use in school to promote foundation math skills. In fact, the slide rule can be seen at the next step since 2 number lines can be placed adjacently to illustrate addition and subtraction, the basic form of the slide rule is two adjacent logarithmically-spaced number lines for use in multiplication and division.

The first place to explore the reason is found in the goals of math and science. **What is the goal of both Math and Science education?** Is it only having an answer to a math problem or knowing a definition of an idea in science? Obviously not. **In both disciplines *it is the process of solving the problem that is the goal.*** As with all good goals, the process of solving a path to it begins and is worked out in one place – the mind. Calculators do not require us to think about what is being done with the numbers or even why we type them in. How many times has a student asked what do you type in when asked a problem?

There are two aspects of this problem – first the need to increase essential math skills and put them into practice and not see them as a luxury or a non-necessity, since nature itself is a set of relations as is our business world and our personal financial world and such.

Second then how to connect fundamental science concepts to our lives, our knowledge base, our mental picture. Looking at tools in the classroom that never come home with the students and these objects do not have a place in everyday life, though colorful and dynamic and fun and such – makes science and its ideas too compartmentalized – to removed from everyday life. Imagine instead I say to you that super ball in your junk drawer at home, using a flexible measuring tape in a sewing kit, and having a free timer stopwatch on your phone (and who cannot do that these days I wonder) you can explore the notion of Radioactive Decay graphically, the Elasticity of the Super Ball and have a numeric representation for it, and still further compute using Galilean relations of distance and time the Earth's Acceleration due to Gravity – Imagine that. The super ball is now not just a toy from 6 years old, but a link to key scientific concepts we all should master – and by the way – take measurements, analyze the data using basic math skills as found in a slide rule. I have crafted a series of books noted as Galaxy Girl's Inquisitive Pioneer Adventures which looks at such things as this and even allows you to not only learn the slide rule but make your own paper version of one that has the computing power of a regular scientific calculator right in the book to construct – so Math, Science, Art, along with Kinesthetic approach of hands on learning using Everyday Items – with added elements of reading about the ideas and the people such as Galileo behind them. You become the scientist, the mathematician, the imaginative artist, and the explorer.

I have Activities to explore Newton's 2nd Law and the relation of Force – Mass – and Acceleration using skates or a skateboard, marble, string, and the stopwatch once again. There is Pi being found with a dart board and darts in one experiment and another simpler one using lids, a flexible measuring tape, and a making a simple graph. How about those marbles being used rolling through a long plastic or cardboard tube at an incline on a desk and exploring Acceleration on an Incline as did Galileo? Next take that device and explore the notion of Energy and Conservation of Energy (one of several central and key concepts in all of ALL the Sciences) – Gravitational Potential Energy – and Kinetic Energy with measurements of distance and time only in another experiment. Use the same marbles and now some water, a set of plastic cups and a fish scale (ie spring scale) and look at Buoyance and Archimedes Principle. Instead use some cool paperclips, string, nuts or small bag with a handful of marbles and a hanger, a ruler and learn the concept of how a Twin Pan Balance Works, the Concept of Torque, the Concept of Balance of these Torgues, and such. You can use your makeshift Balance to determine unknown masses even – Imagine that! How would you like to use a couple of Paper Clips – a few Refrigerator Magnets – a D sized battery – some

Alligator Clips – some Battery Wire – a piece of Styrofoam block – and viola make a functioning Electric Motor?! Wow – Be Brilliant as I always say. Why not measure the weight of your car or family car with only a Ruler and a Tire Pressure Gage. Wouldn't it be amazing to use a slice of cheese or a chocolate bar (your choice), a ruler, and two toothpicks along with your Microwave and determine the Speed of Light right in your kitchen – Science becomes amazing, almost magical only here the Magic is that we know the Science and hence can do it. How about taking two Styrofoam Cups – a Measuring Cup – a Ruler – a Kitchen Grade Thermometer – a Piece of Cardboard – some Water – and an Ordinary Watch – and with measured observations plugged into a carefully orchestrated set of equations then determine the Energy Given Off By our Sun in One Year striking the Earth! Awesome Sauce!

Let's first look at the value of the SLIDE RULE in each of the Computation Skills areas and other critical ideas to use the slide rule in the classroom as a visual tool for acquiring math skills:

1. The Calculator does not promote the knowledge of Estimation, Basic Operation calculations and the like. In order to even properly use a Slide Rule, one must know : Decimal Points and their placement. They must have a good working knowledge of Value Estimation and have sharp memory of the Basic Times Table to even use the device. Unlike the calculator where you press buttons and answers are flashed back at you (we often give it to the student and want only the answer), the slide rule user must interact with the device to find the answer and then compare it to your knowledge, your expectations. One might say then why use it if I have to approach it with skills in place? In the class, even in the elementary school level students could be presented a simple two-scale slide rule to perform basic calculations involving multiplication and division. The tool is visual – one sees the answer as compared to many of the present-day tools in the classroom such as plastic blocks, sticks and other assorted colorful items. They may help too, but the slide rule has the answer directly on it and only required basic skills to operate it. Also one sees the changes in variables as one reads it (see following arguments concerning 3-variable functions). The device encourages mental math with a direct answer (unlike other manipulatives)– unlike the calculator which merely reaches the answer and flashes it at the user. Though one cannot see the value of the slide rule here, with guided practice from a teacher the visual tool in the student's hands can act as the bridge from the tangible world to the conceptual world.
2. The value of estimation cannot be emphasized enough. One cannot use a slide rule without estimating the answer unlike the calculator which totally erases the process of estimation. When computing the area of a floor, walls and the like for tiling, carpeting, painting, and wall papering we round values to no more than 3 significant figures and can mentally estimate where the answer falls. Even in cooking, we estimate costs, amounts needed and so on. How many of us are comparative shopping these days with Cost per Unit Ounce for example – a

simple Slide Rule – pocket sized one (I carry one whilst at the store) answers the question for me – which is the better deal? When employing the slide rule we have to keep track of the decimal place, we need to have reasonable skills in multiplication through practice so that we can read it and determine the answer, and develop deeper computational skills.

3. A slide rule has the numbers from 1 to 10 only, yet because of scientific notion we can find any and all numbers that exist on it! Imagine showing a student that power. Regardless of the number one thinks of and multiplying or dividing it, one can find both values on their independent of magnitude and discover the answer in one's mind and reinforced by the slipstick, the stick with numbers on it. A simple Rule it is : If the values are Multiplied, then Add the Exponents, if the Values are Divided then Subtract the Denominator (bottom one) Exponent from the Numerator Exponent (top one) and the outcome is always the answer – that is all there is to it. Imagine all of Math is reduced to knowing the standard Times Tables from 1 to 9 and then doing a bit of addition and subtraction for the final answer. Cosmic Wow.
4. The Slide Rule is designed to fully illustrate the world of simple ratios and multiplying factors – both of which come from Simple Formulae (A=B/C). One scale is one variable, the other scale is the other variable and the answer (the third variable) is the relation of the two found easily. This makes visualizing the relationship all the more tangible. With this tool, the variables are there in hand. A calculator does not provide this at all. Even as an aside Teachers can use it to construct tables as needed for data and graphing too ! Plus now students can do the same and examine data tables for relations. Take a given speed problem. Place this speed over the index value of 1 on the slide rule. Now all values on that upper scale are distances while the lower scale represents time – just like the equation itself! This works not just for whole numbers, but any decimal value too. A teacher can ask how long did it take to travel some given distance or how far did one travel in x amount of time. The student reads along the scale, the greater the distance along the scale represents a greater amount of time at a given speed. Notice how this reflects the number line and its use to promote number placement. In fact, like one could use two rulers as number lines and place one over the other and it behaves in a slide rule manner. Instead of multiplying and dividing, the rulers can be used to teach adding and subtracting. This extension then takes Elementary school education to the Middle School and the High School.
5. Since #2, then the Slide Rule is a natural tool for examining the effect of the change of one variable with respect to another very easily. One could look along the scale for changes with respect to other variables that a number might be multiplied for example. With a C1 scale (inverse) one could then look at dividing too! One would have to retype the problem time and again to examine the possibilities on a calculator.
6. The idea of one variable over another is especially true when placing a value in question over '1' and looking along the scale at 9, 8, 7, et al – which are now

90%, 80%, 70%, and the like. This is not only useful for math and science, but also business and everyday life calculations. Other calculations, such as increases in percentage on an item and calculating tax on an item are easily done too.

7. This idea extends to the fact that most formulae can be seen as proportions – even 3 variable ones! Take v=d/t. Once we place d over t on the side rule look over the index '1' to find v. Hence it is v over 1 equals d over t. The same is true for all conversions, say 1 ft = 12 inches. Place 1 over 12 and look along the latter scale – this is now the inches and look above to find the feet or vice versa! All conversion scales are really slide rules. Imagine seeing all answers to every question simultaneously when converting. Place 1 representing inches over 2.54 representing centimeters. The upper scale is any and all inch values while all numbers on the lower scale are centimeters. It can be read back and forth for any value. Try that with a calculator! – Notice here the doorway to all of Science itself – as noted earlier all of Nature herself is a set of relations of one variable to another and we can connect the numbers and their relations to what we read and discuss hence strengthening and deepening our understanding of key ideas and their direct applications to everyday life – especially when we connect these explorations and experiments to everyday life items we see and use commonly – we become the explorers, the thinkers, the planners, the experimenters, the data measurers, the data analysts, and the cosmic story tellers of our whole lives where we take our childhood curiosity and make it central to our lives and we become students of life – be willing to think creatively, imaginatively, and connect ourselves to the universe from the atomic to the galactic in scale through basic calculations, ordinary items, and simple relations – we are the Creative, Imaginative, Scientific, Mathematical, and greatly Empowered People We Can Be.
8. In the realm of fractions, their equivalents, and decimal conversions this idea of the proportion is critical. Take any fraction, say 7/8 and place it on a slide rule. Look along and find all other equivalents to it – 14/16 and the like. On the scale of '8' look for the index '1' that has a value above it (the one to your right) and find 87.5 or 0.875 the decimal equivalent of the fraction! Again notice the visual power of the slide rule. It would reflect how it is written in a text book (with one number over another) and makes a good picture when a teacher is discussing this verbally with the class.
9. Another critical area that is related to fractions is proportions. Virtually all formulae, conversions, and the like are proportions. One can solve a proportion on a slide rule as fast if not faster, plus as accurately as anyone using even a scientific calculator. This is because all one has to do is set up the known ratio in the proportion and then find the other known variable so that the value either above or below it is the answer.
10. One might quickly object and say that a Slide Rule gives typically about 3 figures for an answer normally. Clearly a calculator gives many more numbers. This is the fallacy of the calculator. All Science Teachers know this. How many figures are in the answer depends on how many numbers are in the values in the

question. Typically this is no more than 3! Think about it, Reality itself is often this number! What do I mean? When measuring a room or anything for than matter (7ft 9 in, 688 cm, 14 gal, so many dollars and cents, time on a clock, and the like). We rely on 3 figures at most in our daily lives. This is all that is needed. Realize that it is the process of calculation and not just the answer than matters. Think of how often the students when using calculators want to write all the decimal places. Please someone show me with your standard ruler how to measure that 6th or 7th decimal place sometime. This may be needed when I purchase paint, tiles, or wall paper when working on my home. (all sarcasm intended).

11. Since a person using a Slide Rule must keep track of the decimal place, it becomes easier to then discuss the Powers of 10 Math Rules. How. Go grab two ordinary rulers. When you place the rulers alongside each other so that you can read both scales (either English or metric – I recommend the metric) you already have the means to Add and Subtract. Place a number in line with the '0' of the other ruler. Say, for example, 4. Now looking along the ruler with the '0' go to 7 and look back on the ruler with the '4'. It is on '11'. Obviously this can be read in reverse. But wait. I promised exponents, not simple math. Bear in mind that the rulers allows a simple means to teach slide rule use. What if these numbers were Powers of 10 instead of regular numbers and not added but are multiplied. $10^4 * 10^7$ is 10^{11}. This means that all of the multiplication and division rules for exponents are at your fingertips. Side Note : This makes the Slide Rule the natural extension of the Number Line, which is the idea expressed here.
12. Other added features of the Slide Rule : the Slide Rules require no batteries ! Interestingly, despite the fact that Slide Rules may be short or long, or even circular, they have all the same operating rules and employ the same scales – given names like : A,B,C,D,L,S,T. This means there is little adaptation needed as compared to calculators with different keys and the like.
13. Unlike calculators there are no games on slide rules, such as graphing calculators. It is a focused device. Plus it would be harder to cheat with a slide rule as compared to a shared calculator.
14. A basic 9 scale slide rule can perform virtually any and all calculations that a scientific calculator can. (Persons such as Einstein and von Braun relied on the same type of slide rule). Students can start with a basic instruction of 2 scales in elementary school and move on to more scales in middle school and high school using the same 9 scale form they started with. Notice the lack of training needed, the reduction in costs of materials, and the easing of tension to acquire new skills.
15. In the Elementary and Middle Schools there are two points to be made. First, ask yourself what is used for all students in many if not all cases to teach counting, addition, and subtraction. Simple – the Number Line. All of us have used one. Most of us would argue that it is a very useful and tangible way to learn the process. Guess what? The Slide Rule is a Number Line only in a different scale (logarithmic) and can be used for multiplication and division.

Why then not use a number line here too? Imagine making the transition from Elementary to Middle to High School and having to only go from the number line to the logarithmic number line and learn any and all functions that are necessary to master all of the ideas in math and science.

16. The second point when discussing school and tools used, do we need technology simply make the learning faster and easier? If technology is always the answer why do people still use wood and charcoal burning grills instead of microwaves for cooking? Technology does not necessitate usefulness. Also in Elementary and Middle Schools there is a trend to have physical manipulative tools to help the student grasp the idea. The slide rule is the next logical extension of this idea. Perhaps it can even find its way into those arenas. There are even books on paper ones to teach these ideas. A company, TOPS, even wrote a 3 book series involving math and sciences. It is useful for the high school, but some of the simpler ideas could be used in the middle school environment. One book, Far Out Math, is only about Slide Rules and logarithm use. One may think this is an outdated source. It was written with NASA and published in 2006. It uses the idea of a number line and extends to the world of the slide rule (more on number lines next). It is interesting to note that education provided the number line to all then separated the groups later on and provided a different number line for one group and none for the other. Did the slide rule make the others 'smarter'? By no means. But did it grant access to promoting the use of 'mental math' which is the chief argument presented here. Yes it did. Look at any text of the 1960s in physics. Are the problems that more complex and requiring a slide rule? Rarely. Also have the problems changed so significantly that the slide rule would be displaced? The obvious answer is no – the rules of science and math have not changed.
17. Students can even make their own slide rules when given a template and glue them on wood or plastic pieces. These may become projects that tie the science and math departments to art and shop departments in a school. Realize that a personal tool may have less of a tendency to be forgotten, broken, and the like, plus imagine the excitement and power one feels holding such a tool in their hands and accomplishing mathematical prowess with it. (A list of web sites follows at the end of this discussion).
18. Unlike the calculator, the slide rule was a part of history in terms of effecting science (in the areas of astronomy, geology, chemistry, and physics), math, engineering, and the like for nearly 350 years. In fact it can be said that it is the most powerful device for math per unit volume or mass ever devised before the age of the computer. It was in the hands and was the primary tool that lead to the creation of virtually all mobile and immobile engineering structures known. Recognize a wonderful historical fact – the planet Neptune was found through Calculations first – involving Log Tables and tools such as the Slide Rule then it was sought and found – Imagine that predicting its existence from measurements and then finding it. Also there are many famous names involved with it, such as Isaac Newton, Albert Einstein, James Watt, William Oughtred,

Amandee Mannheim, and others. Also virtually all mobile and immobile forms in the world, such as the Panama Canal, Hoover Dam, the Empire State Building, the Golden Gate Bridge, the F16 fighter, and the Apollo spacecraft all have this tool in common as one that was a part of their design. It could be argued that the Slide Rule's history is connected to these people and items in the hands of mathematicians, scientists, engineers, which resulted in the two major Industrial Revolutions (both the Steam and the Electric) which affected the historical shift in society and economics the world over. (And note those earlier Industrial Revolutions changed the concept of work, affected our relations with each other as well as Nature, and paved the way to the eras of Electronics, Electronic Communication (TV, radio, phones, cellular phones, and many more), the ongoing era of the Computer and the Internet and its impact on us all today. We are the extensions of living history and forever connected to these people, these things and to a simple notion such as Numeracy opportunities found in the Slide Rule and Nature being so relational hence connecting the realm of numbers and relations to our need to be Scientifically Literate.

Still skeptics might argue what has been called the 'antique' aspect of the slide rule as compared to other devices, like the ubiquitous calculator. Let's look at some parallels and ask if technology is always the answer.

First consider the microwave. It came into the home in a similar time as did the calculator. Did it replace the gas or electric stove? No. Despite its compact size, much higher power (and power use I might add), we mostly use it to heat up a cup of cold coffee, pop some popcorn or warm up a sandwich at most. Few, if any actually 'cook' with it. Note all of the cooking shows today and see if the microwave is their central tool. Why the gas stove. Much like the slide rule it is a tool that forces the user to be involved with the process of what is going on. The microwave like the calculator is far away from us.

Yet here I have and can illustrate how one can use it and a slice of cheese to find the Speed of Light and the calculation – measure the distance on the cheese melted spots and multiply by the frequency of the microwave (marked on all of them and readily found online) and viola – the Speed of Light. So a tool we let control ourselves and do not use effectively now has imagination, fun, scientific exploration (see my Activity online at www.galaxygirl-brianaandromedapurcell.com or my book on Amazon has it with many many more wonderful ideas : Galaxy Girl's Inquisitive Pioneer Adventures by Briana Andromeda Purcell).

What of Velcro shoes versus those with laces? If Velcro is so inexpensive and all the rage, then why should we even teach kids to tie their shoes at all? They are antiques. The same could be said of having clocks with faces on them. Why not just hand a digital clock to a child in school and have them tell time. Why do schools still use cardboard and foam rubber face clocks? They too are antiques.

What about measuring tapes? Talk about antique. Why not just have students use the infrared beam measuring systems instead. Oh, by the way, try to measure the length of a table, or its height, or your height for that matter, let alone your waist size with one of those beam devices. I guess the antique measuring tape might just have a use after all. But look to my Activities and the range of uses of them – Rulers and Yard Sticks and one can estimate the height of a tall building or a tall tree.

Another skeptical argument might think that the calculator has been proven to be the champion in the classroom at helping students. First note that most studies are about subjective answers to using the calculator and were sponsored by the manufacturers of the devices. In considering the results of these studies, most, if not all had students operating with a calculator and few if any groups with no calculator use, modest calculator use and excessive calculator use in their groups. What is missing? An alternative to the calculator use! One could say this is the no calculator use group studied. No it is not! If we wish to study the success of acquisition of these skills, we need to compare it to other things that can help acquire these skills too. This will help to determine the strengths, weaknesses, and rates of change for both. This is what is lacking in studies of what works versus of what does not in the classroom. This is where the slide rule is left out of the picture wrongfully.

In fact, a math panel to the US government in 2008 even noted that the calculator has not had sufficient short-term, nor long-term studies on it done, plus the ones on the record books (which are over 20 years old) show no real gains or changes in skill levels. The best that can be said of those reports are that students when first introduced to the new calculator (at that time in the 1970s) they found it exciting and interesting. Well, that is often true of most new things. Oddly, though, in a 2005 anecdotal report in the Oughtred Society publication called the Journal of the Oughtred Society in volume 14 number 1 of 2005 there is not one but two teachers, at two different school levels (middle school and high school) in two different states, who introduced their students to the slide rule and found there interest levels go up considerably and participation sharply rose as well. It is critical to realize from this that this presented hypothesis should be investigated and not overlooked. To turn away from a hypothesis based on no information or some internal prejudice is not science at all.

Finally to close out this argument – the use of the slide rule in the classroom – return to the emphasis of education – the goal being the acquisition of skills to solve problems. In history the problem solvers were the ones who took the slide rule from the originators of the tool, such as William Oughtred the mathematician and later those like James Watt, who was what might be called a entrepreneurial engineer, to shape the tool to achieve ends desired. The slide rule has been always placed in the hands of those in education seeking higher education through 350 years. The Slide Rule as noted promotes all of the critical skills to be a problem solver. One must practice these in order to use it properly. In one's hands is a tangible tool to visualize the relationships between the variables in question. One scale is one variable, the other scale is yet another variable. The Slide Rule is the bridge to becoming the computer and not seeing other things as the problem solver. We have forgotten that we solve the problems and only use tools to do that. This property of variable relation is not found on a common calculator and yet embodied by the slide rule. This is the primary goal as noted at the start of the argument for the slide rule in the classroom.

The final skeptical argument is their lack of existence. Today however, there are a number of websites devoted to the slide rule, its history, its use, and there are printable templates for them along with virtual ones to use online. This list comes at the end of the article. Also some of these groups even have slide rule classroom loaner programs that are in place. If a teacher were to want a collection in their classroom, with some time, a little money, and effort a number of inexpensive ones could be found to create a permanent collection for a classroom. Personally I have two sets, a 9-scale plastic set and even a more advanced aluminum Pickett N500 set.

Am I saying a world without calculators, maybe even computers in the long run? No, exactly the opposite. They should be the tool of last resort once the skills are mastered. This is why I promote the slide rule as the tool of choice to enhance one's skills. Note the main goal of education – to have people acquire problem solving skills and then sharpen them. The calculator promotes speed not acquisition of the critical skills. One must think to acquire the skills. Presently we are allowing the 'mystery box' to answer the question.

What then of Scientific Literacy which is the partner to Numeracy? Clearly we operate only in a world based on Science and is the product of several Industrial turning into Informational Revolutions and these are all interconnected, require understanding and maintenance as well as a deep appreciation and understanding of Nature herself – consider our impact on the World and the movement towards Global Warming and Climate Threats and the possibilities of mass extinctions of key and keystone species which could result in the collapse of the entire ecosystem, make food harder to grow and yield smaller yields as well as mass migrations to find and secure water these days plus the out of control weather at a given moment since our deviation from our past trajectory before all these climatic issues which we contributed to. So there is both the dire need to know the science so as to seek solutions – alternative energy forms such as solar and wind and others, distributing energy systems, better decisions on making life more sustainable and not just stuff we make as well as protecting and promoting species safety, growth, and saving them from certain doom – consider the need to eliminate plastic – this is all science – we may be able to use bamboo, kelp, mushrooms, among other ideas proposed – reducing carbon emissions and such – all needing us to be scientifically literate and having some numeracy skills so that we can all be part of the ideas, the opportunities, and other roads. Plus there are other plus sides to all of this and knowing science well – jobs aplenty, new businesses to flourish, new ideas to explore, new solutions to old problems, and the like. We can have both and we can do both – the path is clear – increased Numeracy skills and deeper learning and knowing of key science ideas in a world that needs a Scientifically Literate populace. Always remember – there is no Planet B. As reminded in terms of paraphrasing of Carl Sagan looking at the small pixel sized dot he notes – see that dot there – that is us, that is home – every one you can name in history in the present is or was there – all lifeforms are there – all of our history is there. We are all interwoven and are connected to the Cosmos itself and are made of star stuff – we are the universe aware of itself consciously. We have no other home, no place to go. We have the choice – either we live in harmony with the world or we end up destroying it and ourselves. This burden is our responsibility since we are everywhere, we control the massive number of resources and affect the systems all about us and we have the ability to look forward and grasp the scale of our actions and/or inactions which may affect the planet and its denizens – only we can do that and we are that notes in a paraphrased manner, plus noting we are not special and we all – meaning all lifeforms here – have rights to be here – as noted by David Attenborough.

As a historical aside note that in 1960 a dictionary said that a 'computer' was a person who did computations. This is not true today in terms of that term in the dictionary and more importantly the kids today do not view it in the 1960s way. They see a world that provides them with the answers and not one where they find and determine the answers themselves but instead it is found in machines. The Slide Rule is the tool that can open that door to personal success and mastery of problem solving skills. It is tangible, it forces one to know the basic times tables and decimal placement, make estimations, and try to visualize the relation of the variables in question. This then is the value of the Slide Rule. It is a bridge and a tangible tool to show relations and necessitates involvement. It empowers the user. Notice how it can be employed with students doing the work and then checking it with the slide rule. This philosophy follows what we are told always : measure twice, cut once. Look at the movie Apollo 13 where a calculation at a critical moment done by Lovell (Hanks character) is verified by three slide-rule bearing engineers. Before considering that only a movie, check out the slide rule history website and find a power point presentation about slide rules which shows Michael Collins of Apollo 11 fame with a slide rule in his pocket and a quote from an engineer of those times who noted that he did not even own nor use a calculator until sometime into the early 1970s and the slide rule was a common tool.

Think on this, pick up a slide rule and try it for yourself. Then decide. Next look to my Basic and Essential Everyday Items Science Activities as found in Galaxy Girl's Inquisitive Pioneer Adventures – note the connection all the way to childhood – the bridge to the essential and the needed knowledge put into everyday language, use, and doing math that empowers the student of life both in the world of mathematics and the realms of science. Imagine the ability to look in a drawer and see items you grew up with and instead of them only being forgotten and passed over items – they are tangible bridges to all the key concepts you need to learn and know. Hold up a super ball and realize you can determine the Acceleration due to Gravity for the Earth – something Galileo would be excited by and approve of. Be on the path to increase your Numeracy and Science Literacy skills – both go hand in hand – expand your boundaries, your imagination, your understanding, your hopes and dreams, and makes the whole of the Cosmos within reach.

Always be the Artist, the Scientist, the Student of Life and this will be the greatest and grandest of journeys ever undertaken. Let your imagination be the fire for passion towards new goals and frontiers, use skills of communication, commitment, compromise, and compassion with compassionate rationalism and tools such as Numeracy and Scientific Literacy to embark on such an adventure. Be Brilliant as I always say and invoke in my journey.

List of websites :

Information :
www.oughtredsociety.org

Virtual Slide Rules :
From Derek's Virtual Slide Rule Gallery –
www.antiquark.com/sliderule/sim/

Information, Virtual Slide Rule, Slide Rule power point presentation on how to use the slide rule, and printable scales for making a slide rule
http://www.sliderulemuseum.com/

My website for Slide Rules, Activities and even Cooking Ideas and Crafts
www.galaxygirl-brianaandromedapurcell.com

Other slide rule plans:

Scientific American magazine reference from May 2006 article on slide rules by Cliff Stoll
http://www.scientificamerican.com/media/pdf/Slide_rule.pdf

Luis Fernandes, Dept of Electrical & Computer Engineering, Ryerson University
http://www.ee.ryeson.ca/ elf/ancient-comp/sliderule.pdf

Circular Slide Rule by Dr. Charles Kankelborg, Dept of Physics, Montana State University
http://solar.physics.montana.edu/kanel/math/csr.html

Galaxy Girl's Inquisitive Pioneer Adventures

Let's go onto some Science and Math Adventures and become Inquisitive Pioneers!

Enjoy and Explore
Be Brilliant
Resolve – Solve – Evolve

Ms Briana Andromeda Purcell

Ch.II
Basic Data Analysis with a Slide Rule

Data Analysis Math Tool Alternative Consideration

We looked strongly in the Introduction and the first chapter at two things : The needs for hands-on science activities and my argument for the use of the slide rule as a tool to analyze math related issues. It is agreed upon today that students need to have connections to the ideas they learn and hands-on activities are the first critical step. In these activities it is worthwhile to have communicative skills developed to therefore have an amount of scientific literacy employed and also hence numeracy skills since numbers in science are involved. The next step then is for them to take their measurements and find a way to connect the numbers to concepts. One of the most important goals of science is to **analyze data** to reach mathematical conclusions and find relations between the variables.

The question then becomes : Is there a different way to examine data? An interesting approach would be one where we are not only acting as the scientists taking measurements, but also as the mathematician analyzing their measurements. **The answer is the tool, the 'stick' with numbers on it.**

What if, we were to use only low-cost basic tools (rulers, meter sticks, string, thermometers, small masses, marbles, a personally constructed incline made of meter sticks, stopwatches, mass scales, etc) for measurable labs. With basic tools the students take measurements themselves and then with the help of the laws of mathematics and through the use of a 'stick' with numbers on it, the students come to discover and find the relations that they can then read about in their texts?

Even in the case of non-measurable labs where we as students are merely supplied, straight-forward data, the students can use the very same mathematical 'stick' and find their relations through some graphing and basic computations.

What 'stick' is this? It is the common slide rule!

Why this tool? The **slide rule** is a tangible and visual bridge connecting numbers to the measured real world. It can be seen as an extension of the use of number lines in their early school journey where they were used for adding and subtracting, only here the slide rule is now used for multiplication and division. The slide rule can also act as a motivation for reasoning and mastery of math.

To use a slide rule, one must first estimate answers mentally, know what and why the measured values used are, sequence the mathematical steps of the problem, and understand their place values through scientific notation of both the variables and the answer. Learning to read the graduations on the slide rule, (along with learning to use a new tool for calculation) is useful in itself. *Hence, the student becomes the measuring scientist and the computing mathematician simultaneously once again, like those long ago who used such tools.*

The most critical present-day problem, then, is to find such a tool. The references at the end of the article note the International Slide Rule Museum web site, where there is a student-loaner program. For the cost of about $11 per semester, a teacher can be loaned a classroom set of slide rules. There is a power point on how to use a slide rule, along with ideas on its history, and a way to have medals for slide rule competitions as well. Also in the references is a web site, Cosmic Quest Thinker, for suggestions for many science and math activities using slide rules. Each of these has further links to other web sites for virtual slide rules, printable slide rules, publications, even places to assemble one's own classroom set of slide rules and the like.

This book, Galaxy Girl's Inquisitive Pioneer Adventures has paper template ones you can make copies of and use for your own slide rule too. Imagine that – this book is not only a summary of basic everyday formulae to use (Changing Recipe numbers, Numbers of Gallons of Paint needed, and the like) as well as the Science Ideas for a range of Science Topics and Concepts central to all of Science, but also has a Calculator in it too that can be used for any and all of the Activities that you make for yourself! Imagine that.

Data Analysis with a Slide Rule :

In the following paragraphs I illustrate ideas of analyzing given data and employing the slide rule and discuss its ease of use, how this can be a benefit to the user to acquire increased Numeracy skills in his or her data analysis.

In any and all lab situations or even tables of data cases, we can see ourselves as what I will call the students and say we take recorded (or given) data and then merely convert the values into log values of these numbers (read the log value on the L scale from the data value on the D scale on a slide rule). Now they proceed to graph a log-log plot of each of the variables, such as :

- log(displacement) vs. log(time) for constant acceleration cases;
- log(period) vs. log(distance) for pendulums or planets (Kepler's 3rd Law);
- log(Force or Intensity) vs. log(distance) for inverse-square laws (such as gravitational or electrostatic forces or light intensity), et al).

In this new log data set on its graph, draw a best fit line through these points, and then find the slope of the line. The slope taken as the ratio of two simplified whole numbers will show the exponential relation between the variables and the exponents involved.

For example, in the case of constant acceleration (dropped objects or masses on inclines), the ratio of displacement to time will have a ratio of 2 to 1, hence $d \sim t^2$. This means that the graph is of the form $y = x^2$, which is, indeed, a parabola.

Once the variable relation is found, the slide rule can then be used to then check it as well as explore the relation. Continuing the above example, graph now displacement vs. time-squared as well as the log values of each and again for each draw a best fit line and determine slope. The former determines acceleration while the latter should demonstrate a slope of 1.
In the case of inverse-square laws (gravitational, electrostatic), the slope on a slide rule has a ratio of -2 to 1 for Force to Distance. The negative slope is a negative exponent, so it can be seen as $F \sim \frac{1}{d^2}$. This inverse-square law idea applies to light intensity as well.

Even a situation as complex as Kepler's 3rd Law can be examined this way and one finds what Kepler found (using logarithms, no less) that the period-squared is proportional to the distance-cubed for a planet ($P^2 \sim D^3$).

Note that each of these and many more calculations can be done with as simple a tool as a common 9-scale slide rule! This very tool is as powerful as a conventional scientific calculator today.

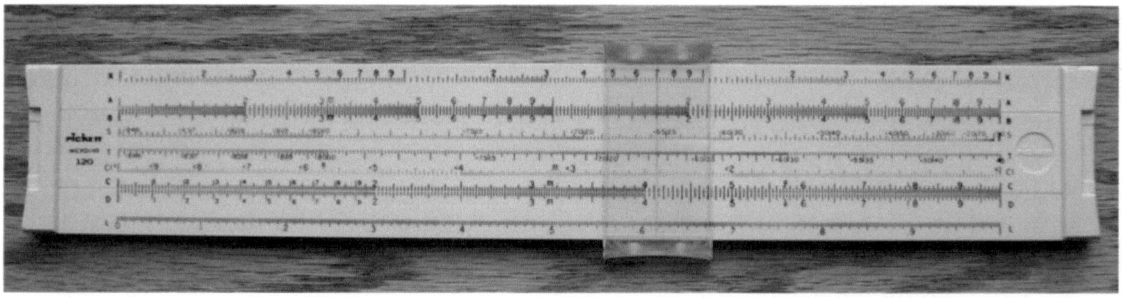

Other mathematical reasons for the slide rule : The slide rule can also be used to illustrate the idea of the *laws of logarithms*, such as the product rule for logs where the log of the product of two values is the sum of the logs of each of the values in question.
(log(A*B) = log(A) + log(B)). Students can compare given values and reach a conclusion here. It is a means to visually conceptualize ideas, such as what happens to variables when one of them changes.

The scales themselves become the variable under consideration.
Take, for example we as students given the density of a pure substance. We then place this value on the C scale opposite the left index of the D scale. Now as they read along the C scale, these values represent mass, the numerator of the equation for density ($\rho=\frac{m}{V}$), while the adjacent D scale is the corresponding volume value for that given amount of mass so as to always end up back at the material's density! Other types of data analysis can be done this way.

Average Speed is similarly done on the slide rule too. Here Distance is the C scale while Time is the D scale. For any determined average speed, as one reads along the C scale, one has driven farther, hence more time (D scale) too. This visual reference where distance and time go hand-in-hand is a concrete visualization that translates into a way of mentally seeing the problem and the solutions. In other cases, Students can be given data here as well to examine, if preferred.

Because *all similar ratios are set up instantaneously,* this same tool can also be used to easily convert fractions into decimals as well as solve any and all proportions even faster than one can on a calculator. Here tables of information can have blanks to be filled in where students can use the slide rule to find the answers. This can be useful for scaling drawings and maps, calculating changes in recipes, determining cost per unit volume or mass, finding unknown sides of similar polygons, and calculating conversions. Most formulae we need to consider to have a good understanding of basic science concepts and applications of math to everyday life have typically no more than 3 variables and if 4 often can be seen as a proportion (even a 3 variable formula can be thought of as a proportion too really). In the book here Galaxy Girl's Inquisitive Pioneer Adventures there are not only a cornucopia of basic science explorations but even activities of practical everyday use – looking at how many gallons of paint are needed for that room? How much is the estimated water bill? What is our average miles per gallon? And so on. Plus all of the science activities. **The applications are limited to one's imagination and mathematical skill.**

Notice how this idea extends to a simple activity connecting ideas in math and science in the understanding of the value of π. You can measure circumferences and diameters of common circular objects and find the ratio on the slide rule. It will show π (3.14) if done correctly and since all similar ratios are set up, for a given diameter (or circumference) they can predict the circumference (or diameter). This application applies to any known ratio.
Other scale explorations of the slide rule allow for examination of squares and square-roots (A & B scales), cubes and cube-roots (K scale), as well as trigonometric relations of sine, tangent, and cosine (S & T scales). In combination these scales are all that is needed for virtually all formula in science and math through school. These can all be done as given tables or through measurements depending on the resources and time.

With this approach, using the slide rule, the goal of having the students do the work and discover the outcome is achieved here. The goal of data analysis is achieved. When they do a lab and take the measurements, they now take the data and find the relations using math reasoning when using the slide rule. The students here engage in the art and act of discovery through actually doing the math. The students come to find the various relations either through measured or as given data tables. Along the way, they connect the numbers to real-world phenomena.

Also a startling notion develops – *all values measured can be represented as a number between 1 and 10,* as does the slide rule and this promotes the use of scientific notation. Image their surprise when they realize they are holding infinity in one's hands! The use of the slide rule is just an alternative and a way to inspire a path to mathematical reasoning and understanding. Also consider that the slide rule has a sufficient level of precision with 2 or 3 significant figures, which is all that is needed. The tool helps in reinforcing this idea.

Does this mean the end of the computer or calculator? No. *In fact, the calculator and the computer can act now as a follow-up to check the answers. Instead of being the source of the answers, they are the checking system for the student's work as a follow up.* .

What of the use of logarithms and the need to explore them? This can be done in the science or in math class, if the students are at that level for understanding. Otherwise, letting them know that logarithms are a tool to uncover such relations may be sufficient at this time. As noted here, this idea can be extended to any and all other variable relations they encounter in various science classes as well as math classes.

This exploration can be a cool math tool adventure. We as Students mentally and mathematically examine data themselves to find the answers. They use tools that help them visualize the concepts and make finding answers a personal responsibility and journey. The slide rule promotes math skills acquisition. As an aside, we as the students can also be introduced to and connected to history through the role of the slide rule. The slide rule was in the hands of numerous scientists, mathematicians, and engineers and used for nearly 350 years (1620-1970). It has a history of being part of the making of the Panama Canal, the Empire State Building, the Golden Gate Bridge, along with development of the steam engine, the discovery of oxygen, and the determination of the density of the Earth. Both Einstein and von Braun used the same 9-scale model themselves – one from the realm of theoretical physics while the other in the practical realm of applied physics to rocket engineering, where he built the Saturn V, the largest human-made device to leave the Earth carrying aloft Apollo astronauts to the Moon, each carrying a Pickett 600 slide rule.

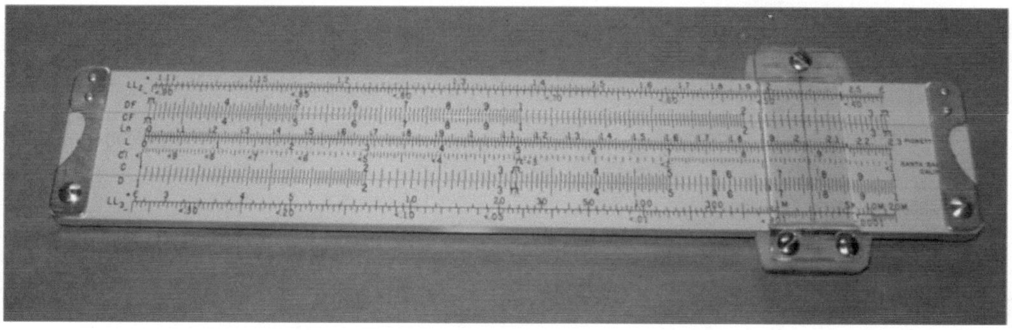

Resources :

Slide Rule Loaner Program, Directions for Slide Rule Use, Make your own slide rule :
http://sliderulemuseum.com
Many Classroom Ideas for Slide Rule use :
www.cosmicquestthinker.com

Ch.III
Brief overview of the history of the Slide Rule

Slide Rule History

There were several key historical developments that directly or indirectly fed into the creation of the **Slide Rule**. Europe of the Middle Ages (1400 on) grew in many ways but what affected the Slide Rule were in the areas of the mind, in particular, many new ideas in **math** with the expansion of algebra, geometry and the birth of calculus and logarithms, a growing empirical, objective and **scientific approach** to the world, an increase in speed, volume and efficiency in **communication** with the development of the printing press to allow the flow of new ideas from one place to another, and **transportation**, particularly shipping at first as well as later on in the first Industrial Revolution items such as the steam engine.

All of the aforementioned paved the way to the second Industrial Revolution which was primarily centered on electricity, wireless communication (radio and television), construction of massive objects, as well as a faster growing numerical characterization of the universe (stars, galaxies). Each of these named in terms of their calculations were connected to logarithms and the primary tool to stem from them, the Slide Rule!

Scientifically-minded people, such as **Galileo**, were the primary catalyst to examine the world in objective empiricism and connect the natural world to the realm of mathematical expression. He is renowned for advancing the scientific method, the basis of kinematics, the foundations of observational astronomy, as well as inventing items. One of his items of interest was the **Sector**. He did not invent it, but improved upon its design which included graduations on movable calipers that could be used for computations involving reciprocals, squares of numbers, and trigonometric formulae. The Sector was used in navigation for the next 300 years.

The foundation of the Slide Rule comes in the early 1600s with **John Napier** (1550-1617), who invented the idea of logarithms which he published in **1614** in his book, <u>A Description of the Marvelous Rule of Logarithms</u>. His idea is not the same base as it is commonly used today (typically 10 or *e*) but the idea is the same basis where the value of the logarithm is the exponent of a base number that when the base is raised to this value, it results in the number in question who log is sought.

$$\mathbf{Log_{Base}(\ Number\ ) = Exponent}$$

$$\mathbf{Base^{(Exponent)} = Number}$$

At first, it seems that complicating numbers to have more calculations to have these results, but it turns out that the properties of logarithms, primarily multiplication, division, raising powers, and taking roots, is actually easier with logarithms. For example :

$$\mathbf{Log(\ A * B\ ) = Log(A) + Log(B)}$$
$$\mathbf{Log(\ A / B\ ) = Log(A) - Log(B)}$$
$$\mathbf{Log(\ A^N\ ) = N * Log(A)}$$

Notice that most math expressions, despite complexity, can become nothing more than simple addition, subtraction, or straightforward multiplication. Imagine that! How this can change the world – from the time of Napier and Oughtred and moving forward the invention of Calculus bu Newton and the former two made the concept of Logarithms hence using powers Science and Math could and did move forward rapidly. In the next 200 years past the publication of Log Tables their were experiments to find Newton's Universal Gravitational Constant which then leads to the mass of the Earth, the mass of the Moon, and then the Mass of the Sun even. How Brilliant – imagine the scale of these numbers. Not to mention in that time period Avogadro develops Avogadros constant and many other calculations involving new numbers of enormous size and scope and scale. Also it turns out a log-log plot of two variables that one might be considering to have a relation to each other, will have a slope that is a ratio of the powers of the variables in question. With a Log Table, a whole series of multiplications can be done quickly with only addition, and this is not only faster, but can have less error. Tables are created by **Henry Briggs** in **1617**, who employs the use of base 10 in the work : Logarithmorum chilias prima. These come to be called **Briggsian logarithms** and are move within twenty years into Europe and are used the world over for many years. Briggs even gives us the terms describing the components of the logarithm such as 'characteristic' (the values before the decimal point) and 'mantissa' (the values past the decimal point).

Logarithms are the primary basis of math (along with Calculus) to advance the realms of math and science from this time forward and their value cannot be overstated. Logarithms are employed directly and through their visual tool, the Slide Rule to be a part of the process of understanding in science areas such as astronomy, chemistry, physics, as well as the direct application of ideas in navigation, construction engineering for virtually all major immobile and mobile items of the last 3 centuries, and even in the world of business and finance.

The first scientific use of logarithms is by **Johannes Kepler** who employs them (as well as proving them as being mathematically valid in other works) in his discovery of his 3rd Law of Planetary Motion, relating the period of a planet's revolution with is distance from the Sun. This rule is still employed today for bodies orbiting each other, such as stars and now planets found to be orbiting distant stars as well.

The first physical manifestation of logarithms as a line of numbers appears on what comes to be called the **Gunter Scale** by **Edmund Gunter** (1581-1626) in **1620**. It is like the Sector, but only a single board about 2 feet in length with various lines inscribed on the wood, including lines of numbers with logarithmic spacing. Using dividers (aka calipers) the spacing between the numbers can be spanned so that addition and subtraction can be done, which is really multiplication and division. This is not a Slide Rule, but only one line of what would become a Slide Rule.

Around **1622** (as best as it can be determined), **William Oughtred**, a mathematician, aligns two logarithmically-spaced lines so that they can move with respect to each other, hence has what is called the first Slide Rule. Not only does he develop the linear model of the Slide Rule, but also the circular form, he referred to as his Circles of Proportion.

An important change comes from **Seth Partridge** in **1657** where the tool now is given a fixed stock body with a slider, similar to the form the linear Slide Rule is in today.

Even the famous scientist-mathematician, **Sir Isaac Newton** played a part in the history of the Slide Rule. It is noted in his works that while he was working with cubic equation solutions in **1675** he made use of three parallel logarithmic lines and suggested the use of a hairline, what we call a cursor today. This idea was 'rediscovered' a couple of more times through time before it became a regular feature in the later 1800s.

Though more than 50 years since the invention of the Slide Rule, very few people have them, and most are mathematical curiosities. At this point, though the Slide Rule makes a transition into the public sphere primarily through Henry Coggeshall and Thomas Everard.

Henry Coggeshall in **1677** develops specialized Slide Rules in the form of the timber and carpenter's rule. In **1683, Thomas Everard** uses a specialized one known as a gauging rule that is used to determine the tax on barrels of wines and spirits (which was actually designed by William Oughtred). This latter form was used long into the 1800s. He is also the first to name the object what we call it today, the **Slide Rule!** In **1722 John Warner** adds scales for squares and cubes so that these can be determined but so too can square roots and cube roots.

At this point, two things are needed : a more accurate Slide Rule and most importantly communication of its effectiveness and usefulness. This is accomplished by **James Watt** and **James Boulton**, of Industrial Revolution improved steam engine design and development which was assisted by the Slide Rule. These two create what came to be called the **Soho Slide Rule** in **1790**. With regards to the Slide Rule they sought to promote its use by scientists, engineers, and mathematicians. What could be called Technical Journals of the time, such as William Nicholson's "Nicholson's Journal" actually had what would be called Slide Rule ads today.

Innovations to the Slide Rule and accolades for it occur at this time. In Nicholson's Journal, the creation of the folded scale, what is now CF, is noted. In **1797, Henry Wollaston** puts in an inverted scale, today known as C1 (which was reintroduced after 1900), and in **1815 Peter Roget** crafts the idea of a log log slide rule, which would not be used for more than 50 years later and then carry on to the 1900s as the central math tool for higher-end engineering. **Joseph Priestley**, the discoverer of oxygen notes in his work that he used a slide rule in his calculations.

Greater standardization comes from **Amendee Mannheim** in **1851** who arranges 4 scales, two double scales, A & B, and two single scales, C & D in a manner that still carries his name for this style, Mannheim arrangement. He also includes a cursor which becomes a more regular feature. Mannheim was in the military where the Slide Rule could be used. The Slide Rule was used a great deal by many nations in both World War I and II and is used in navigation and artillery range determination.

In **1891, William Cox** patents what is known as the duplex Slide Rule model with scales on both sides of the tool and are able to be read in conjunction with each other from the cursor placement since all of the scales were based on the C & D scale placement. This patent is held by K&E for some 40 years.

Slide Rules were primarily wood bodies with marks engraved directly on them. In 1886 Dennert and Pape and Faber take wood bodies and place laminated celluloid marked scales on them, which becomes the standard physical form for most of the next century and is one of the most common Slide Rule. In the early 1900s, the Sun Hemmi company is formed and chooses another material for Slide Rules, namely bamboo. It becomes the most common material for Slide Rules in the Pacific Rim arena of Slide Rule production and has great presence in American and European markets as well. In the mid 1900s, with the development of plastic, it became a cost-effective material to mass produce Slide Rules from, especially by the 1960s. Also from the mid 1900s, aluminum becomes a cost-effective tool and is greatly used by Pickett for its Slide Rules.

Though the linear form of the slide rule is the dominant type, there are varieties of them, such as pocket-sized (5-6 inches), regular (10 inches), and 20 inch models. They could also be one-sided (box-type) or duplex. The number of scales could range from 2 to over 30! Also in the family of slide rules are Circular Slide Rules, which can be metal with glass faces, such as the Fowler line, plastic, or metal. They too can be single-sided or double-sided as well as varying in size from a 1 inch disk to over 4 inches in diameter. Other varieties include the helical-scale design such as on the Otis King or the Fuller Calculator, where scales can be measured in feet.

The varieties of types, sizes, and materials with the large number of international companies involved in their production (Nestler, Faber-Castell, K&E, Pickett, Sun Hemmi, Fowler, to name but a few) plus their changes and filed patents for their form, (there were only 40 models noted in the span of 1625 to 1800, yet 1800 to 1899 saw the introduction of 250 more types) illustrate their importance and wide-spread use through time and history.
Also included is how the Slide Rule was developed in Japan in the first place. In the late 1800s, 1894 Japanese ambassador **T. Kondoue** and industrialist/scientist **Dr. R. Hirota** toured Europe during the 2nd Industrial Revolution and one of the key ideas they returned to Japan with and developed was the Slide Rule. It comes to be the largest Slide Rule company in the Pacific, that is Sun Hemmi, and has sales over 1 million in the later days of the Slide Rule in the 1960s. It is **Jiro Hemmi** who uses bamboo as the core material for these Slide Rules. Companies like Sun Hemmi and Pickett played strong roles in placing Slide Rules in school with demonstration models and promotions for such things. Sun Hemmi also pushed for making the use of the Slide Rule mandatory in the middle schools and have tests for competency in the high school and civil service exams (Some nations like France had done this in the mid 1800s too). The Slide Rule became a world-wide tool found in England, France, Germany, Italy, America, Japan, Australia, U.S.S.R., and many others. It is also interesting to note, though, than an estimated that 40 million Slide Rules were produced during the 1900s.

The Slide Rule had an incredible history with its development and use by many famous people, such as **Albert Einstein** and **Wernher Von Braun** who both used the Nestler 23, a 9-scale model and represented two ends of the physics extremes from the theoretical to the engineering practical with the Redstone Rocket. The Slide Rule also has connections to many famous items directly, such as **the Panama Canal, Hoover Dam, Empire State Building, the Golden Gate Bridge, and the jet engine,** et al). In the realm of avionics there is **Frank Whittle** with the jet engine and the **F-16**, which was probably the last major jet designed using the Slide Rule. The entire **Mercury, Gemini, Apollo** era of NASA was greatly developed by Slide Rule-bearing engineers, often having a 10 inch model hanging from a belt. The Apollo astronauts even had, as a sort of back-up tool, a 6-inch Pickett N-600 Slide Rule. The Slide Rule then is the almost invisible tool not fully if ever present in pictures of such famous human endeavors, ideas, and creations, yet it was a tool to help extend the thoughts, feelings, minds, and imaginations of all these scientists and engineers in these journeys.

The Slide Rule became not only the tangible tool derived from logarithms but also became the visual tool that connects our minds to the universe at the large as well as small scales while also being the primary first tool to employ in the construction from the mind to the form for many mobile and immobile structures in its span of time. This tool, the Slide Rule, is special and becomes all the more brilliant by the genius and creativity of the user and the math journeys undertaken in the cosmos. Pick one up sometime and go on an adventure. Fare thee well.

Notice the history which involves so many important people many of us have heard so many other stories about and now we can include a tool with numbers on it, based on a mathematical idea, that helped shape the minds, imaginations, the experiments, the creations of so many in history – and we can use one too. This is part of the idea expressed here in my book :
Galaxy Girl's Inquisitive Pioneer Adventures – learn by reading, learn by measuring, learn by doing, and learn by reasoning. We become part of the history, the adventures, the ideas, as we learn and grow. Come on – Be Brilliant!

Ch.IV
How to Use a Slide Rule

How does a Slide Rule Work

The Slide Rule is a mathematical tool that enables the user to perform mathematical calculations of a great variety and obtain a reasonable answer. Though it can be used to add and subtract, it is better just to do this oneself. The Slide Rule is best suited for multiplication and division primarily. But with additional Scales the Slide Rule can be used for other things like squaring, taking the square root, cubing, taking the cube root, determining the power or root of a given expression, determining the log of a given number and the inverse of this by finding the number for a known logarithm and in trigonometry can be used to find the values of the sines, cosines, and tangents of given angles along with their inverses where one has the sine of a given angle and needs to find the angle itself. The range of application mostly depends on the user, her math skills, creative but mathematically sound approach to a problem and the number of scales the Slide Rule has to ease the outcome of sought after answer.

With regards to the slide rule, Notice one could call it a magic stick or wand as it were – and more importantly all of the magic – the ability to use it – the mastery of the numbers and their relations all comes from you. I see my B1 as just such a magic scepter and it makes me a queen.

The basic parts (and these are noted for the linear model) are the following : The top and bottom strips are called Stators and respectively are referred to as the Top Stator and the Bottom Stator. These pieces are also called Stock in some books. The moving piece between them is the Slide. The Moving Cursor is simple called the Cursor or sometimes called the Indicator or Runner while the cursor line is also known as the Hairline (it was a hair long ago and first suggested by Isaac Newton).

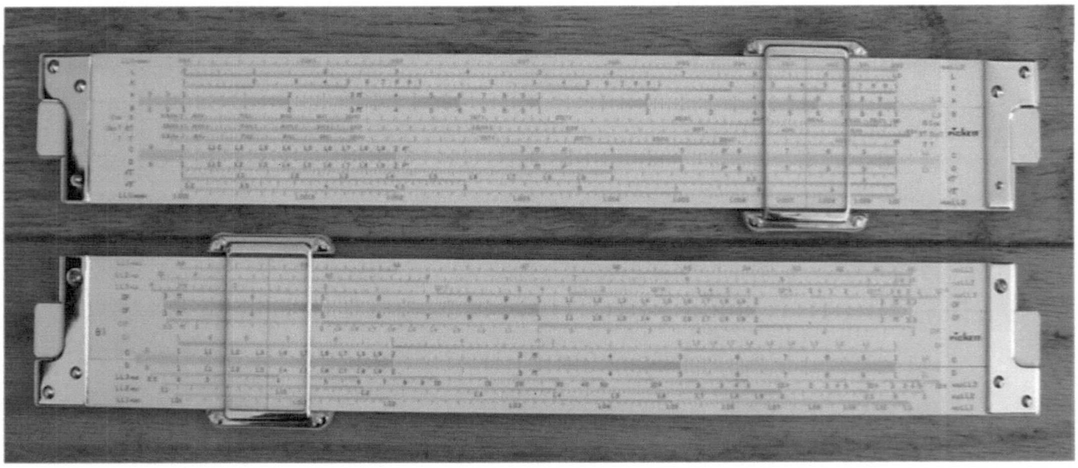

The accuracy of the Slide Rule's answer depends on the Precision of its Scales. Most often the average 10" is effective in its precision of values on its Scales for calculations involving 2 significant digits, but the 3rd sig fig can be estimated as well. The number of gradations on the Scales depends on length, so the longer the Scale the larger the number of gradations, hence a greater level of precision can be found. However, it is not as simple as it might first appear. A 20" slide rule merely has a marginally larger number of gradations hence only a slight increase in estimating the 3rd significant figure. To illustrate : between the 9 and right index 1 of the 10" slide rule there are a total of 10 secondary marks, so values like 9.1, 9.2 are easy enough. Between each of these marks in an ever -narrowing gap (they are spaced logarithmically) on the 10-inch rule there is a mid-point mark, so this would be 9.15, 9.25, et al. It is easy to conclude that one can estimate values to the nearest 0.01, so values like 9.12, 9.37 can be determined where the 3rd digit is estimated while the first two (assuming the slide rule is accurately built and the user is mathematically adept) are reliable. In the case of the 20" rule the same primary and secondary marks exist, but since the scale is longer the tertiary marks now are not 0.5, but instead 0.2. One now has a greater certainty in determining whether the value is indeed 9.13 or 9.14. Notice, however, there is no gain in the number of significant digits though the length of the slide rule has doubled! In general, however, most calculations in the everyday world require only 2 significant digits, so the 10" is sufficient.

The Slide Rule is not a measuring tool, like a ruler, but instead has Scales of Numbers on it that are spaced based on the C & D Scale values which are numbers (such as 1, 1.2, 2, 3.5, et al) placed at a distance from the leftmost number '1' value (aka the Left Index) that corresponds to the logarithm of that given value and multiplied by the size of the scale (in a standard rectilinear rule about the size of a ruler this would be 10" or 25 cm). For example, the number 5 is at a distance from '1' that is log(5)*scale length of the slide rule. This means that all the values are logarithmically spaced from each other on the Scale. What value does having this spacing of values have?

To explore why the numbers on the key base scales (C & D) are logarithmically spaced, first we need to explore logarithms themselves. The reason that it makes the Slide Rule a math tool for multiplication and division comes from the properties of Logarithms, which are :

$$\text{Log}(A*B) = \text{Log}(A) + \text{Log}(B)$$

$$\text{Log}(A/B) = \text{Log}(A) - \text{Log}(B)$$

What this means is this : The log of the product of any two values is merely the sum of the logs of each of the values independently. The log of the division of any two values is the difference of the logs of the respective values as well. What this means is that multiplication can be turned into addition and division can be turned into subtraction. So all we would need then is a table of the log values for any set of values we wish to multiply and we merely add them and find this sum on the table and that would be the product of our values! But the Slide Rule is far easier than this. As noted in the History of the Slide Rule section William Oughtred explored the idea of logarithms and placed the numbers from 1 to 10 at distances from each other that corresponded to the logarithms of the numbers on this line. Knowing the property that the product of any two values is the sum of their logs, all one has to do is combine the distances, hence add them that separate the numbers physically for any given product and the overall total distance will then end up on the number which is the product of the two numbers. Here is a simpler illustration. If I wanted to find the product of 2*3, I merely take the log(2) which equals 0.301 and the log(3) which equals 0.477 and add them together to obtain 0.778 which is the log(6). On the slide rule I merely have to place the starting point, the Index beneath the number in question on another logarithmically-spaced scale – so I place for example the Right Index 1 on the C scale above the 2 on the D scale (go ahead and do this). Now read along the C scale to 3 and look back to the D scale and what do you find? 6, of course. You have traveled the distance of 2 on the D scale and the distance of 3 on the C scale – adding these values together is the same as the product of the values since they are logarithmically spaced so we are to the answer of 2*3 or 6. Read more thoroughly the section on Using the Slide Rule to gain greater insight into using the slide rule for basic and more advanced skills as you learn more and do more with it. Notice we do not need to know logarithms themselves, how to derive them, find tables of them, and with a slide rule in hand, we do not have to construct our own math tool to do this.

Reading the Scale on a Slide Rule :

From here on it is best to have your slide rule available – see it and use it – follow the directions – reread them as needed and practice. A great Activity to turn to once some mastery is well underway is the Everyday Life Calculations first one because there I not only have basic everyday calculations we should all know and use, but walk through examples with a slide rule in a step by step manner to arrive at the answers.

- All slide rule forms (linear, circular, et al) will have the same basic method in reading the scales
- A Scale is the logarithmic-based spaced numbers or a related line to the base scales (C & D)
- Each mark on a Slide Rule can represent any value as needed through Scientific Notation. So only the values 1 to 10 are needed.
- Watch when doing division or multiplication with scientific notation – see those rules if needed.
- Essentially when values are in scientific notation and are multiplied, then the exponents are added. When values are in scientific notation and they are divided, the divisor (the denominator) is subtracted from the quotient (the numerator).
- All scales are related to the base scales (the C and D scales)
- A base scale has the values from 1 to 10, logarithmically spaced.
- However, each mark has a particular reading based on its place on the particular length slide rule.

- The Numbers on the Scale are the Primary Marks. The next major sets of marks are located between them are called Secondary Marks, and when marked, the marks between these are called Tertiary marks.
- When there are 10 marks between any marks, they are tenths (0.1, 1/10)
- When there are 5 spaces between major marks these are two hundredths (0.02, 2/100)

- When a cursor is between any set of corresponding marks, this value becomes the estimated digit and can be done to the level of 0.1 of the major marks involved at best. If very small distances, then it is best to assume only 0.5 of the value between marks.
- The Primary Mark on all scales is the Index which is the number 1. There is a Left one and a Right one on linear slide rules.
- Linear scales (C & D) have 2 Indexes while circular ones have 1.
- The Indexes on a linear slide rule are referred to as the Right Index and Left Index.
- The majority of slide rules have capacity for 3 significant figures in their calculations. (the range is 2 to 4)
- Best General Rule : Use Scientific Notation in computations
- A, B, C, D, C1, CF, DF, C1F, K, and R scales are all logarithmically spaced values. Some are one time (C & D), others (A & B) are double, some are triple (K), some are reverse (C1 & C1F), some are common square roots (R), some are folded (i.e start at a different point) (CF & DF – starting at π)
- Trigonometric Scales include : S, T, ST
- Log Scales include L, LLN(+ & -) Scales – L is the log value (usually base 10) of a value while the LLN scales are representative of natural log based powers of a C or D scale value to a given exponent (see LLN scales for more information).
- Note, unless noted basic math is done with C & D scales
- Scales written in black are from left to right, while in red are right to left
- All basic scales begin with 1 (except for folded, trig or Log scales).
- Scales are aligned to read across them.
- When a Slide Rule is a duplex (two sided), the cursor can be read for both sides as needed.
- The last digit is estimated in reading a slide rule.
- It is necessary to keep track of the decimal place in calculations.
- Some of the most critical skills in using a Slide Rule are these :
    1. Always mentally keep track of an Estimated Answer (some sort of range for the answer)
    2. Keep track of the Decimal place in using the slide rule, since it has infinity contained in 1 to 10
    3. Mentally visualize the formula under consideration and project this onto the slide rule so as to keep track of which scales to read

Multiplication with a Slide Rule :

- Quick summary of rule (explained step by step below) : Set one index of the C scale to the multiplicand on the D scale. Next, set the cursor of the runner to the multiplier on the C scale. Finally, read the Answer on the D scale under the cursor hairline.
- [Note that the answer can be found on either the C or D scale and it depends where one starts and which index is used – it is up to you, but is determined by whether you go off scale – If the calculation is not present, then use the other Index].

- To multiply with a Slide Rule set the Index of one scale (C) over the first number to be multiplied on the opposite scale (D). [Think of this as X from Q = X*Y]
- Slide the cursor to the second value as it appears on the index scale (C). [This is Y from the equation]
- Read the answer under the cursor's hairline on the opposite scale (D). [This is Q]
- Numerical Example :
- Place the Left Index of the C Scale over '2' on the D scale.
- Now read along the C scale to the value of '3'.
- Opposite 3 on C is 6 on the D scale, which is the answer of 2 x 3. This is because you have moved the scales relative to each other both the distance of log(2) and log(3) when added they are the log(6) distance, hence 6 is the answer.
- What is being done is the addition of the distances of the logs of the values in question
- (Log (X*Y) = Log (X) + Log (Y))
- Though illustrated with C & D scales, note it can be done in reverse (that is alternate C for D and vice versa) and this action can also be done on any paired scale-slide combination, such as A & B or if preferred CF & DF.
- With both multiplication and division (plus other functions) keep track of the decimal point. Using Scientific Notation is the best choice.
- <u>Scientific Notation Method :</u>
-
- <u>In Scientific Notation</u>, when multiplying the exponents are added.
- <u>Special Rule :</u> If the slide projects to the Left when performing multiplication, Add One to the Exponent Value for the correct answer.
- (Note : This is for each calculation, so if done twice, then add two for example)
- (Note : Also very important these rules apply to only the use of these scales in use, C & D, for example and does not apply to reading across to other scales, such as CF & DF for example)
- In Scientific Notation, when dividing the exponents are subtracted. (divisor exponent – dividend exponent).
- One must note, however, whether or not the coefficients when multiplied exceed 10, etc.
- <u>Special Rule :</u>
- Also if the slide in division projects to the right, then the answer from scientific notation exponent needs the subtraction of one from the answer to make it correct. – Note this is for C scale as the Numerator and D scale as Denominator. If the scales used are done in reverse of this, then so too is the rule, hence if it projects to the left in that case then subtract one. (The idea here is the same as it was for multiplication – the number of times the slide extends to the right, for each time – subtract one from the exponent total).

-
- Key to calculation is know your decimal placement from products and ratios and using scientific notation it is always one to the right or left of number on the slide rule!

Division with a Slide Rule :

- Quick summary of rule (explained step by step below) : Set the cursor hairline of the runner to the dividend on the D scale. Then slide the divisor on the C scale under the cursor hairline. Finally, read the Answer on the D scale under one index of the C scale.
- [It is important to recognize that for both multiplication and division the use of the scales can be reversed. Here in step one, C is the numerator while D is the denominator. These roles can be reversed].
- To divide with a Slide Rule set the divisor value on one scale (C) over the dividend value on the opposite scale (D).
- Read the answer above the active (D) scale index (1) on the opposite scale (C).
- If the answer falls outside the range of the scale (such as when multiplying or dividing numbers from opposite sides of the scale) then the other index needs to be used in a linear slide rule.
- (Note : This does not occur with a circular slide rule).
- Numerical Example :
- Place '6' on the C scale over '2' on the D scale.
- First note that we are creating a ratio of 6 over 2.
- Next realize that we need to use the Index. In the case of division, always use the Index on the Scale that is the Denominator – here that is the D scale for the example, but it can be the other way if you wish.
- Now notice that the Right Index of D is not opposite any value, so we need to examine the Left Index of D, which is opposite the value of '3' on the C scale, the answer to 6/2.
- What if we wanted 2 divided by 6?
- Here the denominator is the C scale, so again find the useful Index, which is the Right one in this case.
- Since 2 < 6, we expect our answer to be less than 1.
- The right Index of C is opposite 0.333 on the D scale.
- Notice that 3rd digit – it comes from the fact that each division between 3.3 and 3.4 is 0.02 and the cursor is in the mid point between 3.32 and 3.34. Finally following the rules of scientific notation and decimal placement, since the slide went to the left and each of the values has an exponent of 0, so 0-0 is 0 and then we subtract one from that and have -1 for our exponent, so our answer of 0.333.
- What is being done is the subtraction of distances of the logs of the two values in question
- (Log (X/Y) = Log (X) – Log (Y))
- The best way to consider this operation is think of it as a proportion : the Answer is over 1 while along the scales the Divisor is over the Dividend.

-
- The Proportion concept is useful for calculations involving conversions and manipulation of 3-variable functions.
- With all slide rules where one scale can move while the other is immobile all answers are present in both multiplication and division as the scales are read. This makes it a parallel calculator – all relations are instantly set up and visual.
- Though the Scientific Notation method works best for values, be wary and if the slide projects to the right and you are reading the Right Index, then in using the Scientific Notation system for the values, subtract one from the answer in all cases. If it projects to the right, then it is alright for division, yet remember that for Multiplication you add one to the total exponent value!
-
- <u>Alternative Multiplication and Division Method :</u>
- <u>Characteristic Method (which is akin to the Scientific Notation method, only here the characteristic is the exponent) :</u>
- For any given set of values, write the characteristic (the portion in front of the decimal when written as the log of the value in question (recall the notation : characteristic.mantissa – Note only the characteristic is needed, we are not looking up the mantissa, which, by the way, can be found on the L scale of your slide rule if needed) –
- Be sure to be wary of positive and negative values in this case!
- Sum up these characteristics.
- Now perform the Multiplication or Division with the Slide Rule as you normally would. That is to say, each number is merely a value from 1 to 10 only. Note that the answer may end up on the range of values before 1 to 10 (i.e. be 1/10 as large) or on the range after 1 to 10 and be in the range of 10 to 100.
- For multiplication, each time the slide , with the C Scale, extends past the left index of the D scale, add one (+1) to the Sum of Characteristics.
- For Division, subtract one (-1) from the Sum of Characteristics.
- The revised total is now the characteristic (i.e. the power of 10) for the answer to use with the value showing on your slide rule.
- Return to our 2 x 3 and 6/2 examples for a moment. Each has a Characteristic of 0. In the case of multiplication, it went to the right, so the sum is again 0, while in division the sum is 0, but we subtract one yielding an exponent of -1 from these rules here.

<u>Combined Operations with a Slide Rule :</u>

- Combined operations are multiple operations in one problem
- In this list there are many examples and notated use of the Slide Rule when it comes to fractions, ratios, proportions, and applications of these ideas.
- Let's say we have the situation for continued products where Q = A x B x C x ---
- Here set the hairline of the cursor indicator at A on the D scale.
- Next move index of C scale under the hairline.
- Next move hairline over B on the C scale.
- Now move the index of C scale under the hairline.
- Next move the hairline over C on the C scale
- Now move the index of C scale under the hairline.
- Continue moving hairline and the index alternately until all the numbers have been set in this case and come to the answer.
- Second scenario :
- If the problem reads (N x M) / R then
- Place N (C scale) over R (D scale)
- Slide the cursor along the D scale to M (on D)
- Find the Answer on the C Scale
- Repeat this process as needed for more than this set up
- Be sure to keep track of Decimal Point as noted above in the Multiplication and Division Rules.
-
- In essence, combination problems can be seen as proportions which are extensions of ratios, which slide rules are good at.
- For example, any two values over each other is a Ratio or Fraction and once set all similar ratios are automatically established instantaneously. As well opposite the Index of the Divisor is the decimal equivalent of the ratio as well!
- Convert the decimal to a fraction -
- <u>What of converting from a Decimal to a Fraction?</u>
- Place the decimal value on the C Scale over the Right Index of the D Scale and then search along the C and D Scales for a ratio of numbers to represent them.
- Keep in mind this decimal is the percentage value, but you need to multiply in your head by 100 to see it as a percentage.
- For example, 3 on the C Scale over 8 on the D Scale has the Right Index of D under 375 on the C Scale.
- Since 3 < 8, the value is less than one and should be read as 0.375
- What of the Scientific Notation Rules. Each initial value has the exponent 0 and in division, 0 – 0 is 0. But the slide is projecting to the right so we subtract 1 from the result and have -1 for an answer. This means to move the decimal one to the left, and the answer is read 0.375
- As a percentage, multiply by 100 and the answer is 37.5%
- <u>One of the largest uses of fractions is for Sales!</u>

- Place the Price of the Item on the C Scale over the Right Index of the D Scale.
- Read backwards along the D Scale to 9, 8, etc. Each of these is read as N x 10% (such as 90%, 80% and so on). Above it is the Price at that percentage!
- Of course, keep in mind, this is not the percentage off, but what you are paying. To see how much is saved, just use the Left Index of D Scale under the Price instead and read to the right instead of left.
- For example, 25% is found at 2.5 and the value above it is 25% of the price. (Of course what you pay is found at 7.5 or 75% instead).
- Sales tax and total cost can be found in a similar manner :
- Place the Cost of the item over the left index of the adjacent scale 9 say C over D as we have been doing)
- Read along the D scale to the sales tax expressed as a decimal value (this should not be too far, 4% is 0.04, 6% is 0.06, et al)
- The value above on the C Scale is the total cost including sales tax!
-
- Other Fractions or Ratios in Everyday Life :
-
- Mpg = miles per gallon
- Take the ratio of miles driven over the number of gallons used sometime to determine just what gas mileage you are getting!
- Mph = miles per hour
- What was your average rate of travel for a trip?
- Take the ratio of the Miles driven to the amount of time (in hours) to find the average speed of the trip!
- What if you are some fraction of the distance there and have determined the average speed, then
- The question becomes how much longer 'to grandma's house?'
- Take the remaining distance and divide by the average speed to estimate the number of hours to complete the journey.
- Further, from the mpg calculation you could take the remaining distance divided by the mpg and find the number of gallons needed for the trip in order to decide whether to fill up again or not!

- <u>The fraction as a Slope in Algebra</u>
- A ratio might not be just two numbers, but instead represents the ratio of two differences of numbers :

- $m = \frac{\Delta Y}{\Delta X}$

- This is the slope of a linear line, one of the largest topics in algebra.
- The slope is the rate of Rise Over Run. The larger the value, the steeper the line.
- The sign of the slope determines whether the line is moving up or down when examining it from left to right.
- You can consider slope as a 3-variable formula (see below).

- The Fraction as a Rate :
- In many practical applications of the fraction it is seen as a **Rate**. The numerator is one value (distance, gallons, cubic feet of gas, amount of
- growth, temperature change, etc) while the denominator is some other value (typically time).
- The rate could be determined or it may be known in a given problem. If to be determined, the amounts of the other two variables are known
- If the rate is known, then clearly we are missing either the amount in question flowing or the amount of time needed to do this.
- A very good example of a Rate is in Cost per Unit Ounce (Volume), etc. This is a valuable tool for the slide rule in comparing items when shopping for their comparative costs to find the better deal.
- The flow rate is how water and natural gas consumption is measured and billed. For water and natural gas are Ccf (100s of cubic feet).
- In order to estimate the Cost, one has to merely read the meter at the beginning and the ending times, subtract to arrive at an amount used and multiply this by the cost per unit. This is true for all of the meters : Water, Electrical, and Gas.
- Rates and Ratios are very common in Science :
- In Physics there are many ratios, such as Density (amount of substance per unit volume), Pressure (force per unit area), Power (Joules per second), etc.
- Also in Physics, rates can be speed (rate of change of distance with respect to time) or acceleration (rate of change of velocity with respect to time) or from Newton's 2nd Law acceleration is the ratio of Net Force to the mass of the object undergoing the net force.
- Chemistry has many ratios such as Molarity (the number of moles of solute to liters of solution), Molality (the number of moles of solute to kilograms of solvent), Percent by Volume [Mass] (the volume [mass] of Solute to the volume [mass] of solution), the Law of Definite Proportions where the Percent by Mass (is Mass of the Element divided by Mass of the Compound), as well as the Law of Multiple Proportions (these could be looked at in the Proportion section obviously), and so on.
- In Chemistry rates are seen in things such as rate of reaction (is the negative of the rate of change of reactant to change of time) plus many more.

- <u>Conversions :</u>
- In Conversions there are two basic methods that can be used with the Slide Rule :
- First, treat the two items as a Fraction that is to be multiplied by the Number in question.
- Typically the ratio of the items is taken where the units one is 'in' are in the denominator, while the units one wants to convert into are in the numerator.

- $\frac{\textbf{Units to Convert Into}}{\textbf{Units to Convert From}}$ **or** $\frac{X}{Y}$

- This results in this situation :
-
- Beginning Value in Initial Units* $\frac{\text{Desired Units}}{\text{Initial Units}}$ = Final Value

- $N * \frac{X}{Y} = M$

- The best way to handle this on a slide rule is to :
- Place N on the C Scale over Y on the D Scale on the slide rule
- Read along the D Scale to X, the Desired units and read the answer, M on the C Scale.
- Looking at the prior discussion, it is easy to see that the formula presented can be read as :

- $\frac{M}{N} = \frac{X}{Y}$

- All one has to do is take the ratio of Convert Into Units on the C Scale over the Convert From Units on the D Scale
- Now read along the D Scale to the Beginning Value (N) and find the answer above it on the C Scale (M)
- Use the following Table and look elsewhere for everyday conversions. It is best to keep a small list and memory of these as needed :
- **1 inch = 2.54 cm**
- **12 inches = 1 foot**
- **3 ft = 1 yard**
- **1 mi = 5,280 ft**
- **1 minute = 60 seconds**
- **365 days = 1 year (rounded)**
- **1 solar day = 24 hours**
- **1 cup = 8 fluid ounces**
- **1 gallon = 4 quarts**
- **1 pound = 16 ounces**
- **1 kilogram = 2.2 pounds**
- **1 ounce = 28.3 grams**
- **1 liter = 1.06 quarts (rounded)**

- Plus look up whatever you may need !

- For each of the above and any others your find a need for, simply place one value over the other as described above it for use.
- Conversions are needed in many applications :
- One basic unit to another (inches into feet),
- Changing one unit type into another (English to Metric),
- Currency Conversions, and others.
- Computing Costs
- The reason for conversion depends on the needed outcome. For example feet into yards is commonly used when computing the area of a room in square-yards for carpeting by dividing by 9.
- In the case of painting, the Slide Rule readily calculates the area of a wall with Length times Width, but what of the number of gallons of paint needed?
- Take the total area to be covered and use the Slide Rule to divide by 300 if the walls are unpainted or rough – if smooth and already painted divide by 350 – to determine the number of gallons of paint needed!

- In Math, angles are often converted between Radians and Degrees :

- $180^\circ = \pi$ radians

- In the Sciences, there are numerous conversions not only of the aforementioned units, but also of mixed units :

- Such as km/hr to m/s is very common.
- For that calculation the ratio for it is 3.6/1. Check it yourself!
- In Chemistry there are conversion commonly found in the amount of substances and how it is expressed :
- For example take the number of grams of a substance and divide by its gram molecular weight to determine the number of moles present.
- In Science and Math there are many other conversions depending on the situation at hand.
- In Math in the realm of trigonometry since the C and D Scales are the values for the Sines and Tangents as read from the S and T Scales (keeping in mind where the decimal falls), this makes it easy to multiply or divide by the sine or tangent of a value when and where needed.
- Slide the sine value for a given angle read on the C Scale from the S Scale over a given value on the D Scale.
- Read in one way the sine is in the numerator and in the opposite direction it is in the denominator.
- What if you want to find the Sine or Tangent value and are given the sides of a triangle?
-
-
-

-
- Obviously this is again a ratio : For example :
- $$\sin\Theta = \frac{\text{Length of Side Opposite}}{\text{Length of Hypotenuse}}$$
- Also other angular measures are available and can be used for determining distance or size, such as in the Radian measures :
- $$\textbf{Radian measure} = \frac{\text{Arc Length of Circle Portion}}{\text{Length of Radius}}$$
- If our protractor measures $1/10^{th}$ of a radian, then the apparent size of the object in question is $1/10^{th}$ the measure of the distance from us!
- Another interesting one in Math is the conversion from one base to another in terms of logarithms.
- The standard slide rule has base 10 logs, but let's say you want a number in another base, say 2 or the natural log base e (2.71828*)?
- For any positive numbers, N, A, and B with A ≠ 1 and B ≠ 1,
- $$\log_A N = \frac{\log_B N}{\log_B A}$$
- From the question the natural log of any value is the ratio of the log base 10 of the number divided by the log base 10 of the natural log value.
- To illustrate, here is a similar example question :
- What is the log base 2 of 6, for example. Log_2 (6)
- Look up reading from the d Scale to the L Scale both the log of 2 and the log of 6. (0.301 and .778 respectively)
- Now divide on Scale C and Scale D 778 by 301 - We find it to be 258
- Where does the decimal go?
- Since 6 is much greater than 2, it will have a characteristic (here 2 and the rest is the mantissa 0.58) so the answer is 2.58
- So $2^{2.58}$ = 6 (try it and see, rounded off of course)
- The idea of the ratio, fraction goes on to any and all applications.

- **<u>The Proportion :</u>**

- As noted in the prelude, the proportion is two ratios set equal to each other.

- In the conversions section above, it is easy to see its value in use.
- The basic proportion can be expressed as :

- $\frac{M}{N} = \frac{X}{Y}$

- All one has to do is take the ratio of known values - M on the C Scale over N on the D Scale

- Now read along the D Scale to the other known value (Y) and find above it on the C Scale the answer (X)
- It is easy to see that all one needs to know is any 3 of the variables and the 4th is the one to find.
- By using the scaled as fractions it is easy to place one value on one scale and the other on the opposing scale.
- Try this for yourself to find that solving a proportion on a Slide Rule is indeed much faster than one can solve one on a Calculator!
- Also here there are no complex rules, like the calculator, such as cross-products –
- In the case of the Slide Rule the natural form is maintained which is the equivalence of two ratios.
- This is an invaluable tool in Algebra for proportions as well as in geometry for any and all similar figures to determine unknown sides!
- Proportions can be used to determine height or distances - In this case we are using similar triangles.
- For example, hold up a ruler at arm's length to measure the apparent height of a distant object, say a picture on the wall.
- If you read the apparent height, measure your arm's length, (this is the first ratio)
- now measure the distance to the wall,
- Set the first ratio of your measures to the ratio of the unknown height on the wall to the distance to the wall.
- Then the height of the picture can readily be determined.
- Though simple, this technique is used in surveying regularly and is used in the wilderness for distances across rivers and gorges before the advent of electronic equipment.
- The easiest way to envision it is when you are outside and you as well as a tree or a flag pole casts a shadow on a sunny day.
- The ratio of your shadow length to your height is equal to the shadow length of the flag pole to its actual height.

- $\frac{\textbf{Length of your shadow}}{\textbf{Your Actual Height}} = \frac{\textbf{Length of flagpole shadow}}{\textbf{Actual Height of flagpole}}$

- A more thought out activity involves determining the size of the Moon :

- Use a meter stick and place a small card vertically with a hole punched in it at a distance
- Look along the stick through the hole so that when viewing the full Moon it fully fills the diameter of the hole (i.e. move the card until you have proper alignment)
- The ratio of the diameter of the hole to the distance that the hole is from your eye equals the actual diameter of the Moon to the Moon's distance from you. (Here, we assume we know the distance to the Moon). Hence the diameter of the Moon can be determined!

- Proportions can also be used in changing the scale of a recipe :

- $$\frac{\textbf{Recipe Requirement for Material}}{\textbf{Recipe Number of Servings}} = \frac{\textbf{The amount of Material Needed}}{\textbf{Number of Servings}}$$

- Here the ratio of how much is needed to the number of servings is set equal to the amount of unknown material and the number of desired servings. The amount needed is readily found.

- **In Math there are many applications :**

- The very nature of the proportion stems directly from geometry and the relations found in similar figures.
- These can be the aforementioned triangles but also includes any similar polygons, such as squares, rectangles, and the like where one is known, the other is partially known and there is a missing side.
- Still other examples are considered :
- *What if one wants to find the circumference of a given diameter of a circle (or vice versa)?*
- Simply put π on the C Scale of the slide rule over the Index on the D Scale. – Note with CF & DF scales this is already done since these scales are set at π as their beginning point over the C & D Indexes! See CF & DF scale use
- Now read along the known scale. The C Scale is the Circumference, while the D Scale is the diameter.
- *What about the diameter if the area of the circle is known?*
- Take the Area on the A Scale over p on the B Scale.
- The diameter-squared is found on the A Scale opposite the B Index.
- The diameter is then found below reading from the A Scale to the D Scale to read the square root of the value on A.

- Other interesting things can be done with gauge marks (special marks on slide rules for conversions, multiplications, et al like π) as well as personally derived values :

- On some Slide Rules there is a mark, c, on the C & D Scales (1.273) which is $\frac{4}{\pi}$ and comes from $\mathbf{A} = \pi * \frac{d^2}{4}$
- Place this C mark or value of the C Scale over the index of the D Scale
- Slide the cursor over the size of the diameter of a considered circle on the D Scale.
- To find the Area of a circle read the answer on the B Scale!
- What if you want to do this by knowing the radius (of course we could simply multiply by 2 to use the former method, but give your mind a chance to explore the Slide Rule !)
- Look up the square root of π by first finding it on the B Scale and noting its value on the C Scale.
- Slide the square root of π value on the C Scale to the Index of the D Scale
- Now read along the D Scale to any desired radius value for a circle.
- The answer in this case is read on the B Scale for the Area of the Circle in question!

-
- What about the Volume of a Sphere?
- Now place the value 1.61(2) on the D Scale over the index of the C Scale.
- Read along the D Scale, for the radius of a given sphere you are considering.
- The Volume of this sphere is found on the K Scale!
- This comes from $\left(\frac{4*\pi}{3}\right)^{1/3}$ from the formula $\mathbf{V} = \frac{4*\pi*r^3}{3}$
- There are many other problems in Math and Algebra in the area of problem solving:
- For example, if a given material costs so much per ounce, pound, ton, how much will another desired amount cost ?

- $\frac{\textbf{Cost}}{\textbf{Unit Amount}} = \frac{\textbf{How much does it Cost?}}{\textbf{Amount Wanted}}$

- There are numerous ratios of values that can be found or derived from many other references that can be used in proportions to find the answer to many a question one might encounter in a math and or science text:

- $\frac{\textbf{Diameter of Circle}}{\textbf{Side of Inscribed Square}} = \frac{99}{70} = \frac{\textbf{Diagonal of Square}}{\textbf{Side of Square}} = \sqrt{2}$

- Here is a general value for the pressure one feels with depth :

- $$\frac{\textbf{Pounds per Square Inch}}{\textbf{Feet of Water}} = \frac{\textbf{26}}{\textbf{60}}$$

- Even more complex proportions can be solved :

- A classic algebra question might read :
- *If it takes 4 people 7 days to accomplish a job, how much time is needed (assuming the same work rate) for 6 people to do this task?*

- $$\frac{\text{Initial Workers}}{\frac{\text{1 Task}}{\text{Amount of Time Initally}}} = \frac{\text{Workers in case 2}}{\frac{\text{1 Task}}{\text{Amount of Time needed}}}$$

- $$\frac{Wi}{\frac{1}{Ti}} = \frac{Wf}{\frac{1}{T2}}$$

- $$\frac{4}{\frac{1}{7}} = \frac{6}{\frac{1}{X}}$$

- You could go through and first simplify it and then take the ratio of the numbers on one side once the variable is isolated, but the slide rule allows for this to be solved as is!?

- Take 4 on the D Scale and slide 7 on the C1 Scale over it. Start with the cursor here.
- Read along the C1 Scale to 6 and look below on the D Scale to find the answer 4.66 days.

- **In Physics, for example, say you have a balance beam.**

- *If on one side of the balance you have 26 g a distance of 32 cm from the center,*
- *how much mass must be placed on the opposite side at a distance of 20 cm from the center in the opposite direction so that it balances?*

- cw is clockwise, ccw is counter-clockwise

- $$\frac{\textbf{Mass cw}}{\textbf{Mass ccw}} = \frac{\textbf{Distance ccw}}{\textbf{Distance cw}}$$

- $$\frac{\textbf{26 g}}{\textbf{X g}} = \frac{\textbf{20 cm}}{\textbf{32 cm}}$$

- X = 41.6 g

- This idea can be applied to problems in chemistry too.

- Take for example, conservation of mass in a problem where one has to determine the mass of a reactant product in a total mass size where one is only given a small sample for testing.

- $$\frac{\textbf{Mass of reactant in sample}}{\textbf{Mass of Sample}} = \frac{\textbf{Mass of reactant in total mass}}{\textbf{Total Mass}}$$

- This list applies to any and all sciences and is limited only by the imagination in the questions being asked.

Using the CF & DF Scales (πC & πD) :

- The CF and DF Scales are called Folded Scales since they do not start at the Index, but instead are at a chosen point, namely here being π.
- Why π?
- Simple – there are many calculations that involve π that extend from Circles, such as the circumference of a circle :
-
- Take any value on the C or D scale and now look to the cursor value on the corresponding CF or DF scale. It is p times greater. This is the formula for the Circumference of a Circle : $C = \pi * d$, where d, the diameter of the circle is being read from the C or D scale and its circumference is then found on the CF or DF scale.
- For example – put the cursor of your slide rule on 3 on the D scale and find the DF scale. We are assuming we have a circle with a diameter of 3 units, so the question is : what is that circle's circumference? On the DF scale we read the answer of 9.42 units
- Note that the reverse is true as well. If we know the circumference of a circle, we can read it on the CF or DF scale and find its diameter quickly on the C or D scale.
- Another important role that a folded scale has is this : Since it starts at another point on the line, it is easier to do multiplication and division without having to change up Indexes as often.
- For example let's try 3 x 6
- We might first be inclined to slide the Left C Index over 3 on the D scale before realizing we should have first chosen the right C Index.
- But no bother –
- Read the value of 6 on the CF scale and find opposite it on the DF scale our intended answer, 18!
- What of the decimal point rules here, however?
- Going back to our original rules,

- First estimate the answer : Anything past 3.x will result in a value greater than 10, since by 3x4 we are at 12 already.
- Next, if we were to only use the C and D scales, then we would have had to use the Right Index instead of the Left one, hence we would have added one to our exponent total, or 1 in this case (since the values started with exponents of 0).
- Another way to think of this problem is this : Since we have gone past the end of the scale we are on and moved on to the next one in sequence, it is 10x larger, so instead of being 1 to 10, it is now 10 to 100, so the value of 1.8 is really 18 in this case.
- Note : You can do multiplication and division with the CF and DF scales and the same rules apply for them as they did for the C & D scales presented previously. Realize that the index is still 1.

<u>Using the C1 Scale</u> ($\frac{1}{x}$) :

- This is the Inverse Scale. – See an example of use in combination problems – this is a common use of the C1 scale.
- It is basically C scale in reverse.
- It is written right to left (regular are left to right)
- The inverse of any number can be found by aligning a cursor over a given value of C and on C1 is the inverse. (Be sure to keep track of the decimal!)
- For example if reading '5' on the C scale, on the C1 it must be 0.2
- To multiply with the C1 Scale set one of the numbers on C1 over the other number on D.
- Read the Answer on the D scale under the Index on C1 scale.
- Any problem with a fractional component can be examined more easily with C1. – It can be a fraction or be the denominator of a given expression.
- Especially useful in combination problems. – see prior for example in the combination section under proportions.
- Use the rules for multiplication and division and read the proper index for a solution
- One of the other important and common uses of the C1 Scale is to be the scale when reading for a value from the tangent scale angles above 45° (45° to about 80°)

<u>Using the A & B Scales (Squares X^2 & Square Roots $\sqrt{X}$) :</u>

- The A & B Scales are double scales and are the squares of C & D Scales –
- That is to say it is double logarithmic (1 to 10 and 10 to 100)
- Conversely C & D are the square root values of A & B
-
- <u>Rules for Square Roots of Numbers :</u>
- Note Left side is Left Index to Middle Index, and Right side is Middle Index to the Right Index

- Odd Number Digit Rule :
- For Whole Numbers > 1
- If the number of digits left of the decimal point in the value being considered for square rooting is odd, Read the value from Left-Hand side of A
- For Numbers 0 < x < 1
- If the number of zeroes to the right of the decimal is odd then Read the value from the Left-Hand side of A
-
- Even Number Digit Rule :
- For Whole Numbers > 1
- If the number of digits left of the decimal point in the value being considered for square rooting is even, Read the value from Right-Hand side of A
- For Numbers 0 < x < 1
- If the number of zeroes to the right of the decimal is even then Read the value from the Right-Hand side of A – Note this includes no zero at all (just .X)

- To summarize the rule (and make reading a R scale easier) :
- The Left hand side of the A & B scales is for Odd Number of Digits or the Odd Number of Zeroes in a Number (this corresponds to the R1 scale)
- The Right hand side of the A & B scales is for Even Number of Digits or the Even Number of Zeroes in a Number (this corresponds to the R2 scale)

For 0 < X < 1									
No. of zeroes in between X and the decimal	0	1	2	3	4	5	6	7	Continue the pattern of even then odd values
Which Side (L or R)?	R	L	R	L	R	L	R	L	Continue pattern R,L,etc
$\sqrt{\ }$ Answer and the no. of zeroes between ans. And decimal point	0	0	1	1	2	2	3	3	Continue Pattern 4,4,5,5,etc

Number to have a square root taken	Answer on slide rule
0.2	0.447
0.02	0.141
0.002	0.0447
0.0002	0.014

For X > 1									
No. of whole digits in value X	1	2	3	4	5	6	7	8	Continue the pattern of even then odd values
Which Side (L or R)?	L	R	L	R	L	R	L	R	Continue pattern R,L,etc
No. of Digits in the answer $\sqrt{\ }$	1	1	2	2	3	3	4	4	Continue Pattern 5,5,etc

Number to have a square root taken	Answer on slide rule
2	1.41
20	4.47
200	14.1
2,000	44.7
20,000	141.

Using the K Scale (Cubes X^3 & Cube Roots $\sqrt{X}$) :

- The K Scale is a triple scale and is the cube of the values on C & D
- That is its range is 1 to 10, 10 to 100, and 100 to 1000.
- Rules for Taking Cubes :
- For the Rules consider the K scale divided into 3 sections from one index to the next.
- The sections are Left, Middle, and Right
- For All Values under consideration for cube root extraction :
- Divide the Number into groups of 3 (starting at the decimal point) and go left or right of the decimal point as needed in creating these groups
- Look at the Left-most group with non-zero digits in it
- If the group has 1 digit then use the Left portion of the K scale (for example 1-10)
- If the group has 2 digits then use the Middle portion of the K scale (for example 10-100)
- If the group has 3 digits then use the Right portion of the K scale (for example 100-1000)
- To continue further, continue counting in 3's.
- An easy way to find where a value falls if <1 :
- Let right-most 1 be 1 and go backwards as powers of 10 for any given decimal value :
- Example 0.00X is in the thousands place
- The right-most '1' is 1000/1000,
- The next '1' left of it is 100/1000,
- The next '1' left of it is 10/1000
- And the left-most is 1/1000
- If there are more restart at right-most and continue (Note the scale wraps around then)

- The number of groups left of the decimal point determines where the <u>decimal point</u> in the <u>answer</u> falls :
- If there are two groups (of 3) left of the decimal point (complete or not !) (means values 1,000 – 999,000) then there are 2 figures left of the decimal point in the answer.
- Values : X,XX0 to XXX,000
- If there is one group of 3 (values 1-999) then the answer has one value left of the decimal point
- Values : X.XX to XXX.
- If there are 3 groups to the right of the decimal point (where one or more falls in the tenths-hundredths-thousandths columns) then the answer has 3 figures right of the decimal point starting just past the decimal point
- Values : 0.XXX to 0.00X XX-,---
- Values : 0.000 XXX to 0.000 00X XX0 ---
- If the number has four groups to the right of the decimal point where the first group is all zeroes and at least one value is in the ten-thousandths, hundred-thousandths, or millionths place, then the answer will have 4 figures to the right of the decimal point where the first is a zero (0).

Decimal value for X 0 < x < 1		Number of Zeroes in value		
Which portion of K scale to read?	R, C, L	R ,C ,L	R ,C ,L	R ,C ,L
Q number of zeroes in value X	0,1,2	3,4,5	6,7,8	9,10,11
Number of zeroes in answer to $\sqrt{\ }$	0	1	2	3

whole values for X X > 1		Number of Digits in value		
Which portion of K scale to read?	R, C, L	R, C, L	R, C, L	R, C ,L
Q number of digits in value X	1,2,3	4,5,6	7,8,9	10,11,12
Number of whole digits in answer to $\sqrt{\ }$	1	2	3	4

Using the L Scale (Log $_{\text{Base 10}}$ (N)) :

- The L Scale is the Log value of the C & D scales
- It is effectively used for determining powers and roots for a wide range of values. Note that the powers do not have to be whole numbers as well as roots can be any fractional value one is interested in determining
- The L Scale provides the log of values 1 to 10 with no changes
- If the Number is <1 (includes <0) or >10, the log value has both the Characteristic (the value before the decimal point) and the Mantissa (the value after the decimal point found on the scale)
- Note that the mantissa will be the same for a given value independent of the decimal point.

-
- For example the log (2) mantissa is 301, as is for log (20), log (200), etc – the only difference is the Characteristic, so log(2) = 0.301, log(20)=1.301, log(200) = 2.301, et al
- When the number is <1, can use scientific notation to determine the value to add to the log of the number in question. Log(0.2) = Log(2×10^{-1}) = 0.301-1 = -0.699
-
- **Procedure below for logs in general :**
- First Look up the Log value for any given number treating it in Scientific Notation format so that it is >1 and <10
- Locate the characteristic of the scientific notation value on the C or D Scale and look below to the L Scale
- This L value is the Mantissa
- Add the Mantissa to the Exponent of the Scientific Notation exponent
- This Sum is the Final Answer
- The L Scale is useful for X^N and $X^{1/N}$
- To solve – Look up X on C and find its log value on L

- Then Multiply or Divide as needed the N or 1/N value involved in the problem using the rules for multiplication or division. Be sure to watch the decimal point.
- This new value, Q, is computed
- Now Search for the Q Mantissa on the L scale
- Note that if there is a Characteristic, it becomes the power of 10 for decimal placement of the answer!
- Alternative to finding a log for values between 0 and 1
- If a desired value for log(X) has $0<X<1$ as a decimal in the tenths place, the C1 scale can be used. Reading 2 on C1 the log value is the log of (0.2) for example.

Using the S Scale ($\sin(\theta)$ & $\sin^{-1}(\theta)$) :

- The Sine (S) Scale has values representing the angles from approximately 5.5° to 90°
- When the cursor is on any value of the S scale (the left number printed in black typically) its sine value can be found under the cursor on the C or D Scale
- Note : Since Sine functions range from 0 to 1, all values read from C or D are decimal and begin as 0.XXX (for most models)
- Note : Some Box-style Post Slide Rules have a cursor on the backside where the S scale often is and with a value on it the sine & tangent (T scale) of the angle is found on the B scale!
- For angles 0° to 5.5° there is often an ST scale and represents both Sine and Tangent functions as these have approximately the same values for small angles such as these (also the decimal is 0.0XXX)
- Many Sine Scales have a second set of numbers written in red to the right of the values on the scale –
- These are the complimentary angles hence they represent the Cosine of those angles listed in red
- Recall $\sin(\Theta) = \cos(90^{\circ}-\Theta)$
- To multiply by the Sine of a given angle simply place the angle over the appropriate D scale index
- Next read along D to the value to be multiplied by
- Find the Answer on the C Scale above that point on the D scale
- Remember to watch the decimal points since the C value is a decimal value
- To divide by the sine of an angle simply place the angle on S scale acting as the dividend over the divisor on the D scale
- Find the Answer at the C scale corresponding D scale Index
- If the square to the sine of an angle is needed since the sine value is on C, the square of the sine value is found on A.
- To find the log of an angle for a trigonometric function such as sine, look up the sine of the angle on C scale, then read on the corresponding L scale for the log of the sine of this angle.

Using the T Scale ($\tan(\theta)$ & $\tan^{-1}(\theta)$) :

- The Tangent (T) Scale has values printed in black from 5.5° to 45° when read left to right
- The values (typically red) read right to left are 45° to 84.5°
- Like the directions for Sine (S) Scale,
- To read the tangent (Θ)
- place the cursor on the desired angle
- If the angle is from 5.5° to 45°, the Tangent to this angle is found on the C or D scale and is read as a decimal 0.XXX
- Recall that tan(45°) = 1.00
- In fact this value lines up with the Right Index of C & D
- If the angle is 45° to 84.5° read the T Scale from Right to left to find the value and place the cursor there
- Answer to the Tangent of this angle is found on the C1 Scale
- The values on the C1 scale are read as whole numbers as the line appears
- (Note there are slide rules that have two T lines, hence are read from left to right on the C scale and one must keep in mind that the C values start at 1 (tan 45°)
- Like the S scale since the tangent (like the sine value) value is on the C scale the rules for multiplying and dividing are the same as noted in the S Scale section
- Note since $\tan(\Theta) = 1/\cot(\Theta)$ one can compute these as needed

Using the ST Scales :

- If your Slide Rule has a ST Scale it can be used for small angles for both the sine and tangent function as these are very close to each other in value between 0.6° and 5.5°
- Much like S Scale and T Scale readings the angle is on the ST scale and the reading comes from the C Scale.
- Important Reading Scale Note : Here however the readings range from 0.01 and 0.1, so the reading has 10^{-2} as its exponent.
- <u>What if there is no ST Scale :</u>
- In approximations :
- We can use : $\sin(x) = \tan(x) = \frac{x}{\frac{180}{\pi}} = \frac{x}{57.3}$
- Note that this approximation also is the radian value for that angle!
- Here x is the small angle and using the C and D scales then provides a reasonable value for the sine or tangent of the small angle (within the same range).

Using the Log (LLn Scales) :

- The LLn Scales are used to raise a number in question to a power or find a root of that number.
- Each line is a power of e (2.718...), some models it is the power of 10, of the next line in the list (LL1, LL2, et al).
- That is to say each line is e to some power (from -10X (LLO3) to +10X (LL3))
- The Table of Names and Powers for LL scales are below showing both the power and the range of values to be found on those scales. :
- Looking at the tables, it is clear that the spacing needs to be examined carefully when reading the scales. Be sure to look at the two primary values that your values is between (as an example) and then look at the number of secondary divisions to determine the appropriate value.
- These scales are indispensible when looking for any given power or root as needed (not necessarily whole number ones either).
- Also since each is 10x the line before it, for example LL3 is 10x LL2, and LL2 is 10x LL1, and so on, hence LL3 is 100x LL1. This means that for any power of 'e' which is used for the scale when reading from the C or D scale each is 10x the others.
- For example e^2 is found by looking at the D line for 2 and reading from the e^{1-10x} line, which is LL3 and is approximately 7.4
- Better still, what if you wanted the inverse of e^2 or 7.4? Simple read its value on LL03 – Be careful in reading the scales, these are in reverse first off and have decimal values – so read it from right to left
- Here the inverse of 7.4 is between 0.1 and 0.2 and reads 0.135 – try this yourself and practice reading the scale properly.
- What about $e^{0.2}$, that is found on the LL2 scale, (yields a value nearly 1.222)
- and $e^{0.02}$ is on the same cursor line with the cursor the whole time on 2 on D.
- This is one of the great values of the LLn scales. Any value from 1.001 to nearly 100,000 can be used from LL0 to LL3 and its inverse is present as well if needed. That would be 10^{-5} to 0.999 found on LL00 to LL03. It is typically written in this range though, due to the effectiveness of the slide rule : having a range of values (up to 80 inches in length on these LLN scales) of running from 0.00005 to 20,000.
- Looking at a value on say LL0 scale and reading the value below on LL1 scale it becomes that number raised to the power of e 10x more. So an increase in number is 10x the prior line.
- The LL0 value if read with the LL2 scale will then be that value raised to the power of 100.
- Going from a higher LL scale number to a lower one means that it is $1/10^{th}$ power per each jump.
- To raise to a negative (–) exponential power : This can be done directly or find the positive value then its inverse on the corresponding negative exponent line.

- If raising negative exponent to positive exponent values :
- If one wants to find 10x the value in power simply read from one line to the next, such as going from LL1 to LL2 in essence multiplies the power by 10 if needed.
- Going in the reverse direction divides the power by 10, of course.
- For example read a value on –LL1 scale and find that value to the 10th power on –LL2 scale.
-
- Note : the –LL Scales are normally written in red. And the values increase from right to left (as do all other inverse scales)
- Recognize that –LLX scales are the inverse of LLX scales correspondingly. So the inverse of the values found on LL1 are found on LL01 also known as –LL1.
-
- **<u>Using the LLn Scales in general :</u>**
-
- To find X^N :
- Look up the value X on the LL Scale.
- Slide the Index of the C (1) scale over it.
- Move the cursor along the C Scale to the power (N) and
- Read below the answer on the needed LLX scale
- For example, 2 is squared, 3 is cubed, 5 is the 5th power – but you can do any power in between too – such as 4.7th power for whatever reason is needed :
- (be sure to estimate since it may have gone from one of the LLX Scales to the next one in line)
-
- **<u>One could find roots as well :</u>**
- For a given value, X find this on the needed LLX scale
- Place the needed root over it (N) on the C scale
- Note : Recognizing the this is read as 1/N mentally
- Now read in conjunction with the C scale Index the value on the appropriate LLX scale the root value
- Essentially taking the root is the reverse of the process of determining a power – much like doing multiplication and division with the slide rule in terms of directions.
- Note : this process works for –LL Scales as well.
- LL Scales are not good for numbers very close to 1, such as 1.001 or 0.999.
- There is an approximation for this value for small values of 'n' with :

  $(1 + n)^p = 1+d^p$

Name	Power	Range	Name	Power	Range
(+)LL3	$e^{+1.0x}$	e to 22k	(-)LL03	$e^{-1.0x}$	1/e to 1/22k
(+)LL2	$e^{+0.1x}$	1.1 to e	(-)LL02	$e^{-0.1x}$	0.91 to 1/e
(+)LL1	$e^{+0.001x}$	1.01 to 1.1	(-)LL01	$e^{-0.001x}$	0.990 to 0.91
(+)LL0	$e^{+0.0001x}$	1.001 to 1.01	(-)LL00	$e^{-0.0001x}$	0.999 to 0.990

Note : in table, the + and – denotation is used together, such as –LLN and +LLN, on some slide rules, while on others it is LLN and LL0N denotation depicted

<u>Tricks & Tips in Slide Rule Use :</u>

- Check alignment of scales and adjust as needed
- Always be sure of the level of precision of the scales for proper reading. Most 10" Slide Rules are quite good to 2 significant digits with even a 3rd estimated significant digit being possible.
- In terms of all calculations undertaken, Be sure to first estimate the answers ahead of time mentally
- Often it is a good idea to convert most values to scientific notion both for ease of calculation and determination of decimal placement.
-
- **Summary of Scientific Notation Rules :**
- **<u>The Rules for Decimal Placement :</u>**
- 1) Always first estimate the Answer.
- 2) Convert all values into Scientific Notation.
- 3) For Any Multiplied Values, Add up the Exponents and For Any Divided Values, Subtract the Exponents.
- 4) If for any single Multiplication operation the Slide moves to the Left, then add +1 to the exponent total (Why? Because in essence, we have gone off this scale and are adding two values that extend beyond the scale in front of us to a next one in line, which is 10x the line before it)
- 5) If for any single Division operation and the Slide moves to the Right, then -1 from the exponent total (Why? Because in essence, we have gone off this scale and are subtracting two values that place the answer on the scale to the left of the one in front of us, which is 1/10th the line before it)
- 6) Mentally treat any two values independently (just as whole numbers now) but keep track of what the exponent will be through the rules.
- 7) That is to say answer the question as to where the Slide has moved and factor in that addition or subtraction as needed for each operation independently.
- 8) Take this final answer and use the exponent figure arrived at to determine the decimal placement!

- **Alternative Decimal Place Method :**
- **Summary of Decimal Rules using the Counting Digits Method found in some slide rule books :**
- The key to the digits method is to tally the number of digits in a number.
- The basic rule for this is this :
- For numbers X > 1 the number of digits is simply the number of digits in the number.
- For example : 7 has 1 digit, 70 has 2 digits, and 70,000 has 5 digits, etc.
- What if the number is greater than 1 but has decimals, though?
- For example the number is 23.45, how many digits does it have?
- It has only 2.
- In the case of X > 1, all decimal values are overlooked in the digit count.
- So the next question then is, what of values 0 < X < 1 then, what is their digit count?
-
- In the Digit Count Method all values 0.1 to 0.9 expressed as nothing more than 1/10ths values have 0 (zero) digits.
- So 0.6 has 0 digits, for example.
- With each decimal place it is like a reverse number line :
- 0.01 has -1 number of digits,
- 0.001 has -2 number of digits,
- 0.0001 has -3 digits,
- And so on...
- To help remember this the value of digits is essentially the negative number of zeroes past the decimal point.
- Summary of these ideas on Counting :
- For numbers greater than 1, count all the numbers up to the decimal and treat these values as positive. For numbers less than 1, count the number of places up to the first nonzero number and view these as negative values. Sum up these numbers in multiplication and when the slide extends to the right, subtract one from the sum.
-
- With the number of digits in all the values you have in your problem now look to the following rules when it comes to multiplication and division :
-
- <u>Multiplication with C & D Scales :</u>
- 1) If the slide projects to the right of the stock during multiplication, the digit count for the product is one less than the sum of the digit counts for both the values in your calculation (called the multiplicand and the multiplier).
- 2) If the slide projects to the left of the stock, the digit count for the product is equal to the sum of the digit counts for the multiplicand and the multiplier.
- <u>Division with the C & D Scales :</u>
- 1) If the slide projects to the right of the slide rule stock during a division, the digit count for the quotient is one more than the digit count for the dividend (the numerator) minus the digit count for the divisor (the denominator).
- 2) If the slide projects to the left of the stock, the digit count for the quotient is equal to the digit count for the dividend minus the digit count for the divisor.

-
- Multiplication with the C1 & D Scales :
- 1) If the slide projects to the right of the stock during multiplication, the digit count for the product is equal to the sum of the digit counts for the multiplicand and the multiplier.
- 2) If the slide projects to the left of the stock, the digit count for the product is one less than the sum of the digit counts for the multiplicand and multiplier.
- Division with the C1 & D Scales :
- 1) If the slide projects to the right of the stock during division, the digit count for the quotient is equal to the digit count for the dividend minus the digit count for the divisor.
- 2) If the slide projects to the left of the stock during division, the digit count for the quotient is one more than the digit count for the dividend minus the digit count for the divisor.
-
- Other Ideas :
-
- Examine the problem carefully, working from the inside out, and use the best scale for that calculation
- Recognize that values can be found easily with given scales – squaring, square rooting, cubing, cube roots, sine values, etc. When moving the cursor its alignment may find it at a place to make use of these as needed.
- Proportions (which also conversions and 3-variable functions can be treated as) are very straightforward and easy to set up and solve on a slide rule.
- Note that most things are Ratios or Proportions and the Slide Rule is the best tool for these calculations since it sets up all similar ratios instantly with any setting!
- Create a needed scale! Take the activity where we used the C & D scales with the C1 scale. If there is no C1 scale, simply invert the slide in the slide rule and use the reversed C scale as if it were a C1 scale now.

- **How to Add Numbers on a Slide Rule :**
- Suppose we have two numbers X & Y so that we want their sum (X + Y) and yet use a slide rule!
- Let's first rearrange this expression : $X*(1 + \frac{Y}{X})$
- With a slide rule simply set the Y value on C over the X value on D.
-
- If Y > X read answer above the D index from the C scale as whole number with decimal and add one to it. Note that the decimal placement rules also must apply however.
- OR
- If Y< X read answer at D index on the C scale as decimal value and add one so that it is 1.***. Note that the decimal placement rules must also apply however.
-
- Now Take the result and find it on D and place this value under the left C index.
- Now, Regardless of Y > X or Y < X, now read along the C scale to the X value and find the answer for the sum of X + Y on the D scale below! Be certain to

have an estimate of your answer and employ the proper decimal placement rules for reading the slide rule

-
- **<u>Other Things to find on the Slide Rule :</u>**
- There are many ratios that can give good estimates in situations :
- Set up these ratios and find the known value and opposite it is the sought after answer -

$\frac{\text{Circumference of Circle}}{\text{Diameter of Circle}} = \frac{355}{113}$ $\frac{\text{Feet}}{\text{Meters}} = \frac{82}{25}$ $\frac{\text{Atmospheres}}{\text{Feet of Water}} = \frac{23}{780}$

$\frac{\text{U.S Gallon}}{\text{Cubic inches}} = \frac{1}{231}$ $\frac{\text{Feet of Water}}{\text{Pounds per sq inch}} = \frac{60}{26}$ $\frac{\text{Side of a Square}}{\text{Diagonal of a Square}} = \frac{70}{99}$

$\frac{\text{US gallons}}{\text{Liters}} = \frac{14}{53}$ $\frac{\text{US Gallons}}{\text{Imperial gallons}} = \frac{6}{5}$ $\frac{\text{Inches of Mercury}}{\text{Feet of Water}} = \frac{15}{17}$

$\frac{\text{Yards per Minute}}{\text{Miles per Hour}} = \frac{88}{3}$ $\frac{\text{Pounds per sq yard}}{\text{Kgs per sq meter}} = \frac{46}{25}$ $\frac{\text{Weight of fresh water}}{\text{Weight of sea water}} = \frac{38}{39}$

$\frac{\text{US Gallons of Water}}{\text{Weight in pounds}} = \frac{3}{25}$ $\frac{\text{Ounces}}{\text{Grams}} = \frac{6}{170}$ $\frac{\text{Diameter of Circle}}{\text{Side of equal square}} = \frac{79}{70}$

$\frac{\text{Inches}}{\text{Centimeters}} = \frac{26}{66}$ $\frac{\text{Pounds}}{\text{Kilograms}} = \frac{75}{34}$ $\frac{\text{Area of Circle}}{\text{Area of Inscribed Square}} = \frac{322}{205}$

Ch.V
Gauge Marks and Scales of the Slide Rule

The Scale

The primary use of the slide rule is seen in our activities for multiplication and division. This was done by using non-linearly divided scales that are divided instead in a logarithmic manner (that is to say the numbers were not evenly spaced like on a ruler but here by logarithmic distances).

Why use the logarithmic spacing of the numbers ? The most basic linear form utilizes two scales where the numbers are logarithmically spaced. Due to the mathematical properties of logarithms, the spacing of the numbers on these lines allows for easy and rapid multiplication and division with the same set of rules despite the type of slide rule used. (This idea is explored in the history section in more detail discussing the discovery and importance of logarithms to history which the slide rule can be seen as the physical visual manifestation of this math form).

Many slide rules have more than 2 scales, and these can be used for many other mathematical operations as we have seen, such as : squaring, square roots, cubing cube roots, raise to various powers, taking a various root of a number, common and natural logs, and trigonometric values such as sine, cosine, and tangent.

Standard scales spaced in a logarithmic fashion, like C and D are said to be **Single Scales**. That is, there is one run from 1 to 10 in the distance for the slide rule, such as 25 cm (i.e. 10 in). If there are two scales running from 1 to 10 in the same distance, it is said to be a **Double Scale**, such as for A or B scales. Of course if there are 3 runs of 1 to 10, which is the case for the K scale, this is a **Triple Scale**.

Another type of scale is the **Folded Scale**. Instead of starting at the Index point of 1, it begins at or near another point on the logarithmic line. The usual choice is at pi (π). The choice of pi was two-fold. One it allowed for using pi as an Index, so that calculations involving circles and cylinders could easily be done and two, by cutting the original C or D scale at a different point, made alignment more convenient for times when multiplication involved numbers from each end of the regular fundamental scale.

The C and D scale are the primary or fundamental scales of the Slide Rule and all others are based on them. For example, we see from the activities, the tangent (as well as the sine) scale is read in relation to the C or D scale. This idea is further illustrated in the table which lists in alphabetical order the scales, their mathematical relationship to C or D and a brief description of that scale.

Table of Scales and Uses

SCALE	Mathematical Relation to C or D	DESCRIPTION and Other Notes
A	X^2	Square Values of Fundamental D scale Double scale, Opposite to B scale
A1	$1/x^2$	Reciprocal of A scale, Reciprocal of square of D scale
B	X^2	Square Values of Fundamental D scale Double scale, Opposite to A scale
C	X	Fundamental (Single) Scale On Slide opposite to D scale
CF	πx	Folded Fundamental Scale, starts at π Opposite to **DF scale**
CI	$1/x$	Reciprocal of Fundamental Scale On Slide – Basically a Reverse Scale
CIF	$1/\pi x$	Reciprocal of CF scale Reciprocal of Folded Fundamental Scale C
D	X	Fundamental (Single) Scale On the Stock opposite to C scale
E	e^x	Log-log scale – see LL3
K	X^3	Cube Values of the Fundamental D scale Triple Scale
L	$Log_{10}x$	Mantissa of the common logarithm of Fundamental D scale value
LL	Ln(x)	Mantissa of the natural logarithm of the Fundamental D scale value
LL0	$e^{0.001x}$	Scale yields 'e' raised to 0.0001*x power, where x is read from the fundamental scale. Positive Log-log scales are used to raise a number to some exponent or find roots >1 LL scales are to enable one to raise values to a power and take roots very readily
LL1	$e^{0.01x}$	Scale yields 'e' raised to 0.01*x power, where x is read from the fundamental scale.
LL2	$e^{0.1x}$	Scale yields 'e' raised to 0.1*x power, where x is read from the fundamental scale.
LL3	e^x	Scale yields 'e' raised to x power, where x is read from the fundamental scale.
LL00	$e^{-0.001x}$	Scale yields 'e' raised to (-0.001*x) power, where x is read from the fundamental scale. Negative power means $1/e^{-0.001x}$ Negative exponent log-log scales are used to raise numbers to a power or find roots <1

LL01	$e^{-0.01x}$	Scale yields 'e' raised to (-0.01*x) power, where x is read from the fundamental scale. Negative power means $1/e^{-0.01x}$
LL02	$e^{-0.1x}$	Scale yields 'e' raised to (-0.1*x) power, where x is read from the fundamental scale. Negative power means $1/e^{-0.1x}$
LL03	e^{-x}	Scale yields 'e' raised to (-x) power, where x is read from the fundamental scale. Negative power means $1/e^{-x}$
P	$(1-(0.1x)^2)^{1/2}$	Pythagorean Scale. Cosine of $\sin^{-1}$ D scale
R1	$\sqrt{\ }$	Square root of Scale D value Scale twice the length of D scale R1 runs 1 to 3.2, R2 runs 3 to 10 Aka: W1, W2 and Sq1, Sq2
S	$\text{Sin}^{-1}x$	**Scale D** value is the sine of the angle in degrees read on the **S scale** Runs 5.7° to 90°
ST	Sinx, Tanx	Sine & Tangent of small angles 0.58° to 5.73° Same and S&T scale Used since sine and tangent are similar at these angles
T	$\text{Tan}^{-1}x$	**Scale D or C and CI** value is the tangent of the angle in degrees read on the **T scale** Runs 5.7° to 84.3° D scale read for angles 5.7° to 45° increasing CI scale read for angles 45° to 84.3° (printed in red and in reverse order)

The table above is by no means all of the scales, nor does it trace the differences in names for the same scales as they changed through time. It is a representation of the most common scales and represents a list that most slide rules commonly have a subset of.

As noted earlier, one could simply use a 9 scale form (such as in the activities) and be able to compute the overwhelming majority of problem types that even a regular scientific calculator is capable of today.

This does not mean that the other scales were merely for show, though there was probably some pride and showing off in the office if one had a more advanced model I imagine. The other scales were applicable to various disciplines, such as chemistry, physics, electrical engineering, and the like.

This idea is particularly true when one considers not just the right scale combination form, but also relevant and useful gauge marks that may be on the scales. Some are listed and described below in the accompanying table. Like the scale table, this is not complete and represents a cross-sectional view of them.

Table of Gauge Marks on the Slide Rule

Besides the numbers on a slide rule perhaps you have noticed some out of place marks or letters? These are gauge marks. These marks are at specific places corresponding with their value and have a given prescribed mathematical value to the user of the slide rule.

The table below lists some of the more common ones. This table is very far from complete but illustrates some of the possibilities. Note that not all makers use the same letters nor do they have all of the same marks on their various models. Some were probably put on there since those models may have been marketed to a particular set of professions where that mark would have value and use.

Look at some of the examples such as symbols for the weight of copper conductors, watts in one horsepower which clearly had applications in the electrical industry. Other constants as gauge marks can include the acceleration due to gravity for calculations in physics.

More common ones include the conversions of radians to degrees and vice versa as angular measures in calculations were commonplace.

Probably as no surprise the most common gauge point is π, since it can be used for all circular calculations. Using π we realize the value in this. Knowing where it is at allows quick calculations of either multiplication or division as needed.

Gauge Marks	Meaning	Value
C	Square root of $4/\pi$ Circle area calculations	1.128
G	Acceleration due to Gravity (metric)	9.8 m/s^2
G	Acceleration due to Gravity (English)	32.2 ft/s^2
L	Natural log of 10 Convert $\log_{10}$ & $\log_e$	2.3026
R, p, or r	$180/\pi$ Convert radians & Degrees	57.3
p′	Minutes in a radian	3438
Q	Radians in one Degree	0.01745
	$\pi/4$	0.7854
π	Pi – ratio of circle Circumference to Diameter	3.1416
W	Weight of copper	111000
-	Watts in one Horsepower (hp)	745.47 (746)

There are also common conversion marks and value marks with no special symbols on various slide rules, such as on the Fowler's Circular Calculators. Many of them have value marks for square-root of 2, square-root of 3, along with pi, and conversions for inches to centimeters, kilograms to pounds, square centimeters to square inches, and the like.

Activity #1
Using Math (and the Slide Rule) in Everyday Life

Grade Level : Middle School
Math Level : Calculating

Everyday Life Calculations with the Slide Rule Activity

Hello Inquisitive Pioneers – Ms Briana Andromeda Purcell, the Galaxy Girl here and here is our first adventure and it involves only the everyday! There are so many parts of life that involve numbers if we let them be a part of it, so that math becomes a part of our thinking process. We might ask ourselves how many miles per gallon is my car getting? How many gallons of paint would I need for this project – painting the walls in a room say? If I change the number of people in the recipe how do I change the recipe I plan to use ? Connecting ideas to the numbers that are involved and practicing mental math and even learning the slide rule for fun and to encourage good mental math practices strengthens our skills, reasoning, and understanding and in the case of these everyday life calculations can be empowering. Have fun! This first Activity is the mathematical exploration of everyday life – such as determining miles per gallon, cost per unit ounce and the like, but all using a slide rule. Even if you don't, though I recommend it, these are basic things we should all know and engage in mentally to some degree. It is good to practice common math sense! :)

- All of these calculations require either a question on the part of the person to consider the hypothetical or the actual items at hand to work on for calculations.
- The materials needed depend on what is being done.
- Know the basic rules for multiplication and division on a slide rule (they are summarized below and if needed in the first question – miles per gallon).
- When scales other than C & D are used the rules are explained.
- In most cases, only the C scale and D scales are needed.
- Always estimate an answer and watch decimal placement.
- Not always needed but is handy is a small pad of paper.
- In the practice to learn the slide rule – run through a problem a couple of times to see if it is the same result and in the early go of it check your answers on a calculator to see how well you are doing.

Slide Rule Basics

1) For almost all calculations in this Activity, the C & D scales are used.
2) If considering a Ratio or Fraction, It is best to see the C Scale as the Numerator since it sits atop the D Scale and the D Scale as the Denominator
3) To Multiply, Place the Right or Left Index as needed of a given scale, say the D Scale below the first number to be multiplied on the C Scale.
4) Next, read along the D scale to the other number to be multiplied by and read above it the answer on the C scale!
5) If the number is not available, use the other index (Right or Left) and start the process over as needed.
6) If Dividing simply place the Numerator value on the C Scale over the Denominator Value on the D scale.
7) Read the Answer opposite the Index of the D Scale on the C Scale.

8) What about the use of Scientific Notation?
9) It is best to consider all values in Scientific Notation in all calculations. This is done by setting the decimal point past the first non-zero number in the value under consideration and multiplying by 10 raised to the appropriate power.
10) The power is determined to be positive if the decimal needs to move to the right to obtain the original number (2000 is 2×10^3 for example), and negative if the decimal needs to move to the left to achieve the original number (0.002 is 2×10^{-3} for example).
11) When multiplying with values in Scientific Notation, simply add the exponents (Result = Operand 1 Exponent + Operand 2 Exponent).
12) When dividing by Values in Scientific Notation, simply subtract the Denominator Exponent from the Numerator Exponent (Result = Numerator Exponent – Denominator Exponent).
13) Note the special case rules for linear slide rules :
14) If Dividing and the Slide Projects Right, Then Subtract One from Your Exponent Total to achieve the correct decimal placement.
15) If Multiplying and the Slide Projects Left, Then Add One to Your Exponent total to achieve the correct decimal placement.
16) Here is a full summary of decimal placement :
17)

- **<u>The Rules for Decimal Placement :</u>**
-
- 1) Always first estimate the Answer.
- 2) Convert all values into Scientific Notation.
- 3) For Any Multiplied Values, Add up the Exponents and For Any Divided Values, Subtract the Exponents.
- 4) If for any single Multiplication operation the Slide moves to the Left, then add +1 to the exponent total
- 5) If for any single Division operation and the Slide moves to the Right, then -1 from the exponent total
- 6) Mentally treat any two values independently (just as whole numbers now) but keep track of what the exponent will be through the rules.
- 7) That is to say answer the question as to where the Slide has moved and factor in that addition or subtraction as needed for each operation independently.
- 8) Take this final answer and use the exponent figure arrived at to determine the decimal placement!

<u>Miles per Gallon Measured</u>

1) This is a good place to start – the miles per gallon – here we have in our minds not just how effective is our car at using its fuel but in the back of our minds with this one is the cost per gallon too. At the end of this exercise take time to take the miles per gallon and them look at it from the point of view of cost (the price of a gallon of gas) per gallon too.

2) For this calculation it is best to begin with a full tank of gas. Once full write down the number of miles on the Odometer [called the starting miles] (or reset the trip odometer to zero).
3) Drive for some period of time (it can be a day, but if you have a regular work schedule it may be better to go for 2 or 3 days).
4) Needless to say you put more gas in some time later. At that time write down 2 numbers.
5) First the current reading of the Odometer now [called the ending miles] (or look at the trip odometer and write down the number of miles driven).
6) Second write down the exact amount of gas put in this time. This is the <u>number of gallons used</u>.
7) The 10" slide rule will give reasonable results despite the number of digits in your numbers, but it is best to round off as needed to facilitate the calculation.
8) If you did not use the trip odometer, determine the difference between your starting miles and your ending miles by subtracting them. This is your <u>trip total</u>.
9) Quick review for Division and Multiplication on a Slide Rule :
10) To Divide place the numerator read from the C scale over the denominator on the D scale and read the answer opposite the D scale index on the C scale.
11)
12) To Multiply place the left (right if needed) Index of the C scale over the first operand read from the D scale. Read along the C scale to the other operand and read opposite it on the D scale to find the Answer.
13) To determine the <u>average miles per gallon</u> :

$$\textbf{Mpg} = \frac{\textbf{trip total}}{\textbf{number of gallons used}}$$

14) This value is important in many ways – On long trips and the car is gassed up, you can now estimate the range the car will go before filling up by *multiplying the Mpg by the number of gallons.*
15) On a map, knowing the distance to some place an estimate of the amount of gas needed can be determined by *dividing the Trip Miles by the Mpg.*
16) Also estimated costs can be found by taking *the number of gallons used and multiplying by the average cost of gas at that time.*
17) This calculation can go along with the Average Speed one as well.

<u>Average Speed along with Distance and Time Determinations</u>

1) To find the average speed what is needed is the total distance travelled and the total time taken for the trip.
2) This calculation can be done for cars, bikes, walking, boats, or any moving object under consideration.
3) If by car, either determine the total distance to be traveled by examining a map (paper for those who like to figure it out themselves or on the internet and a map system like Google Maps) OR
4) Actually travel the distance and either reset the trip odometer before the trip or write down the current odometer reading (starting miles) and then at the end of the trip write down the new odometer reading (ending miles).

5) In either case of the trip, it must be traveled and times.
6) For the traveled trip be sure to not only keep track of miles but use a watch or a timer. On a watch mark down the start and end times.
7) Determine the miles traveled by subtracting the starting miles from the ending miles (Miles Traveled) and
8) Do the same for the amount of time to travel by subtracting the start time from the end time (Travel Time).

9) The Average Speed is found by :

$$\mathbf{V = \frac{Miles\ Traveled}{Travel\ Time}}$$

10) Note that one does not need the total miles and total times to determine an average speed for part of a trip.
11) If you are on a long trip and had the start time and start miles down, then at any point in the trip the average speed for that portion of the trip can be determined.
12)
13) This might be useful in the case of a very long drive where if the total miles to cover is known and the average speed is determined for some point in the trip, then the amount of time needed is found by :
14) First subtract the miles covered from the total miles. *Take the remainder and divide this value by the average speed. This will give the amount of time in hours left for the journey.*
15) Estimated times can be determined for an entire trip by first estimating the average speed and dividing it into the total miles of the trip. This gives the number of estimated hours for the trip.
16) If there is a situation where the average speed is known and the amount of time is also known then *the distance traveled is simply the average speed times the amount of time traveled.*
17) Always watch to see that units match up as needed! If not convert units to agree with the problem at hand.
18) A good exercise is to convert units for those who like math practice. For example, convert mph to miles per second or feet per second.
19) Remember to keep decimal placement in mind here when using the slide rule – be sure to look to the summary of rules at the start of the Activity if needed
20) A good way to test your skill at using the slide rule is check your answer with a calculator too
21) Miles per Second = mph / 3600
22) Feet per Second = Miles per Second * 5280
23) Also why not try the metric conversions as well, such as mph to km/hr and then convert these to m/s.
24) Kilometers per Hour (km/hr)= 1.6(09) * Miles per Hour
25) Meters per Second = (Km/hr) / 3.6
26) These ideas apply to taking a flight too. Know the distance flown and time the trip to determine the average speed of the plane.

27) Note that distances do not have to be miles, it can be any measurement unit. A bicycle can be looked at from the number of houses it passes in a given time, for example. A crawling insect might be done in centimeters/minute.

Cost per Unit Known (Mass, Volume, Weight, Item Number)

1. All of us should be comparative shoppers – it helps us stretch our dollar and find the best deals. This next exercise helps in this :
2. Most stores today will provide the cost of an item per unit ounce, per pound, per unit number of the item. The first question then to determine is this : Is it correct?
3. The next goal of this activity is to compare it to other brands or the same brand on the shelf only packaged in a different size.
4. Note that often the larger sizes are not labeled in the same manner of cost per unit as the smaller ones.
5. Be sure to read the labels carefully since the unit cost provided might be something like cents per ounce yet the material is labeled in dollars and pounds.
6. Therefore it is a good idea to keep in the back of one's mind basic conversions, such as :

- **1 pound = 16 ounces**
- **1 cup = 8 ounces**
- **1 gallon = 4 quarts**
- **1 quart = 2 pints**
- **1 pint = 16 ounces**
- **1 kilogram = 1000 grams**
- **1 kilogram = 2.2 lbs (pounds)**
- **1 ounce = 28.3 grams**

7. With these in mind it is best to choose base units for calculation and comparison, for example turn all units to ounces for mass or volume considerations.
8. Reading a label first determine the total number of ounces if not given.
9. To calculate the cost per unit ounce (CPO):

10. Cost per Unit Ounce (CPO) = $$\frac{\textbf{Cost (\$)}}{\textbf{No.of Ounces}}$$

11. For all other types of costs per unit whatever it is simply the cost divided by the number of ounces (number of pounds) (number of items) and the like.
12. In all cases these are rates and a rate is a ratio or fraction.
13. The best way to treat all ratios is to let the Numerator (here Cost) be the C scale and the Numerator be the D scale (here the number of Ounces). The answer is read opposite the D scale index on the C scale.
14. Once you know the cost per item it is easy to determine the total cost of a purchase by taking that ratio and multiplying by the number of items you intend to buy.
15. Be an effective shopper!

Computing Sales Tax or Determining the Tip

1) This activity is not a mistake though these two things are different circumstances. It turns out they are calculated in the same way on the slide rule.
2) For both you are mathematically taking a cost and adding to it a percentage of the cost.
3) Read the Cost on the C scale and place the left index of the D scale beneath it.
4) Now read along the D scale to the known sales tax rate or the desired amount of tip.

5)
6) Here each mark on the slide rule represents 1% or 1/100.
7) For example, The second mark is 2%, the fifth mark is 5%.
8) The easy mistake is misreading the D scale. Be sure to keep track of the decimal point here.
9) For example if the cost were $20 the first mark past the left index on the D scale has a reading of $20.20, the fifth mark is $21.00
10) Yet if the cost had been $200, then the first mark has a reading of $202 and the fifth mark has a reading of $210.
11) This illustrates the power of the slide rule since all numbers can be represented on this logarithmically-spaced number line and it is merely the decimal point that needs to be tracked.
12) When you read to 1.1 this is 10%. The value above it on the C scale is 10% greater than the original value.
13) If you want the amount of tax or tip you have to subtract the original cost from this new value to see how much the tip is.

Calculating Cost of a Sale Item (given percent off)

1) Many of us have been to stores proclaiming some given percentage off and large tables on cards describing what one pays.
2) With a slide rule this is a very easy task.
3) First, we need to acknowledge the trick in advertising. If it is 10% off, we are still paying 90% (100%-10%), and we pay 75% if it is 25% off (100%-25%) and so on.
4) Knowing this makes the task of finding our actual cost very easy and we can even find our savings too.
5) Read the Original Price on the C scale and place it over the Right Index of the D scale.
6) Now read from right to left along the C scale and find the percentage you are paying.
7) For example, if you have 10% off, read to 9 which is now considered 90% and read above it on the C scale your price.
8) If you have 25% off read to 7.5 since this is 75% and read the price above on the C scale.
9) What if you want the amount of savings? Instead of the right index, place the Price over the Left Index of the C scale.

10) Reading from left to right find the percentage off and find its value on the D scale above.
11) Note that you may have to use the Right or Left Index differently if the values go off the scale!
12) Notice the speed of this reading. In a single moment one can determine what any percentage is for any given value. Try that one with a calculator for those who think that faster!
13) By the way go back to the sales tax activity above if there is tax on it to determine the total cost if needed.

Computing Pay Amount from Rates of Pay

1) For estimated gross pay, take the rate of pay (R) on the C scale
2) Place this value over the index on the D scale
3) Read along the D to find the Number of Hours worked (H)
4) Find the answer on the C scale above (G)
5) Notes : First note that the pay rate has to be in the same time unit as is multiplied. Second, this is the gross without any withholding. To factor that in take an estimated percentage of pay withheld for taxes and use the percent off calculation to find an estimated Net Pay. Third, if there is overtime pay, be sure to calculate those hours at that pay rate and add that to the base gross pay.

G = R*H

Estimating the Electric (Water) Bill from Meter Readings

1) This activity here examines directly the process of reading the meter, taking the values there and finding an estimated bill (excluding taxes and other fees on the bill).
2) Depending on the meter you have for the Electric and Water Meters, this will affect how it is read.
3) There are some that give a direct reading of the current level of consumption, and if present then you can proceed with the calculation.
4) If the number is not directly on it, then you need to start on the first day of the month if you want to monitor it for the month. (If for a week, pick any day).
5) Read the dial carefully. Most have dials and these go in opposite directions with each dial, first clockwise, then counterclockwise, etc.
6) When the arrow is between numbers be clear on how to determine what value it is. The arrow is always the value it is coming from and until it reaches the next value on the dial.
7) If the meter is in a dial pattern, be sure to start with the largest place value and work your way around the dial for the other digits.
8) In both the Electric and Water Meter cases, you need the starting value and then an ending value some time later (keep track of the time between your readings – a month is a good time frame).
9) Simply take the Final Number you record and subtract the First Number you record. This Net Amount Used is what you consumed.

10) In the case of Electricity it is normally in kilowatt*hours (kWh) while in the case of Water it is 100s of cubic feet of water (Ccf).
11) Strangely, the Water Company typically charges for Ccf and not gallons. For those who want to know the gallons, take the Ccf number and multiply by 748 as there are 7.48 gallons in a cubic foot)
12) In both cases, one can call or look up on line the average cost per kWh or Ccf and then simply multiply your use by these values respectively to arrive at an estimated cost.

Cost = Net Amount Used x Rate

13) Note that this cost is estimated, since it does not factor in any type of tier-pay system for cost changes for different amounts, any taxes, fees, and other costs that may appear on the bill.
14) Fun to note, here in the book, Activities 3 & 4 help us explore doing some readings using some simple tools to even look at where these total usages come from by looking at individual items and their measures – for example the amount of water used in a shower.

Home Project Needs for Painting and its Cost

1) The most basic of calculation is simple enough, first measure the Length (L) and Height (H) of a wall in a room in the chosen units (ft or m) and calculate the Area (A). (**A = L * H**)
2) To make the process simpler and avoid mistakes round all values up to the next whole number value. Most walls are 8 ft tall for example.
3) Be sure to sum up all of the walls. TA = total area (**TA = Sum of all Areas**)
4) Not Recommended , But For those who like some level of precision, in the case of inches divide the inches past the feet measure by 12 and tack this decimal value onto your feet measure, instead of the rounded up values used.
5) For still greater precision, you can subtract out non-painted areas, such as windows and doors (A = L*W) if you like.
6) Whatever the case, take the final number TA and divide by 300. This gives the number of gallons needed for rough, textured, or unpainted wallboard.
7) If it is smooth walls instead divide TA by 350.
8) Note that these are estimates. Always overestimate and round up. For example you are not going to buy 3.3 gallons, buy 4.
9) The final calculation is cost, simply take the Cost per gallon on C scale place over 1 on the D scale and read along the D scale to the needed number of gallons. The Cost is found on the C scale above.

The next several calculations look at Home Projects and Recipes – clearly it is about how much of a resource is needed for that given project.

Home Project Needs for Wallpapering and its Cost

1) The most basic of calculation is simple enough, first measure the Length (L) and Height (H) of a wall in a room in the chosen units (ft or m) and calculate the Area (A). (A = L * H)
2) To make the process simpler and avoid mistakes round all values up to the next whole number value. Most walls are 8 ft tall for example.
3) Be sure to sum up all of the walls. TA = total area (TA = Sum of all Areas) This is the Wallpapering Area.
4) For still greater precision, you can subtract out non-papered areas, such as windows and doors (A = L*W) if you like, instead of rounding up. Subtract the total of the non-papered areas from the papered ones
5) Use the chart below to find the Usable Yield Value. This is divided into the Wallpapering Area to determine the Number of Single Rolls Needed.

$$\textbf{No. of Single Rolls} = \frac{\textbf{Wallpapering Area}}{\textbf{Usable Yield}}$$

6) The final calculation is cost, simply take the Cost per roll on C scale place over 1 on the D scale and read along the D scale to the needed number of rolls. The Cost is found on the C scale above.

Pattern Repeat (Drop)	Usable Yield (American rolls)	Usable Yield (European rolls)
0 to 6 in.	32 sq ft.	25 sq ft.
7 to 12 in.	30 sq ft.	22 sq ft.
13 to 18 in.	27 sq ft.	20 sq ft.
19 to 23 in.	25 sq ft.	18 sq ft.

Home Project Needs for Carpeting and its Cost

1) The most basic of calculation is simple enough, first measure the Length (L) and Width (W) of the room in the chosen units (ft or m) and calculate the Area (A). (A = L * W)
2) To make the process simpler and avoid mistakes round all values up to the next whole number value.
3) *Not Recommended But For those who like some level of precision, in the case of inches divide the inches past the feet measure by 12 and tack this decimal value onto your feet measure, instead of rounding up.*
4) Determine the Number of Square Yards Needed by taking the Total Area (in square feet so far) and divide by 9.

$$\textbf{Number of Yards Needed} = \frac{\textbf{Total Area in sq ft}}{\textbf{9}}$$

5) Note that this Calculation is for Vinyl Flooring too! Follow the same procedure noted above down to the cost here below.
6) The final calculation is cost, simply take the Cost per yard on C scale place over 1 on the D scale and read along the D scale to the needed number of yards needed. The Cost is found on the C scale above.

Determining Recipe Needs through Proportions

1) These directions addresses the question of what to do when a recipe does not match your materials on hand and/or the question of how much is needed when you are changing the scale of the recipe to a larger or smaller yield.
2) This is the type of activity that the slide rule does very well since it is a proportion, which slide rules excel at.
3) The idea to solve this comes from noting that the ratio of the needed amount for a given ingredient in the recipe of a given number of servings will equal
4) The amount you need to have for your batch divided by the number of servings you wish to make!

$$\frac{\textbf{Recipe Requirement for Material}}{\textbf{Recipe Number of Servings}} = \frac{\textbf{The amount of Material Needed}}{\textbf{Number of Servings}}$$

5) This type of logic can be used for any and all variations of the same theme here. Remember to let the Numerator be the C Scale while the Denominator is the D Scale.
6) First place the known values of the Recipe over each other and look along the D Scale to the Number of Servings you wish to make and find the amount of the unknown needed material above. Be sure to employ the proper use of scientific notation when needed.
7) Keep in mind, you can determine how many servings you are making by looking in the opposite direction and going from what you have and looking back on the D Scale at how many servings it will make!
8) Go have fun in the kitchen with your recipes!

Basic Conversions :

1) The Slide Rule is the best conversion system around. This is because this operation is similar to the percentage calculation and the recipe calculation above.
2) Here, in this activity we can convert fractions to decimals or vice versa. We can also find equal fractions as well.
3) Also, if we need to convert one unit into another (inches into feet, feet into yards, ounces into pounds, inches into centimeters, and even complex ones like mph into kph) this is considered here.
4) In the case of fractions simply set the C Scale as the Numerator and the D Scale as the Denominator. Reading along the C Scale one finds all of the similar ratios.

5) (For example, if 2 on C is over 3 on D one finds 4 on C over 6 on D, and 6 on C is over 9 on D, et al. –
6) Not only that, but above the D Scale Index we find the decimal equivalent of the fraction, 0.66)
7) This means for any fraction or decimal we can find the other easily.
8) What is needed is first a ratio of what is known, say one unit over another. (For example 1 inch = 2.54 cm, 8 ounces = 1 cup, 1 foot = 12 inches)
9) Set up this ratio on the slide rule. It can be in either order where one is the C Scale value and the other is the D Scale value.
10) Now look for the other known value along the line that is known. On the opposite Scale will be the sought out answer.
11) (For example if Inches are on the C scale and Centimeters are on the D Scale and we wish to know how many centimeters are 4.5 inches, we read along the C Scale to 4.5 and find the answer below on the D Scale).
12) What this implies is that with a simple list of conversion factors, one can readily perform rapid calculations.
13) Conversions can include currency, one set of units into another, and the like.
14) For Currency conversions : 1 unit (say the dollar) equals N units of another currency. Set the Left Unit (1) over the equivalent on the D Scale. Reading along C is the number of dollars and below is the number of corresponding units in the other currency!
15) Another fact is this : Recall our calculations previously for miles per gallon, average speed, and cost per unit ounce.
16) Notice that each of these is a 3-Variable Formula where :
17)
18) What this means is that if the ratio is known (X & Y) then the answer is always found above the D Scale Index (N).
19) Also if we know the outcome value (N), we can estimate answers more readily.
20) (For example if our average speed is 45 mph, how long will it take to travel 280 miles. With 45 on C over 1 on D, read along C to 280 and below it on D is the answer : 6.2 hours)
21) This is how any and all 3-Variable equations can be treated!
22) Some of the basic set of 3-Variable functions needed are as follows :

$$\textbf{Rate} = \frac{\textbf{Cost}}{\textbf{\# Items, etc}} \qquad \mathbf{v} = \frac{\Delta \mathbf{d}}{\Delta \mathbf{t}} \qquad \mathbf{a} = \frac{\Delta \mathbf{v}}{\Delta \mathbf{t}} \qquad \mathbf{V = I*R} \qquad \mathbf{P = V*I}$$

23) We have used the first 2 here in this activity amongst others (area, etc) while the last 2 are Ohm's Law concerning voltage, current & resistance, and the final one is electrical power. The middle one is the formula definition of acceleration.
24) Note that there are many more and this is a small list. The Activities have these to learn from and use. Some involve the amount of food one eats and the graphical breakdown of these items. Another Activity calculates the amount of energy used and power exerted in exercising.
25) The key to this is your use of the Slide Rule as the means to connect math to the real world.
26) The Proportion Formula can be used in a number of cases (see other Activities involving its use) such as in the height of an object or distance to it.

27) For example if you know your height and measure the height of your shadow on a given sunny day, and measure the height of a shadow of another thing (tree, light pole, house) you can determine its height.

$$\frac{\textbf{Your Height}}{\textbf{Size of Your Shadow}} = \frac{\textbf{Height of Object}}{\textbf{Size of Shadow}}$$

Home Economics - Simple Interest on a Loan or Savings Account

1) This calculation is a quick look at the Basic Interest Equation and not the more complex one of being compounded daily or with some other time value.
2) Here, the Principal (P) is known, and perhaps the Interest Rate (R), along with the time period (T).
3) From these variables we can find the amount of accrued interest for a total time period (I).
4) This is a multiple step problem, so follow the directions.
5) First place P on the C Scale over the Index of the D Scale.
6) Read along the D Scale to R and find the answer to this on the C Scale, which we will call 'V'.
7) Now move 'V' to the D Scale Index and again read along the D Scale to the variable T.
8) Above T find the Answer I now on the C Scale.
9) Be sure to rearrange the formula as needed if solving for some other variable.

I = P*R*T

More Complex Loan Equation Calculations

1) This next excursion is the maximum use of math for this activity and requires concentration. Here we are going to examine the compounded interest rate on a loan or savings account.
2) First look at the equation :

$$A = P*(1+\frac{r}{n})^{n*t}$$

3) Here : P = Principal value, r = Annual nominal interest rate expressed as a decimal value, t = the number of years, n = number of times the interest is compounded per year, and A = Total Amount
4) All of the variables are known here except A. We want to determine the total amount owed or earned from the Principal at these given rates and conditions.
5) First compute with your slide rule r / n by using 'r' on the C Scale over 'n' on the D Scale. Call this X. Jot it down for reference.
6) Add 1 to X and multiply this by P on the slide rule.
7) For example place P on the C Scale over the D Scale Index and read along the D Scale to (1+X) and find the answer, Z, on the C Scale above.
8) Next multiply n*t on the slide rule. Call this M and jot it down.

9) Rewrite Z in Scientific Notation.
10) Reading Z (see step 7) on the D Scale, find the log (Z) on the L Scale. Recognize that this is the characteristic and the mantissa is the exponent from the Scientific Notation value. This new value is Q.
11) Now multiply this value, Q, by M (step 8) and find the value W.
12) Look up the characteristic (all of the numbers past the decimal) of W on the L Scale and find its corresponding value on the D Scale.
13) The decimal place is determined by the mantissa of W (the numeric value before the decimal).
14) With proper placement of the decimal, the value read on the D Scale is the Total Amount, A.
15) This takes time to do and to understand, but with patience, practice, and determination, you can succeed. As Slide Rulers say, Keep On Computing!

Well – there are many more of these ideas but the good thing this Activity is a good beginning at learning and using the slide rule plus noting all of the basic math relations in our daily life that have a use. Along the way you may find other formulas for other things in the everyday world and the nice thing is that it resembles the form of one of the formulae we have used here and guess what? You can then employ the slide rule in a similar fashion, only changing the variable names to match your situation and you have a template here to use as a tool.
As Slide Rulers say – Keep on Computing! I also add – Be Brilliant!

Activity #2
The key to the Slide Rule use - Proportions

Grade Level : Middle School
Math Level : Calculating

The World of Ratios & Proportions using the Slide Rule Activity :

The first thing to say is this, the Slide Rule was made for fractions, decimal representations, and proportions. The sliding scales allows one value to be placed over another readily to generate the needed relation. This is what this Activity explores. Let's Go!

What of the importance of fractions, decimals, and proportions, though? Obviously not all things are considered simply as whole objects and frequently a given object is analyzed in parts of the whole. A couple of quick examples are Earth's atmosphere (78% N_2 – nitrogen, 21% O_2 – oxygen, 1% all other materials), another is the Sun's overall composition (73% Hydrogen, 25% Helium and 2 % other materials), and so on illustrates the beginning point for a need to understand fractions and their expression as a decimal. This idea leads to the concept of examining parts of whole, which is the simplest view of what a fraction is.

A more precise way to consider a fraction is this. It is any two values A & B such that B ≠ 0 (zero) and they are expressed as a ratio of A:B or :

$$\frac{A}{B} \quad \text{or} \quad \frac{\text{Numerator}}{\text{Denominator}}$$

Fractions and Ratios are of great use in everyday life. One of the most common is the Scale. A map might say 1" = 2 mi (1 inch equal 2 miles). This is a ratio and can be rendered as 1":2mi. Also, the English measurement system has a large number of fractional uses for parts of an inch (1/64, 1/32, 1/8, ¼, and so on) which also applies to tools and parts such as drill bits, socket sizes, and the like. Also they are present in cooking with measuring cups marked off in fractions of 1 cup, such as 1/8, ¼, and ½.

All decimal values are really fractions in another form. One-tenth expressed decimally is 0.1 and fractionally is 1/10. The decimal system is a 10-based system, so that any decimal placement is that number divided by a power of 10 associated with that place value. 0.01 is 1/100.

The 10-based system makes conversion of decimal to percent very easy as well. Take any decimal value and multiply by 100 to convert it to percentage. For example, the decimal value 0.34 is 34% and is 34/100 or 34 parts of the 100 parts of the whole. What is important here is not only the simple conversion, but how to get the value in the first place. How does one find the fractional value of one part to the whole? This is the process of constructing a pie chart.

An important application is sales in stores which note things as either fractional values off (¼ off, 1/3 off) or use percentages (such as 20%, 25%, etc). Decimal values have an enormous number of connections to our everyday lives in the form of interest rates, such as sales tax. This idea is summarized both here and in the Everyday Life calculations using the slide rule to find these values.

Beyond the economics of the ratio of two numbers is the fact that many formulae can be seen as ratios (average speed is total distance divided by total time for the trip, density is mass per unit volume, pressure is force per unit area, and so on). The ability to compute with these formulae is important, but also to understand the relation of the variables to each other and when graphed, their slope determines the outcome values. In fact the whole idea of slope the rise over the run or m = $\frac{\Delta}{\Delta}$ is a fraction itself. The concept of slope is a **rate** or ratio and the slide rule is very functional for this idea (examples of rates : cost/unit volume, mass, or item – miles per hour – gallons per day – cubic feet per second – inches per year - et al). Some of these are considered in the activity to illustrate the use, but the full use stems from the slide rule user, while others have been addressed in other activities (Everyday Life Calculations, for example). In fact, it can be said that rates are common (interest rates, rent and other types of payments are seen as monthly/weekly/daily rates, and even the news about people and things is often presented as ratios too such as '3 out of 5 people do this or like that').

Much of Science can be seen as the change of one variable with respect to another (as seen in many of the activities in the book) hence it is the Rate at which something happens.

Obviously having a good mental model of dealing with numbers as ratios is a great tool to increase not only speed but accuracy as well as abilities to understand the outcome when one variable is manipulated what happens to the outcome. This is at the heart of a good science experiment for example. We control the Independent variable – typically the x axis value and we see and measure its effect on the Dependent variable – here the y axis. We are looking for patterns of change – this ratio as it were is the heart of good science.

In the case of regular everyday formulae this ratio mental model is an useful tool. For example look at average speed ($v = \frac{\Delta d}{\Delta t}$) . If the trip were to take the same amount of time, yet the distance covered was doubled, what then happens to the average speed? It too is doubled. If instead the distance is constant but the time is doubled, what then happens to average speed? It is half as much as the original value.

Beyond the realm of the fraction, its decimal equivalent, and its numerous applications to formulae is when there are two ratios set equal to each other, which then is a Proportion.

Proportions are relationships between things or sometimes parts of things which have comparative magnitude or quantity. By definition it is the equivalence of two ratios.

Any time one converts from one form of units into another one is using a proportion. For example consider the question "How many centimeters are in 18 inches?" It would be set up this way on paper:

$$\frac{\textbf{18 in}}{\textbf{X}} = \frac{\textbf{1.0 in}}{\textbf{2.54 cm}}$$

On the slide rule simply set 1.0 Index on the C scale over 2.54 on the D scale. Read along to 1.8 on the C scale and there below is the answer on the D scale (45.7). Notice how now you can convert any and all values instantly once the primary ratio is set up. All the others are now in place. Better still it is merely the movement of the decimal for any and all other conversions! Wow – how fast and easy is that!?

Conversions can include units from one type to another and these can be one unit (such as distance, time, mass, etc), mixed units (speed, acceleration, force, etc), and can include such things as monetary conversions too amongst other things. This will be further explored in the activities with a few other examples. Keep in mind, the limit to the slide rule is only set by your exploration of math and your imagination!

Proportions are very useful in recipes to scale them up or down from the directions given. (Find this is Everyday Life Calculations Activity) Also, it turns out in nature there are a number of laws and rules that can be seen as proportions, such as the ratio of sides in one triangle to another in a similar triangle. (Note this applies not only to triangles but any two similar geometric forms).

$$\frac{\mathbf{A}}{\mathbf{B}} = \frac{\mathbf{C}}{\mathbf{D}}$$

$$\frac{\textbf{Recipe Requirement for Material}}{\textbf{Recipe Number of Servings}} = \frac{\textbf{The amount of Material Needed}}{\textbf{Number of Servings}}$$

The goals of this project is to examine the ratio, the fraction, its decimal equivalents, conversions, and other relevant proportions and illustrate that the slide rule is designed for just such an exploration and makes the objective of evaluation easy and efficient. One might find that they can actually solve a proportion faster on a slide rule than on a conventional calculator. This is because the slide rule is made to have two scales to move in relation to each other. One scale can act as the Numerator of the Fraction and the other by default is the Denominator. Once any fraction is set up all similar fractions are instantly established.

For example, place 1 on the C scale over 2 on the D scale. Now simply read along the C scale and find 2 there over 4 on the D scale, 3 on C over 6 on D, and so on. Though this is a simplified view and easy to know, this works for any and all fractions and their similar ratios as needed.

A fascinating side note, as it were, to these fractions is their use in angular measurements. This idea is fully expressed in terms of numbers and their uses in the Altitude / Distance Activity where similar triangles are employed to determine the unknown side of a triangle when a similar triangle is known and there is only one corresponding sides known on the similar triangle.

In the case of angular measurement, this idea of the ratio of two numbers can be related to a circle, where the numerator is the span of the arc of the circle and the denominator is the circle's radius. If in the same units this creates a ratio that measures in radians!

One radian is where the ratio is one (the arc length equals the radius). The natural question for most of us is, what angle is this in degrees?

Consider this, let's say we had an initial angle of 180°. Also let's make the radius 1 unit of length. The arc length numerator will be p (3.14***) units in length! Hence there are π

radians in 180°. Back to the question of 1 radian. To find the answer, place 180 of C Scale over π on the D Scale. Read above the D Scale Index and find that this is an angle of 57.3°. This is one radian!

Why are radians useful? One of the simple answers to this is that all one needs to know is the ratio of arc length to radius to find an angle that an object subtends (i.e. its angular size) [this is used in astronomy all the time]. Also this ratio (given limitations of measurement and other considerations) can be seen as the side of a triangle as mentioned previously for similar triangles and hence unknown sides can be uncovered.

What about 3-variable functions and the use of proportions ? In the case of a formula like Density ($\rho = \frac{M}{V}$), one scale is Mass and the other is Volume. One can visually see the relation and what happens as the scales are moved. Density, itself is found on the C Scale over the Index of the D Scale.

$$\frac{\textbf{Density}}{\textbf{1}} = \frac{\textbf{Mass}}{\textbf{Volume}}$$

Better still is the notion of any 3-variable function being seen as a proportion. All one has to do with formulae like density or average speed is take the main variable (density or average speed) and place it on the C Scale over the Left or Right Index of the D Scale. Treat the Scales as the ratio of the other two variables (for density this is mass for C and volume for D) (for average speed this is distance traveled for C and time for travel for D). If one knows any two of these variables, the third is found rapidly. Say you have an estimated speed of travel and the total distance for the trip to be traveled. Setting up the proportion as described one finds below the distance on the D Scale the amount of time this will take.

This is the value of the slide rule as a visual tool for learning the relationship of variables. In the case of speed once set on the C Scale and now reading along the C Scale for distance and D Scale for time one visually senses how the relation results in answers. For any given speed it requires more time for a greater distance traveled, hence we read further along the scale. One must read along the scale and can visually see and consider the distance and time considerations.

The slide rule acts as a sort of tangible bridge to begin the building of our mental models and the relations between variables in 3-variable functions and basic proportions. Wow this is awesome sauce! This then is one of the fundamental strengths of the slide rule. It acts as not only a means to perform a calculation but becomes an extension of our ideas as we take the physical world and consider it mathematically and wonder about the relations between values, variables, and changes in them.

Purpose : To examine Ratios, Rates, Fractions, Decimal Equivalents with the slide rule.

Purpose : Exploring Proportions with their applications in similar figures, scale values, changing amounts for recipes, and other related formulae,

Purpose : Calculating common Conversions in a proportion-manner with a slide rule from many areas

Purpose : To use the slide rule as a visual tool to examine basic 3-variable functions as a proportion (such as speed, density, et al)

Materials :

- Proportion Formulae,
- Example Story Problems with proportions,
- Quick Activities for illustrative purposes,
- Slide Rule

Procedure : In a simple term – Practice!

Basic Rules in Reading along with Multiplying & Dividing with a Slide Rule :

1) In most exercises the use of the C & D Scales is done, but other Scales will have basically the same rules (such as the A & B).
2) First note that most if not all of the Scales are read with respect to the C & D Scales.
3) For example the square of a given value from the D Scale is found on the A Scale.
4) The cube value of the same value on the D Scale is found on the K Scale.
5) And the Log Value of the value on the D Scale is on the L Scale.
6) To Multiply, Place the Index of one of the Scales (for this description we will use the C Scale) over the first of the multiplicands on the other adjacent scale (here the D Scale).
7) Now read along the index scale (C Scale here) to the other multiplicand and find the answer on the adjacent scale (the D Scale).
8) For Division, place one value over another on the sliding scales (for the example, place the Quotient as the C Scale value over the Divisor on the D Scale).
9) Read along the divisor scale (here the D Scale) to the Index that is adjacent to the answer on the other scale (here C Scale).

The Fraction :

1) Place any value on the C Scale over any other value on the D Scale. You now have a Numerator over a Denominator and hence a Fraction.
2) Fractions can be made for a variety of reasons :
3) Try for example 1 on C Scale over 2, then 3 and so on. Realize that the slide is moving to the left. The Right Index of D is moving closer to our fraction, hence the number above it is a smaller value, so ½ is larger than 1/3 and this is larger than ¼ and so on.
4) Similar Fractions :
5) One of the most common is to find other related fractions. To do so, once you have done step 1, now read along the C Scale and find where any two numbers, one on the C Scale and one on the D Scale, line up! This is a similar ratio or fraction.
6) For example for ¼ you will find 2/8, 3/12, and so on!
7) **The next most common thing is to find the decimal equivalent of a fraction.**
8) Again after step 1, now look to the D Index where it is below a value on the C Scale.
9) When treating the C & D Scale as a Fraction : If the value of C Scale < the value of the D Scale, then the value on the C Scale above the D Scale is a decimal value. Otherwise it is not.

10) **What of the decimal point?**
11) Recall the rules for Scientific Notation and the use of the Slide Rule (these are found in the chapter on using the Slide Rule and the first Activity too):
12) **In summary, when in Scientific Notation, Add exponents when Multiplying figures, and Subtract exponents when Dividing figures.**
13) **The key to the rules is not only employing the Scientific Notation method, but knowing when to add or subtract one from the total.**
14) **If Dividing, Subtract one if the Slide Projects Right, and**
15) **If Multiplying, Add one if the Slide Projects Left.**
16) Here is a full summary of decimal placement :
    - **The Rules for Decimal Placement :**
    - 1) Always first estimate the Answer.
    - 2) Convert all values into Scientific Notation.
    - 3) For Any Multiplied Values, Add up the Exponents and For Any Divided Values, Subtract the Exponents.
    - 4) If for any single Multiplication operation the Slide moves to the Left, then add +1 to the exponent total
    - 5) If for any single Division operation and the Slide moves to the Right, then -1 from the exponent total
    - 6) Mentally treat any two values independently (just as whole numbers now) but keep track of what the exponent will be through the rules.
    - 7) That is to say answer the question as to where the Slide has moved and factor in that addition or subtraction as needed for each operation independently.
    - 8) Take this final answer and use the exponent figure arrived at to determine the decimal placement!

17) Note that the Numerator can be larger than the Denominator and this still works on the Slide Rule.
18) If still uncertain follow the direction on the 'How To use a Slide Rule page'.
19) In fact, the ratio of the two numbers will provide not only the whole number in front of the decimal, but also the following decimal equivalent of the remainder.
20) What if you do not want the remainder in decimal form?
21) Convert the decimal to a fraction -
22) <u>What of converting from a Decimal to a Fraction?</u>
23) Place the decimal value on the C Scale over the Right Index of the D Scale and then search along the C and D Scales for a ratio of numbers to represent them.
24) Keep in mind this decimal is the percentage value, but you need to multiply in your head by 100 to see it as a percentage.
25) For example, 3 on the C Scale over 8 on the D Scale has the Right Index of D under 375 on the C Scale.
26) Since 3 < 8, the value is less than one and should be read as 0.375
27) What of the Scientific Notation Rules. Each initial value has the exponent 0 and in division, 0 – 0 is 0. But the slide is projecting to the right so we subtract 1 from the result and have -1 for an answer. This means to move the decimal one to the left, and the answer is read 0.375
28) As a percentage, multiply by 100 and the answer is 37.5%
29) <u>One of the largest uses of fractions is for Sales!</u>
30) Instead of step one, place the Price of the Item on the C Scale over the Right Index of the D Scale.
31) Read backwards along the D Scale to 9, 8, etc. Each of these is read as N x 10% (such as 90%, 80% and so on). Above it is the Price at that percentage!
32) Of course, keep in mind, this is not the percentage off, but what you are paying. To see how much is saved, just use the Left Index of D Scale under the Price instead and read to the right instead of left.
33) For example, 25% is found at 2.5 and the value above it is 25% of the price. (Of course what you pay is found at 7.5 or 75% instead).
34) Sales tax and total cost can be found in a similar manner :
35) Place the Cost of the item over the left index of the adjacent scale 9 say C over D as we have been doing)
36) Read along the D scale to the sales tax expressed as a decimal value (this should not be too far, 4% is 0.04, 6% is 0.06, et al)
37) The value above on the C Scale is the total cost including sales tax!
38) Other Fractions or Ratios in Everyday Life :
39) Mpg = miles per gallon
40) Take the ratio of miles driven over the number of gallons used sometime to determine just what gas mileage you are getting!
41) Mph = miles per hour
42) What was your average rate of travel for a trip?
43) Take the ratio of the Miles driven to the amount of time (in hours) to find the average speed of the trip!
44) What if you are some fraction of the distance there and have determined the average speed, then

45)
46) The question becomes how much longer 'to grandma's house?'
47) Take the remaining distance and divide by the average speed to estimate the number of hours to complete the journey.
48) Further, from the mpg calculation you could take the remaining distance divided by the mpg and find the number of gallons needed for the trip in order to decide whether to fill up again or not!

49) The fraction as a Slope in Algebra

50) A ratio might not be just two numbers, but instead represents the ratio of two differences of numbers :

$$\mathbf{m = \frac{\Delta Y}{\Delta X}}$$

51) This is the slope of a linear line, one of the largest topics in algebra.
52) The slope is the rate of Rise Over Run. The larger the value, the steeper the line.
53) The sign of the slope determines whether the line is moving up or down when examining it from left to right.
54) You can consider slope as a 3-variable formula (see below).

55) The Fraction as a Rate :
56) In many practical applications of the fraction it is seen as a **Rate**. The numerator is one value (distance, gallons, cubic feet of gas, amount of growth, temperature change, etc) while the denominator is some other value (typically time).
57) The rate could be determined or it may be known in a given problem. If to be determined, the amounts of the other two variables are known
58) If the rate is known, then clearly we are missing either the amount in question flowing or the amount of time needed to do this.
59) A very good example of a Rate is in Cost per Unit Ounce (Volume), etc. This is a valuable tool for the slide rule in comparing items when shopping for their comparative costs to find the better deal.
60) The flow rate is how water and natural gas consumption is measured and billed. For water and natural gas are Ccf (100s of cubic feet).
61) In order to estimate the Cost, one has to merely read the meter at the beginning and the ending times, subtract to arrive at an amount used and multiply this by the cost per unit. This is true for all of the meters : Water, Electrical, and Gas.
62) Rates and Ratios are very common in Science :
63) In Physics there are many ratios, such as Density (amount of substance per unit volume), Pressure (force per unit area), Power (Joules per second), etc.
64) Also in Physics, rates can be speed (rate of change of distance with respect to time) or acceleration (rate of change of velocity with respect to time) or from Newton's 2nd Law acceleration is the ratio of Net Force to the mass of the object undergoing the net force.

65)

66) Chemistry has many ratios such as Molarity (the number of moles of solute to liters of solution), Molality (the number of moles of solute to kilograms of solvent), Percent by Volume [Mass] (the volume [mass] of Solute to the volume [mass] of solution), the Law of Definite Proportions where the Percent by Mass (is Mass of the Element divided by Mass of the Compound), as well as the Law of Multiple Proportions (these could be looked at in the Proportion section obviously), and so on.

67) In Chemistry rates are seen in things such as rate of reaction (is the negative of the rate of change of reactant to change of time) plus many more.

Conversions :

1) In Conversions there are two basic methods that can be used with the Slide Rule :
2) First, treat the two items as a Fraction that is to be multiplied by the Number in question.
3) Typically the ratio of the items is taken where the units one is 'in' are in the denominator, while the units one wants to convert into are in the numerator.

4) $\dfrac{\textbf{Units to Convert Into}}{\textbf{Units to Convert From}}$ **or** $\dfrac{X}{Y}$

5) This results in this situation :
6) Beginning Value in Initial Units* $\dfrac{\text{Desired Units}}{\text{Initial Units}}$ = Final Value

7) $N * \dfrac{X}{Y} = M$

8) The best way to handle this on a slide rule is to :
9) Place N on the C Scale over Y on the D Scale on the slide rule
10) Read along the D Scale to X, the Desired units and read the answer, M on the C Scale.
11) Looking at the prior discussion, it is easy to see that the formula presented can be read as :

12) $\dfrac{M}{N} = \dfrac{X}{Y}$

13) All one has to do is take the ratio of Convert Into Units on the C Scale over the Convert From Units on the D Scale
14) Now read along the D Scale to the Beginning Value (N) and find the answer above it on the C Scale (M)
15) Use the following Table and look elsewhere for everyday conversions. It is best to keep a small list and memory of these as needed :

16) **1 inch = 2.54 cm**
17) **12 inches = 1 foot**
18) **3 ft = 1 yard**
19) **365 days = 1 year (rounded)**
20) **1 solar day = 24 hours**
21) **1 cup = 8 fluid ounces**
22) **1 gallon = 4 quarts**
23) **1 pound = 16 ounces**
24) **1 kilogram = 2.2 pounds**
25) **1 ounce = 28.3 grams**
26) **1 liter = 1.06 quarts (rounded)**

27) Plus look up whatever you may need !
28) For each of the above and any others your find a need for, simply place one value over the other as described above it for use.
29) <u>Conversions are needed in many applications :</u>
30) One basic unit to another (inches into feet),
31) Changing one unit type into another (English to Metric),
32) Currency Conversions, and others.
33) <u>Computing Costs</u>
34) The reason for conversion depends on the needed outcome. For example feet into yards is commonly used when computing the area of a room in square-yards for carpeting by dividing by 9.
35) In the case of painting, the Slide Rule readily calculates the area of a wall with Length times Width, but what of the number of gallons of paint needed?
36) Take the total area to be covered and use the Slide Rule to divide by 300 if the walls are unpainted or rough – if smooth and already painted divide by 350 – to determine the number of gallons of paint needed!

37) **In Math, angles are often converted between Radians and Degrees :**
38) **180° = π radians**

39) In the Sciences, there are numerous conversions not only of the aforementioned units, but also of mixed units :
40) Such as km/hr to m/s is very common.
41) For that calculation the ratio for it is 3.6/1. Check it yourself!
42) In Chemistry there are conversion commonly found in the amount of substances and how it is expressed :
43) For example take the number of grams of a substance and divide by its gram molecular weight to determine the number of moles present.
44) In Science and Math there are many other conversions depending on the situation at hand.
45) In Math in the realm of trigonometry since the C and D Scales are the values for the Sines and Tangents as read from the S and T Scales (keeping in mind where the decimal falls), this makes it easy to multiply or divide by the sine or tangent of a value when and where needed.

46)
47) Slide the sine value for a given angle read on the C Scale from the S Scale over a given value on the D Scale.
48) Read in one way the sine is in the numerator and in the opposite direction it is in the denominator.
49) What if you want to find the Sine or Tangent value and are given the sides of a triangle?
50) Obviously this is again a ratio : For example :

51) $$\sin\Theta = \frac{\text{Length of Side Opposite}}{\text{Length of Hypotenuse}}$$

52) Also other angular measures are available and can be used for determining distance or size, such as in the Radian measures :

53) $$\text{Radian measure} = \frac{\text{Arc Length of Circle Portion}}{\text{Length of Radius}}$$

54) If our protractor measures $1/10^{th}$ of a radian, then the apparent size of the object in question is $1/10^{th}$ the measure of the distance from us!
55) Another interesting one in Math is the conversion from one base to another in terms of logarithms.
56) The standard slide rule has base 10 logs, but let's say you want a number in another base, say 2 or the natural log base e (2.71828*)?
57) For any positive numbers, N, A, and B with A ≠ 1 and B ≠ 1,

$$\log_A N = \frac{\log_B N}{\log_B A}$$

58) From the question the natural log of any value is the ratio of the log base 10 of the number divided by the log base 10 of the natural log value.
59) To illustrate, here is a similar example question :
60) What is the log base 2 of 6, for example. Log_2 (6)
61) Look up reading from the d Scale to the L Scale both the log of 2 and the log of 6. (0.301 and .778 respectively)
62) Now divide on Scale C and Scale D 778 by 301 - We find it to be 258
63) Where does the decimal go?
64) Since 6 is much greater than 2, it will have a characteristic (here 2 and the rest is the mantissa 0.58) so the answer is 2.58
65) So $2^{2.58}$ = 6 (try it and see, rounded off of course)
66) The idea of the ratio, fraction goes on to any and all applications.

The Proportion :

1) As noted in the prelude, the proportion is two ratios set equal to each other.
2) In the conversions section above, it is easy to see its value in use.
3) The basic proportion can be expressed as :

4) $$\frac{M}{N} = \frac{X}{Y}$$

5)

6) All one has to do is take the ratio of known values - M on the C Scale over N on the D Scale
7) Now read along the D Scale to the other known value (Y) and find above it on the C Scale the answer (X)
8) It is easy to see that all one needs to know is any 3 of the variables and the 4th is the one to find.
9) By using the scaled as fractions it is easy to place one value on one scale and the other on the opposing scale.
10) Try this for yourself to find that solving a proportion on a Slide Rule is indeed much faster than one can solve one on a Calculator!
11) Also here there are no complex rules, like the calculator, such as cross-products –
12) In the case of the Slide Rule the natural form is maintained which is the equivalence of two ratios.
13) This is an invaluable tool in Algebra for proportions as well as in geometry for any and all similar figures to determine unknown sides!
14) Proportions can be used to determine height or distances - In this case we are using similar triangles.
15) For example, hold up a ruler at arm's length to measure the apparent height of a distant object, say a picture on the wall.
16) If you read the apparent height, measure your arm's length, (this is the first ratio)
17) now measure the distance to the wall,
18) Set the first ratio of your measures to the ratio of the unknown height on the wall to the distance to the wall.
19) Then the height of the picture can readily be determined.
20) Though simple, this technique is used in surveying regularly and is used in the wilderness for distances across rivers and gorges before the advent of electronic equipment.
21) The easiest way to envision it is when you are outside and you as well as a tree or a flag pole casts a shadow on a sunny day.
22) The ratio of your shadow length to your height is equal to the shadow length of the flag pole to its actual height.

23) $$\frac{\textbf{Length of your shadow}}{\textbf{Your Actual Height}} = \frac{\textbf{Length of flagpole shadow}}{\textbf{Actual Height of flagpole}}$$

24) A more thought out activity involves determining the size of the Moon :
25) Use a meter stick and place a small card vertically with a hole punched in it at a distance

26) Look along the stick through the hole so that when viewing the full Moon it fully fills the diameter of the hole (i.e. move the card until you have proper alignment)
27) The ratio of the diameter of the hole to the distance that the hole is from your eye equals the actual diameter of the Moon to the Moon's distance from you. (Here, we assume we know the distance to the Moon). Hence the diameter of the Moon can be determined!

28) Proportions can also be used in changing the scale of a recipe :

29) $$\frac{\textbf{Recipe Requirement for Material}}{\textbf{Recipe Number of Servings}} = \frac{\textbf{The amount of Material Needed}}{\textbf{Number of Servings}}$$

30) Here the ratio of how much is needed to the number of servings is set equal to the amount of unknown material and the number of desired servings. The amount needed is readily found.

31) In Math there are many applications :
32) The very nature of the proportion stems directly from geometry and the relations found in similar figures.
33) These can be the aforementioned triangles but also includes any similar polygons, such as squares, rectangles, and the like where one is known, the other is partially known and there is a missing side.
34)
35) Still other examples are considered :
36)
37) What if one wants to find the circumference of a given diameter of a circle (or vice versa)?
38) Simply put π on the C Scale of the slide rule over the Index on the D Scale.
39) Now read along the known scale. The C Scale is the Circumference, while the D Scale is the diameter.
40)
41) What about the diameter if the area of the circle is known?
42) Take the Area on the A Scale over p on the B Scale.
43) The diameter-squared is found on the A Scale opposite the B Index.
44) The diameter is then found below reading from the A Scale to the D Scale to read the square root of the value on A.
45) Other interesting things can be done with gauge marks (special marks on slide rules for conversions, multiplications, et al like π) as well as personally derived values :
46) On some Slide Rules there is a mark, c, on the C & D Scales (1.273) which is $\frac{4}{\pi}$ and comes from $\mathbf{A} = \pi * \frac{d^2}{4}$
47) Place this C mark or value of the C Scale over the index of the D Scale
48) Slide the cursor over the size of the diameter of a considered circle on the D Scale.
49) To find the Area of a circle read the answer on the B Scale!

50) What if you want to do this by knowing the radius (of course we could simply multiply by 2 to use the former method, but give your mind a chance to explore the Slide Rule !)
51) Look up the square root of π by first finding it on the B Scale and noting its value on the C Scale.
52) Slide the square root of π value on the C Scale to the Index of the D Scale
53) Now read along the D Scale to any desired radius value for a circle.
54) The answer in this case is read on the B Scale for the Area of the Circle in question!
55) What about the Volume of a Sphere?
56) Now place the value 1.61(2) on the D Scale over the index of the C Scale.
57) Read along the D Scale, for the radius of a given sphere you are considering.
58) The Volume of this sphere is found on the K Scale!

59)

60) This comes from $\left(\frac{4*\pi}{3}\right)^{1/3}$ from the formula $\mathbf{V} = \frac{4*\pi*\rho^3}{3}$
61) There are many other problems in Math and Algebra in the area of problem solving:
62) For example, if a given material costs so much per ounce, pound, ton, how much will another desired amount cost ?

63) $$\frac{\textbf{Cost}}{\textbf{Unit Amount}} = \frac{\textbf{How much does it Cost?}}{\textbf{Amount Wanted}}$$

64) There are numerous ratios of values that can be found or derived from many other references that can be used in proportions to find the answer to many a question one might encounter in a math and or science text:

65) $$\frac{\textbf{Diameter of Circle}}{\textbf{Side of Inscribed Square}} = \frac{99}{70} = \frac{\textbf{Diagonal of Square}}{\textbf{Side of Square}} = \sqrt{2}$$

66) Here is a general value for the pressure one feels with depth :

67) $$\frac{\textbf{Pounds per Square Inch}}{\textbf{Feet of Water}} = \frac{26}{60}$$

68) Even more complex proportions can be solved :
69) A classic algebra question might read :
70) If it takes 4 people 7 days to accomplish a job, how much time is needed (assuming the same work rate) for 6 people to do this task?
71)
72)
73)
74)
75)
76)

77) $$\frac{\text{Initial Workers}}{\frac{\text{1 Task}}{\text{Amount of Time Initally}}} = \frac{\text{Workers in case 2}}{\frac{\text{1 Task}}{\text{Amount of Time needed}}}$$

78) $$\frac{Wi}{\frac{1}{Ti}} = \frac{Wf}{\frac{1}{T2}}$$

79) $$\frac{4}{\frac{1}{7}} = \frac{6}{\frac{1}{X}}$$

80) You could go through and first simplify it and then take the ratio of the numbers on one side once the variable is isolated, but the slide rule allows for this to be solved as is!?
81) Take 4 on the D Scale and slide 7 on the C1 Scale over it. Start with the cursor here.
82) Read along the C1 Scale to 6 and look below on the D Scale to find the answer 4.66 days.

83) In Physics, for example, say you have a balance beam.
84) If on one side of the balance you have 26 g a distance of 32 cm from the center,
85)
86) *how much mass must be placed on the opposite side at a distance of 20 cm from the center in the opposite direction so that it balances?*
87) cw is clockwise, ccw is counter-clockwise

88) $$\frac{\textbf{Mass cw}}{\textbf{Mass ccw}} = \frac{\textbf{Distance ccw}}{\textbf{Distance cw}}$$

89) $$\frac{\textbf{26 g}}{\textbf{X g}} = \frac{\textbf{20 cm}}{\textbf{32 cm}}$$

90) X = 41.6 g

91) This idea can be applied to problems in chemistry too.
92) Take for example, conservation of mass in a problem where one has to determine the mass of a reactant product in a total mass size where one is only given a small sample for testing.

93) $$\frac{\textbf{Mass of reactant in sample}}{\textbf{Mass of Sample}} = \frac{\textbf{Mass of reactant in total mass}}{\textbf{Total Mass}}$$

94) This list applies to any and all sciences and is limited only by the imagination in the questions being asked.

Summary :

Wow! Clearly there are many applications and uses for fractions, ratios, proportions and the idea of looking at various formulae as ratios and proportions both in the everyday life world and in the academic worlds of science, math, and engineering. This is the great thing about the slide rule! It can be used in all of these areas and in essentially the exact same way and it is the matter of the slide rule user to be the Inquisitive Pioneer – to reflect on the numbers, consider their relationships, use the slide rule as a tool to help in finding a path to the solution! In all, there could be nearly an infinite number of pages written here for the use and application of the slide rule to Ratios, Fractions, Decimals, Conversions, Rates, and Proportions. This is where the slide rule will be referred to in all of our Science Activities in this book, Galaxy Girl's Inquisitive Pioneer Adventures. Come back to these Activities (one and two) when and where needed for added ideas and practice if using the Slide Rule as your problem solving tool. Explore and Enjoy! Be Brilliant!

Activity #3
Calculating the Cost of Water Usage at Home

Grade Level : Middle School
Math Level : Calculating

The Cost of Water Usage with the Slide Rule Activity

A common question arises and grows into a set of questions – such as how did this water bill become this? Clearly it is the total amount used for that time, but are there places and things we do that add the most to this bill – surely. So why not make it an activity or analytical adventure – a sort of detective story for ourselves and investigate the sources of usage areas!

In the Activity, we will monitor the rate of water usage and estimate the amount of water used for the month and from that the cost of it on the bill. Note that this clearly has many other costs added often, such as sewage and other fees and taxes. This Activity will explore direct measurements unlike the Everyday Life Activity (Activity #1) that examines the meter reading for both water and electricity. The first goal of the activity is to explore the units that are used to describe water consumption (100 cubic feet or Ccf).

First we need to consider how they measure water usage. Most Water Usage is measured in Ccfs? This is a terms that means 100s of cubic feet of water. The capital letter 'C' at the start means 100. A family of 4 can use approximately 0.5 to 1.0 Ccf per day so that in a month this can be 15 to 30 or more Ccfs (this is a lower end figure and does not factor in outdoor water use – other figures may be 50 to 60 Ccfs per month). Also, many water bills may be for a time period of several months (3 usually) so that it will have the total usage for that time. The rates will vary but often are over $1 per Ccf.

Where does water use come from? The list is fairly straight-forward :

1) toilet use,
2) shower and tub use,
3) laundry,
4) dishwashing and dishwasher use,
5) hand washing, teeth brushing and other sink uses,
6) outdoor use – watering the lawn, wash car, etc

In order to successfully measure the consumption of water use in the home, we need to develop simple measuring tools – using a marked-off bucket and gallon jugs which will be used to measure the rates of water usage to find the gallons per minute. Then with simple estimates of rates, measure the amount of time of use to determine the total number of gallons which are then converted into cubic feet (1 cubic foot is 7.48 gallons).

The primary goal of this Activity is to monitor, measure, and calculate the usage and costs of water consumption. This activity extends from the everyday life calculations where the meters are read, and in that case, the total usage is known. Here it is the examination of the items that use the water usage.

With the monitoring of life style, appliance usage, one becomes more conscious of usage and hence this can lead to considerations of reductions or alternatives if applicable. For water, it could be turning them on less often and at lower intensity as well as valves to allow on-and-off while in use, such as in shower heads.

We compile lists of those items that use water and construct pie charts (and/or bar graphs) for comparison purposes.

Activity : Computing the Water Usage & Cost

Purpose : To estimate the cost and amount of usage of water usage in daily living for a month and computing their costs.

Materials :

- Gallon plastic jug (2),
- 2-3 small plastic cups or tubs (butter, et al),
- Timer,
- Bucket,
- Slide Rule

Notes to Pre-Activity :

1) The first primary note is this : have parental permission and help in the activity!
2) The key to the activity here is timing. You do not necessarily need a stopwatch, but this can be used.
3) Recognize this only gives an estimate value for both items since you are not measuring all of the sources of use. Also the bills for each are based not only on consumption, but also taxes, fees, and other items.
4) For the Water Usage Activity, it is best to use both a plastic bucket and one or two empty plastic gallon jugs. These will be used to determine the rate of flow when the water is on. For example, fill the gallon jug with water and pour this into the bucket. Now use a permanent marker to mark this level. If possible do this for a second and maybe a third one too. With the bucket marked and ready you can continue with the water usage activity.
5) Note a better way to have greater accuracy is to use the bucket but use measuring cups to mark off 0.25 gallon marks as well. This way estimation is easier if and where needed (such as determining the amount of water in something like a toilet or a washer).

Procedure :

1) See Notes for Pre-Activity above for prepping the bucket for use.
2) Use the bucket and measure the amount of flow rate for various water outlets in the house, such as the faucets and the shower.
3) You may consider doing it more than once and taking the average so as to have an accurate measure of the amount of water used.
4) Record each of these items and flow rates in the table below.

5) For items that are harder to determine, consider first looking them up on the internet or looking in the manuals for them (such as for the washer, dishwasher, and toilet).
6) If permitted (with adult supervision and permission), the toilet can easily be done by shutting off the water, removing the top and using a small plastic cup (such as a butter tub) and filling your marked off bucket.

7) The same measurement can be done for a washer (again with adult permission and supervision). Here run a regular load without any clothes or soap since we are only measuring the water use.
8) For the toilet and washer, merely measure the amount of water per use and the number of uses in a given time period (week or month). Note if a time period is not one month, you need to convert it to one month's reading (multiply by $\frac{\text{Number of Days in the Month}}{\text{Number of days monitored}}$).
9) When it comes to hand-washing, running the sink for dish-washing, water for one reason or another, merely keep track of time after the rate has been measured.
10) Turn the total number of gallons used by all items into 100s cubic feet. Then calculate the estimated cost of water usage.
11) Create a bar graph and/or a pie chart of these items.
12) If you have the rate of cost use the formula C = NCcf * R (where C is total cost and R is the Rate) to find the cost.
13) As part of this Activity, look up the rates for your water consumption which is needed for the formula.
14) Note : Though the Slide Rule is a recommended tool, all of these calculations can be done with a regular or scientific calculator. Some scientific ones even have built-in averaging formulae. For those who like spreadsheets, the data can be typed in and the formulae then also be typed in its own cell where the formula references each of the measured variables in their respective cells (for example in cell B1 the total minutes is typed and a later cell, say B3 has the formula where it occurs).

Data :

Item	Flow Rate (GPM)	Time Used (min)
Faucet in kitchen		

Item	Amount Per Use	Number of Times Used
Toilet		
Washer		

Tables for Use IF you cannot obtain rates or amounts :

Note that these are estimates and can vary

Item	English Amount	Metric Amount
Bath	30-40 gallons	100-150 L
Shower	5 gal/min	20 L/min
Washing Clothes	3-7.5 gal/load	12-30 L/load
Toilet flushing	3-5 gal/flush	10-20 L/flush
Dishwasher	15 gal/load	50 L/load
Cooking	8 gal/day	30 L/day
Watering a Lawn	10 gal/min	40 L/min

Calculations :

Be sure to use your Slide Rule!

Gallons Per Minute :

$$\mathbf{GPM} = \frac{\mathbf{60}}{\textbf{Time in Seconds for 1 Gallon of Water}}$$

Gallons Used (GU) = GPM * Mins of Water Running (min)

$$\textbf{Cubic Foot of Water (cf)} = \frac{\textbf{Gallons Used (GU)}}{\mathbf{7.48}}$$

Number of 100 cubic feet :

$$\mathbf{NCcf} = \frac{\mathbf{cf}}{\mathbf{100}}$$

Time Usage :

$$t\ (\text{hrs}) = \frac{\text{Total minutes}}{60}$$

Graphing Formulae :

Decimal Value of Usage :

$$\mathbf{D} = \frac{\textbf{Part Amount}}{\textbf{Total Sum of All Parts}}$$

Pie Chart Piece :

$$\mathbf{P = 360^{o} * D}$$

Conclusion :

So how did the detective work and investigation go for you?
The best way to examine this data is to look at the overall costs, where the highest costs come from, and if possible then examine ways to reduce these areas.

Summary :

The overall goal of this activity is to monitor ones consumption rates of water to help isolate the areas of maximal use and this first makes one focus on these uses and next on possible solutions for reductions.

Extension Activity :

Other Fun Measurements & Challenges !

One could do things such as determining the amount of water needed for a pool for example. The volume of a rectangular pool is simply V = L*W*H, while for a circular pool is $V = h*\pi*\frac{d^2}{4}$. Measured in feet this gives the cubic feet, which when multiplied by 7.48 equals the number of gallons needed. All one has to do is measure the rate of water discharge from their hose which is used to fill the pool (by timing how long a gallon bucket fills to then find the Gallons Per Minute for flow rate) and divide this into the number of gallons which will yield the amount of minutes to fill the pool!

Activity #4
Calculating Cost of Electricity Usage at Home
Grade Level : Middle School
Math Level : Calculating

The Cost of Electricity Usage at Home with the Slide Rule

Another critical use of resources is electricity as a cost generator for our lives and this activity has the same detective aspect to investigation to it as did the water usage activity in Activity #3. Here we are monitoring, tabulating and calculating the usage and cost of electrical power use.

The first goal of the activity is to explore the units that are used to describe electrical power usage (the kilowatt*hour or kWh) and then to estimate the use of electrical power and calculate an estimate cost.

Let's start with Electricity. Many of us have heard terms like Volts, Amps, Watts, and many others, but how are these related to determining the electrical needs and use of a given appliance. The American home has 120 V AC source for all connections. In reality, most appliances are rated for 110 V to 120 V. The appliances we plug in are designed for this voltage, but through their electronics, they may use different amounts of current (I), or the Amps. It turns out that the product of Voltage and Current is Power. Power is measured in Watts. (P = V*I). So, if an appliance does not have its rating in Watts, but has the Amps listed, then one has to only multiply this by 120 V to find its power rating in Watts.

What is a Watt? Obviously from the formula it is a volt*amp, but what is that. Volt units are Joules per Coulomb while Amp units are Coulomb per Second. When canceled out correctly this leads to Joules per Second. Joules is an amount of energy and Seconds is the amount of time that this energy is transferred. So 1 Watt means that for every 1 second of operation there is a transfer of 1 Joule of energy. For example, a 75 W bulb is converting and using 75 Joules per second of operation! Recognize though that a Joule is a small amount of energy. Lift a cup of water weighing 1 N (about ¼ lb) 1m in 1s and this requires 1 J of energy, to illustrate the point.

The wattage is not the use of the item, however. We are not paying for electrical power, but for electrical energy usage. Energy is Power times Time. So in the case of Electrical Power Usage Cost comes from the Rate per kilowatt*hour times the number of kilowatts a device uses.

So the kilowatt *hour is a unit of energy, but how does it compare to other forms of energy? 1 kilowatt*hour of electricity is 3,412 BTUs or 860,369 calories. These latter units are NOT food calories, those are 1,000 of these calories listed. (How about other forms of energy? 1 cubic foot of natural gas has 1,000 BTUs or 252,164 calories of energy available).

Notice that it is kilowatts and not watts as well. This means for that 75 W bulb, we have to divide by 1000 first to make it 0.075 kW. If we run it for 10 hours then it would use 0.75 kWh. Typical costs range and depend on where one lives, the power source, cost for peak hour usage, and may also be in some sort of tier-cost system, but ranges from 12 cents to 30 cents per kilowatt*hour. If we take our bulb example and have a cost of 20 cents per kWh, then the 75 W bulb at 10 hours yields 15 cents of total cost.

Realize that that is one bulb and on for a short time, plus the fact that there are many other costs factored into bills such as taxes, fees and the like. Also for a given number used there may be one rate, but as usage increases the cost per unit time can be higher (a step up in price). The number of kilowatt*hours used has a wide range and depends on the usage of electrical items, their power and amount of time, plus the number of household members using them. It can go from several hundred kilowatt*hours in a month (typical home value used ranges 600 - 1100) to a couple thousand kilowatt*hours for large-scale use in the home and even thousands of kilowatt*hours use in large-scale business.

Of course, any operating electrical device contributes to the electrical costs, but we need to recognize that just because a label has its wattage on it, it does not mean you are using that amount when it is on. The label is rated at its maximum, so that a microwave on lower power settings will use less electricity for example. The opposite can be true as well. Speakers may say 40 W, but this is the rating of the sound energy output and not the power needed to generate this, which could be double this amount. Also there are small devices that run a lot even though not 'on', such as the electrical clocks on VCRs, DVD players, and others with displays.

The primary goal of this Activity is to monitor, measure, and calculate the usage and costs of electrical power consumption (which is similar to the water usage in that Activity). This activity extends from the everyday life calculations activity where the meters are read, and in that case, the total usage is known. Here it is the examination of the items that use the electrical power.

With the monitoring of life style, appliance usage, one becomes more conscious of usage and hence this can lead to considerations of reductions or using alternatives if applicable (solar panels, et al).

What is fun and fascinating about this activity is that it is not only personal, but can lead to surprises – maybe something uses more or less power, for example, than one previously thought. For example a 5 W night light on for 8 hours a day for 30 days uses 0.12 kWh. That amount of electrical power consumption is equal to a microwave using 1200 W for 6 minutes of time.

We compile lists of those items that use electricity and water and construct pie charts (and/or bar graphs) for comparison purposes.

What is important to note here is that you have parental permission and supervision. Do not move large electrical appliances, do not tamper with their plugs or wiring. In most cases you can find the information about a given appliance from the guides provided with it, on-line, or on a label on the device (only if it is accessible)

Notes to Pre-Activity :

1. The Electrical and Water Activity have the same set of ideas in mind – monitor those things that use either electricity or water for its usage. Keep in mind your safety and operate in a safe manner. Have parental permission and supervision.
2. For the Electricity Activity, the best device to use for common objects, like televisions, microwaves, computers, stereos, and refrigerators for determining the actual usage is a device called a Kill-A-Watt. It is plugged into the device and then it is plugged into the wall. Decide on the amount of time it will be plugged in for and monitor it regularly. Record the results in the table below.
3. If you do not have the Kill-A-Watt device, then monitor the amount of time a given device is running and use tables (such as on this Activity or in the product's guide book or on-line) to determine the kilowatts that the device uses.
4. Obtain for Electrical Power consumption the rate(s) for kilowatt-hours used. Notice here that there may be more than one rate, what could happen is a tier-payment system where for a given range of usage, there is one cost, and it changes at greater amounts of usage.
5. Note that this activity can be considered an extension of the Everyday Life Calculations Activity where the reading of the Electric and Water Meters is done and the cost is calculated.
6. The key to both activities (Electrical Power Usage and Water Usage) here is timing. You do not necessarily need a stopwatch, but this can be used.
7. Recognize this only gives an estimate value for both items since you are not measuring all of the sources of use. Also the bills for each are based not only on consumption, but also taxes, fees, and other items.

Activity : Computing the Electrical Usage & Cost

Purpose : To estimate the cost and amount of usage of electrical power in a home during a month through measurement and calculation of kilowatt*hours for given appliances for cost determination.

Materials :

- Power Consumption Device, like Kill A Watt monitor,
- If no power consumption device, use Tables or Guides,
- Timer,
- Everyday home electric appliances,
- Slide Rule

Procedure :

1) Realize there are some items that you merely time after recording their wattage rating. This includes lights for example. You can also include any of the items in the table below. Note that this table is generalized and may not be your actual devices. Further research may be needed.
2) The major thing to do is to plug the Kill A Watt device into the wall and attach a given appliance, such as the refrigerator or television. Decide how long of a time it will be monitored. If for a week, realize you need to change that into the month by assuming that the usage for that time period will be constant for the other days. Rotate the device to other appliances that are not easy to monitor. – If using the Kill-A-Watt device, be sure to read the directions and follow them. It works with some items and not others.
3) Important Note : that the Kill A Watt will not work with high power appliances, like electric stoves, washers, dryers, and anything else of more than 120V. There are some power meters in the works for higher power ones, but none currently. Use the Table below for these ratings for cost estimates. In these cases, measure the amount of time it runs and multiply by the listed power rating.
4) If not using the Kill-A-Watt, then use tables-guides-internet information along with measured amounts of time the appliance is running to determine its power usage. Take for example the table given below on this Activity. You may have to estimate the power rating of your given device, since this table provides a range for some items.
5) Make a list as long as you want, but try to monitor the major appliances so that it reflects the greater electrical power usage.
6) Take all of your time measurements and turn them into hours.
7) Multiply the hours by the power rating (in kilowatts) for the item. This determines the Electrical Energy Usage.
8) Sum up the Electrical Energy Usage.
9) Determine the percentage and pie chart piece size for each appliance in your list.
10) Create either or both a Bar Graph and/or Pie Chart of Electrical Power Usage for each category of appliance.

11) This Electrical Energy Usage is then multiplied by the Cost (X cents per kilowatt) to find the cost.
12) Though the Slide Rule is a recommended tool, all of these calculations can be done with a regular or scientific calculator. Some scientific ones even have built-in averaging formulae. For those who like spreadsheets, the data can be typed in and the formulae then also be typed in its own cell where the formula references each of the measured variables in their respective cells, for example B1..BN has the measurements and values used in the equation while BN+1 has the formula for all of these variables (why not the A cells? Simple – use them to label you variables)

Data :

Appliance	Power Rating (W)	Time Used (hr)	Kilowatt*hours Total (kWh)	Total Cost ($)

Table of Common Power Ratings for Appliances :

Note : these are estimated ranges

For those with a range, you may have to do further research in your guide books or on the internet.

Appliance	Power Rating (W)
Electric Furnace	8,000 – 25,000
Electric Space Heater	600 – 1,200
Gas Furnace blower	750
Central Air Conditioner	3,500
Window AC	500 – 1,300
Ceiling Fan (speed & size)	25 – 100
Electric Clothes Dryer	4,400
Washer (cold/cold)	300
Washer (hot/cold)	2,800
Dishwasher (heated water)	3,600
Electric Oven	2,000
Desktop Computer & Monitor	150 – 340
Laptop Computer	45
Plasma TV (42" – 56")	100 – 500
LCD TV (32" – 42")	90 – 250
CRT TV (19")	55 – 90
Microwave or 4-slot Toaster	1400
Coffee Maker	900
Range burner	800

Calculations :

Be sure to use your Slide Rule!

Time Usage :

$$t\ (hrs) = \frac{\text{Total minutes}}{60}$$

Electrical Power :

$$P = V*I$$

Electrical Energy usage :

$$W = \Delta E = P*t$$

(P in Watts, t in hours)

Cost of Electrical Energy Usage :

$$C = \Delta E*R$$

(C = cost, E = Energy, R = Rate of electrical cost)

Graphing Formulae :

Decimal Value of Usage :

$$D = \frac{\text{Part Amount}}{\text{Total Sum of All Parts}}$$

Pie Chart Piece :

$$P = 360°*D$$

Conclusion :

Well detectives – how did the investigation go and what did the data tell you? As with all data look at and think about the results – perhaps even do some internet research to see if this is on par with what is regular – also look for solutions to excessive uses.

Examine the Total Costs of a given appliance, its amount of usage, and consider are there alternatives or ways to reduce use. Also examine the pie chart and recognize where most of the electrical bill comes from and consider ways to decrease the areas of largest use.

Activity #5
Some Indirect Measures determined

Grade Level : Middle School
Math Level : Calculating

Indirect Measures and the Slide Rule

Imagine being handed a ruler and a single sheet of paper and then asked to measure not is length or width which are easily found with the ruler (plus which are marked on the pack of paper if needed) but instead you were asked to measure its thickness!

What at first seems impossible is actually possible within a certain amount of error. How? Instead of a single piece, what if there were a stack of exactly 100 pieces? Could you measure the thickness of that? Sure. Now divide that by 100 to obtain the thickness of a single sheet! We are making some assumptions. First assumption, that all sheets have exactly the same thickness and second, that there is no appreciable gap between any of the pages.

Well Inquisitive Pioneers – this is one of the most important areas of science – how to estimate with a reasonable accuracy numbers that cannot be directly measured – the mass of a whale or a mountain, the number of stars in a galaxy, or even the number of leaves on a tree (there is an Activity in this book for that too in this case – wow) and so on.

Is this real science? Definitely. Most measurements in science are actually indirect. For example, realize that we cannot see an individual molecule, let alone an atom (as of yet – being worked on), yet from the behaviors of a large number of them characteristics of the individual members can be determined. (There is a related activity to read on how the size of an atom is determined – see A Marble's Small Size Determination Activity #20).

Also this technique works in both directions. Instead of taking a large number and inferring the characteristics of a single member, there are times when the values for a small sample of a population are extended to the whole. Samples are taken from a group in question and statistical mathematics is applied to reach reasonable results within a certain margin of error. Take for example even examining the characteristics of the human population. Look at the simple idea of counting the number of people on Earth. Even if one counted a number per second, it would take 100 years to reach a number far less than 4 billion. Keep in mind that the Earth is suspected to have 7.1 billion presently. That number is large but even larger is the number of stars in our home galaxy the Milky Way Galaxy. It is estimated that there are 200 billion stars in it and that there are some 100+ billion galaxies in the universe. (For scale, the number of stars in the Milky Way is 15x just Earth's current population and the number of stars in the universe is in the neighborhood of all the grains of sand on all the beaches in the world).

Could one really measure the exact mass of a mountain, an ocean, even a planet? What of the population of the insects in a forest or the number of leaves on a tree? From Astronomy consider the sheer number of stars in a galaxy. (see the Determination of the Mass of a Brick Wall in the Density of Various Items Determination Activity #13 – the volume determination method used in Experiment IV here (the volume of a paper clip) is used for a marble too).

Still other indirect measures come from the fact that we cannot have the item in our labs, such as the masses of large objects like whales or even the mass of a star located thousands of light-years away. All of these require indirect measures. They are inferred from our data and understanding of the known rules of science which is the fabric of the universe.

So Inquisitive Pioneers – here we will measure a number of simple things in this Activity (and there are a couple of others that rely upon Indirect measures as well – estimating pi by using darts, counting the number of leaves in a tree, measuring the density of a brick in a wall from a small sample, and the amount of charge on a pith ball and so on plus the other representative ideas here in this Activity – wow – awesome sauce! Be Brilliant – Explore and Enjoy – along with Ms Briana Andromeda Purcell – the Galaxy Girl.

Purpose : To use common items and measure them with tools that have limitations outside the range of these items physical properties such as mass, length, and volume.

Materials :

- pack of standard paper,
- box of paper clips,
- marbles,
- piece of cardboard,
- ruler,
- water,
- graduated cylinder,
- stack of pennies,
- mass scale,
- Slide Rule

Procedure :

Experiments Ia & Ib : Thickness & Mass of a Piece of Paper

1) Open a fresh pack of paper and remove the plastic wrapper. Be sure to not ruffle or separate the papers as much as possible.
2) Record the number of sheets in the pack as recorded on the package. (This needs to be verified later).
3) Place the stack at the edge of a table so that a ruler can be placed alongside the stack and begun at zero '0'.
4) Measure the thickness of the stack at two points along this side.
5) Rotate the stack and measure a third time along another side.
6) Record all of these values and take the average of this number.
7) Use a scale that can read the mass of the entire stack.
8) First place a stiff piece of cardboard on the scale and zero it out. – this is here to act as a plate or bridge to be able to set things on as needed.
9) Place the stack on the scale and measure its mass.
10) Verify the number of pages in the stack by counting them.
11) Calculate the average thickness and mass of the individual pieces of paper.

Experiment II : Mass of a Paper Clip

1) First place a cup on the scale and zero it out.
2) Start placing paper clips in the cup counting them as they are put in.
3) When the scale has read at least 1.0 oz (or about 30 grams), then record the number of paper clips used.
4) Record the mass reading on the scale.
5) Calculate the average mass of the individual paper clip.

Experiment III : Mass of a Marble

1) Note the directions below assumes that the scale is not sensitive enough for 1 marble. If so then take its mass, otherwise use these directions.
2) First place a cup on the scale and zero it out.
3) Start placing marbles in the cup counting them as they are put in.
4) When the scale has read at least 1.0 oz (or about 30 grams), then record the number of marbles used.
5) Record the mass reading on the scale.
6) Calculate the average mass of the individual marble.

Experiment IV : Volume of a Paper Clip

1) Use a 50 mL graduated cylinder and fill it to 20.0-25.0 mL with water. Record this initial value in the table. (Be sure to begin with enough water to cover the full volume of the paper clip)
2) Note that other sizes can be used but always use enough water to cover the paper clips put in.
3) Begin counting and placing paper clips in the graduated cylinder and watch as the volume changes.
4) Add between 8 and 12 (10 is a good number here). A smaller graduated cylinder can use 5.
5) Record the new volume reading.
6) Determine the change of volume and then the volume of a single paper clip.

Experiment V : Density of a Penny

1) Using a ruler, measure the diameter of a penny and record this in centimeters (cm).
2) Note that you will have to read the rule as accurately as possible (i.e. 3 sig figs).
3) Take a known number of pennies (10, 20, or 30)
4) Measure the height of the stack of these pennies to the same level of accuracy as the diameter and record this.
5) Record the mass of the stack of pennies from the mass scale.
6) Calculate the Volume of the penny stack.
7) Determine the density of a Penny.

Data :

Experiment Ia : Thickness of a Sheet of Paper
Experiment Ib : Mass of a Sheet of Paper

Thickness Measurement 1	
Thickness Measurement 2	
Thickness Measurement 3	
Average Thickness Measurement	

Number of Sheets in Stack :	
Average Thickness of Stack :	
Mass of Stack :	

Experiment II : Mass of a Paper Clip

Number of Paper Clips on Scale	
Mass of Paper Clips on Scale	

Experiment III : Mass of a Marble

Number of Marbles on Scale	
Mass of Marbles on Scale	

Experiment IV : Volume of a Paper Clip

Number of Paper Clips in the Graduated Cylinder : ________

Final Graduated Cylinder reading	
Initial Graduated Cylinder reading	
Change in Volume	

Experiment V : Density of a Penny

Number of Pennies in Stack : ____________

Mass of Pennies in Stack : ___________ g

Width (Diameter) of a Penny (d) : ___________ cm

Height of Stack of Pennies (h) : ___________ cm

Calculations :

Be sure to use your Slide Rule for each calculation. Since it is a ratio, it is merely taking one value over another on the scales. The C & D are good choices here.

For each of the experiments the calculation is merely the ratio of total mass (or thickness) to total number of items involved. This yields the average value for an individual item.

$$\text{Average mass (thickness)} = \frac{\text{Total Mass (Thickness)}}{\text{Total Number of Items}}$$

Volume of a Penny Stack :

$$V = h*\pi*\frac{d^2}{4}$$

Density :

$$\rho = \frac{m}{V}$$

Extension Activity :

Beyond using the Metric System for measurements here, one could do conversions as well into other units, such as ounces or inches.

For the teacher, questions can be posed since the average value has been calculated, questions about 'how many would be needed to reach' a given value are possible.

Another possibility is to redo the activity but use different numbers of items and see how similar the results are.

In using the graduated cylinder one could estimate the number of pieces of cereal in a box, the number of grains in a box, and so on. Simply count how many pieces fit in a known volume you have decided upon. Next calculate the volume that it came in, such as the volume of the cereal box (realize that it does not fill the entire box – can you estimate what volume the cereal takes up versus the whole of the box? . Divide these two figures and find how many estimated pieces should be in there!

Conclusion :

The first basic question is this : Do these results seem reasonable or not? Exploration on the internet or comparison to other repetitions of the activity or others performing the activity can answer this question.

Hopefully there was fun, curiosity, and some unexpected results in your findings.

Summary :

This activity is useful from a number of perspectives. The first is the use of tools to measure things. Proper use of tools and understanding how to read them is fundamental. Next the precision and the level of significance from the outcome of measures is important here as well.

Also it can be seen that questions that at first seem unanswerable without the use of unique tools is not always the case. Simple assumptions, which are common in science, are used here as well.

Activity #6
Finding Pi using everyday objects and by graphing

Grade Level : Middle School
Math Level : Calculating

Finding Pi with Measures and a Slide Rule

Ms Briana Andromeda Purcell here and looking at a Math Activity though it involves objects. We will do in the next couple of Activities an exploration and examination in finding a value for Pi, a remarkable number with a long history and a number of uses. Its main fame of course is that it is a constant value (though an irrational number, meaning that it continuously goes one indefinitely with no pattern or ending) that we find from the ratio of the Circumference of a Circle (the distance all the way around) divided by the Diameter of the same Circle. This value comes from any and all circles as long as they are circles and not out of shape.

The following is a short article containing a simple exercise that can be done with a simple slide rule and becomes a great launching point for discussion with a class for a teacher or parent(s) with their son or daughter. It is the 'discovery' of pi in natural by using human-made objects which can be found as the ratio of the circumference to the diameter of a given circle which is the constant, yet irrational number pi (3.14...). Here we take measurements, graph the data, and doing some basic calculations. In the end with a good sample size, good measurements, we discover the value of Pi using Lids, a piece of graph paper, a sewing measuring tape, a ruler, and a slide rule of course.

A classic and easy to do classroom Activity for discovering Pi (π), which will be referred to as the Pi Project, is to have a collection of circular objects (coins, coffee-soup-tuna cans, plates, cups, circular lids from many common items, pot-pan lids, et al) and a set of measuring tools (rulers, measuring tapes, calipers) and to have students take measurements for a given object of their circumference and diameter in an organized way by creating a table of this information. Note that in the absence of too many measuring tools, and only having rulers, for example, use string to span the distances (both the diameter and the circumference) and then measure the string measurements with the rulers.

Note the easy way to find the measure of the span of the diameter of a given circle. For any circular item place the string or ruler at any point on the edge of the circle and move it across the circle until it spans its maximum distance. The greatest span is the diameter of the circle.

The next step is the analysis of this information. A good idea is to plot the circumference for a given item as the y-axis and the diameter of the same object as the x-axis on a Cartesian coordinate plane. Next draw a best fit line and then find the slope (with good measures, it is close to and should be pi (π), of course). A good calculation should result in 3.14 as an answer here.

If graphing skills are still not in place for those involved, then taking the ratio of circumference to diameter is the way to find pi as well. In both cases, however, we face a problem : Our measurements are typically accurate to 3 significant figures at best, and in both cases we are performing calculations of the slope or ratios and they often have numerous figures in their answers, most of which are not significant.

Instead of simply telling them to stop after the hundredths place, why not use a tool that already does so? Namely the Slide Rule is the solution. A classic linear slide rule (5 inch or 10 inch) has two common scales, the C scale and D scale, of values logarithmically spaced so that multiplication and division are easy to perform (due to the laws of logarithms). The Slide Rule has only up to 3 significant figures available, which is all that is needed not only in this case, but in most cases in real life applications. Also, the Slide Rule is designed for variable relations that are Ratios and Proportions. Here, for example :

$$\frac{\pi}{1} = \frac{C}{d}$$

With minimal instruction on operation, such as the C scale is the Circumference of the circle (C) and the D scale is the Diameter of the circle (D) (nice how the letters correspond to the scales), placing one value over the other, interesting observations can be made. The first is that for all of the circular objects measured, the ratio of said variables now appears (or at least is very close depending on the measurements) to being established on the Slide Rule immediately. This is because the Slide Rule is a parallel computer which solves all similar ratios simultaneously. Hence if you were to ask a slide rule user to predict a circumference from a given diameter (or vice versa) they could now do so, without even knowing of 'pi'.

The second observation is the answer to the question of the ratio of the circumference to diameter for a given circle, which is found opposite the left D scale index on the C scale, and the answer, of course, is pi (π). Now there is a tangible reason for the answer and the form of the formula for the circumference of a circle and its relation to its diameter. Pi (π) becomes no longer just a symbol, or even just a number, it stems from the physical measurement of circular objects in real life and the computation of the ratio of circumference to diameter.

Here it is best to create a table with 3 columns – the first is the name or description of the circle you are measuring – a blue lid, a red disk, or some other designation. The next two columns are labelled at the top Circumference and Diameter and in each space corresponding to the given circle its measurements are placed.

It is often best to have at least 3 items measured for a reasonably good graph, but you can have as many as you like. The more the merrier but also typically more accurate in the results too in this case.

You can use a spreadsheet, such as Excel to place your data just as you have written it and then have the program graph the data as points with the y-axis being Circumference and the x-axis being Diameter. Draw a best fit line through it with the software and look at its slope – what does it tell you? Wow – more than likely it will read 3.14... - Isn't that amazing!?

You could have just graphed it yourself using graph paper if available, plotting each dot and then draw a best fit line and then determine the slope of that line. The slope is a ration too : It is the change of Y over the change of X (this means that two points you have picked on your best line, you read its coordinates and take the difference of them (Y1 – Y2 and X1 – X2)

$$\frac{m}{1} = \frac{\Delta Y}{\Delta X}$$

Realize each and all of these calculations (barring subtraction) can be done on the Slide Rule. With every set of measurements – you can readily check the ratio of Circumference to Diameter and see the result – it should be Pi (wow). But why use this tool here?

The first question concerning the Slide Rule is that Slide Rule scales only run from 1 to 10, so what about measures greater than this? The answer depends on the slide rule users level of understanding of decimal places. With no prior knowledge of decimal place, now is the best time to illustrate how to not only read a ruler (both in common and metric units) but also the slide rule as well. Knowing graduation marks on various scales is a benchmark in itself. Also note, how amazing it must be to come to realize that any and all numbers can be represented in a single scale from 1 to 10! In handing the Slide Rule to the student of life and one with curiosity, one can note that they are holding infinity in their hands. For more advanced students, the Slide Rule allows for easy use of Scientific Notation if they are aware of this.

The major question and concern is the access to the Slide Rule. None are to be found in the stores. A first source is on-line with virtual Slide Rules (www.antiquark.com/sliderule/sim). For those who want something more tangible, there are free paper ones on-line to download as well (www.scientificamerican.com/media/pdf/Slide_rule.pdf). Imagine a fun project where students create their own personal calculator, such as the 9-scale, 6-inch wonder noted which is as powerful as a scientific calculator, yet made of paper! With some creativity, the project can involve cardboard, slats of wood, some glue, et al. This idea as well as a paper one is explored on my web site : http://galaxygirl-brianaandromedapurcell.com/index.html.

Also at the end of this book, Galaxy Girl's Inquisitive Pioneer Adventures is a set of paper template slide rules to make use of. Think of it – a paper calculator that is essentially as useful to one or two significant figures in calculations that is as useful as a basic scientific calculator – wow. Right here in this book!

Finally, what if a one-time or short-time adventure is needed, there are even web sites offering complete classroom sets of Slide Rules to be borrowed for up to a semester (http://www.sliderulemuseum.com/SR_Loaner.htm). The Slide Rule Loaner program is not only in America, but is international in scope. Their web site offers a Power Point on the history and how to use the Slide Rule plus details on how to obtain medals to award for Slide Rule competitions as well. Other web searches can find a wealth of information on the Slide Rule too. The side question is cost. For the virtual and the paper ones, there is none, other than district access to the internet. In the loaner program the only cost is return shipping (presently $10.30 USA Priority Mail Flat Rate) and pictures of the use of the

Slide Rules in your classroom. Requests for this are sent on school letterhead. If a student asks (with teacher permission) that student can receive a Slide Rule to keep.

The key to using the Slide Rule is that it is a great visual aid in connecting the world of math and numbers to real-world measures and their formula-based relations. It empowers the user to be the 'computer' and not some external device. One has to focus on skills such as estimation, decimal placement, the relations of the numbers, and a practiced effort at basic math skills. The Slide Rule also connects to history since most mathematicians, scientists, and engineers used them for 350 years (such as Isaac Newton and James Watt) and were used by others who created such works as the Empire State Building, Hoover Dam, and the Golden Gate Bridge with them. This lends itself to stories and inspiration in the classroom.

Also, beyond the pi project, as noted, the Slide Rule is a natural tool for Ratios, hence fractions, similar fractions, including fraction to decimal conversions which can readily be done with it. Notice that it clearly distinguishes the numerator and denominator in its arrangement. A direct real-life application is ideas such as cost per unit measure. Students can investigate costs and amounts for items as sold by mass, weight, volume, or number. The Slide Rule quickly yields the ratio of cost to unit measure for comparison. Also the Slide Rule works well with any and all formulae that can be treated as Proportions that are found in math, science, and everyday life as well. Some simple examples are : converting common and metric units or formulae for average speed or density (the scales become the needed variables once again as in the pi project. For average speed, the C scale is distance, the D scale time. For a given speed, one must read along the scale, a greater distance, meaning that the trip takes a greater time). Also proportions for similar sides to geometric figures, which can for example, be applied to everyday life – such as the height of a person over their shadow length is proportional to the height of some unknown object, like a flag pole, over its' shadow length). A simple case is parallel to the pi project here, and answers the question : how many centimeters are in an inch (2.54 cm = 1 in.)? Have students measure a given set of items in both types of units and take their ratio to find the conversion factor! This can be done for any unit conversion situation. The historical math tool, the Slide Rule, connects too many of the foundations in math that start early and move through all of one's education. The classic 9-scale model is as powerful as a common scientific calculator today. It is a visual tool that can spark imagination, creativity, and promote stronger math skills. The list of ideas for its use is only limited by the imagination of the user of the Slide Rule in its applications.

Hopefully this was fun, exciting, and helps you learn many new skills – reading various measuring tools, thinking creatively, learning graduation marks on these tools, using a slide rule and finding pi for yourself.

Realize the importance of Pi as well – it is found not just in this relation here (note there is a story of Archimedes and his work at giving us the regularly quoted and used 3.14 value from over 2000 years ago – found in Activity # as well as noted in another Activity about him as well Activity #) but Pio is also in various formulae in Physics and Chemistry in examining Gravity, and other phenomena in science – too cool.

Explore and Enjoy – Be Brilliant!

Activity #7
Determining Pi (π) through Probability Activity

Grade Level : Middle School
Math Level : Calculating

Purpose :

Archimedes, historically is one of the first recorded persons to approximate a value for π (pi) at 3.1416. His value has been used for centuries and essentially is quite useful even today. We will use it to see how well our method for pi determination is. For more information on Archimedes on this matter and many others read the Preludes to Archimedes Density Activity, as well as the Pi Activity, and the Archimedes Buoyancy Activity.

For all Circles the Ratio of the Circumference (the measure of the distance around the circle) to its Diameter is a constant and is a value we call pi which is represented by the Greek letter of the same name π. We found this in Activity #6. In that Activity for measuring pi we simply measured the Diameter's and Circumferences of already made or personally made circles and use this data by graphing the data and determining the slope of a line which should yield the value π (3.14). In this Activity interestingly we will use statistics to arrive at pi, however.

In this Activity we will draw a circle in a square piece of foam core board, reverse it so we cannot see it and with eyes closed toss darts at it numerous times (the larger the number the better it should turn out). It turns out if we draw what is classically called a mandala – essentially a circle that fits perfectly inside a square where the radius of the Circle is say 's' and this makes the side of the Square 2s. Then if we create on a posterboard (thick foamcore board is best since we will throw darts at it)this mandala on the posterboard and turn it around so we cannot see our possible target and then randomly throw darts at the board and keep count of the numbers of those that had landed inside of a circle and those that did not – say between the circle and the square. If we throw enough darts – that is – increase our pool size of measurements then through a simple calculation we will find Pi – wow.

It turns out that the ratio of the Area of the Circle (πs^2) to the Area of the Square ($4s^2$) is equal to the ratio of the number of dart impacts in the Circle to the dart impacts to the square which are both equal to $\frac{\pi}{4}$.

Wow – Think about it – to find Pi by throwing darts and counting essentially hits and misses at one (or more - see the alternative mandala model with 4 instead of 1) mandala and using a very basic formula to find Pi! Math is amazing isn't it?!

Materials :

- Foam Core Poster Board,
- Meter Stick,
- Ruler,
- Thumb Tack,
- String,
- Medium Point Marker,
- Scarf,
- Compass,
- Darts (one or two sets of standard darts),
- Slide Rule

Notes on Safety : First be sure to have parental permission and supervision in the usage of darts. At no time are they used inappropriately or unsafely. Be sure to have your target board against a wall surface that will not be damaged by the darts should they go astray.

Procedure :

1. Creation of Target Board :
2. Purchase to use a standard foam-core poster board to act as the target for the darts thrown. Cut the board to be a square – it is best to use one of the dimensions (the shortest side) as the standard, for example 24 in. by 24 in..
3. Use the meter stick in one of two ways to find the center of the square : either A) measure half way along each side and lightly draw a line from side to side so as to have a point at the center or B) use the meter stick to draw two diagonals from corner to corner to create a point at the center.
4. Secure the thumb tack at the center and cut a loop of string a bit longer than twice the distance to the side and tie it off so that when a marker is in the loop and it is stretched from the thumb tack to the marker it reaches the side. This way you can create a circle that touches the four sides of the square when going around. (see photo)
5. Now set target board on the target wall (one that is approved by parents for use when using darts since there may be a stray dart). Note that the circle is facing the wall so all the thrower would see if looking at the target is a blank board. Measure a reasonable standard distance for throwing the darts (say 8 ft for example). Tape may mark the throwing line if needed and approved.
6. Note : There is an alternate design to test if you want to for a different approach. Cut out the foam board square as originally done.
7. Next mark the half-way point along each edge and draw 4 boxes by connecting the midway points across from each other.
8. Within each of the four squares create a circle – perhaps use a compass if you have one that spans from the center of the box to the box edge. If not employ the string method as described in the circle creation method first noted in step 4.
9. This might be fun to try to see if the results are different from the other method. So here you have 4 mandalas instead of one.

10. Throwing the Darts :
11.
12. Choose a number of tosses of the darts to be used. Note that the larger the number the better the results (but this can be tested by tabulating the results as one does the experiment to verify such a claim!). A good number is 100+ to use.
13. With each throw first look to see where the board is and set yourself at the line. It is a good idea to move a little to the left and right with each throw and let yourself have a certain amount of randomness in your throw (to the right or left, up or down, or even just straight on). To help this using a scarf as a blind fold or closing one's eyes is helpful. Throw hard enough to penetrate the board.
14. Be sure to keep track of the count of throws until completed with the task.
15.
16. Computing results :
17.
18. First take down the board, and before turning over determine your rule for any dart hole that lands on the circle (does it count for the circle or not and be consistent with each trial you do)
19. Now turn over and count the number of dart holes in the circle and record the result to your count (C).
20. Now count the remaining dart holes that are not in the circle and add these to the number of dart holes in the circle – this total is the number of dart holes in the square – record this number (S).
21. Notice that if for some reason a given dart did not hit any surface of the square, then it is not counted in the total!
22. Calculate your estimate for the value of Pi (π) using your results.
23. Note that you can stop at any time your efforts to see how the number of throws affects the outcome. You should find with a larger sample size, the value approaches and approximates pi (π).

Photo :

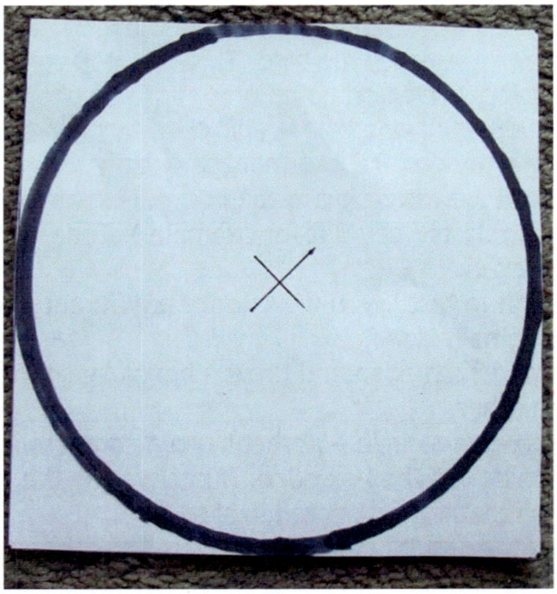

Data :

Number of Dart Holes in Circle (C) : ________
Number of Dart Holes in Square (S) : _______

Calculations :

Be sure to use your Slide Rule!

________________________ = _____________ = –

$$\pi = 4*(C/S)$$

Conclusion :

How did your exploration go – how close were you in your results to Pi? Did you find your results giving a reasonable estimate of Pi? Can you think of changes to your study to affect these results? Did you try 4 mandalas instead of 1 and did this affect the outcome? Why do you think the fewer the results would miss the mark and the Pi value instead of more trials with more times the dart is thrown? Fascinating isn't it that mathematical models and the use of other math types, here Statistics can lead us to a value like Pi!

Activity #8

Similar Triangles projected and determined exploration

Grade Level : Middle School

Math Level : Calculating

As in the prior Activity #6, Activity #7 here is an article containing an Activity along with the usual argument for why the slide rule can be used in the classroom. Though this may seem redundant, I am writing each of these separately and assume that one reads what they find interesting, so he or she may not know of the other documents, so pardon my repetition. The Activity is based on looking to examining an everyday life application of similar triangles. Here we know three measurements, such two legs on a smaller triangle and only one of the corresponding legs of the larger triangle, then we calculate the length of the other corresponding leg can be found. This technique is useful in navigation, map making, and astronomy.

Explore & Enjoy!

Be Brilliant!

Have fun! :)

Quest for a 'New' Tool in the Classroom?!

How can an available tool, namely the slide rule, be effectively used for in the classroom today ?

The introductory question has actually 3 questions to be explored, which this paper does. The first is : What can the slide Rule be used for in the classroom? Second : How can the slide rule be an effective tool in the classroom? And third : How available is the slide rule today?

Of all the subjects in school, math ranks as number one as being the most difficult and abstract. American students today have been shown to be lacking in both math and science. Numerous reports illustrate that critical math skills can diminish very early and by college there is a growing trend towards many of the incoming students needing remediation as well as other reports that illustrate the lack of skills for most people at most ages. (the Final Report of the National Mathematics Advisory Panel in 2008 from the U.S. Dept of Education amongst many others which denote the same findings such as the Harvard University's Program on Education

Policy and Governance sponsored study by the Journal Education Next). More alarming and importantly, math is the main subject that separates most socioeconomic levels. The greater one's math skills, the more adaptable the job skills base, as well as the greater a person's income potential.

In Mathematics Education, there then, is a need to find ways to connect the realms of numbers and operations to conceptual ideas in math as well as practical everyday life applications so that math is accessible to all. Hence, there is a need to find a set of tools to act as bridges from one's focused thoughts to the concept or the application. In math, the general trend today is to find everyday life connections so that math becomes more tangible. Also it is important to find concepts that span all of the years in school since these can act as foundations to build upon over time.

The candidate topic to be explored in this paper is ratios, which can be extended to fractions, decimal equivalents, and proportions. The concept of ratios as applications in everyday life (miles per gallon, cost per unit ounce, et al) and are a critical foundation for all of math through the years. There are numerous benchmarks in national and state levels concerning each of these topics and they range from middle school to high school.

Unfortunately, most of the tangible tools in use are items that only allow for a visual reference (plastic pie shaped pieces) but offer little in the way of seeing the numbers or thinking about the numbers literally and more importantly in use, such as in multiplication and division. In fact, most tangible math tools are only for young people and are too simplistic. Typically there are blocks, bricks, and other plastic pieces representing fractions at best. An interesting fact is that these tools start and stay in the elementary school. What is needed, then, is a tool that can begin as early as late elementary school or in the middle school and extend to the high school. Also the tool needs to act as a bridge for calculation, to spark the notion of conceptualizing the problem and connect numbers to the real world.

The tool chosen to be considered is the Slide Rule because it has numbers and acts as a portable number line (only in logarithmic form instead of a linear one) which can perform any calculation considered and only depends on the imagination and math skills of the user.

Before a quick dismissal of such an idea, since we live in an age of overly electronic, recognize that slide rules and calculators were never studied side by side in any study. Recognize that there are many educators out there that emphasize the need to do the math first and check the answer with the calculator after the fact. Finally, two points : one, is that the employment of math needs to be mastered by the person (and not a machine) and two, why not try a new idea to see if there is a place it works, even if it came from the past?

In most of our early math experiences we encountered the Number Line. If we were to use two Number Lines opposite each other, we can illustrate addition and subtraction. These are clearly visual tools and have been considered by all as very effective. One problem is that they are useful only for addition and subtraction, but not multiplication and division.

What, then, if there were scales that were not linear, say logarithmic, so that when two of them when placed adjacent to each other can show ratios? These scales would be the analog of the number line. The property of logarithms is such that $\log(A*B) = \log(A) + \log(B)$ and log(A/B) = log(A) - log(B). Simply place the numbers (A & B) at logarithmic distances from each other on a logarithmic number line. Have two of these lines, like the linear number line case. Moving one with respect to the other allows for multiplication and division since the distance moved can be added and subtracted hence the numbers are multiplied or divided!

Also these lines can be seen to represent variables. One scale could represent 'miles' and the other 'hours', for example. The answer would be read opposite an index (1 in this case) which would be the 'miles per hour' answer. The description of that item is a Slide Rule, invented by William Oughtred in 1625 which arose from John Napier's invention of Logarithms in 1614. With just 2 scales on a slide rule, the classic C scale and D scale, all of the aforementioned ideas can be found there in one tool. The majority of the argument presented here uses only 2 scales, the C & D, but with access to a 9-scale slide rule, one has a tool that has the power potential of a scientific calculator!

Beyond the tool, the question to consider is the area of best application of the Slide Rule as a tool for connecting the virtual thought world of math to the tangible world of reality. The best concepts to be explored are from the scales themselves and their arrangement – one value placed over another value, the basis of the Ratio. The Slide Rule, unlike the calculator (except for high price high-end ones) shows the numbers as they are shown in textbooks – a numerator over a denominator. Here, then, let's look at the realm of fractions, ratios, decimal equivalents, conversions, and proportions since they are all interrelated ideas. Each of these is a separate stand alone math standard, but are easily interconnected to each other as well as to science and everyday life.

From math, along with various ratios and fractions, we have ideas like slope ($m = \frac{\Delta y}{\Delta x}$), and properties of circles ($\pi = \frac{C}{d}$). The realm of probability is a ratio

($P(\text{Event}) = \frac{\text{Number of Events}}{\text{Total Number of Occurences}}$). Other math ideas have ratio formulae :

$\sin\Theta = \frac{\text{side opposite length}}{\text{hypotenuse length}}$ and $\tan\Theta = \frac{\text{side opposite length}}{\text{side adjacent length}}$ are but a few amongst many.

In these latter areas, such as science, most formulae are 3-variable functions where two of the variables are a ratio equal to the other variable. For example consider these formulae : average speed ($v = \frac{d}{t}$), density ($\rho = \frac{m}{V}$), Newton's 1st Law relating acceleration to force ($a = \frac{F}{m}$), and Pressure ($P = \frac{F}{A}$), amongst many others. Many of these are quite useful and are benchmarks in most curricula today. Note too here in my book Galaxy Girl's Inquisitive Pioneer Adventures each of these formulae are used in a number of Activities both directly or as part of the overall Activity – there is an Activity involving Pressure for example and there are two other Activities that involve Pressure – one in which you measure the weight of your car with a ruler and a Tire Pressure Gauge! Wow!

Also the use of the common 3-variable formula is also abundantly found in everyday life : For example, Any rate idea (cost per unit ounce-volume-mass-pound, miles per gallon, pay rate, flow rates of material, et al).

A Quick Exercise with the Slide Rule for Fractions, Similar Ratios, and Decimal Equivalents

Take a slide rule and place two values opposite each other on the C and D scales, say 3 on C and 4 on D. Read along C and find at 6 on C that it is over 8 on D. Better still opposite the right index of D (the rightmost 1) it is beneath 75 on the C scale, which we know as 0.75. The slide rule is the supreme tool for showing similar fractions, yielding or converting to or from decimal equivalents, and providing the least common multiple ratio for a given ratio in one simple movement.

Imagine placing a decimal equivalent on the D scale opposite the C scale index. For example place the right index of C over 125 on the D scale. Read along the D scale, it is now the numerator while the C scale is the denominator. 1 on D is over 8 on C, Next place the right index of C over 125 and now read 2 on D is over 16 on C, 3 on D is over 24 on C. Even a shift in the decimal can be considered.

The first skeptical question is this : Why not just use the calculator? Unless one has the resources (i.e. money) and time (spent on training and maintaining) very high-end calculators such as used in later high school for graphing in advanced math classes, there is no simpler tool than this. Also the slide rule is a parallel computer (i.e. all variations to the answer are shown simultaneously) as compared to the clumsy calculator which is a serial computer (i.e. one result at a time).

Beyond the obvious math class use noted above for fractions, decimals, consider where ratios are needed for answers. When students work on simple linear equations until there is a variable with a coefficient equaling a number, at that moment, the slide rule then becomes the tool at hand to solve for the answer. In doing this, we emphasize the internal work by the student. The student should even estimate the answer before even using the slide rule. In this way, any multiplication and division can not only be done but also checked with a slide rule, so as to emphasize mental calculations. Then the final answer can readily be found as the ratio of the two values in the final step.

A second quick skeptical question might be : What if the numbers exceed 10? For example how can I find speeds of 25 mph when the regular scale only reads from 1 to 10? Imagine teaching students that any and all numbers can be found from 1 to 10 regardless of size and it is all based on decimal placement! To hold infinity in one's hands in a tangible manner with a slide rule. Perhaps numbers and their mastery in math becomes all the more tangible with a slide rule then! The world of numbers, though infinite, now seems manageable even tangible.

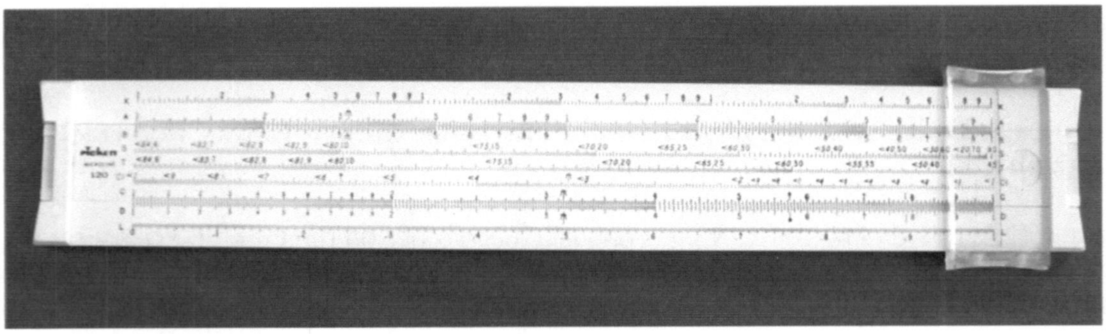

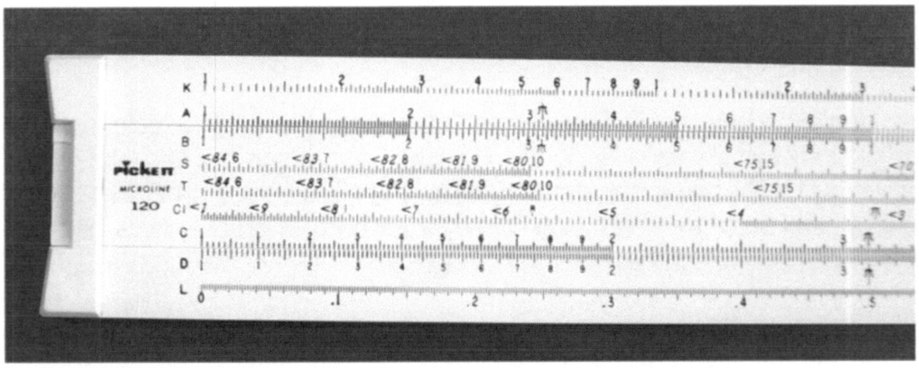

Notice what this use of the C & D scales does though. One has to practice decimal placement and more importantly, be able to estimate answers. (Noting that estimation is actually a benchmark in math as well). Also unlike a calculator, where numbers are merely pushed in and there is no other mental effort needed on the part of the user, the slide rule forces the user to consider one scale as one idea and the other as yet another along with forcing a needed knowledge and practice of estimation, decimal placement, and basic operations such as multiplication and division skills.

Let's return to our speed consideration. Let's say we place 25 (mph) and let it be our speed on the common 'C scale' over the left index (1) on the 'D scale'. Look along the top C scale now, it is the number of miles. Below it, on the D scale, is the number of hours needed to travel that distance. One must look further along the slide rule as one travels further both in distance and in time. Read along D to '2', it is below '5' on C. This represents 50 miles. Above '3' on D is '75' on C and so on. Even decimal values such as 2.5 hours can be found – the answer of 62.5 can be easily read. Instead of repeatedly typing in information over and over, the entire set of answers is in front of the student. The same can be done for being given the distance and the amount of time to travel, and then finding the speed. Not only can questions be quickly answered, the distance and time traveled along the slide rule becomes all the more a tangible concept. The slide rule is a visual tangible tool illustrating how much one variable changes with respect to another.

Any 3-variable relation can be considered this way (like the ones mentioned above). Take, for example, density. With a given mass and volume the density can be readily found. Also if given any of two of the variables, such as density and volume, mass can be readily found. For example, let's say you have a 30 cm^3 sample of aluminum, what is its mass? A student would need to look up the density of aluminum from a table (2.7 g/cm^3). Before grabbing the slide rule mentally estimate the answer : If the metal had density of 2 g/cm^3 then the mass would be 60g (2*30) and if the metal had density of 3 g/cm^3, the mass would be 90g (3*30). Our estimated answer is between these. Placing 2.7 on D under the C scale index and reading along C to 3 (now thought of as 30) the student finds the answer below, 81 g (in the expected range). What if you now cut the piece in half, what is the mass? Unlike a calculator where a recalculation is done, with the Slide Rule, without moving the scale now read along to 1.5 (read as 15) to find 40.5 g.

Here estimation, decimal placement, significant figures, and concepts like density can be conveyed. The realm of numbers becomes connected to science and everyday life quickly and efficiently. Also notice that the distinction of mass and density can be readily noted here, which is often a student misconception. The ratio of mass and volume is density. For a given material, the density of a material remains consistent while the mass will be related to the volume. With a known density, any value for mass and its corresponding volume can be readily found.

Everyday Life Calculations with the Slide Rule

Beyond the science consider the 3-variable use in everyday life – cost per unit (mass, volume, number, et al). Go to the store and find the cost and the number of ounces of a given can of some item (beans, vegetables, soup, etc). The C scale is the cost and the D scale is the number of ounces or some other unit of measure. The answer for cost per unit ounce (mass, pound, item, et al) is found opposite the D scale index. Any and all items can be quickly compared to find the best deal. I personally do this regularly either with a circular or 6 inch model in my pocket I may have on hand.

This same idea applies to using the scale as a line to find percentage of cost for items with a given percentage off. Place the cost of the item on the C scale over the right index of D. Next read back along the D scale where now each of the numbers is the percentage one is paying, 9 is 90% (or 10% off), 8 is 80% (20% off) and so on. For example a $74 dollar item says there is a 25% sale, what is the cost. 74 on C over 1 on D and read above 75% (25% off) on D to find 55.5 or $55.50 in cost for the item! This is because this is where the slide rule is used for proportions, its main power. Here, the Price is set over 100% while the cost at a given percent is found above the noted percentage.

The Slide Rule, Estimation, and Scientific Notation :

Emphasis on Estimation (a required math skill since it is the person and not the machine that is the 'computer' (look at definitions before 1980 in a dictionary)) becomes paramount for computation. Learning to keep track of the decimal point makes the transition to Scientific Notation all the easier and more natural instead of sudden. Instead of just typing values and not paying too much attention to their order, their placement in the equation to be solved, and their magnitude, with the slide rule these become the norm of operation and one cannot succeed without practice.

As an aside, consider the ease once one masters the art of scientific notation when it comes to multiplication and division. Estimation and problem solving are quite easy (most of us can quickly do calculations of numbers between 1 and 10) and then all we have to do is add exponents for multiplication and subtract exponents for division problems. Hence the Slide Rule is a natural for this transition.

The Slide Rule and Significant Figures :

A third new question of the skeptic then arises, what about answers requiring 3 or more significant figures? The answer is this question – how often do we need more than 3 figures in our everyday lives? Take values such as calculating the area of a room or miles per gallon, and the like. Each of these only requires 3 significant figures at best. Besides, if one has mastered the skills at 2 significant figures, why add only more work? Also the use of the slide rule can allow students to better understand the use of significant figures.

This brings up the next interesting side of using the Slide Rule. Students must learn to read the scales and the graduations. Understanding the markings on a tool are critical skills by themselves. Here is a tangible tool to bring this up (much like using a measuring tape, ruler, or graduated cylinder).

This leads to summarizing the value of the Slide Rule in the classroom. As the aforementioned ideas note, 2 simple scales are indispensable tools that has numerous applications. The slide rule can then go from early middle school all the way through to high school as can be seen in the ideas discussed here. It is an extension of the regularly used number lines and act as a visual tool for analyzing fractions, ratios, 3-variable formulae, and proportions. Properly used, one becomes more accustomed to estimation and scientific notation use. Also it avoids the excessive non-significant digit use that calculators promote (try 19 over 29 on a calculator and a Slide Rule. The slide rule shows 0.655 and if these were measurements we round to 0.66 while the calculator goes on and on like students often do in their answer). All of this stands in sharp direct contrast to the calculator which does not help in these areas, which some today are wondering if it is diminishing the overall math skills of the students today.

Take the classic 9-scale model. It can be used for all those things mentioned along with squares and square roots, cubes and cube roots, sine and tangents of angles, the logs and anti-logs of numbers. The log scale (L) allows one to take any power or root as desired, and not even whole number ones! This model has been the staple of the slide rule world and has the equal calculating power of a scientific calculator! Even more important is the fact that slide rules were once the tool for the mathematically gifted and driven. It was taught to the students headed to college and in college when moving towards advanced degrees in mathematics, the sciences, engineering, and medicine. Were the educators of yesteryear wrong so long ago to teach the brightest this tool?

The Main Power of the Slide Rule – Used for Proportions

$$\frac{A}{B} = \frac{C}{D}$$

Since the slide rule has these two scales moving with respect to each other and ratios can be readily created and since they are equivalent, the realm of proportions, a mainstay math national benchmark, can be utilized.

Returning to the C & D scales and the realm of ratios - Notice, too, how the slide rule can be used in a ratio fashion that utilizes the proportion for scales in models and maps. Let's say a map has a scale that 0.8 in = 1 mi. Place the right index of C over 8 on the D scale. Since we are using the map we will measure with a common ruler and determine distance using the

scales : $\frac{\text{Number of Miles}}{\text{Number of Inches}} = \frac{\text{1 mile}}{\text{0.8 inches}}$. The C scale is miles while the D scale is the map

measurements. We measure 2 inches, how far is it? Read 2 on C and find 2.5 miles on D above. 3 inches on C yields 3.75 miles on D, 4 inches on C is 5 miles on D and so on.

Take this proportion idea to the realm of conversions, such as English to metric units :

$\frac{\text{Number of centimeter}}{\text{Number of Inches}} = \frac{\text{2.54 cm}}{\text{1 inch}}$. Place 2.54 on the C scale over 1 on the D scale.

Depending on what is known, find that value while reading along that scale. Of course, this can be applied to many other conversions.

Those interested in geometry in particular (and algebra as well) quickly see the connection and use here. Similar figures of all types can be examined. For example, one triangle has two sides 5 and 12. On the corresponding triangle one side is 9 and the other is X. So here:

$\frac{5}{12} = \frac{9}{X}$. Place 5 on the D scale over 12 on the C scale and read along to 9 to find 21.6

for a final answer. Also the geometric mean is a proportion hence can be solved this way too.

The use of proportions is by far the greatest advantage of the Slide Rule – any and all proportions (which many formulae can be expressed as) can be found very quickly on a slide rule and at a rate even faster than a high end graphing calculator. Here any of the variables can be the unknown and the other three are known, so the unknown can be found efficiently and effectively.

Return to the example of speed or density. If the main variable, speed or density is considered over 1 on the C and D scales respectively then the other two variables involved here (distance & time, mass & volume) are the ratios along the scales. Other areas, such as chemistry have a number of formulae that are or can be treated as proportions, such as when studying a sample and determining the percent composition of a given material in it and generalizing it to the whole of the source to find the amount of that material. Also the gas laws frequently are computed this way (such as Charles' Law). Another similar use comes in biology and measuring the population of a species in a given region. Take a sample and mark them. Return them to the region and recollect at a designated time, place, and method later. The ratio of the number originally caught to the population is proportional to the number of marked ones caught the second time to the total number captured.

Think, too, of other real life examples for the proportion – such as in cooking. For example we have a recipe which has directions for making 6 servings, requiring ¾ stick of butter. We wish to make 20 servings. Place 0.75 on the C scale over 6 on the D scale. Read back along the D scale to 2 (now considered 20) and find on the C scale 2.5 sticks of butter.

A Proportion Activity – Making Measurement Tangible with a Slide Rule

Since the Proportion is a natural for the Slide Rule, why not employ it with an Activity. Make an Activity out of Proportions using geometry and similar figures. The one noted here is about Triangulation and is a useful Indirect Measurement Activity for both Math and Science. It is used in map making, surveying, meteorology, and astronomy.

This is a great adventure for all Inquisitive Pioneers! Ms Briana Andromeda Purcell finds this one quite exciting in an unique way – measure tall objects, such as a tree with a ruler and a measuring tape for example – too cool.

If there is a poster or painting on a wall, one can claim to be able to measure the height of it with a ruler when held up at arm's length at a measured distance from the wall without ever actually measuring the poster at all. This is due to the geometry of the situation and the similar triangles. Use this proportion :

$$\frac{\textbf{Apparent Height of Poster via ruler}}{\textbf{Distance from eye to ruler}} = \frac{\textbf{X height of poster}}{\textbf{Distance to Poster}}$$

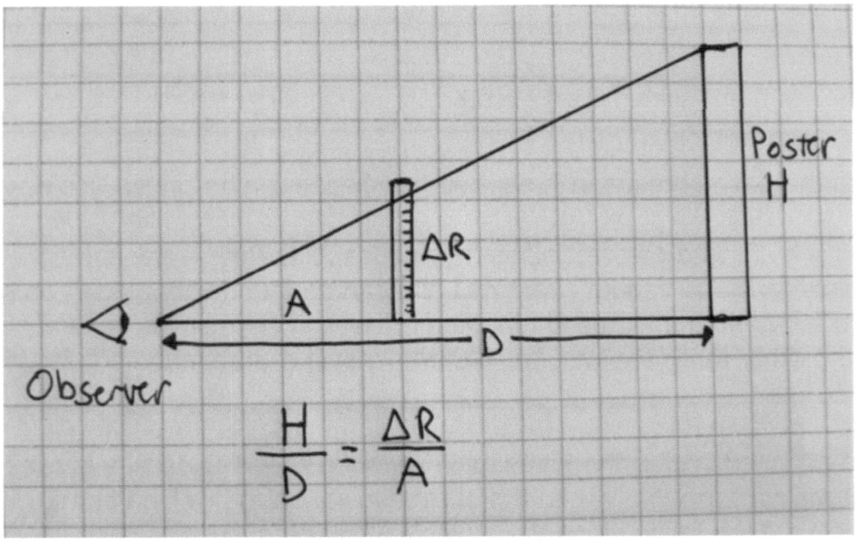

The basic tools needed are a ruler, measuring tape(s), a Slide Rule, and the object in question. Notice too that calculation comes after the thoughts of what one is doing and taking measurements.

This activity can be rearranged in numerous ways with different questions : For example, How far is one standing from an object of known height? Here the object is not the question, but the distance from it is.

Alternatives to the Activity!

A teacher can even take this outside to find the height of various objects outside (flag pole, telephone pole, backboard and/or rim, house or building, tree, etc).

Also one does not have to use the Ruler! The observer can use their height in addition to the heights of the ***shadows*** of the object and the observer to find the unknown object height! Here the measures are the shadows and the height of the one observing.

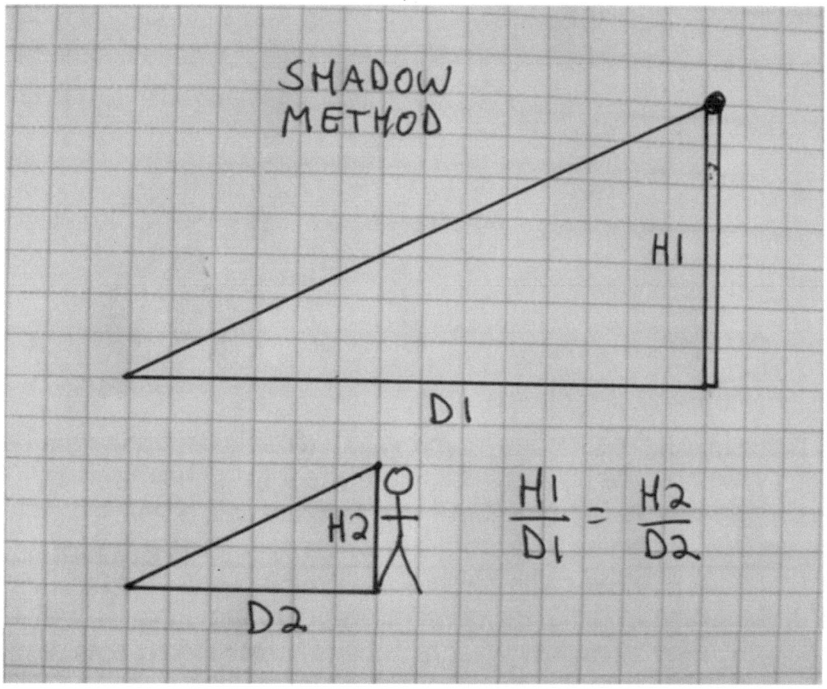

Another method for comparison (or if not too sunny) is to have students work in pairs. The one observing crouches low to the ground and uses the other as the sighting stick. At a given distance the observer notes the alignment of the partner and more distant unknown height object now aligned. With the distance to the partner from the observer (D2) and the height of the partner (H2) along with the distance from the observer and the object (D1), it is easy to find the height of the object (H1). Note that the sighting stick can be any known measured tool (meter stick).

$$\frac{H1}{D1} = \frac{H2}{D2}$$

Still Other Alternatives :

With only a tape measure and a **mirror**, object heights can be found both indoors and outdoors. Here the mirror is between the observer and the object in question (height of wall, flag pole, etc). On the mirror are tape crosshairs. The observer stands at a measured distance (D2) from the mirror so that the top of the object in question matches up with the crosshairs. The distance from the object to the crosshairs is also measured (D1). The observer's height from the floor to the observer's eyes is measured (H2) The proportion above solves for the unknown height (H1).

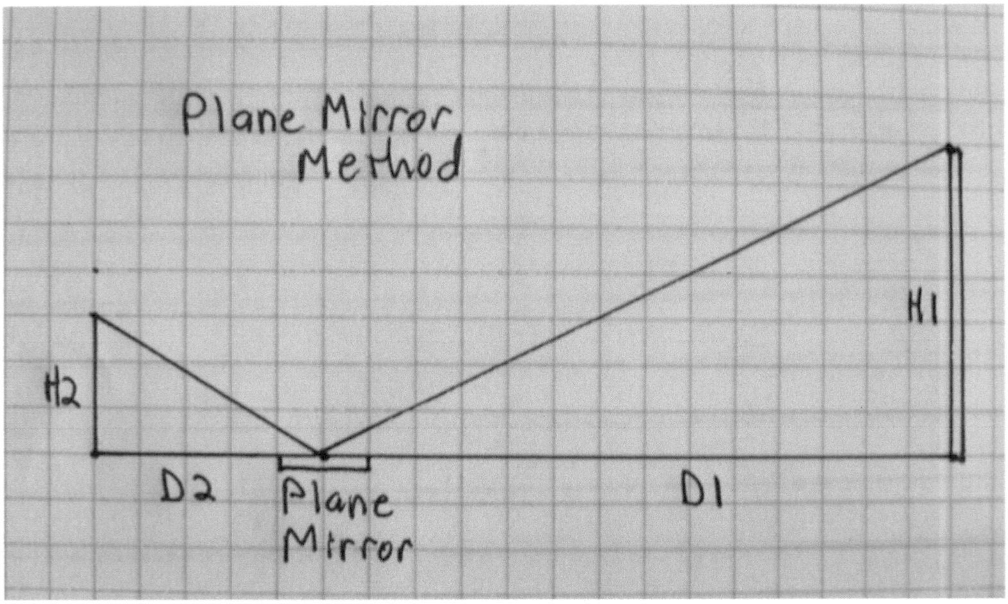

A more advanced mathematical idea can be done by trying to indirectly determine the distance to an object, such as a light pole or a fire hydrant across the street from the observer. The observer is to stand directly across from the object at some unknown distance (X). Then moving a measured distance (D1) perpendicular to the line connecting the observer and the object, the observer makes an observation point which is marked by a stick or small flag or marker (call this point P). Returning to the initial observation point, the observer walks a measured amount (N) directly away from the object and then repeats the process of moving a measured distance (D2) perpendicular to the observer and object so that when the observer now views the object it is in line with the first marked point (P). This proportion has two unknowns, but with algebraic manipulation can readily be solved and with the use of a slide rule! Use the following relations from the aforementioned description :

$$\frac{X}{D1} = \frac{X+N}{D2}$$

Solve for X :

$$X = \frac{N*D1}{(D2-D1)}$$

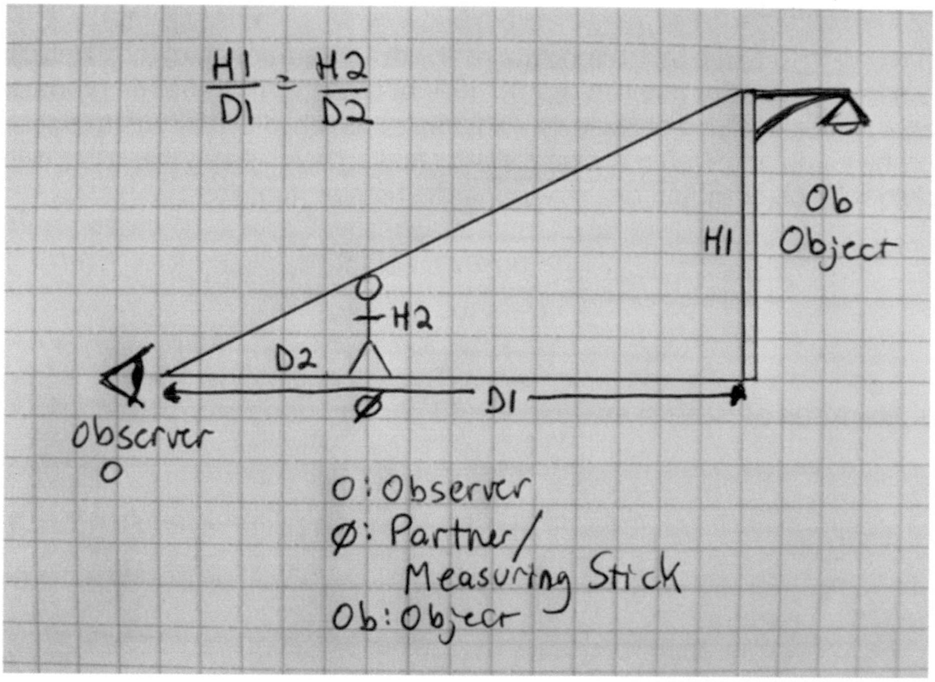

H1/D1 = H2/D2
Ob
Object
H1
H2
D2
D1
Observer
O
O: Observer
Ø: Partner/
Measuring Stick
Ob: Object

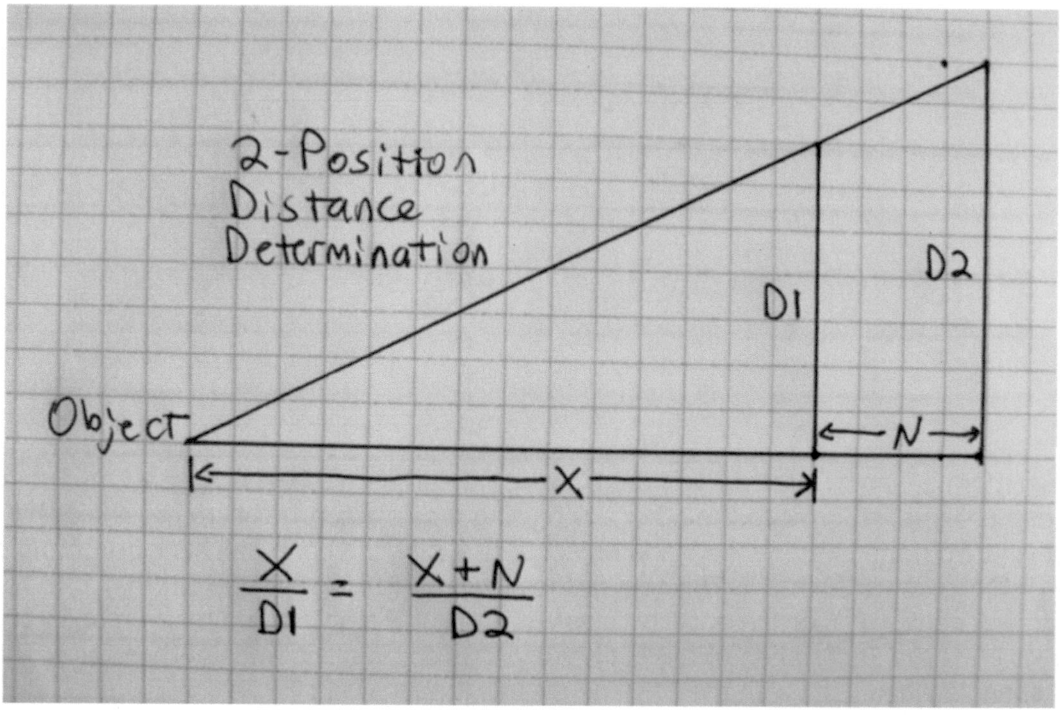

2-Position
Distance
Determination
D2
D1
Object
N
X
X/D1 = X+N/D2

The Geometry and Trigonometry use of the Activity :

An alternative allows one to find either the height or the distance from an object. Here, in place of the ruler, one uses a protractor and can use the idea of tangent. In using the protractor attach a straw as a sight and string at the axis with a mass attached allows one to measure angles subtended by an object. With a known distance and a known angle measure, using a classic 9-scale slide rule, one can find distance using the tangent function :

$$\text{tangent of angle subtended} = \frac{\text{unkown height}}{\text{distance to object}}.$$ Beyond the other method,

with some work it can be reconfigured to connect to Parallax and how astronomy measures distance to stars which can be done in an Activity outdoors and using proportions.

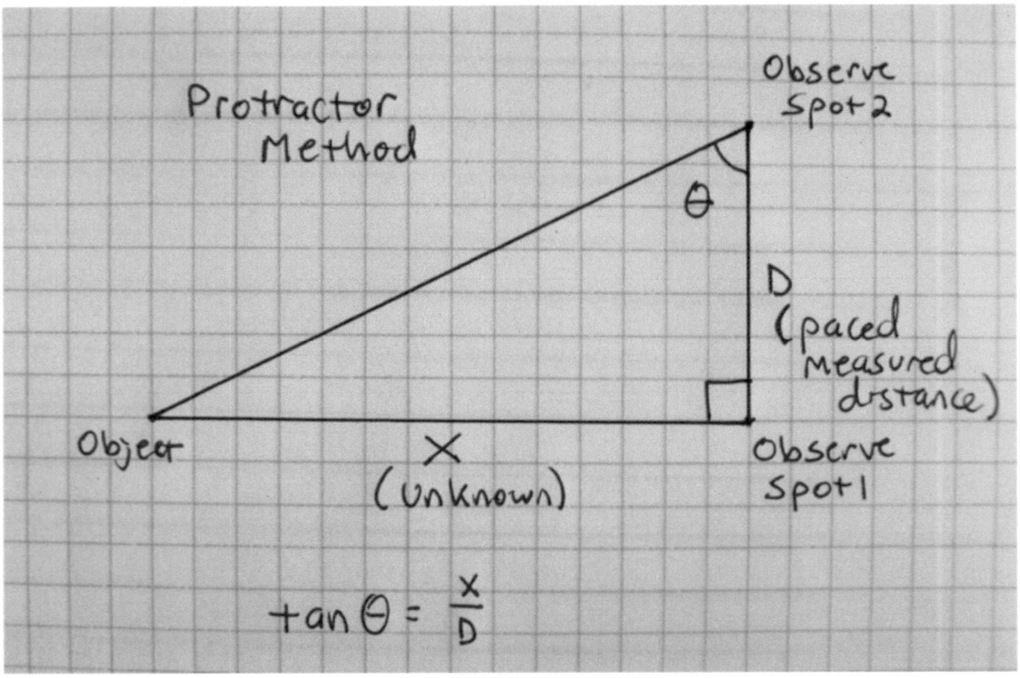

An even more advanced method has the observer move along a baseline (B – the distance from one observation point to the other) and using a protractor, measure the angle that the object is at from the baseline. (A good object to use is a ping pong ball hanging from a string some distance down the hall). The sum of these two angles are subtracted from 180° to find the parallax angle (P), or the apparent angle of shift of the object due to the motion of the observer (like the Earth when observing nearby stars at different times of the year). This parallax angle when used with the baseline creates the relation :

$$\mathbf{D} = \frac{\mathbf{B}}{\mathbf{2 * \tan (P)}}$$

With the relation, the distance to the 'star', D, can be determined.

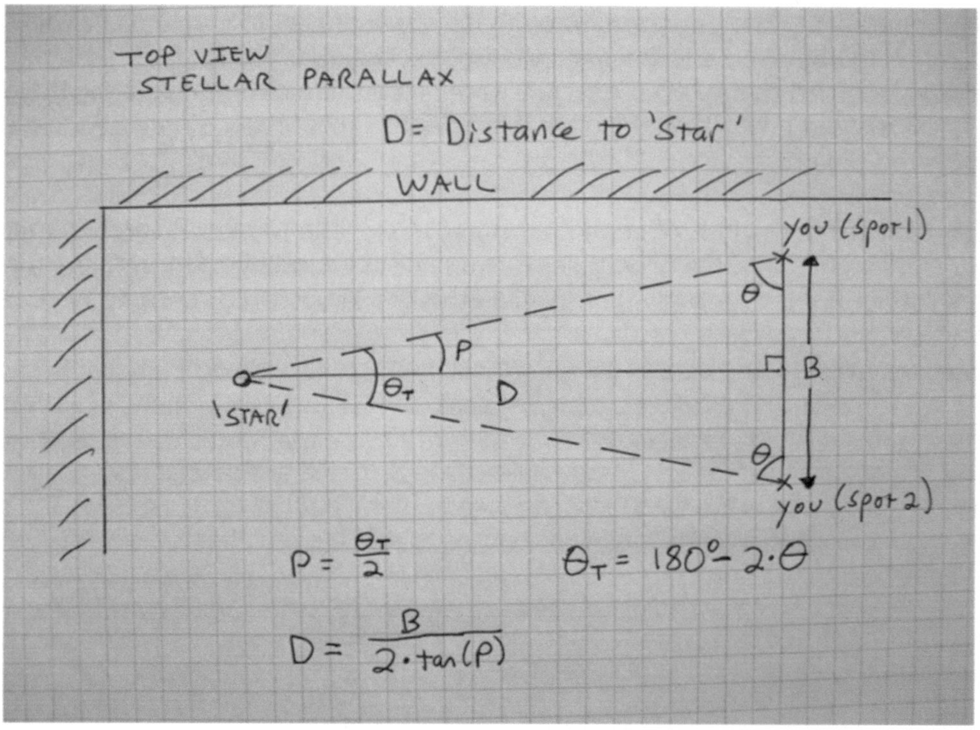

There is more exploration of this very Height and Altitude Activity in Activity #10 in this book – look there for specific directions for doing this beyond this overview.

Notice the range and evolution of the Activity, where the tools do not change too much, but the level of complexity can. The extension to geometry and trigonometry is the best illustration of this. The ratio of tangent as the length of the opposite side to the length of the adjacent side is readily found on a Slide Rule and with a simple 9-scale one, the tangent scale is there as well. No need to change the tool that the student started with in middle school and now in high school.

Why the tool and the Activity? The students do all the work and connect math to the real world. The students can test their ideas and find how close to true it is. The activity promotes simplicity yet sophistication. It promotes algebra, geometry, and practical engineering in the real world. The two triangles in ratio formation can be visualized on the Slide Rule readily. Also it shows a good illustration of determining the measure of something without directly measuring it (which is common to much of science).

Other Reasons for the Slide Rule in the Classroom :

Besides being the tangible visual tool as a number line for multiplication and division, along with being the natural launching point for estimation, mental calculation, and scientific notation, the Slide Rule has many other elements to make it a tool for the classroom. Other fascinating facts about Slide Rule use :

The Slide Rule has a 350 year history of use, unlike the calculator. Slide Rules were used by scientists, mathematicians, and engineers through history and used to create such items as : the Panama Canal, the Empire State Building, the Golden Gate Bridge, the F-15 jet fighter, and the Saturn V which carried the Apollo spacecraft to the Moon. We often think many things began with tools. Instead it was the curious and imaginative mind exercising mathematically and often using the slide rule.

The Slide Rule came from the idea of Logarithms in 1614 by John Napier. It can be argued that the logarithm and its offspring, the Slide Rule, was one of many catalysts for not just the first Industrial Revolution, but the second one as well in the late 1800s which became world-wide.

The Slide Rule was used by and altered by people such as Sir Isaac Newton and James Watt (who used it in his crafting of a better steam engine), as well as Peter Roget (of Roget's Thesaurus fame). Even Albert Einstein and Werner Von Braun used one (both incidentally used the same 9 scale Nestler model). One could argue that one of the important tools of the advances in technology and ideas came from both logarithms and its manifestation, the slide rule. Imagine the inspiration, stories, and connections to real life with this.

The slide rule requires no batteries and can play no games – low maintenance costs and efforts for the teacher (i.e. no hard resets, no major difficulty in teaching how to use)

The slide rule is nearly impossible to cheat with even when students exchange them during a test or quiz – what is the answer and which of the indices is the answer opposite from?

One can personally make a slide rule from paper templates readily found here in this book and on-line. This allows for an interesting potential of creative projects. Imagine making your own scientific calculator – only from paper! The template from Scientific American – a 9 scale 6 inch model – is a mini scientific calculator (www.scientificamerican.com/media/pdf/Slide_rule.pdf) (A couple of others : Luis Fernandes, Dept of Electrical & Computer Engineering, Ryerson University http://www.ee.ryeson.ca and Circular Slide Rule by Dr. Charles Kankelborg, Dept of Physics, Montana State University http://solar.physics.montana.edu/kanel/math/csr.html).

Interesting to Note : The same slide rule can be used from middle school to high school and the rules for using one are all the same, unlike various models of calculators. This includes the fact that the rules to the slide rule do not change essentially while there are differences between more advanced and basic calculators.

Where to find the Slide Rule :

The final critical question of the skeptic, then, is this : what about the availability of such a tool as the slide rule? A quick reaction might see it as impossible and at best few and far between. The truth of the matter is quite a pleasant surprise.

The first place to go is to sources for information and simple materials.
This is one of the great elements of my book,

Ms Briana Andromeda Purcell's book :
Galaxy Girl's Inquisitive Pioneer Adventures

– there is a paper slide rule template at the back of the book so you can make as many and as decoratively slide rules as you wish.

There are also many websites that describe the slide rule and its use. Some give history, photos, and power points on how to use a slide rule that can be downloaded, such as the International Slide Rule Museum and the Oughtred Society (www.sliderulemuseum.com , www.oughtred.org , and many others with a google search) There are even ones for paper templates for slide rules which can be downloaded, printed out, and used. (the aforementioned slide rule museum is a good one as well as others on Wikipedia under slide rule). As noted earlier, think about having students build their own calculator. With some creative effort, the scales can be attached to thin slats of wood and a student will have their own powerful calculator. Imagine crafting your own personal and powerful tool. Beyond this, there are even websites for virtual slide rules! (the best is www.antiquark.com/sliderule/sim). Consider having students on computers and able to manipulate a very effective slide rule. Once they have learned to read it, they have a virtual yet fascinating tool that forces the user to practice good math skills – knowing basic times tables, estimating answers, and keeping track of the decimal and most important here for the virtual one is the reading of scales, which is a benchmark itself.

Still, what about a tangible slide rule to place in the hands of the students? There is even a web site for a Slide Rule Loaner Program. The web site is the international slide rule museum noted before. Here the program is as follows : At the web address http://www.sliderulemuseum.com/SR_Loaner.htm is the description, pictures, and details on how to borrow slide rules for your classroom! This is not only being done in America, but also in other countries, such as Germany, Netherlands, Australia, and Japan. Notice this is not just an American idea and there are other schools in other countries examining this idea.

The Slide Rule sets can be kept for up to a semester. There is an article in the Oughtred Society's publication, Journal of the Oughtred Society, Spring 2010 Volume 19, Number 1 detailing the use of them in the class. In the same issue is a more extensive article for the argument for the use of the Slide Rule in the classroom. The web site notes that it even has medals for competitions for slide rule use that a teacher can obtain. Think of forming a Math Club where tools of the past are talked about and physically examined! Skilled competitions in math ability (perhaps even the ability to letter in it?).

Those with resources can look on websites, personal and ebay for low cost Slide Rules and develop their own classroom collection. (Sphere Research, ThinkGeek, amongst others) I have created two sets of my own. There are many websites and people willing to work with teachers in this area. Slide Rules can still be purchased, such as from Concise in Japan (http://www.concise.co.jp/eng0731/slide.html).

There are ways to connect these ideas to English and literature classes. The mention of a Slide Rule has happened in numerous classic science fiction of the past. Robert Heinlein, Isaac Asimov and others are noteworthy here (Have Spacesuit, Will Travel – Space Cadet – and Starman Jones by Robert Heinlein and Caves of Steel and I, Robot by Isaac Asimov). There are even good non-fictional historical pieces such as : A History of the Logarithmic Slide Rule by Florian Cajori. A good straightforward overview of the slide rule can be found in the Oughtred Society's book : Oughtred Society Slide Rule Reference Manual.

The history-crafted tool, the Slide Rule exists and can be used to tangibly connect ideas to reality – to act as a visual tool that extends from early childhood into adulthood. The Slide Rule promotes proper significant figure use, estimation, mastery of basic math computation, and the use of scientific notation. Also the single idea presented here, the proportion, can be extended too many realms and regions for application as noted. This, then, is the power of the Slide Rule, to have the user be the computer and not a machine. To use one, one has to do the work, think about what, why, and how the numbers are relating to each other, quite unlike a calculator user. It only remains to be seen if it is utilized to act as a spark to ignite creativity, imagination, and exercise mathematical focus in the modern world to build strong math skills which are grounded first in thought and then placed into practice by the innovative teacher and the student as the Inquisitive Pioneer.

Explore & Enjoy!
Be Brilliant!

Activity #9

Right Triangles Calculations with a Slide Rule

Grade Level : High School

Math Level : Calculating

Historically Geometry is one of the oldest math forms and used extensively through time and finds a lot of uses even today – such as designing and building buildings and such as well as working with floors and walls for covering in one fashion or another. One of the most used and studied items in Geometry is the Right Triangle – one in which there is a 90o angle and the other two will add up to 90o (since all angles in any and all triangles add up to 180o).

One of the more interesting and important uses of the Right Triangle comes from Physics particularly in Kinematics (the study of motion). Galileo pointed out that one could consider both horizontal and vertical motions separately – meaning mathematically – over 400 years ago! His simple thought experiment involved two people, two equal masses and a horse. Each person has a mass in hand and are at the same height off the ground (the one person not on horse is on a stage or table perhaps). The horse rides off with the other person on it With a signal each lets go of the mass in hand and it falls. They each hit the ground simultaneously Galileo reasoned correctly so. The one falling from the hand of the person on the horse will take a parabolic arc since the vertical motion will be accelerated (by Earth's gravity) and the horizontal motion will be constant – the same as the horse and rider (barring friction). Needless to say each motion – the horizontal and the vertical can be resolved and looked at independently through mathematics!

This is what this Activity explores – the examination of the motion of object(s) in two dimensions (x & y) through using Right Triangles and resolving them to reach solutions. This is also called Vector Resolution in Physics as well.

The great thing about looking at a particular case, here the boat going at right angles to the current across the river – we can apply these ideas to other situations – remembering to employ what is needed in those cases (such as if the situation were vertical we need to factor in Earth's gravity too).

This little overview activity with solution in it is then followed by a series of Right Triangle questions all on their own with the values where one finds the missing value.

Right Triangles, Vectors and the Slide Rule Activity

- First read the description, then look at the diagram pictured at the end of this description and think about the problem.
- We are going to use the classic physics question of the determining of the speed of the boat moving at a constant speed directly across a river where the water runs at right angles to the boat.
- **In our case we watch as our friend motors her boat toward us from the opposite shore after pushing off from the opposite shore directly across from us at a constant speed of 6.0 m/s.**
- **Also, the river flows directly across her path and from left to right from our point of view on the shore at 12. m/s**

- **Additional information in the question : The river is 200 m across.**
- The question we are asked is this:
- ***What is the relative speed of the boat as compared to us on the shore?***
- A quick briefer on physics, vectors, and combining similar things (both here in the question are speeds): mathematically when dealing with things *that are at right angles to each other* – the short of it – we can add vectors by using the Pythagorean formula $A^2 + B^2 = C^2$.
- Here A can be the boat's speed, B can be the river's speed, and C is what we are trying to determine - the relative speed of the boat from our point of view on the shore. (see diagram in photo)
- This is how this question is both a physics 2-dimensional vector problem and a geometry problem where we are solving for the hypotenuse of a right triangle.
- So the geometry question here is : What are the lengths of the sides and the angles of the right triangle?
- We first need to find the squares of each of the speeds. Though these are easy, let's use the slide rule to show its usefulness.
- Look up 6 on D scale and look for D^2 by looking directly above at A scale. The answer is 36. (Note : A & B are Scales on the slide rule and are not directly the extensions of the Pythagorean formula)
- Do the same for 12. Note that 12 is located at 1.2 on D scale and look above on A scale and find the cursor is just short of 145 on A scale. We know 2 x 2 is 4, so it ends in 4, so this answer is 144.
- Add these together : 36+144 = 180. Note that this number is C^2 and has 3 places so we read the left side of the A scale for 180! (Refer to the How to Use the Slide Rule page or recognize that the A & B scales are really double scales, they go from 1 to 10 then from 10 to 100 in the same distance span that the C & D scales run).
- Now look up 180 on A and using the cursor read the square root of it on the D scale.
- This answer is 13.4 m/s
-
- (note since our problem has only 2 significant figures, we should leave the answer at 13 m/s)
- If our values had been given to us as 12.0 m/s and 6.00 m/s then we could use the 13.4 m/s
- How do we know the decimal point? Recall 10 x 10 = 100, and our value we are looking up is 180, so it must have 2 digits before the decimal point and be greater than 10.
- On your own try different river and boat speeds and find the resultant speed as seen by the observer on the shore.
- What about the angle that the boat makes with respect to the shore in its travels?
- Here we can use the Tangent Scale (T).
-
- $\text{Tangent}(\Theta) = \frac{\text{side opposite length}}{\text{side adjacent length}}$
- $\text{Tan}(\Theta) = \frac{12}{6}$

-
- First note that on the tangent scale, the tan(45°) = 1.
- Our value is greater than one, (12 over 6 on our C over D scales yields an answer of 2) so the angle is greater than 45°
- With the answer of 2, this means we must read the Tangent scale in reverse, from right to left.
- Look for 2 now on the C1 scale which is written right to left. Place the cursor on it.
- On the cursor line, reading from the tangent scale and reading the numbers marked in some manner to indicate they are read in this direction (either with a < symbol or written in red or italics).
- It should read 63.4°.
- This is the angle east of north that the boat is moving at from our perspective on the opposite shore.
- Notice something else. The '2' on C1 is over 5 on C. Notice that if we were to compute the tangent of the other angle in the triangle it would be 6 over 12 or 0.5.
- In this case, since it is less than 1, we read from the C scale and would have put our cursor on the same spot.
- We would have read the complimentary angle (now reading from left to right) and find 26.6°.
- With regards to all of these calculations, now try them in a scientific calculator and find how many decimals there are. Our slide rule has given more than enough information to solve this vector addition (or geometry) problem.
- Depending on what the question is asking for, we can write the final answer a number of ways :
- The boat is heading at $13._4$ m/s at 63.4° East of due North.
- OR the boat is heading at $13._4$ m/s at 26.6° North of due East (i.e, with respect to the shoreline). The angle is from (90° – 26.6°)
- Note : For the geometry question addressed here, what are the lengths and sides of the triangle, we have found all of the parts with these values. – The sides are : 6, 12, $13._4$, the angles are 90°, 26.6°, and 63.4°.
-
- This physics problem can continue, such as how far down stream will the boat land on our side of the river ?
- Recall that the river is 200 m across. Our friend's boat is traveling this distance at a rate of 6 m/s, so the time of travel is found from the average speed formula (use the C & D scales – a good strategy here is to let 200 be the D scale and 6 be the C scale, so that you do not have to move the cursor and only the slide to the answer for the next calculation for distance):
-
- $$t = \frac{d}{v} = \frac{200}{6} = 33.3 \text{ s}$$
-
- This is the amount of time our friend will travel at 12m/s east along the river. (d = v*t)
- You should have an answer of 400 m (You might consider to start running!)
- Do our answers make sense?

- We would expect the final speed to be greater than the speed of the boat and the speed of the river and using vectors it must be between the minimal and maximum vector sum of these two vectors : 12+6 = 18 and 12-6 = 6. Our answer of 13.4 is between these two values.
- With regards to distance traveled East along the river, notice that the speed of the river is twice that of the boat. So logically it must travel twice as far as the boat did across the river, which was 200 m, so it will land 2*200m or 400m east of our spot across from our friend in her boat. This matches our answer perfectly.
- ***ASIDE***
- *An interesting side note here is the fact that one can solve Pythagorean Triangle problems so easily. Given any two sides, all you have to do is square the values and either add or subtract these values as needed (add if they are the sides A and B, subtract if given C and one of the sides such as A). Then take the square root of that number and viola you have the unknown side!*
-

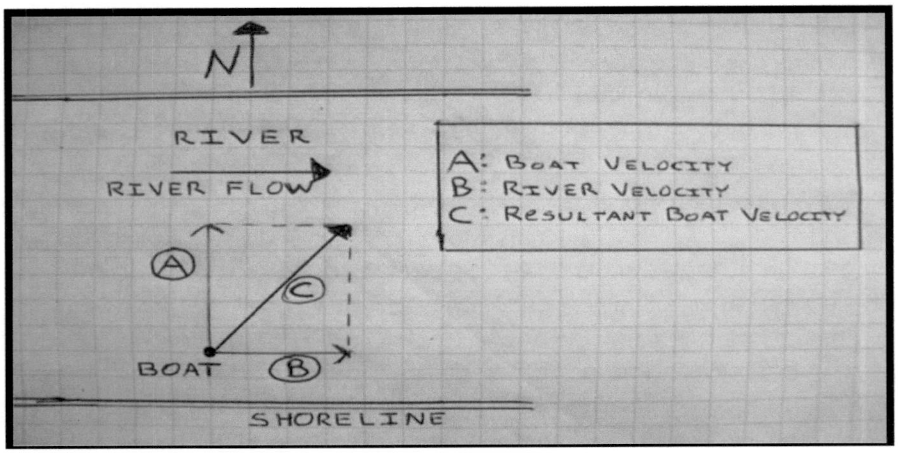

Geometry : Right Triangles, Pythagoras & Trigonometry

Here is a walk through of solving Right Triangles and afterwards are the formulae needed and a table of Right Triangles with a missing puzzle piece to find and solve and the answers are given after that too! Learn by going through and doing the examples – this will help you not only use the slide rule but also gain skills at other scales other than C & D which we have been mostly using thus far in the book (Galaxy Girl's Inquisitive Pioneer Adventures) til now. This Activity will help you become a master Slide Rule User like Ms Briana Andromeda Purcell myself. Here you will use C, D, A, B, S, T scales – wow!

Purpose : To be given a Table of data of a series of Right Triangles and find the missing values through use of the Pythagorean Theorem and Trigonometric Relations of Sine, Cosine, and Tangent.

Note : Right Triangles are useful ideas not just in themselves along with physical applications to everyday life (such as architecture), but also in the realm of Physics and the vector addition of vectors at right angles to each other, or when given a vector finding the resolution of it into x & y components. (Refer to the example in the section to learn the slide rule and the boat crossing the river where the speeds are perpendicular to each other).

Scales to be used in this Activity : C, D, A, B, S, T
Note : When you have a S Scale, you can find the Cosine as well.

Procedure Steps with Examples :

1) For all Triangle considerations refer to the picture associated with the Formula below.
2) In the first example, we have side A = 3 and B = 4. All the other values are blank.
3) To find C, look up both the values for A & B on D and read the square of each of these values on the A Scale. 3^2 is 9 and 4^2 is 16.
4) Take the sum of these values from the Pythagorean Theorem. The sum is 25.
5) Find this value on the A Scale. Since it has 2 digits in the number we use the right hand side of the A Scale. Read down below for the square root and find 5 as the answer for C.
6) The $\tan(a) = \frac{3}{4}$.
7) Place 3 on the C Scale over 4 on the D Scale, the ratio's answer is opposite the D Scale's D Index, 0.75.
8) Continue reading to the T Scale and find the angle 36.8°.
9) At this point, you can determine the other angle in a similar manner or simply find angle (b) from 90°-36.8° which equals 53.2°.
10) If you have found the angle by one method, it is often a good idea to check your answer by using another trigonometric ratio, such as for sine or cosine.
11) In another example A = 16 and ∡a = 30°. We can quickly find the other angle, ∡b = 90° – 30° = 60°.
12) To find B and perhaps C, we need to use one of the trigonometric relations.
13)
14) For example $\tan(a) = \frac{A}{B}$.

15)

16) So here $\tan(30°) = \frac{16}{B}$.

17)

18) Rearrange the equation to read $B = \frac{16}{\tan(30°)}$.

19)

20) Use the cursor and find 30° on the T Scale. This will now be over the tan(30°) on the D Scale.

21) Now move the slide to 16 and place it over the tan(30°) value.

22) The final answer for B will be found opposite the D Scale Right Index and it is 27.7

23) Now you can use any method you want to find C.

24)

25) For example, $\sin(a) = \frac{A}{C}$.

26)

27) Or the Pythagorean Theorem can be used instead.

28) Again, it is best to use more than one method so as to practice good math skills and check your answer from another different point.

29) These methods will work for any and all of the triangles in the table.

30) Be sure to note that when given sides A & C, for example, you not only have to take the square as we did earlier, but now you take the difference of them and not their sum, then find the square root.

31) Develop strategies to deal with decimal placement (hint). For example you will find the same answers if the triangle has sides 3,4,5 than if it has sides 30, 40, 50.

32) There also may be cases where angles could be too small to be determined by the S or T Scales. This occurs when the magnitudes of the values being used is less than 0.1

33) When this occurs, we have to use the R marking on the Scale, which is 57.3. This is because the $\sin\Theta \sim \tan\Theta \sim \frac{X}{57.3}$

Formulae to be used :

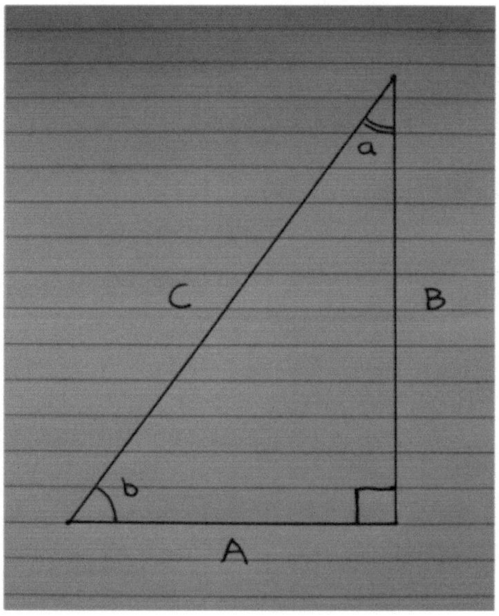

Pythagorean Theorem :

$A^2 + B^2 = C^2$

Trigonometric Relations :

$$\text{Sin}(\Theta) = \frac{\text{length of side Opposite}}{\text{length of Hypotenuse}}$$

$$\text{Sin}(a) = \frac{A}{C}$$

$$\text{Cos}(\Theta) = \frac{\text{length of side Adjacent}}{\text{length of Hypotenuse}}$$

$$\text{Cos}(a) = \frac{B}{C}$$

$$\text{Tan}(\Theta) = \frac{\text{length of side Opposite}}{\text{length of side Adjacent}}$$

$$\text{Tan}(a) = \frac{A}{B}$$

$\text{Sin}(\Theta) = \text{Cos}(90^\circ - \Theta)$

$$\mathbf{Tan(\Theta) = \frac{sin\theta}{cos\theta}}$$

Triangle Relations :

$\measuredangle a + \measuredangle b + \measuredangle c = 180^\circ$

Since $\measuredangle c = 90^\circ$, then $\measuredangle a + \measuredangle b = 90^\circ$

Table of Triangles :

Triangle	A	B	C	∡a	∡b
1	0.3		0.5		
2	6.0	8.0			
3	195	216			
4	5.0		13		
5	0.08		0.17		
6	240		260		
7	9.0		41		
8	65	72			
9		16			14.3^{o}
10	15.0		113		
11	119			44.8^{o}	
12			8.5		25°
13	0.33			30.5^{o}	
14		$112x10^{3}$	$130x10^{3}$		
15	$7x10^{-3}$	$24x10^{-3}$			
16	105			50°	
17		44	125		
18	$3.5x10^{-3}$			45°	
19			84	30°	
20			200	15°	
21	0.4			30°	
22			1240	60°	
23	24	1.7			
24			$3.2x10^{3}$	64°	
25	32			32°	
26		7.5		75°	
27	40			18°	
28		40		18°	
29	14.0	18.0			
30	3.9		8.9		

First try all the problems above -

Answers Below :

Triangle	A	B	C	∡a	∡b
1	0.3	0.4	0.5	36.8°	53.2°
2	6.0	8.0	10.	36.8°	53.2°
3	195	216	291	42.1°	47.9°
4	5.0	12	13	22.6°	67.4°
5	0.08	0.15	0.17	28.1°	61.9°
6	240	100	260	67.4°	22.6°
7	9.0	40	41	12.7°	77.3°
8	65	72	97	42.1°	47.9°
9	63	16	65	75.7°	14.3°
10	15.0	112	113	7.6°	82.4°
11	119	120	169	44.8°	45.2°
12	7.7	3.6	8.5	65.0°	25°
13	0.33	0.56	0.65	30.5°	59.5°
14	$66.0x10^3$	$112x10^3$	$130x10^3$	30.5°	59.5°
15	$7.0x10^{-3}$	$24x10^{-3}$	$25x10^{-3}$	16.3°	73.7°
16	105	88.0	137	50°	40°
17	117	44.0	125	69.4°	20.6°
18	$3.5x10^{-3}$	$3.5x10^{-3}$	$4.9x10^{-3}$	45°	45°
19	42.0	72.7	84.0	30°	60°
20	51.8	193.	200	15°	75°
21	0.4	0.69	0.8	30°	60°
22	1074	620	1240	60°	30°
23	24	1.7	24.1	86°	4.0°
24	$2.9x10^3$	$1.4x10^3$	$3.2x10^3$	64°	26°
25	32.0	51.2	60.4	32°	58°
26	28	7.5	29	75°	15°
27	40.0	123	129	18°	72°
28	13	40	42	18°	72°
29	14.0	18.0	23.0	37.9°	52.1°
30	3.9	8.0	8.9	26°	64°

Activity #10
Parallax – Height & Altitude Considerations

Grade Level : Middle School
Math Level : Calculating

Ms Briana Andromeda Purcell here and this Activity is an essay at first and then explores an idea presented in Activity # but only on the surface – here there are more details and directions in the pursuit and we once again look to the Right Triangle as a tool to uncover Height, Altitude, and Distance considerations we might wish to consider.

Interesting to note is that these Activities here are scaled down versions of what really goes on – such as using Parallax to determine the distance to a star – one of the questions posed long ago – if the Earth does indeed move about the Sun it was reasoned then closer stars should appear to shift in their displacement with respect to stars that are farther away. It took until 1838 to finally see this as it is so small because stars are very far away. This same sort of idea is used by Rangers to determine distances to distant bluffs and the like by looking from two different vantage points and taking measurements.

The essay and the Activity :

The primary goal of science is the application of math models and formulae to the real world in order to do such things as understand cause-and-effect, make predictions when changes in one variable are made to another variable, and to solve for the unknown in a given situation.

One of the simplest applications is the direct application of math formulae for such things as area and volume as well as similar triangles.

If we start with geometry and a right triangle where the base is 4 units long, the height is 3 units tall, then we can find from the Pythagorean Theorem that the hypotenuse is 5 units long ($3^2 + 4^2 = 5^2$).

Now we change the side from 4 units to 8 units long and we also double the height from 3 units to 6 units. What becomes of the hypotenuse? It too doubles to 10! ($8^2 + 6^2 = 10^2$). Though this illustration only employs whole numbers, any multiplying factor for the hypothetical triangle would work. This is because each of the triangles, the original and the magnified one, are what are called Similar Triangles. Similar Triangles are triangles that have the same angles comprising the triangle and their sides are all proportionate to each other. For example from our triangle example, if we take ¾ on the first triangle it corresponds to 6/8 on the other one. (also for these triangles, they have the same set of angles : 90°, 36.9^o, and 53.1^o).

Also realize that similar triangles do not only apply to right triangles, but to any two triangles that have the prescribe description above applying to them.

What value are these similar triangles? Let's say one of the two triangles have the values known for sides and angles, and from the other one we have the angles (as needed) but only one of the two sides – the other is unknown. We could then easily solve for the other side by proportions. Take our example (let's assume we do not know the corresponding side to 3 in the other triangle) :

– = –

Beyond math, this has many applications in the real world. Imagine you hold up a ruler vertically at the end of your arm in your hand. You have measured your arm length (one side of the small triangle) and you look at a tree, house, a telephone pole with the ruler and measure its apparent height (the other side of the small triangle). Now all you have to do is measure how far away you are standing from the object under consideration (the distance to the tree) which gives you the corresponding triangle side length to your arm. The unknown value here is the height of the object in question and is solved like our proportion above.

______________ = __________________

The idea of similar triangles can be reduced to one triangle and the use of trigonometric formulae. All one needs to know is the side of a triangle and the angle associated with it. From this one can determine another side. More importantly instead of the triangle(s) being vertical as noted in the early part of the discussion, this is true even if the triangles are on the ground and we can only determine one of the sides! This is simply because there is no special direction in nature. Look directly across a river at an object, such as a tree. Walk parallel to the river a measured distance and look at the tree again. Using a protractor measure the angle from the line you walked to the tree. Here we can use the ratio of two sides of this triangle, the width of the river to the distance walked is the tangent of the angle you have measured. .

Tan Θ = _________________

Parallax and Astronomy Use :

One of the most important applications of this idea comes from Astronomy. One of the easiest arguments employed by peoples of long ago to argue that the Earth is not moving was this question : If we are moving, why do not the closer stars appear to move due to our motion as compared to the more distant stars in the sky?

How does this question work? Simple – hold a pencil erect at arm's length in front of you and first look at it with one eye and then the other. Next try this closer to you. The pencil as compared to more distant objects, such as furniture along the wall, appears to move more when closer to you than when farther away.

How this applies to the question of the ancients is this – your eyes are to represent the Earth at two different locations in its orbit. The pencil is a nearby star. If we are moving and we look at the nearby stars, we should see their apparent motion. For centuries this was not detected, because stars are so far away, there apparent motion as it is called is very small. Once the apparent angle is measured, the distance the Earth travels, which is the baseline for the triangle, allows us to determine the distance to the star.

This term in astronomy is called parallax. Defined, parallax is the apparent shift or motion of a relatively close object with respect to more distant objects as the location of the observer changes.

Finally in 1838 the parallax of several stars were discovered. It turned out that even the closest stars have a parallax of less than 1" (1 second of arc – Note : 1° has 60 minutes of arc, while each minute has 60 seconds of arc). The distinction goes to 3 different people of 3 different nations : Friedrich Bessel with 61 Cygni, Friedrich Struve with the star Vega, and Thomas Henderson with alpha Centauri. For those interested, the largest is of course the closest star (other than our own Sun) Proxima Centauri at 0.772 arcseconds of parallax, which results in a distance of 1.30 parsecs or (4.2 light-years).

Our Activity will be a series of activities and use each of these techniques in turn to determine the size of objects or distance to objects, including a 'star' we make from a ping pong ball to demonstrate the idea of parallax.

Purpose : To determine the measure of the height (altitude or size) of a distant object by indirect measurements and proportion calculation.

Purpose : To determine the measure of the height (altitude or size) of a distant object by indirect measurements and definition of tangent calculation.
(First a local one to test the method both ways – poster/painting/mirror/fixture on the wall, then to use on other objects such as the height of a tree, telephone pole, or a house and possibly a launched rocket and its altitude outdoors).

Purpose : To determine the distance to or across a region by using the side of a triangle and the tangent of the angle made by 2 observations of a distant object.

Purpose : To measure the distance to an object that represents a star (ping pong ball) so as to illustrate the process of Parallax in Astronomy.

With any and all Activities read through, then set up the process for the Activity and run it without measurement and watch where you may have bumps in the road so as to anticipate these. Be patient – Repeat processes. Once all is well when measuring see to the values – do you notice patterns or trends – what do you predict will happen next as you change one thing or another from one trial in an experiment to the next. Always Explore & Enjoy. Be Brilliant!

Materials :

- Item to determine height of (poster on wall, billboard, tree, wall, house, lamp post, telephone pole, et al),
- protractor,
- string (possibly dental floss),
- nut,
- measuring tape (flexible about 2 m),
- Measuring tape (long about 20 m or more),
- ping pong ball (note this can be Styrofoam, or something easier to poke a hole in so that it is easy to use),
- paper clip,
- launch-capable object (water or air-propelled rocket),
- slide rule

Procedure :

Activity 1 : Determination of the height of an object in the home on the wall through proportion method.

1) For the indoor activities take all measurements in centimeters. (It is a good practice to use the metric system as this is a worldwide measuring system employed in all sciences and engineering)
2) Tools needed for this activity : Object on Wall (can be painting, poster, et al), ruler, regular measuring tape, flexible measuring tape, slide rule
3) Use a measuring tape and from the wall on the floor just below the chosen object to determine its height open and start it there and measure off up to 3m (this depends on room size) perpendicular to the wall away from the object.
4) The goal is to be somewhere between 1m and 3m from the object for measurements so that when holding up your hand at that distance the object appears about the size of your hand at most.
5) With ruler and flexible measuring tape in hand pick a spot to stand at along the measuring tape on the floor. Choose a number that will be the distance to the wall (D) on the table of data values so that a line from your eye to view the object with to the floor is measured from here to the object.
6) Face the object. Look at it with one eye. With the corresponding hand with both the ruler and flexible tape in it raise it so as to hold the ruler perpendicular to your arm. (Note : Your arm is parallel to the floor).
7) Find a comfortable distance to hold the ruler erect at so that the edge can be read (the centimeter side and the numbers going up vertically).
8) Once done, reach with the other hand and let the flexible measuring tape unroll as your free hand draws it from your hand with the ruler back to your face in the area of your eye that is used for viewing.
9) Once there, have this hand hold the value on the flexible tape and hold that tape with this hand.

10) Looking at the ruler in the other hand, try to align the bottom of the object with a whole number on the ruler.
11) Keeping this value in mind read it and the value that appears to be the top of the object from the perspective of the ruler from this view.
12) (Side Note : to optimize the reading in this case align the ruler at the end of your arm with the bottom of the object. This may mean crouching to align).
13) Record your data. The number on the flexible measuring tape goes in the Arm Length box on the data table (A), while the Apparent Height Top (R_{Top}) and Apparent Height Bottom (R_{Bottom}) go in their respective boxes.
14) Calculate the Apparent Height, which is the difference of the Top and Bottom Values, place in box. (R)
15) Use the formula to determine the Actual Height of the Object.
16) Do this 3 times from 3 different distances and take the average of your calculated height value for the object.
17) Measure the actual height of the object. Then calculate the percent error of your indirect value to the actual value.
18) Look at the photo for Activity 3 which applies to this case since this one is only a smaller scale of that outdoor one which shows the Ruler Method of Similar Triangles.

Activity 2 : Determination of the height of an object in the home on the wall through tangent method.

1) Follow Steps 1 through 4 in Activity 1 Noted above.
2) Stand somewhere between 1m and 3m away from the object along the measuring tape. Write down this value in the table.
3) Materials needed here : protractor, string (dental floss is best), lug nut, tape, straw (thin stirring one is best so long as it is a tube and straight).
4) As in the first activity it is best to crouch and align your view from the base or bottom of the object on the wall so that your view of the object is a direct line parallel to the floor.
5) Tape the straw along the base of the protractor so that it goes from 0° to 180°. Attach the string to the vertex point of the protractor and have it extend in the direction of the angular measurements and be long enough to tie the lug nut to it (see photo of Astrolabe). This is called your astrolabe.
6) Use the protractor device by looking through the straw (along the bottom of the straw) first at the bottom of the object where you are directly in line with it so that the nut hanging down is at 90°.
7) Now slowly rotate the protractor so that you are now looking through the straw at the top of the object.
8) At this point use your other hand to hold the string against the protractor.

9) Read the angle measured here. (Θ_2) – Note this is not the angle used. Take the compliment of this angle ($90^\circ - \Theta_2$) to obtain the needed angle (Θ).
10) Calculate the height of the Object using a slide rule.
11) Do this 3 times like the first activity and average the values determined for the height of the object done by this manner.
12) Using the actual value for measured height, compute the percent error of your indirect measurements.
13) See photo of Tangent Activity in photo of same name.

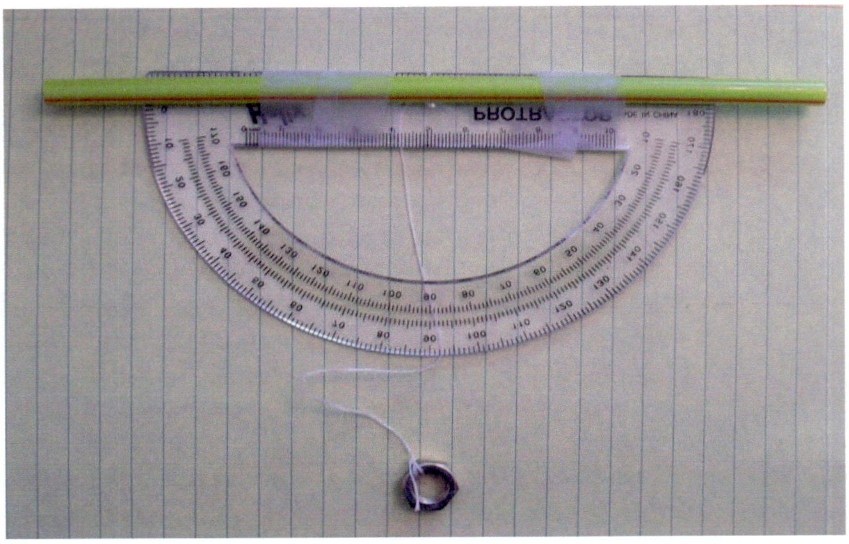

Activity 3 : Determination of the height of an object outside (such as a telephone pole or tree) through proportion method and the tangent method for comparison.

1) For the outdoor activities take all measurements using the long measuring tape but measure in meter.
2) Tools needed for this activity : Object outside (telephone pole, light pole, house, flag pole, or tree), ruler, long measuring tape, flexible measuring tape, slide rule
3) Use a long measuring tape and from the object in question at its base in a direction to stand far enough away so that when holding the ruler erect at arm's length the entire object falls within the scale on the ruler. (This may be as little as 3m up to 18 m distant).
4) Record the distance that you are standing at from the object.
5) Now measure the height to your arm which will be extended and parallel to the ground with the erect ruler in your hand. Record this height (it will be added to the height of the object later on).
6) Next with ruler and flexible measuring tape in hand at your spot to stand and face the object with arm extended with ruler erect and look at the object with one eye closed aligning the centimeter side of the ruler

along the vertical axis of the object starting at its base and going to it's top.

7) With the corresponding hand with both the ruler and flexible tape in it raise it so as to hold the ruler perpendicular to your arm. (Note : Your arm is parallel to the ground).
8) Find a comfortable distance to hold the ruler erect at so that the edge can be read (the centimeter side and the numbers going up vertically).
9) Once done, reach with the other hand and let the flexible measuring tape unroll as your free hand draws it from your hand with the ruler back to your face in the area of your eye that is used for viewing.
10) Once there, have this hand hold the value on the flexible tape and hold that tape with this hand.
11) Looking at the ruler in the other hand, try to align the bottom of the object with a whole number on the ruler.
12) Keeping this value in mind read it and the value that appears to be the top of the object from the perspective of the ruler from this view.
13) Record your data. The number on the flexible measuring tape goes in the Arm Length box on the data table (A), while the Apparent Height Top (R_{Top}) and Apparent Height Bottom (R_{Bottom}) go in their respective boxes.
14) Calculate the Apparent Height, which is the difference of the Top and Bottom Values, place in box. (R)
15) Use the formula to determine the Actual Height of the Object.
16) Redo Activity 2 only as an outdoor exercise using the astrolabe to determine the angular height of the object and knowing the distance from it, you can calculate the unknown height. Use the tangent formula. Look at the Tangent Activity photo.
17) If possible, find records or general information to verify your calculated height as compared to what is known. If impossible, then try the alternate method below which uses shadows and compare results.

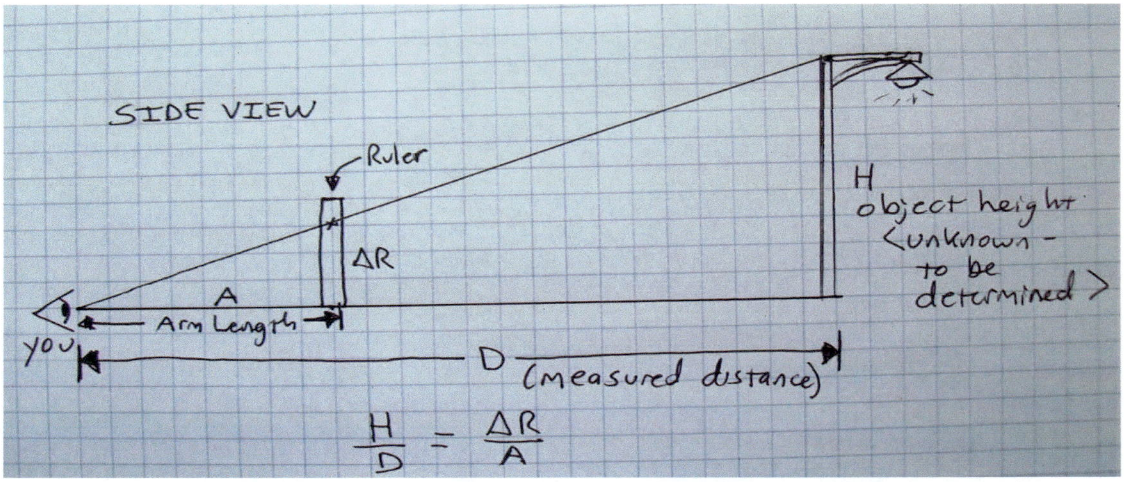

Activity 4 : Determination of the distance across an object outside (such as the distance across the street) through tangent method.

1) Use the Astrolabe you constructed in Activity 2 above (see photo of it) for this next activity.
2) Stand a good distance (over 20 ft, perhaps on the order of 60 ft to 100 ft) but still measurable by your longest measuring tape so that the answer can be determined. (Measure it later, not now).
3) A good way to do this activity is to be on one side of a residential (low traffic volume) suburban street. The best objects to consider are one of the following on the other side of the street from you : A) light pole, B) fire hydrant, C) tree, or D) telephone pole. Typically some sort of object in a boulevard portion of a lawn is best.
4) Stand directly across from the object so that the street runs between the two of you.
5) Walk in a straight line perpendicular to the object a measured distance (10 ft or 3 m) (this is 'd') so as to make the base of a triangle (your path is one leg, the other leg is the distance from your starting point to the object and the hypotenuse will be the your view path to the object from where you end up). (see photo)
6) Facing the direction from which you came, point the straw back at where you started and have the protractor portion pointed away from the object. Swing the straw in the direction of the object until it is in view, all the time sliding the string and nut so as to be able to measure the angle covered.
7) To emphasize : the string and nut point in the direction from which you came and are marking an angle while the straw with you viewing through it now points at the object. Start the string at zero. Angle measure Θ.
8) Record both the distance traveled and the angle measured.
9) Use this information and the formula to determine the distance to the object (x) on the other side of the road from where you originally stood.
10) If possible, measure this actual distance and compute the percent error.

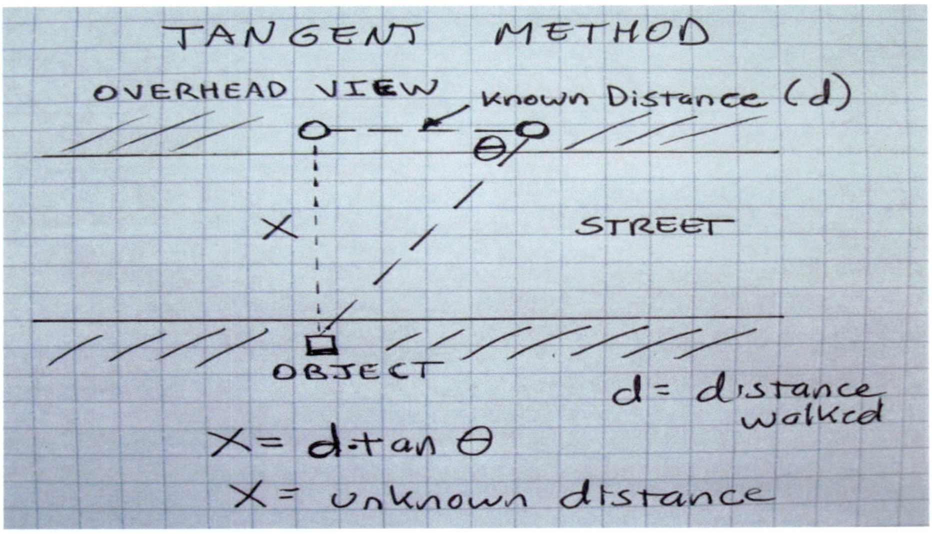

Activity 5 : Determination the altitude an object rises to when launched (such as a water rocket) through proportion method.

1) This activity can be done in conjunction with the water bottle rocket activity in the book.
2) In any case, one needs either a water-propelled or air-propelled rocket for the activity.
3) Also it is best to have at least 3 friends besides the one launching the rocket (though the minimum is 2 people – one to launch and one to measure the maximum altitude).
4) All of those engaged in measuring the maximum altitude need to use the astrolabe constructed in activity 2 above.
5) The rocket launcher readies and activates the rocket. The others performing the measurements are situated in a manner so as to be 3-5 m away from the launch. Whatever the distance, it must be measured and recorded. (This is 'd').
6) When launched, the altitude measurer uses the astrolabe and follows the path of the rocket to its maximum altitude by looking through the straw of the astrolabe and letting the nut and string hand down.
7) Once at the peak the altitude measurer holds the string in place on the protractor so as to measure the angle made.
8) Note that if not measured from zero, you will have to take the reading and subtract it from 90°.
9) The angle and the distance from the launch are used with the tangent formula to determine the altitude of the rocket.
10) If more than 1 is measuring the maximum altitude, have as many as possible measure the maximum height angle simultaneously and average the values for the measured angle.
11) Better still is if there are 5 people, cross out the largest and smallest angles and average the middle three to arrive at the angle to use.
12) In any case, repeat the experiment 2 to 3 times being sure to launch the rocket in exactly the same manner for each trial. Then these results can be averaged as well.

Activity 6 : Determination of the distance of an object (here a ping pong ball representing a star) in the home through parallax method.

1) In a long hallway or in the far part of an open area basement attach a string to a ping pong ball and suspend it from a ceiling so that it is about eye level.
2) Take your measurements at a place some distance from the ping pong ball 'star'. (this may be 3+ m).
3) It is best to have a line to measure from that is perpendicular to a wall that runs in the direction of the 'star' (see diagram).

4) Pick two points on this line so that if a line were to be drawn directly from the 'star' to the line it would bisect it the two chosen points would be nearly equal in distance on either side of the midpoint.
5) For the measurements, you use the astrolabe which was constructed in activity 2 above. For measurements you will be doing them horizontally. Hold the astrolabe parallel to the floor.
6) From each point stand so that you can point the straw first at the other point to be measured from and then moved so as to be pointed directly at the 'star' when viewing it.
7) While moving it from the line of observation to the 'star' hold the string and nut so that it sweeps across the protractor and measures the angle from the measurement line to the star from your point of observation.
8) If the angles are not congruent, choose a place so that they become congruent for best results.
9) The measurements to record are the distance between your observation points and the two angles measured.
10) Find the 3rd angle of the triangle from your two angles when added and subtracted from 180°. Now divide this answer by 2.
11) Use the tangent formula to find the distance to the 'star'.
12) Actually measure the distance with a measuring tape and determine the percentage error.
13) See photo below for the basic set up as well as an idea in a sketch.

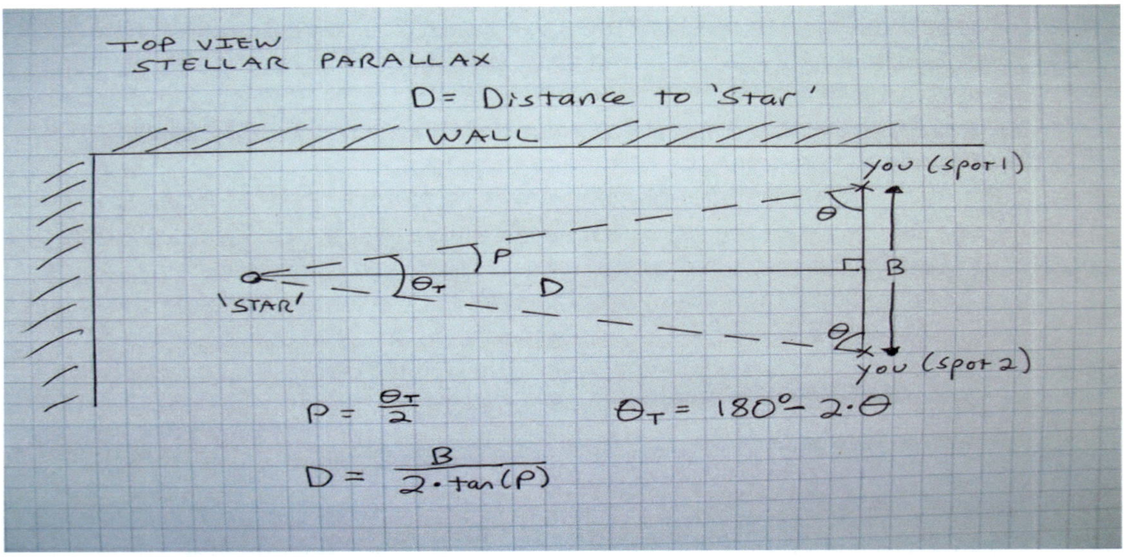

Data :

Activity 1 : Height Determination of Object on Wall

Calculate Apparent Height Table

Trial	Top Height	Bottom Height	Apparent Height
	R_T (cm)	R_B (cm)	ΔR (cm)

Similar Triangles Data Table

Trial	Distance from Object	Arm Length	Apparent Height	Calculated Height of Object
	D (cm)	A (cm)	ΔR (cm)	H (cm)

Activity 2 : Determination of Height through Angle (Tangent) Method.

Trial	Distance from Object	Angular Height	Calculated Height of Object
	M (cm)	Θ (^{o})	X (cm)

Activity 3 : Height Determination of Object Outside

Calculate Apparent Height Table

Trial	Top Height	Bottom Height	Apparent Height
	R_T (cm)	R_B (cm)	ΔR (cm)

Similar Triangles Data Table

Trial	Distance from Object	Arm Length	Apparent Height	Calculated Height of Object
	D (cm)	A (cm)	ΔR (cm)	H (cm)

Activity 4 : Determination of Distance of Object through Tangent Method

Trial	Distance from Object	Angular Height	Calculated Height of Object
	d (cm)	Θ (°)	X (cm)

Activity 5 : Determination of Altitude by the Tangent Method

Trial	Distance from Object	Angular Height	Calculated Height of Object
	d (cm)	Θ (°)	X (cm)

Activity 6 : Determination of Distance of Star by 'Parallax' Method

Trial	Θ^o	$\Theta_T^{\,o}$	P^o	B(m)

Calculations :

Be sure to use a Slide Rule for your calculations.
It is best to use average values,
so do several trials for any given activity in this set.

Formulae to Use :

$$\Theta = 90^o - \Theta_2$$

$$- = \frac{\Delta}{\;}$$

$$X = d * \tan\Theta$$

$$\Theta_T = 180^o - 2 * \Theta$$

$$P = -$$

$$D = \frac{\;}{* \quad (\;)}$$

For known heights (such as a picture or poster)
Determine the percent error from your experimental value :

$$\%E = \frac{[\qquad\qquad]}{\;} * 100\%$$

Conclusion:

Wow – so many applications of the same basic idea! Would you believe?! That is where creative thinking and application comes into play and the employment of basic concepts in math as applied to science – here we are mostly adapting the Right Triangle to our ideas in the real world.

From these activities it is clear that there is a connection between math in the form of algebra and geometry and the real-world when we find the place to use the ideas from math. These techniques are the means to solve the problems since there are many times we cannot verify the answer so readily. Consider the case of a star's distance. The most reliable is the measure of the parallax! All other measuring methods in astronomy have to be calibrated from it.

The key to the activities is patience, proper measurement, doing the work repeatedly and then determining the accuracy of the work when comparing to known values.

Hopefully you found this fun, Ms Briana Andromeda Purcell here really likes this one and it makes you want to create and explore more and find more ways science and math connect to the everyday world. Always be curious!

Explore & Enjoy – Be Brilliant!

Summary :

Notes : From Activity 6 in this Activity , one could argue to use the angles measured from the points along the line and then use the tangent to find the distance. This does work and can be used. I designed the activity this way as parallax for stars is measured by the angle of the star's displacement and not the angle that it is measured with respect to the baseline here at Earth's orbit.

Alternative Outdoor Method :

On a sunny day go outside, measure your height with a measuring tape. With the Sun in place to cast shadows, measure both the height of your shadow and the height of a shadow cast by an object (such as a flag pole, telephone pole, light pole, tree, et al).

Using the proportion method, estimate the height of the object from the following :

$$\rule{5cm}{0.4pt} = \rule{5cm}{0.4pt}$$

Activity #11
Law of Sines Use with Story Problem Solving
Grade Level : High School
Math Level : Calculating

Law of Sines Activity

In this part of the book we are still in the realm mostly of mathematics, but these ideas apply to the real world as we have seen. In the last few Activities we used the Right Triangle as our tool and applied it to many things in creative ways. In this next Activity we look to various angles in the triangle and their relations – one of the key ones in Geometry is called the Law of Sines. Here we will explore ideas such as answering – how long does that cable need to be for a given pole or support – how far away or apart are these two places and the like. As in the Right Triangles Activity where there are solutions walked through you can read along, use your slide rule, learn and learn to anticipate the next thought put down as a direction. This way you are becoming the problem solver. Following those exercises there are a series of questions for you to answer and apply these newly learned and sharpened skills. The answers to the questions, like the Right Triangles worksheet follow the exercise. Plus there is a good summary of key ideas from Geometry of important ideas contained there in the formulas commonly used and basic fundamental rules from there. Remember that these rules and ideas need to be in our creative minds when examining other problems and situations since they may be useful in that case – much like the boat crossing the river situation and using Right Triangles ideas in that Activity. Let's go and explore!

Triangles, like many of the geometric forms, have various relations between their sides and angles. It is these relations that allow for determinations of unknown sides or angles when needed. The Law of Sines is one such relation. The Law of Sines is the relation of the lengths of the sides of any given triangle to the sines of the angles opposite the sides. Each of these expressed as a ratio is equal to any other length of side to sine of the opposite angle value. The sine of an angle in a triangle is the ratio of the length of the side opposite the angle in question to the length of the hypotenuse of that given triangle. In the unit circle, the sine is the measure of the y-coordinate (since the hypotenuse is 1).

The Law of Sines :

$$\frac{\sin(A)}{a} = \frac{\sin(B)}{b} = \frac{\sin(C)}{c}$$

The Law of Sines is also called the Sine Law, the Sine Formula as well as the Sine Rule. The ratio can have the sines of the angles in the numerator or the lengths of the corresponding side lengths as the numerator.

Uses of the Law of Sines :

The Law of Sines can have many uses. When any two angles and a given side between the two known angles are known, then the remaining sides and angles can be determined. It is used in finding the unknown or missing measures in a given triangle when : A) there are two angles and any side known, or B) when two sides and the angle opposite one of them is known. This law is often employed in Triangulation or Surveying in the real world application. Look at the Right Triangle Activity (#9) and the Quest for a New Idea article (Activity #10) for another angle (joke intended) on this sort of problem where similar trigonometric methods are employed to find triangle angles and/or sides (distances).

There are 3 cases to consider when trying to prove the Law of Sines. In this matter only 1 of them will be considered : the Acute Triangle Case.

Proof Using the Acute Triangle :

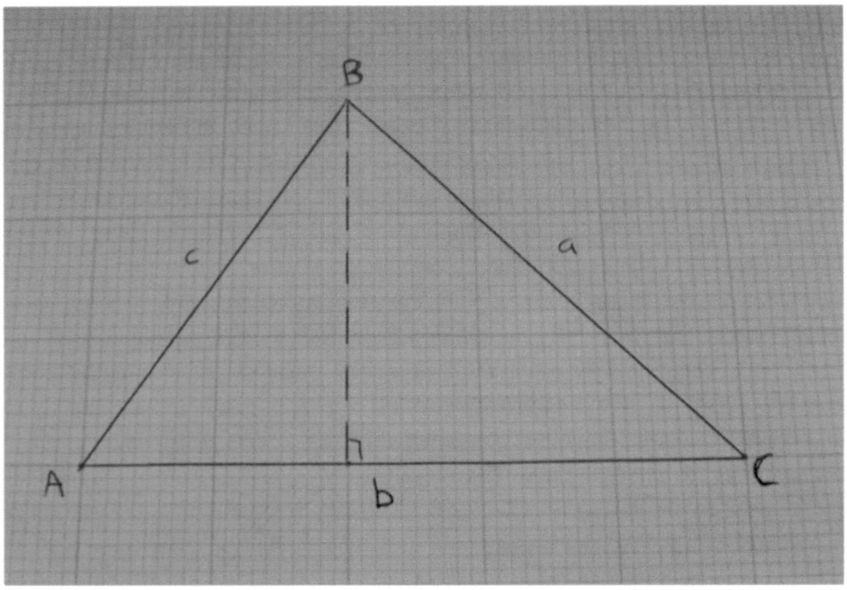

In the above diagram of triangle ABC drop an altitude at angle B and call it 'h'. The Area of this triangle is given by Area = 0.5*b*h.
sin(A) = h/c,
so we can substitute in this Area equation for h = c*sin(A)
and then have : Area = 0.5*b*c*sin(A)
There is no reason to only have an altitude at angle B, this can be done for angle A and then angle C subsequently.
In each case, we can then generate the following equations, all of which are equal to the area of triangle ABC :
Area = 0.5*a*c*sin(B) and
Area = 0.5*a*b*sin(C)
Since they are all the same area, we can set them equal to each other :
0.5*b*c*sin(A) = 0.5*a*c*sin(B) = 0.5*a*b*sin(C)
We can divide each of the expressions by 0.5*a*b*c to then yield :

$$\frac{\sin(A)}{a} = \frac{\sin(B)}{b} = \frac{\sin(C)}{c}$$

Our expected and familiar expression for the Law of Sines!

This is not the only key relation used when searching for sides and/or angles in a given triangle. The other is the Law of Cosines – which is the formula for Right Triangles or the Pythagorean Theorem in a more generalized format. This is expressed this way :

The Law of Cosines (expressed for each side) :

$$a^2 = b^2 + c^2 - 2*b*c*\cos(A)$$
$$b^2 = a^2 + c^2 - 2*a*c*\cos(B)$$
$$c^2 = a^2 + b^2 - 2*a*b*\cos(C)$$

The following Activity illustrates how to use a Slide Rule in solving various and classic Law of Sines questions and Story Problems. Pay careful attention to the examples since it is not only the Law of Sines that one employs to find the solution – there are other facts and relations concerning triangles, geometry and trigonometry that may be needed as well. Use these to then try and solve other questions here as well for you to try. Answers are provided afterwards. Enjoy :)

Purpose : To explore and solve various problems illustrating uses of the Law of Sines in conjunction to other trigonometric and geometric relations concerning triangles and angles while employing a Slide Rule

Materials :

- Ruler,
- Protractor,
- Trigonometric and Geometric Relations,
- Slide Rule

Procedure :

- The Procedure will be an illustration of the type(s) of Law of Sines Problems and How to Solve them with a Slide Rule
- Both the Procedure (solved questions) and the Data section (questions for you to try) will rely on the photos following them and enumerated.
- The S scale will be used a lot and is essentially a table of angle values from about 5° to nearly 90° (effectively about 87° at best)
- The S scale is for the Sine of a given angle. The value of the sine of the angle is not found here, but when the cursor is placed on a given angle, the sine value is then read from the C or D scale. Note however, that all sine values range from 0 to 1, so one has to mentally read the scales as if they are divided by 10 – shifting the decimal place one to the left.
- For example, the cursor on 30° on the Scale is over 5 on the C scale and the sine value for this angle is a value of 0.5 on the C scale, which is correct.

-
- **<u>Example 1 – refer to Figure #1 :</u>**
-
- In triangle ABC, angle A measures 35° , side 'a' is 10.0 units in length, angle C is 20°, What then is the length of side 'c' ?
- Looking at what we have, this is a problem where there are two angles known with a side adjacent to one of them of known length.
- In any given problem we wish to reduce all formulae to a set of known values and only one unknown, the one we are solving for.
- We look to the formula with our given information and see if it can be done with no other work or information – which we will see in later examples.
- It is better to use the portion of the proportion as this : sinC / c = sin A / a
- Here we have angles A and C and side a, and are looking for side c.

-
-
- $$\frac{\sin(A)}{a} = \frac{\sin(C)}{c}$$
-
- $$\frac{\sin(35°)}{10.0} = \frac{\sin(20°)}{c}$$
-
- We can cross-multiply if we want to, the slide rule is a natural tool for ratio and proportions such as this formula.
- We merely set the sin(35°) found on the C scale (read as 0.573)(if your slide contains the Trig scales of the S scale)
- If your slide rule has the Trig scales, such as S, et al on the Stator then it is often best to use the D scale. In all of the examples it is assumed that the trig scales are on the slide.
- Move the slide so that the S scale value of 35° is now above the Right Index of D. The value on the C scale (0.573) is the sin(35°).
- We chose the Right Index first because it is '10'.
- Next we choose between the left '1' and the right '1', we choose the right one because the other angle in our consideration (20°) is less than or to the left of the first value (35°) considered. If the second angle were to the right, or greater than, the first angle we would use the other Index.
- Now, Simply read along and move the cursor along to 20° on the S scale. Beneath it is the sine value for it (0.342), but it is not even needed here. We want 'c', which is now found below on the D scale which reads : 5.96 which is the correct answer!
- Note that in this case we are assuming that the angles given are perfect values (having infinite number of significant figures) and using the 10.0 value as being the one to establish the number of significant figures. If the answer were to only have two then the answer would be 6.0.
- Be sure to realize we do not always need the index, as our next modified example from this problem illustrates.

- **Example 2 – refer to Figure #1 :**
-
- In triangle ABC, angle A measures 35° , side 'a' is 19.0 units in length, angle C is 20°, What then is the length of side 'c' ?
- Looking at what we have, this is a problem where there are two angles known with a side adjacent to one of them of known length. This is the same as the last example, but we have changed the length of side 'a' to illustrate how the answer changes, but also how to use the slide rule in a different situation – that is, we will use the other end of the slide rule.
- In any given problem we wish to reduce all formulae to a set of known values and only one unknown, the one we are solving for.
- We look to the formula with our given information and see if it can be done with no other work or information – which we will see in later examples.
- It is better to use the portion of the proportion as this : sinC / c = sin A / a
- Here we have angles A and C and side a, and are looking for side c.

- $$\frac{\sin(A)}{a} = \frac{\sin(C)}{c}$$
-
- $$\frac{\sin(35^\circ)}{19.0} = \frac{\sin(20^\circ)}{c}$$
-
- We can cross-multiply if we want to, the slide rule is a natural tool for ratio and proportions such as this formula.
- Unless needed the examples will continue to use this proportion method.
- We merely set the sin(35°) found on the C scale (read as 0.573)(if your slide contains the Trig scales of the S scale)
- If your slide rule has the Trig scales, such as S, et al on the Stator then it is often best to use the D scale. In all of the examples it is assumed that the trig scales are on the slide.
- Move the slide so that the S scale value of 35° is now above 19 (which is 1.9 on the D scale). The value on the C scale (0.573) is the sin(35°).
- We do not need the Indexes in this problem first because we are not doing any division (but we can if we want to for whatever reason is needed). We are merely treating the problem as two ratios equal to each other. Once we have one ratio, we find the other by reading along the scales.
- Now, Simply read along and move the cursor along to 20° on the S scale. Beneath it is the sine value for it (0.342), but it is not even needed here. We want 'c', which is now found below on the D scale which reads : 11.3 which is the correct answer!
- Notice that independent of the lengths of the sides of the triangle the sine values of given angles remain constant.

- **Example 3 – refer to Figure #1 :**
-
- In triangle ABC, side 'a' is 10.0 units in length, angle B is 80°, and angle C is 35°. What then is the length of side 'b' ?
- Looking at what we have, this is a problem where there are two angles known with a side between them of known length.
- In any given problem we wish to reduce all formulae to a set of known values and only one unknown, the one we are solving for.
- However, we quickly see a problem here. We want to find 'b' so we need the ratio of sinB / b, but we do not have side 'c' even though we were given angle C – it is best not to use this ratio.
- It is better to set sinB / b = sin A / a
- This makes one look at it in a questioning manner, since we were not given angle A, though we have side a.
- This is where basic background knowledge is assumed on the part of the problem solver.
- For all triangles, the sum of all of the angles is 180°. Since we know two of the angles, it is easy to find the unknown angle A :
- 180° = measure(A) + measure(B) + measure(C)
- 180° = measure(A) + 80° + 35°
- Measure (A) = 65°
- Now we can use the ratio we originally considered :
-
- $$\frac{\sin(A)}{a} = \frac{\sin(B)}{b}$$
-
- $$\frac{\sin(65°)}{10.0} = \frac{\sin(80°)}{b}$$
-
- We can cross-multiply if we want to, the slide rule is a natural tool for ratio and proportions such as this formula.
- We merely set the sin(65°) found on the C scale (read as 0.906)(if your slide contains the Trig scales of the S scale)
- Move the slide so that the S scale value of 65o is now above the Left Index of D. The value on the C scale (0.906) is the sin(65°).
- Simply read along and move the cursor along to 80° on the S scale. Beneath it is the sine value for it (0.984), but it is not even needed here. We want 'b', which is now found below on the D scale which reads : 10.9 which is the correct answer!
- If interested you can find side C as well – Note it is best to place the sin(65°) above the right index of D in this case! (answer : 6.33)
- *A quick and interesting side note, with a little practice on this, it is clearly faster than with a calculator and shows the true form of the formula as a proportion.*
-
- **Example 4 which comes from figure #2**
-
- In triangle ABC from figure #2 side 'c' is 50 m long and side 'a' is 40 m in length. Angle C measured 125°. With this information find the measure of angle A, the measure of angle B, and the length of side 'b'. Round all answers to whole number values.

- In this case we have two sides and an angle.
- Looking at the 125° angle our first concern is the size of the angle if we are to use a slide rule since they have values up to 85+° – what do we need to do?
- As in the prior example of needing to have prior knowledge, we need to realize that the supplement of this angle in the second quadrant will have the same sine value as the original angle (try it on a calculator or better yet look it up, if needed).
- So the angle we can use in this problem will be angle C = 180° – 125° = 55°
- Notice that the question asks us first to look for Angle A – there is a reason for this :
- We do not know angle B nor side b, yet we do know side a, but do not know angle A – we have only one unknown, so this can be solved!
- Now we can use the ratio we originally considered :
-
- $\frac{\sin(C)}{c} = \frac{\sin(A)}{a}$
-
- $\frac{\sin(55°)}{50\text{ m}} = \frac{\sin(A°)}{40\text{ m}}$
-
- Since side 'a' is smaller than side 'c', we would expect angle A to be smaller than angle C – knowing this we be looking to the left once we set up the ratio of the sine of angle C over C on the C and D scales.
- Unlike the other examples, we have to now read along the D scale and not the C scale, since we are dealing with the sides of the triangle.
- Read along the S scale to 55° and place this above 5 (read as 50) on the D scale.
- If interested the answer to this ratio is read opposite the Left D Index (be sure to keep track of the decimal point!) (answer : 0.0163) Also the reading for the sine of 55° is : 0.819 .
- Like before, we do not need to find the sine of the angle (it will be : approx 0.655), but read the answer above 4 (read as 40) on the D scale back to the moved slide on the S scale and find the angle of 41° (rounded from 40.9° – from the directions in the question) for angle A.
- We now can find angle B readily by recalling that the sum of all of the angles in any triangle is 180o. So angle B = 180° – 125° – 41° = 14°
- Realize we have not needed to move our slide after the first move.
- Now read along the S scale to 14° and read down to the D scale to find the final answer of side 'b' which is read as 14.8, but using the directions we round to 15 m as our final answer.
-
- **<u>Example 5 from figure #3</u>**
-
- Here is a real-life example. We have a tower that has two supportive cables attached as depicted in the drawing.
- The Height (H) of the tower is 420 m, angle C measures 60°, angle D measures 40°. The question posed here is this : What is the distance between points C and D (essentially the length of the segment CD)?
- Realize that this problem can have a host of questions, which you can consider and solve for yourself, if interested (such as what are the lengths of each of the cables, what is the angle between the cables)?

- To make the question more exciting we will find these answers along the way to finding the distance CD!
- The first thing to realize is that there are two overlapping triangles in this diagram : namely triangle ABC and triangle ABD. It is recommended you draw each of these separately and label them with the known information :
- Such as Angle B in each of these is 90°, each has the leg H measuring 420 m, we can even determine the unique angle A in each of the triangles (and they are different – if it helps give them different names, like A1 and A2).
- To find each of these A angles, A1 and A2, we recall the sum of the interior angles theorem for all triangles being 180°.
- So A1 = 180° – 90° – 60° = 30°
- And A2 = 180° – 90° – 40° = 50°
- Why start with the pictures and the angles?
- Ultimately we want the segment CD which is the side of a triangle and this triangle has an unknown angle A* (which is not A2 mind you, but is the difference of A1 and A2 (50° – 30° = 20°) and two other unknown sides – we need to simplify our approach.
- The best solution is to untangle the overlapping triangles and treat them as right triangles because the sin(90°) = 1, which makes setting up ratios on the slide rule quite easy.
- Also look at the logic, we can subtract the length BC in the smaller triangle from the length BD in the larger triangle and find the needed distance of CD, the goal of the problem!
- From the first triangle ABC, we can find 'b', the length of the cable from this proportion:
-
- $\dfrac{\sin(\mathbf{B})}{\mathbf{b}} = \dfrac{\sin(\mathbf{C})}{\mathbf{H}}$
-
- $\dfrac{\sin(90^\circ)}{\mathbf{x\ m}} = \dfrac{\sin(60^\circ)}{\mathbf{420\ m}}$
-
- Set the angle 60° on S scale over 420 on the D scale,
- Push the cursor along to 90° on the Scale and read the answer to the problem here on the D scale – be wary of decimal placement : It should be read as 485 m
- By the way, do not move the slide, we can find BC already!
- Push the cursor to 30° on the S scale and find the answer to BC beneath it on the D scale, which reads : 242 m
- Now for triangle ABD
-
- $\dfrac{\sin(\mathbf{B})}{\mathbf{c}} = \dfrac{\sin(\mathbf{D})}{\mathbf{H}}$
-
- $\dfrac{\sin(90^\circ)}{\mathbf{y\ m}} = \dfrac{\sin(40^\circ)}{\mathbf{420\ m}}$
-
- Follow the same directions for this proportion, only here start with 40° on the S scale and slide it over the same initial value of 420 on the D scale
- Under the cursor at 90°, you should find the cable length, y, as 653 m
- Moving the cursor to 50° you find the answer as 500 m for BD

- The final answer is CD = BD – BC = 500 m – 242 m = 258 m
- We have solved all the questions posed and only with one slide of the slide and two movements of the cursor for each of the triangles!
- This problem illustrates a few of the fundamental aspects of a good story problem and its approach :
- First realize what you know and what is looked for
- Next, break the problem into pieces when and where appropriate
- Look for similarity between the pieces
-
- **<u>Example 6 from Figure # 7</u>**
-
- This next problem has some other unique challenges that one might encounter with a slide rule, such as when an angle is not on the scales present!
- From Figure #7, angle B is measured at 70°, side AB measures 500 ft and Angle D measures 3.5°
- With the given information, find all of these things : find the distance AC, the the distance CD, and finally the distance AD. Take all answers to two significant digits.
- Looking over our values, we might first think that this is similar to other examples, only different numbers – and that is so.
- But from the point of view of our tool, the slide rule, there is a problem : the small angle 3.5°
- If you only have a S scale and not a ST scale, it will not have it on there. The ST scale provides values for sines and tangents for very small angles (about 0.6° to 5.8°) – it is both since in this range they are nearly identical in the first two significant digits.
- If you have a ST scale, then you need do nothing else, other than realize unlike the S scale values for sine which are read as decimal values on the C and D scale where it is 0.XY, they now have another place added and are read as 0.0XY – which means a greater emphasis on keeping track of the decimal point!
- If there is no ST scale, then what to do?
- Actually the slide rule is designed to take care of this :
- All small angles, such as those from 0.6° to 5.8° can be determined by first looking up the given number in question, such as angle D at 3.5° here and placing it over a common gauge point on most slide rules at 57.3
- Doing so, such as 3.5° on the C scale and 57.3 on the D scale one reads the answer opposite the Right Index of the D scale to find 0.0610 – try this on a calculator to see if correct (it is).
- Having this obstacle managed, now we can examine and solve the problem :
- In case needed, let's determine angles A1 and A2 –
- A1 = 180° – 70° – 90° = 20°
- A2 = 180° – 3.5° – 90° = 86.5°
- Now let's determine the distance AC :
-
- $$\frac{\sin(B)}{AC} = \frac{\sin(C)}{AB}$$
-
- $$\frac{\sin(70^\circ)}{AC} = \frac{\sin(90^\circ)}{500\text{ ft}}$$
-

- Use your slide rule as spelled out in the other examples to set up the proportion and find the answer to AC :
- AC = 469 (almost 470 – a calculator yields 469.8, so if you use the value of 470 you are quite alright)
- Keep the mark for the answer to AC and now slide our answer to the sin(3.5°) over it since we are now going to look for AD :
-
- $\frac{\sin(C)}{AD} = \frac{\sin(D)}{AC}$
-
- $\frac{\sin(90°)}{AD} = \frac{\sin(3.5°)}{469\ ft}$
-
- AD = 7690 ft (calculators disagree, but we really do not have that many significant digits to work with here, and this is sufficient) –
- A calculator would yield 7682, the answer to 2 significant digits would be 7700, which is what you might be seeing on your slide rule (I look at mine with magnifying reading glasses by the way).
- I will stay with my estimated answer of 7690 and employ it to find the distance of CD.

- We can then move on to segment CD :
-
- $\frac{\sin(A2)}{CD} = \frac{\sin(C)}{AD}$
-
- $\frac{\sin(86.5°)}{CD} = \frac{\sin(90°)}{7690\ ft}$
-
- Notice from the S scale we are heavily estimating the placement of the cursor for 86.5°, but all values for sine over 85° have as their first two decimal places 0.99X
- CD = 7670 ft
- (The calculator shows 7668 using the previous calculator value through the calculation)
- All in all the values are quite good and very close to 'actual' values – I place this in quotes since it does depend on the precision of the measurements in the first place hence the number of significant digits we can use in our final answers.
-
- **Example 7 from Figure #9**
-
- For this example, we are given segment BD as 40.0 inches, angle B measures 35°, angle A2 measures 25° and angle D in the triangle ACD measures 60°.
- There are two questions here : what is the length of segment AD and what is the length of segment AC?
- This is one of the cases where the answer in one case becomes used to find the final answer. This is classic in many story problem situations.
- First let's find segment AD.

- $\dfrac{\sin(B)}{AD} = \dfrac{\sin(A1)}{BD}$
-
- $\dfrac{\sin(35°)}{AD} = \dfrac{\sin(25°)}{40.0 \text{ in}}$
-
- After setting up the proportion, you should find that AD = 54.3 in
-
- $\dfrac{\sin(D)}{AC} = \dfrac{\sin(C)}{AD}$
-
- $\dfrac{\sin(60°)}{AC} = \dfrac{\sin(90°)}{54.3 \text{ in}}$
-
- From this proportion, you should find AC = 47.0 in.
-
- **Example 8 from Figure # 6**
-
- This is a classic story problem. Imagine you are in a building and have a protractor that enables you to measure the angle from horizontal to the top and bottom of a nearby structure (it may be a statue, a building, or a tree – whatever you want it to be).
- Here are the numbers that are known : You are at a height H of 80 m, your measurement of angle A is 20° and the other measurement angle to the bottom of the structure is 30°.
- The obvious question, then is this : What is the height of the structure?
- The first thought might me is that there is insufficient information to solve this problem – but this is not the case.
- Once again, we are expected to know a number of things.
- First is that the overall figure is like a rectangle – extend the top line horizontally and project the center line of the structure in question so that they meet at a right angle.
- Next, we assume that each of these structures are both perpendicular to the ground which is assumed to be flat. This means that the height of the structure, we will call Q and the space above it up to the horizontally projected line, we will call N in sum is the same as the height of the building we are in (80 m),
- so Q + N = 80 m
- Due to Alternate Interiors Angles Theorem in Geometry (see information below) the angle B is congruent to the small angle formed with the ground at the base of the unknown structure, and so it too is 30°.
- With that being known, we have essentially three triangles. We will concentrate on the bottom one.
- It has the angles of 90° in the lower left corner of the picture which is where the building we are in meets the ground, the right hand angle was determined to be 30°, so the other angle in the upper left corner is 60° (180° – 90° – 30° = 60°)
- We want to find the distance on the ground between the buildings, we will call this X. It is the same measure of the horizontal line running parallel as projected from the window we are standing at 80.0 m up
- Given this information we can solve for X : Examine the ratio and make a drawing if needed to follow along :

-
- $$\frac{\sin(30^\circ)}{80} = \frac{\sin(60^\circ)}{X}$$
-
- We have a problem. Placing the S scale 30° over 80 will not allow us to read the answer opposite 60°!
- In this case it is best to cross multiply and solve for X :
-
- X = 80*sin(60°)/sin(30°)
-
- Place the cursor on 8 (for 80 on the D scale)
- Above it place the Right Index of the C scale
- Now move the cursor to 60° on the S scale – the answer to the multiplication is found on the D scale, but we now need to divide it by the sin(30°)
- Leaving the cursor on this running answer now move the slide so that the S scale reading of 30° is above the running answer
- The answer is now found opposite the Left Index of the C scale on the D scale
- You should find X = 138. (perhaps estimated at 139 depending on your reading of the slide rule, it is between them – the key is leave the cursor here !)
- Now turn your attention to the smallest triangle where angle A is one of the angles and in the upper right hand corner is a 90° angle. The other angle in this triangle must be 70°.
- With this information we can find the distance for the space above the unknown structure up to the projected line from the building, or distance N.
-
- $$\frac{\sin(70^\circ)}{138} = \frac{\sin(20^\circ)}{N}$$
-
- We have the same problem as before. Placing the S scale 70° over 138 will not allow us to read the answer opposite 20°!
- In this case it is best to cross multiply and solve for N :
-
- N = 138*sin(20°)/sin(70°)
-
- So instead of the S scale 70° above 138 on the D scale, place the Left Index of the C scale above it.
- Now move the cursor to 20° on the S scale. The answer to the multiplication is now found on the D scale under the cursor.
- Now move the slide so that the S scale reading of 70° is above this running answer.
- Finally the answer can be read opposite the Right Index on the C scale
- We come to find N = 50.4
- Now Q, the height of the unknown structure, can be found :
- Q = 80 m – 50.4 m = 29.6 m
- An important aside : Is there any way to continue the proportion method that we have employed in the other examples? Yes there is.
- In more advanced slide rules with more scales, there are ones with a CF and DF scale set as well. These tend to be duplex rules (i.e. scales on both sides). Since the C and D scales are now split at the value of pi we can move the scale, for example in one of our

calculations, to 70° on the S scale over 138 on the DF scale, then move the cursor to 20° on the S scale and read the answer from the DF scale 0f 50.4!

- With all of these examples, you can do any and all the other problems with some thought and effort. Be the Scientist – Be the Mathematician – Explore &Enjoy – Be Brilliant! :)

Data :

- The Data Section will be a series of classic Law of Sines Problems
- Both the Procedure (solved questions) and the Data section (questions for you to try) will rely on the photos following them and enumerated.
- Be sure to refer to the photos when indicated. The problems are in no particular type of order as compared to the examples detailed above, but be sure to look to them for ideas on how to solve a problem as needed.
- Answers are provided in the Conclusion Section of the Activity.
- In some cases, a general triangle is mentioned, but one could refer to the first diagram in Figure #1 but be sure to label accordingly as listed in the given question.
-
- **Questions :**
-
- 1) In Triangle ABC, the measure of angle A is 52°, side 'c' is 10 ft, while side 'a' measures 15 ft, what is the measure of angle C?
- 2) In Triangle DEF, the measure of angle F is 43°, side 'd' is 16 mm, side 'f' is 24 mm, what is the measure of angle D ?
- 3) In Triangle RST, the measure of angle R is 78°, while angle T measures 39°, and side TS is 19 in. What is the measure of side RS?
- 4) In Triangle JKL, angle L measures 64°, angle K measures 36°, and side 'j' measures 18 m, what is the length of side 'k' ?
- 5) Triangle ABC has where angle A measures 30°, angle C is 50° and side 'b' is 44 units in length. What are the measures of sides 'c' and 'a'?
- 6) Referring to Figure #2 if side 'c' is 38 and side 'a' is 25 while angle C measures 108°, what is the measure of angle A, the measure of angle B, and the length of side 'b' ?
- 7) Use Figure #8 for this question : Two towers are points A and B and are 10 miles apart. There is an emergency signal flare spotted 39° E of N (North is up in the diagram) from tower A and is simultaneously spotted 42° W of N from tower B. The question is : how far is the signal flare from each of the towers?
- 8) Use Figure #8 for this question : Two towers are points A and B and are 15 kilometers apart. There is a fire spotted 32° E of N (North is up in the diagram) from tower A and is simultaneously spotted 48° W of N from tower B. The question is : how far is the fire from each of the towers?
- 9) Use Figure #3 for this question : There is a Radio tower that is 600 ft tall. Two cables, labeled AC and AD attach and secure the tower to the ground and keep it vertically standing. Angle C is 58° and angle D is 44°. Note also that the angle ABC is 90°. You are to find the length of the two cables and the distance between their anchor points on the ground.
- 10) Use Figure #9 for this question : In your diagram, label angle A2 as 23°, angle B as 33°, and angle ADC as 56°. Also, the segment BD is 112 units in length. The question at hand is this : what is the length of the segment AC?

- 11) For this question refer to Figure #6 : You are at the observing platform of a 145′ tower and can see a statue in the distance. The angle A from a parallel horizontal plane to the ground at your observation level measures 13° to the top of the statue and the angle B measured from the same plane now measuring to the bottom of the statue is 31°. With this given information, find the height of the statue.
- 12) Refer to Figure #7 for the following question : If angle A1 measures 3.00°, angle 2 measures 22.0° and segment AD measures 48.0 ft, to 3 significant figures, what is the measure of segment CD?
- 13) From Figure #4, a balloon is held by two cables. Cable 1 is 842′, while angle A measures 65° and angle B measures 4°. Find the distance BC, where C is the point directly below the balloon on the ground.
- 14) In Figure #5 : There is a lighthouse on the shore atop a cliff as illustrated in the figure. There are two boats on the water at points A and B directly in line with each other and in a plane through the lighthouse. At Boat A, the observer there notes that the bottom of the lighthouse subtends an angle 'x' of 23°, while the top of the lighthouse from point A called angle 'y' measures 40°. From boat B's point of view, the top of the lighthouse, angle 'b', subtends an angle of 16°. In these positions, the boats are 1320′ apart. The questions then are these : What is the height of the lighthouse by itself and what is the height of the cliff?

Pictures to reference for Procedure & Data sections :

Figure #1

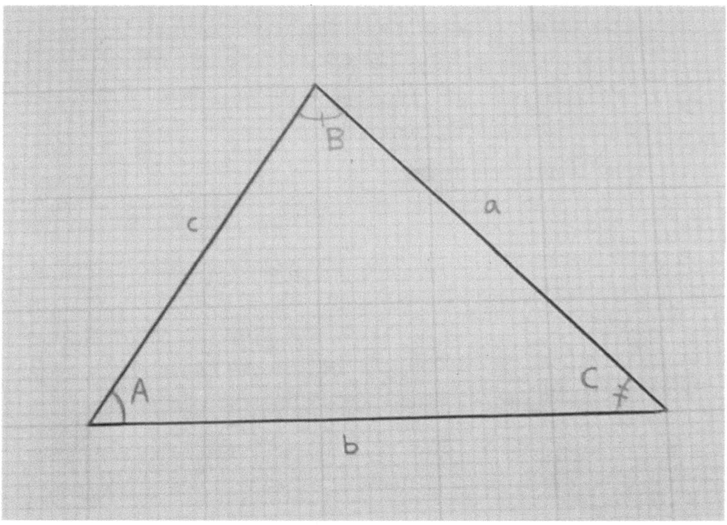

Figure #2

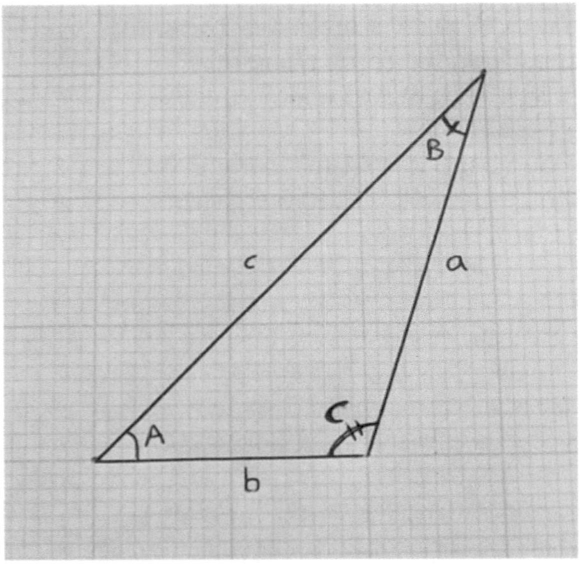

Figure #3

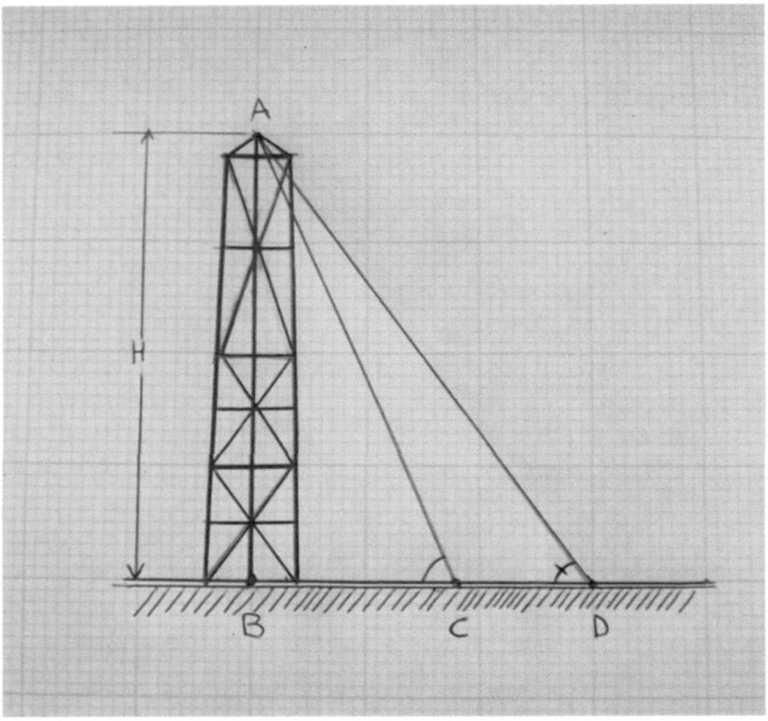

Figure#4

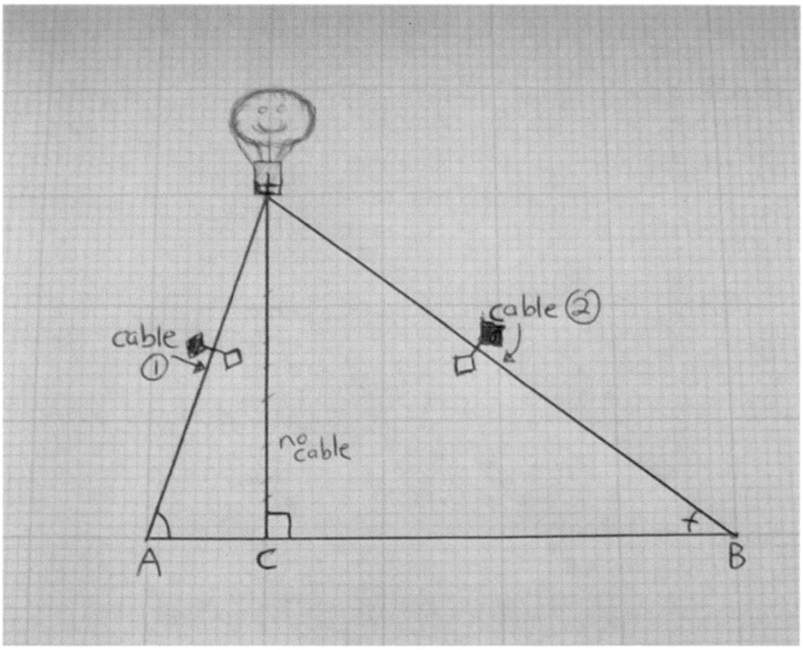

Figure #5

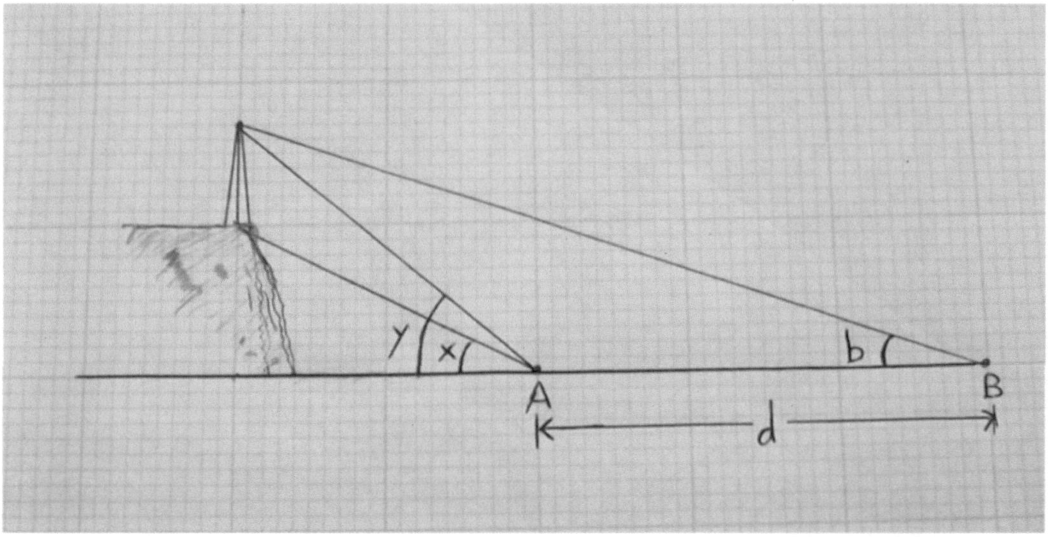

Figure #6

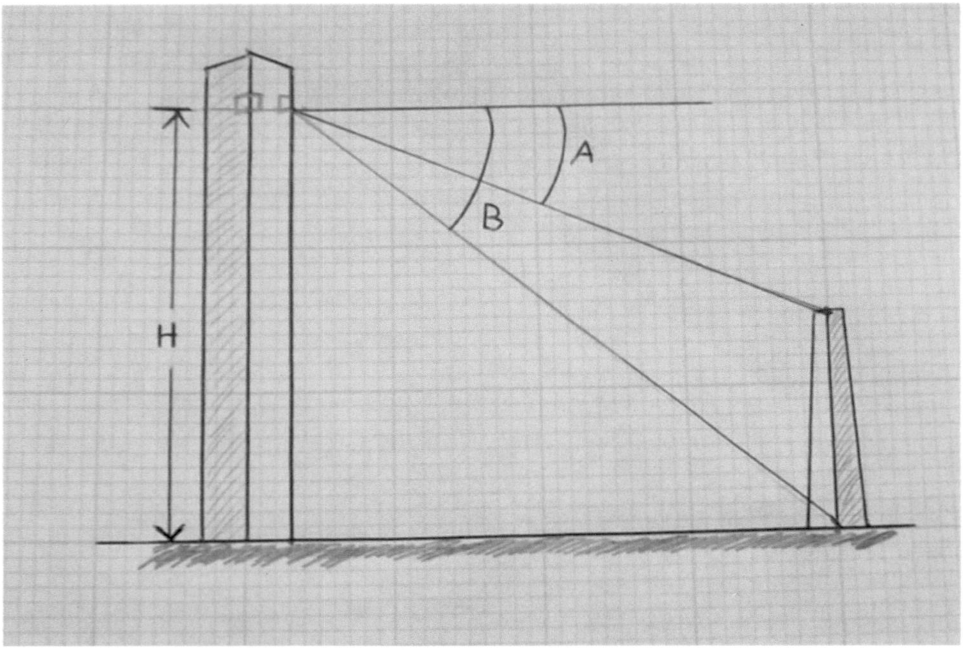

Figure #7

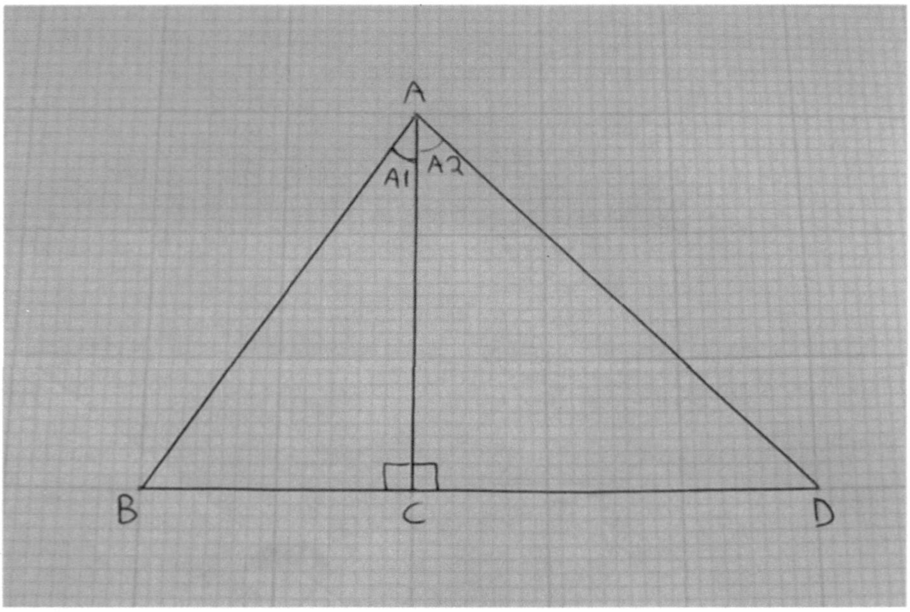

Figure #8

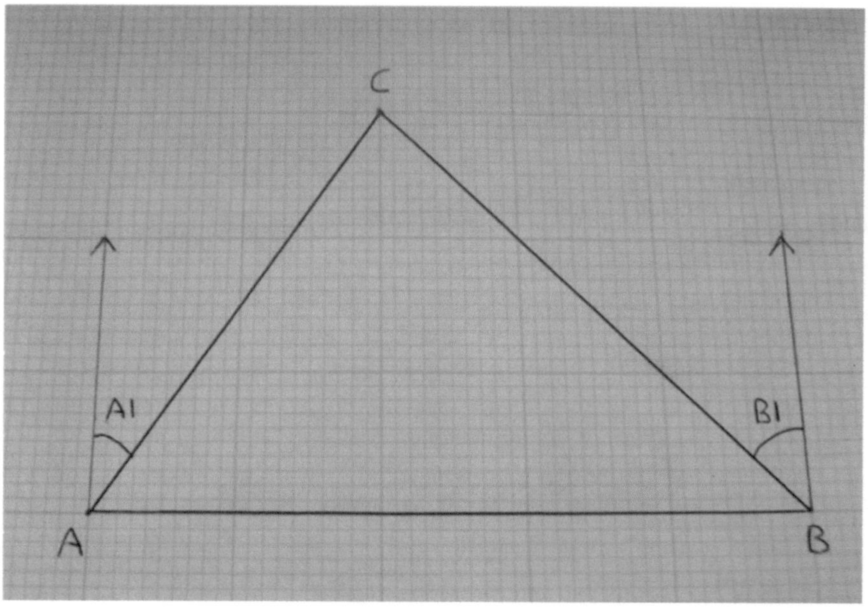

Figure #9

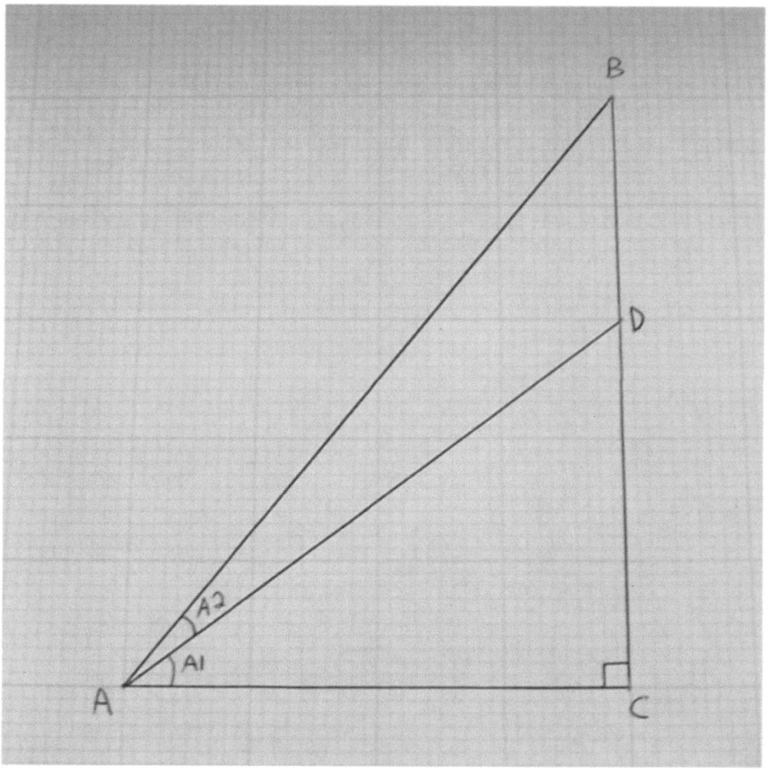

Calculations :

Be sure to use your Slide Rule !

$$\textbf{Sin}(\Theta) = \frac{\text{length of side opposite}}{\text{length of hypotenuse}}$$

Law of Sines : In any triangle the ratio of any two sides is proportional to the ratio of the sines of the angles opposite those sides.

$$\frac{a}{b} = \frac{\sin(A)}{\sin(B)}$$

Law of Sines :

$$\frac{\sin(A)}{a} = \frac{\sin(B)}{b} = \frac{\sin(C)}{c}$$

Other Useful Trigonometric and Geometric Formulae that may be useful :

Concepts, Components of a Triangle and its Relations :

Sine of Angle Θ : $\frac{\text{length of leg opposite } \angle\Theta}{\text{length of hypotenuse}}$ $\sin(\Theta) = \frac{\text{opp side}}{\text{hyp}} = \frac{a}{c}$

Cosine of Angle Θ : $\frac{\text{length of leg adjacent } \angle\Theta}{\text{length of hypotenuse}}$ $\cos(\Theta) = \frac{\text{adj side}}{\text{hyp}} = \frac{b}{c}$

Tangent of Angle Θ : $\frac{\text{length of leg opposite } \angle\Theta}{\text{length of leg adjacent } \angle\Theta}$ $\tan(\Theta) = \frac{\text{opp side}}{\text{adj side}} = \frac{a}{b}$

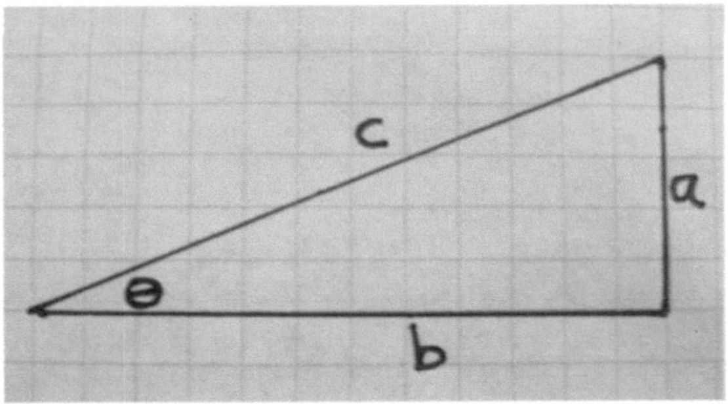

Other Trigonometry Identities and Relations :

Sin (Θ) = Cos (90° - Θ)

Sin(Θ) = 1/Csc (Θ)
Cos(Θ) = 1/Sec (Θ)
Tan(Θ) = 1/Cot (Θ)
Tan(Θ) = Sin(Θ)/Cos(Θ)

$\sin^2(\Theta) + \cos^2(\Theta) = 1$

Triangle Angle Sum : The sum of all the interior angles of a triangle equal 180°.

$m\angle A + m\angle B + m\angle C = 180^\circ$

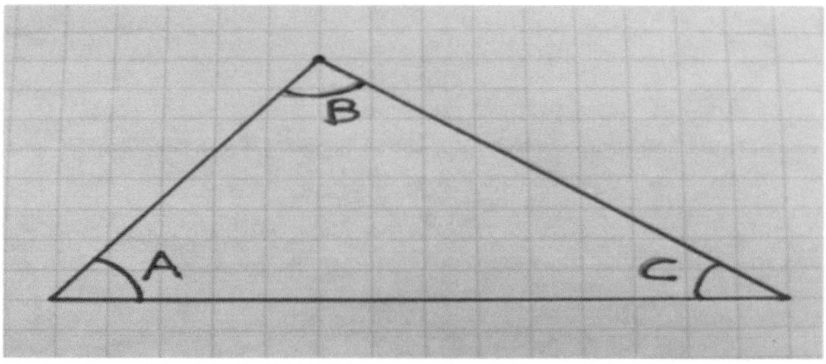

<u>Useful Geometric Concepts :</u>

Complementary Angles : Are two or more (≥ 2) angles that when summed up equal 90°

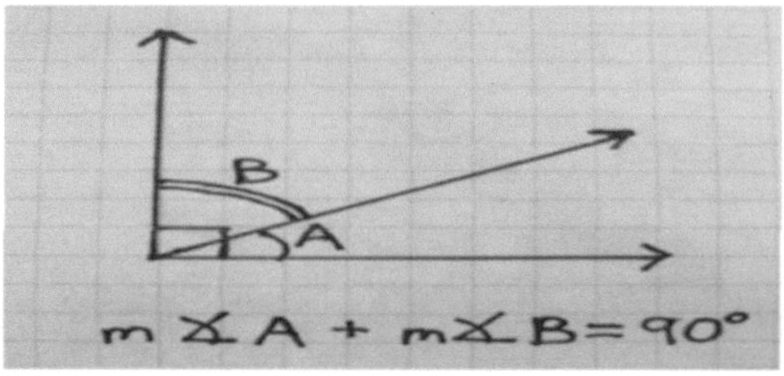

Supplementary Angles : Are two or more (≥ 2) angles that when summed up equal 180°

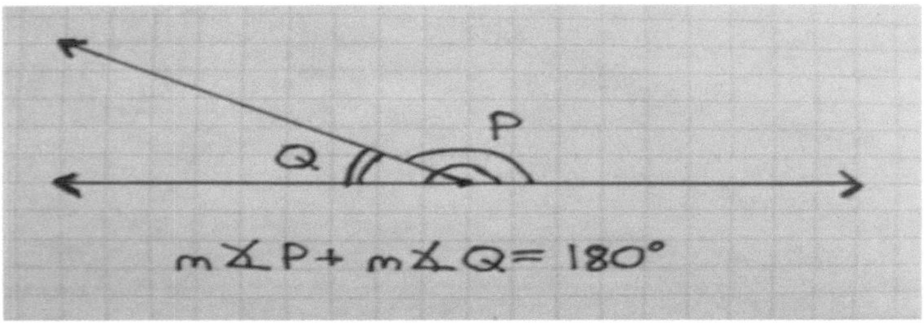

Similar Triangles : Are 2 (or more) triangles where the corresponding angles of each are congruent and the corresponding sides of each figure are proportional.

Vertical Angles : Vertical Angles are formed by any two non-parallel lines crossing at a point. The angles opposite each other are congruent.

Parallel lines intersected by a Transversal : $\neq$

When two parallel lines are cut by a transversal line, several conclusions can be drawn :

A) The 2 adjacent angles on the parallel line both horizontally and vertically are supplementary
B) The corresponding angles between the two parallel lines are congruent
C) The alternate interior angles are congruent
D) The angles on the same side of the transversal cutting the parallel lines are supplementary
E) The alternate exterior angles to the parallel lines are congruent

Pythagorean Formula :

For all Right Triangles :

The sum of the squares of the legs of the triangle is equal to the square of the hypotenuse of the triangle.

$$A^2 + B^2 = C^2$$

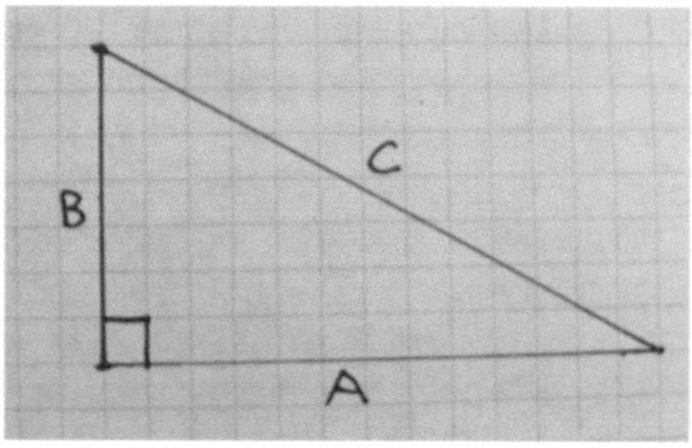

Can also be expresses as the Law of Cosines for any triangle :

$$A^2 + B^2 - 2*A*B*\cos(c) = C^2$$

Conclusion :

Here are the solutions to the Questions in the Data Section :

1) Measure of angle C = 31.7°
2) Measure of angle D = 27.0°
3) Side RS = 12.2 in
4) Side 'k' = 10.7 m
5) Side 'c' = 34.2, side 'a' = 22.3
6) Measure of angle A = 38.7°, measure of angle B = 33.3°, side 'b' = 21.9
7) Tower A is 7.5 mi, while Tower B is 7.9 mi from the signal flare
8) Tower A is 10.2 km, while Tower B is 12.9 km from the fire
9) Cable AC = 708 ft , Cable AD = 864 ft , and distance CD = 246 ft
10) AC = 129.4 units
11) Statue height = 89.3 ft
12) CD = 19.4
13) Distance BC = 10,910 ft
14) Height of Lighthouse = 284 ft, Height of Cliff alone = 291 ft

How did it go? Always go back over your steps if needed – say them aloud if it helps so that if something was off you can consider where to change what you did to arrive at the answer then to now.

Activity #12
Estimate the Number of Leaves on a Tree

Grade Level : High School
Math Level : Calculating

Estimating the Number of Leaves on a Tree Activity

This Activity is a good math activity involving the estimation of a 'Population' size. Think about it – how does someone come up with numbers such as the number of stars in a galaxy, the number of bacteria on our bodies, the number of a given type of fish in the ocean and the like? That being said, the idea here is this question, which many of us may have had at one time or another : How many leaves does a tree have?

This seems an impossible, if not useless, question. But is it? Can knowing an estimated value for the number of leaves indicate some aspect of a tree's health, likelihood to survive, ability to adapt or thrive in a given environment? This sort of data may find a place when used in conjunction with other information. Trees are definitely a critical life form on the planet – providing food, shelter, and oxygen for many others.

To ponder this, consider a small forested area with several tree species. Do trees with the most leaves utilize the most resources? Are they found nearest water sources? Do they have the largest leaves? Do trees of similar size have the same number of leaves, even if not of the same species?

Needless to say, this is an Activity that allows one to look up at a tree and with observational counts and a few basic measurements, one can estimate the number of leaves on the tree. Explore & Enjoy

Purpose : To estimate the number of leaves on a given tree from indirect measurements and estimates in measures of surface area of the tree, the numbers of leaves on a branch, and number of leaves per unit area.

Purpose : To use the number of estimated leaves on a tree to determine an estimate of the surface area of exposure of leaf area to collect sunlight.

Purpose : To use the number of estimated leaves on a tree to determine the mass of the leaves on a given tree.

Materials :

- Tree(s) outside with leaves on them,
- Leaves from the Tree in the investigation,
- Ruler(s),
- Measuring Tape,
- Protractor,
- String,
- Mass Scale (if interested in mass of leaves on the tree),
- Binoculars,
- Slide Rule

Note : One of the best times to conduct these Activities is during the fall, since there are leaves on the ground to use for the calculation (versus having to acquire some during the summer)

Note : Do not acquire leaves without parent permission and/or supervision. It is advised to NOT climb the trees or do any other potentially dangerous actions so as to acquire leaves.

Procedure :

1. For both Estimation Activities, choose 1 (or more) trees for your study.
2. **Activity I :**
3. With all directions it is best to read them through once or twice, think about what to do and develop a plan of action.
4. From the trunk of the tree measure a distance using the measuring tape so that you are far enough to see the full span of the crown of the tree (the widest part of the branches of the tree in its full form).
5. Record this value of distance on the data table (r).
6. Facing the tree, hold the protractor level to the ground so that the 90° mark lines up with your eye on one side of the crown when looking from the vertex point looking across the protractor. (for ex., the right side)
7. Use string as an arm on the protractor measuring from the vertex and sweep out the span of the crown so that when looking from the vertex you are looking along across the protractor first in the direction of 90° and then along the string to the angle mark for the opposite side of the crown (now the left side of the crown). (Be sure to hold the protractor in place while measuring and hold the string on the angle mark).
8. Record this angular measure (Θ) on the data table. Note : That the angle is the difference of 90° and the angle mark on the protractor,
9. With this measure, you might consider doing this 3 times at the same distance (r) from different perspectives and averaging these values.
10. Note that the goal of the prior several steps was to measure the angular span of the crown of the tree.
11. Now gather a small group of leaves from only that tree (Note do not climb or do anything dangerous or destructive). Try to acquire a range of sizes of leaves from that tree.

12. On level ground (or a table), place 2 rulers at right angles to each other so as to form the sides of a 'square'.
13. Use the leaves gathered and place them within the 'square' without overlapping the leaves, leaving as few gaps as possible, and not exceeding the boundaries of the 'square'. Note, do not alter the leaves by folding, cutting, and the like, hence there will be some gaps in your 1 sq ft area.
14. Count and record the number of leaves in the square (N)
15. Redo the square foot area twice more with different leaves from the same tree, recording the number of leaves each time, and take the average of these (N_{ave}).
16. Calculations :
17. After calculating the needed Averages noted in the directions above, do the following :
18. Convert the angular measure to radians (R).
19. Calculate the length of the arc length of the crown of the tree (S).
20. From the arc length (S) we can now determine the radius of the crown arc length (r_T).
21. With the radius we can now Calculate the Surface Area of the Tree (SA)
22. Use your determined values (of N and SA) to calculate the estimated number of leaves on the tree (T_L).
23. Further Calculations :
24. Take a given number of leaves (M) and find their mass on a kitchen mass scale.
25. Determine the average mass per leaf on the tree (M_{ave}).
26. Using the same set of leaves for mass determination, measure both the length (l) and the width of the leaves (w), recording these values in a table you make.
27. Calculate the area of the leaves (A) and take the average of these areas (A_{ave}).
28. Using the Average Mass and Areas values, calculate the estimated total mass of the leaves on the tree (MLT) and the estimate of the surface area of the leaves exposed to sunlight for the tree (SLA).
29. Consider doing other similar species trees or others for comparison.
30. **Activity II :**
31. In this Activity, the Calculations for Estimated Mass of the Leaves and Estimated Leaf Exposure is the same as in Activity 1.
32. The only difference in Activity 2 is in the method for estimating the number of leaves (T_L).
33. In this case, one merely looks up and counts through visual estimate the following (Note this is not done through climbing nor removing branches – it is a visual estimate):
34. The number of main (large) branches that extend from the trunk (S) and record this value.
35. Looking along at least 2 (the more the better for a better estimate from the average of the number used) main branches, count and record the number of smaller branches on a given main branch (M).
36. Look carefully at several of the small branches (from the ground – perhaps using binoculars) and visually count and record the number of leaves on a given small branch (N).
37. Note for all the measures of small branches and the leaves on a small branch, the larger the sample size taken, the better the results.
38. Average the values for the number of small branches and the number of leaves on a given small branch.

39. Use the values of : Number of Main Branches along with the Average values for the Number of Small Branches on a Main Branch and the Number of Leaves on a Small Branch in order to calculate the estimated number of leaves on the Tree (T_L).
40. Use the Estimated Number of Leaves value along with other average values for mass and area to determine the total mass of leaves on the Tree and the Surface Area of the Leaves on the Tree (MLT) (LSA).
41. Note for the calculations done here in either Activity you can do this : Though the Slide Rule is a recommended tool, all of these calculations can be done with a regular or scientific calculator. Some scientific ones even have built-in averaging formulae. For those who like spreadsheets, the data can be typed in and the formulae then also be typed in its own cell where the formula references each of the measured variables in their respective cells, for example B1..BN has the measurements and values used in the equation while BN+1 has the formula for all of these variables (why not the A cells? Simple – use them to label you variables)

Data :

Activity I :

Angular Measure of Tree Crown (Θ) : _______ (°)

Distance from Tree for Angular Measurement (r) : _______ (ft)

Trial	Number of Leaves covering 1 sq. ft. (N)
1	
2	
3	
Average	

Activity II :

Visual Count of Main Branches connected to Tree Trunk : ____ (S)

Visual Count of Small Branches connected to a Main Branch : (M)
(Use Average Value)

Sample	Small Branch on Main Branch Count
1	
2	
3	
Average (M)	

Visual Count of Leaves on a Small Branch : (N)
(Use Average Value)

Sample	Number of Leaves on Small Branch
1	
2	
3	
Average (N)	

Other Needed Data :

Leaf	Length (l)	Width (w)	Mass (m)
1			
2			
3			
Average			

Calculations :

Be sure to use your Slide Rule!

Convert Degrees to Radian Measurement (angular crown size) :

$$R = \frac{\theta^\circ}{57.3}$$

Tree Arc Length of Crown (r is distance from tree) :

$$S = r*R$$

Radius of Crown measurement :

$$r_T = \frac{S}{2}$$

Surface Area of Tree Crown :

$$SA = 4*\pi*(r_T)^2$$

Estimated Number of Leaves on the Tree (Activity I) :

$$T_L = N_{ave}*SA$$

(N_{ave} is the average number of leaves from the tree that covers a 1sq. ft. area)

Estimated Number of Leaves on the Tree (Activity II) :

$$TL = S*M*N$$

(S = no. of main branches connected to trunk,
M = no. of small branches on a given main branch,
N = no. of leaves on a small branch)

Approximate Area of a Leaf :

$$A = l * w$$

Estimated Leaf Surface Area Exposed to Sunlight for the Tree :

$$LSA = TL*A_{ave}$$

(A_{ave} is the average of the sample leaves taken and measured for surface area consideration)

Estimated Mass of Leaves on Tree :

$$MLT = T_L*M_{ave}$$

(Mave is the average mass of a leaf determined through measurements)

Average Values :

$$M_{ave} = \frac{\Sigma m}{n}$$

(Σm is the sum of all the values in question for an average to be determined from, n is the no. of items used to determine the average)

Conclusion :

So cool – to look at the world in terms of Mathematics, apply the science, and do things like estimate 'population' sizes – all in the shade of a beautiful tree! This is awesome. Ms Briana Andromeda Purcell loves here loves it!

First note that in this set of Activities we make a larger number of assumptions and are probably estimating a lower limit of leaves on the tree.

Do the Values determined from Tree Activity I and Tree Activity II match (at least in magnitude)? How reasonable do these results seem. For an idea trees can have easily tens of thousands of leaves into the hundreds of thousands. How do two similar species of trees compare to each other – or how do two different species compare? (Consider looking at two different trees where leaf size is considerable so as to compare the numbers of leaves – do trees with smaller leaves have significantly more or less leaves than trees with larger leaves?)(In the case of the same species, how does size affect the outcome of the number of leaves?)

Activity #13
Density of Various Items Determination

Grade Level : Middle School
Math Level : Calculating

Density Determination
The regular way and the Archimedes way and with a Slide Rule

Ms Briana Andromeda Purcell here and we are on to other fun Activities. Here we will look at the physical property density and even have a bit of a history aside from the famous Archimedes who has one of the most well-known tales about density and its use.

Density is a very important and early-on understood concept in human history for the physical properties of materials. Anyone who has picked up a stone and a sponge of roughly the same volume can testify to as to which is the more massive and therefore the more dense.

Density by definition is Mass per Unit Volume :

$$\textbf{Density} = \frac{\textbf{mass}}{\textbf{Volume}}$$

$$\rho = \frac{m}{V}$$

Density is one of the characteristics of a material to help in its determination (when there are few other means to consider). Metals have large values of density, such as aluminum at 2.70 g/cc, iron at 7.80 g/cc, copper at 8.63 g/cc, and lead at 11.3 g/cc. Other than metals, other objects (gases and liquids too) can have measured density values, such as water (1.0 g/cc), typical glass (2.60 g/cc), common wood (0.70 g/cc), ice (0.92 g/cc). Even other items, like butter (0.94 g/cc), cork (0.25 g/cc), gasoline (0.72 g/cc), sand (2.80 g/cc), carbon (2.60 g/cc), the average person (0.99 g/cc) have density values measured.

In fact, things like wood are even specified and have tables of data, so that they can be recognized in part by this property. For example : Oak has a value of 0.85 g/cc, while white pine has a value of 0.43 g/cc, and maple a value of 0.69 g/cc. In our activity, we will use blocks of wood (rectangular in shape) and measure their mass (via a mass scale), then calculate volume from measurements of length, width, and thickness of the blocks. From this the density can be calculated. (Note : Unless you find some common wooden blocks in toy stores and the like you will have to construct these – this, of course, requires adult permission, supervision, and the adults need to do the work).

Though noted as 'typical glass' above, note that glass is separated since this is a useful crime investigative tool when bits of glass are found at a crime scene. Its density (and refractive index) can reveal the type of glass it is.

As an aside, density is not just a physical property concept. In everyday life, we consider the density of such things as populations (people, plants, animals, insects, et al) when considering the requirements for a given population, its mobility, and access to their needs.

The density of a given geometric form is easily determined once there is a scale for the mass of the object in question and since the object has a regular geometric form, its volume can readily be calculated from mathematical formulae, which we will do in this exercise.

But what if the object's form is not regular? For example it is a lumpy stone. One could approximate the volume and with enough measurements may even find an upper and lower limit to it.

This puzzle is one of the questions posed over 2000 years ago to Archimedes, one of the greatest of mathematicians and inventors of all time.

Archimedes of Syracuse (c. 287 BC – c 212 BC) is best described as a Greek mathematician, inventor, engineer, physicist, and astronomer. Some of his notable claims to fame are his explanation of the lever, the Archimedes Screw to act as a pump, a useful value for pi (π) of 3.1416, an approximation for the $\sqrt{3}$, and in his work, The Method, the basis of modern day calculus.

One of the more famous stories is the basis of this activity and can simply be referred to as the Golden Crown. King Hiero II had a crown made by a goldsmith whom he felt may have cheated him by withholding some of the supplied gold to make the laurel-leaf shaped crown. The problem is how can one determine the crown to be pure gold without melting it back down and then comparing that amount to the amount of gold used. It is said that while attending a bathhouse to bathe Archimedes noted the amount of water being displaced when entering the tub and recognized that the amount displaced is equal to the volume of the item placed in the tub. "Eureka" (Greek, meaning "I have found it!") he shouts as he races sans clothing from the bathhouse.

The exact method of gold determination was probably based on his work in hydrostatics (such as found in his work, On Floating Bodies) and buoyant forces (he gives us the Buoyant Force). (There is even an Activity looking at Archimedes Principle and boats in the book here). Here this method looks for a difference in buoyant force. The more water displaced the greater the buoyant force basically.

His idea of Buoyancy is the basis of why objects sink or float. A ship either made of wood or even metal must displace its weight in order to float. The weight displaced is based on the volume of fluid, here water, which must equal the weight of the vessel.

Hence a floating object has no net force acting on it since the force due to weight is equal to the buoyant force of the fluid it is in. An object sinks when its displaced weight of fluid is less than the object's weight. So a block of metal in air weighs more than one under water. The one under water has a net weight equal to its weight in air less the buoyant force acting on it.

The concept of buoyancy applies to any mass acting in any fluid. Not only is water a fluid, but so is air. So this idea applies to explaining and analyzing why hot air balloons and helium-filled balloons float.

Not only can this idea be examined from the perspective of buoyancy, but conclusions can be drawn about the density of the object and the fluid it is in. If the overall average density of the object, say a metal ship, is less than the fluid it is in, here water, then it floats. If it is greater than the fluid, then it sinks. If equal, then it has neutral buoyancy and can be placed anywhere, such as in the case of a submarine.

Also this is why people tend to float (lower density than that of water) as well as the fact that nearly all forms of wood float in water (since they have density values less than water).

Returning and Examining the Gold Crown story and the use of Buoyancy and Density :

First the crown is balanced on a scale beam with the crown at one end and the needed amount of gold at the other. This balanced system is then placed in the water. If the crown is equally dense the rod would stay straight since both displace the same volume.
If the crown was more dense for some reason it would displace less volume hence a smaller buoyant force so the scale would tip in the direction of the crown. If, however, it was less dense, the crown would displace more water, hence have a greater buoyant force and the scale rod would tip in the direction of the gold mass at the other end.

The other oftentimes reported method noted the amount of displaced water by both the crown and the equal mass of gold. If a greater amount of water for the crown, means a greater volume, hence a lower density.

In either case, the investigation notices the property of density. It is reported that the crown was less dense hence had been blended with other less dense materials and the goldsmith had pocketed some of the gold.

The value of density cannot be overestimated in any case. The physical property of density is one amongst many to characterize a material to begin with. Next, it then can be used to quickly determine its ability to float or not and the level to which it will sit in a fluid when it floats. With the use of density by Archimedes, a major step in the direction of scientific analysis was taken in history.

For example, in the determination of the mass of the Earth by scientists such as Maskelyne and Cavendish, they calculated first the density of the Earth and then when multiplied by the volume of the Earth yielded the Mass of the Earth. Also, by examining the average value of the Earth and comparing it to water and the ground outside it was determined that it must increase with depth and this lead to a greater understanding of the composition of the Earth's interior and later formation and piecing the puzzle together of plate tectonic theory and the rock cycle.

Beyond mass density there are many forms of density where it is a ratio of a quantity of material per unit some other variable. These can be used to describe flow rates, concentrations, and other types of values, like population density.

To illustrate the value of the average density in our activity, when we have the average density of a piece of brick from a wall and calculate its estimated volume, we can then determine the mass of the wall. Think of how difficult it would be to determine the weight of a whale and other large items. This method, like the one to determine the average density of the Earth to find its weight, is the best method for large scale things when there are no scales available for the item.

Purpose : To use a measured mass from an regularly-shaped Object, along with its dimensions and then calculate its volume to determine its density. (wood block)

Purpose : To use a measured mass from an regularly-shaped object (that is non -dissolving) with its volume determined from water displacement to determine its density. (marble & paper clip)

Purpose : To use a measured mass from an 'irregularly-shaped' object (that is non -dissolving) with its volume determined from water displacement to determine its density. (pebbles)

Purpose : To use a measured volume and measure mass from a given quantity of water, graph the mass vs. volume and from slope determine the density of water .

Purpose : To use the density of a given mass and the calculated volume of an object that it is part of in order to determine the mass of the object. (Density of a Brick, Density of a Piece of Brick, and the Mass of a Brick Wall)

Note : It is assumed that classroom style equipment is being employed here, such as using a sensitive mass scale and a precise graduated cylinder. However, if these are not available, the lab volume can be done none-the-less. Use for volume a regular yet graduated large measuring cup. This would mean adding more masses (paper clips or marbles) to the cup to register a change in volume, but the same is true for using a less sensitive scale in this place as well. The mass, however, requires a Scale.

Note : If using a regular graduated cylinder, be sure that the items fit and come out easily. If not find smaller ones or use a larger graduated cylinder. In any case, tip the graduated cylinder when placing items in it so as to not damage the graduated cylinder nor the items.

Materials :

- Kitchen Mass Scale,
- Bathroom Scale,
- Graduated Cylinder (minimum 25 mL size),
- brick wall for measuring,
- bit of brick and cement from wall debris,
- brick,
- small pebbles,
- rectangular wooden blocks,
- water,
- measuring tape,
- ruler,
- goggles,
- small hammer,
- marbles (few),
- Slide Rule

Note : The size of the Graduated Cylinder will determine the maximum size of object that can go into it and be used for the lab. Either the 25 mL size or 50 mL size is readily available, as are Kitchen Mass Scales on the Internet.

Note : Wood blocks can be found as toys already formed, but one can create their own from materials at a store where one safely and properly cuts the wood to hand-sized. Choose wood that does not readily splinter. This should be done with proper tools, wearing goggles, and only adults should do this. A child requires the permission and help of a parent in this matter.

Procedure :

1) With any and all Activities read through, then set up the process for the Activity and run it without measurement and watch where you may have bumps in the road so as to anticipate these. Be patient – Repeat processes. Once all is well when measuring see to the values – do you notice patterns or trends – what do you predict will happen next as you change one thing or another from one trial in an experiment to the next. Always Explore & Enjoy. Be Brilliant!

Activity I : Density of a Wooden Block :

1) Each of the wooden blocks can be found either as already formed toys or if one has the means to create hand-sized blocks by purchasing wood from a store and having a parent cut them safely by an adult.
2) With the given set of blocks at hand, choose one at a time. Each of these is a trial to be noted in the table below in Activity I.
3) Measure its Mass (m) on the scale. (measure in grams)

4) Using the ruler, measure its dimensions L, W, T (length, width, thickness) and place these values on the table. (measure in centimeters – to nearest 0.1 cm)
5) When all of the blocks are measured do the calculations. (Volume & Density)
6) For each block use the volume formula and determine the Volume of the block (V).
7) Now determine Density (ρ) for each block.

Activity II : Marble and Paper Clip Density
Mass of a Marble (later Paper Clip)

1) If using a scale that cannot read one marble mass follow the directions below. If it can then take the mass of one marble.
2) First place a cup on the scale and zero it out.
3) Start placing marbles in the cup counting them as they are put in.
4) When the scale has read at least 1.0 oz (or about 30 grams), then record the number of marbles used.
5) Record the mass reading on the scale.
6) Calculate the average mass of the individual marble.
7) In both cases record the mass of a marble (m)

Volume of a Marble (later Paper Clip)

1. Use a 50 mL graduated cylinder and fill it to 15.0 mL with water. Record this initial value in the table. (Be sure to have enough water to cover the marble when placed in there).
2. Note, one could use a 25 mL grad. cyl. And 1 marble as well or use the alternative method noted in the Note above.
3. Begin counting and placing marbles in the graduated cylinder and watch as the volume changes.
4. Add between 3 and 6 items (marbles in first activity, paper clips if doing this activity with them)(5 is a good number here).
5. Record the new volume reading.
6. Determine the change of volume and then the volume of a single marble.
7. In both cases, record the volume of a single marble (V).
8. Using the Average Mass and the Average Volume to calculate the Average Density of a Marble.
9. The Density of a Paper Clip :
10. Follow Steps 1-7 in the mass section and the steps 1-8 above in this section but replace Marble with Paper Clip. Note that the use of multiple paper clips will probably work since both the mass and volume of a single paper clip may be too small for the devices used here, hence we need to use averages. Be sure to use the same type and approximately the same size ones for this.

Activity III : Mass & Volume of a small irregularly-shaped Object

1) Choose a small Object that satisfies three criteria : A) Non-dissolving solid (small pebble is best), B) massive enough to register on the mass scale to use (i.e. > 0.1 g), C) small enough to be placed into a graduated cylinder.
2) Measure and record the mass of the Object using the mass scale.

3) To determine volume, place enough water in the graduated cylinder that would cover the object if it is placed in there. Make the reading easy, for example 5.0 mL, 10.0 mL or something similar.
4) Record the initial reading of water without the Object in the graduated cylinder.
5) Place the Object in there and record the new volume reading.
6) Determine the volume of the Object (V).
7) Calculate the Density of the Object.
8) Be sure to check the Extension Activity at the ends of this Activity, where an exploration of Silly Putty and its density is explored.

Activity IV : Density of Water

1) Place an empty graduated cylinder on the mass scale and zero it out.
2) Pour into the graduated cylinder a measured amount of water (for example 3.0 or 5.0 mL)
3) Record the Volume of Water (V), and measure and record the mass of the water in the graduated cylinder (m).
4) Add more water to a new level (for example, if you started with 5.0 mL, increase to 8.0 mL)
5) Record the total mass (m).
6) Measure and record the new total mass of the system with the new volume measure (V). Note it is best to add the same amount each time.
7) Continue to add volume and measure mass so as to do at least 4 trials.
8) Graph the date with mass (m) on the y-axis and volume (V) on the x-axis.
9) Draw the best fit line and determine slope of the line, which is the average density (ρ) for water.
10) Compare this value to the generally accepted value for water (1.0 g/cc) and determine the experimental error
11) A great follow up Activity is to create Solutions of Salt Water or Sugar Water by starting with about 150 mL water and adding 1tsp to one solution, 2tsp to a second solution, or more to a given volume of water separately. Stir until totally dissolved. Then determine the density of each solution for comparison in the same manner as noted in these directions for water, but also creating a bar graph for comparison after the line graph for each independently. (Note for salt 1 tsp is about 6g, but measure it on your scale and record the numbers).

Activity V : Mass of a Brick Wall

Note :
Follow all of the directions for the Mass & Volume of a small irregularly-shaped object only now the object is specifically a piece of brick which may have a small amount of cement on it.

1) First let's estimate the mass of a brick wall.
2) Take a brick, if available, and weigh it on a bathroom scale. Record its weight. This will be called the 'Brick Density'.

3) For a given wall count the number of bricks across it for 3 rows. Record each of the values separately and then take the average of them to arrive at the average number of bricks per row.
4) Count how many rows there are.
5) Note : If there is a triangular situation, such as at the top of a house, count the number of rows up as well, but realize that you will have to use the triangle formula instead of the rectangle formula for determining the number of bricks here.

6) Determine the number of bricks in the wall.
7) Multiply this value by the 'brick density' (the weight per brick) to arrive at the weight of the wall.
8) If a brick is not available :
9) Find a piece of brick near the wall in question on the ground. It is best if it has a bit of cement on it (small amount at most).
10) Measure and record the mass of this piece. It must have enough mass to be measurable and small enough to fit into the graduated cylinder.
11) Note : If pieces have to be broken off, have a parent do this with a small hammer. Wear eye protection.
12) Be sure that it is small enough to go into the graduated cylinder and use this piece.
13) As you did with the marble, paper clip, and pebble find the volume of the piece of brick by the Archimedes method of volume determination.
14) Record the final and initial volumes of the water in the graduated cylinder and determine the volume of the brick piece.
15) Use the measurements and determine the density of the brick piece.
16) Next find the area of the wall in question.
17) Use a measuring tape to measure the length and width of the wall in question.
18) Measure the thickness of an average brick in the wall by going to the edge and measuring this.
19) If the wall has openings, such as doors and windows, measure their heights and widths separately.
20) Determine the total area of the wall.
21) Determine the area of the openings separately and subtract each of these from the area of the wall.
22) Determine the Volume of the Wall by multiplying the Net Area of the Wall by the thickness of a brick in the wall.
23) Determine the Mass of the Wall by using the Density and Volume of the Wall.
24) Recognize we are assuming that the average density of the brick is representative of the entire wall, which will be the main source of error in the calculation.

Data & Calculations :

Be sure to use your Slide Rule!

Activity I : Density of a Wooden Block

Trial : ________ (Number of Trials equals the number of blocks)

Mass of Wooden Block : ________ (g)

Dimension	Measure (cm)
Length (L)	
Width (W)	
Thickness (T)	

Formulae :

$$V = L*W*T$$

$$\rho = \frac{m}{V}$$

Side Note :
If you are using a cylindrical object :

$$V = \frac{\pi * h * d^2}{4}$$

Activity II : Density of a Marble (Or Paper Clip)

Mass of a Marble

Number of Marbles on Scale	
Mass of Marbles on Scale	

Volume of a Marble

Number of Marbles in the Graduated Cylinder : ________

Final Graduated Cylinder reading	
Initial Graduated Cylinder reading	
Change in Volume	

Note : Use these above Tables in Activity 2 for the Density of a Paper Clip, only changing the word Marble to Paper Clip!

Calculate Mass and Volume :

Change of Volume (Volume of Object) =
Final Volume in Grad. Cyl. – Initial Volume in Grad. Cyl.

$$\Delta \mathbf{V} = \mathbf{V_f} - \mathbf{V_i}$$

$$\textbf{Average Mass Value} = \frac{\textbf{Total Mass}}{\textbf{Number of Items}}$$

Calculate Density : $\rho = \frac{M}{V}$

Activity 3 : Density of an Object (pebble)

Mass of the Object : ________ g

Volume of the Object

Final Graduated Cylinder reading	
Initial Graduated Cylinder reading	
Change in Volume	

Calculate Density : $\rho = \frac{M}{V}$

Activity 4 : Density of Water

Trial	Total Volume (V)	Total mass (g)
1		
2		
3		
4		

$$\text{Density (Slope)} = \frac{\text{change of mass}}{\text{change of Volume}}$$

Accepted Value : Density of Water = 1.0 $\frac{g}{cc}$

$$\%\text{Error} = \frac{[\text{ accepted value} - \text{activity value }]}{\text{accepted value}} * 100\%$$

Activity V: Density of a Brick & Cement Piece then Mass of Wall
Brick Density Method :

Number of Bricks in a Row :

1) ____________

2) _____________

3) _____________

Average Number of Bricks in a Row : __________

Weight of a Brick (Brick Density) : _____________

Density of a Piece of a Brick :

Mass of the Object : _________ g
Volume of the Object

Final Graduated Cylinder reading	
Initial Graduated Cylinder reading	
Change in Volume	

Calculate Density : $\rho = \frac{M}{V}$

$$\text{Average Number of bricks in row} = \frac{\text{Sum of the Number of Bricks in Rows}}{\text{Number of Rows used}}$$

Number of Bricks in Wall = Ave No. Bricks in Row * Number of Rows

'Brick Density' is the Weight per brick

Weight of Wall by Brick Density Method :

Weight of Wall = 'Brick Density' * Number of Bricks in Wall

Density of Brick Wall by Small Piece Method :

Area of Wall = L * W

Net Area of Wall = Area of Wall – Opening Areas

Volume of Net Wall = Net Area of Wall x Brick Thickness

Mass of Wall = (Net Wall Volume) * (Brick piece Density)

Conclusion :

Wow – what a world of density and this is only the tip of the iceberg as they say. We can explore so many other avenues, but many of the ideas for solids and liquids are explored here. Did any of your results surprise you?

The first Activity is a good warm up exercise in measurements and density determination. It is best to use the same type of material for each of the blocks and they do not have to be the same size. If using the same material, the overall density should be rather close. Minimally the first significant figure should be the same and possibly the second. If you come across a table of density values for a given type of wood, you can compare your results to this to see how close you came. This can be found on the internet.

From the second activity, you can determine the margin of error for a regular geometric object after first measuring and determining the volume of the marble for comparison to the results in this experiment. This would mean taking measurement of its diameter and then using the formula for a sphere $\mathbf{V} = \frac{4*\pi*r^3}{3}$ for comparison to your results. The best tool to use to measure the diameter is a caliper, if available.

From the second experiment, more than one pebble may be done to have more than one sample for consideration. Also in pebbles, the question to consider is : 'why are the values different?' for different pebbles. Also, how do they compare to the density of the wood, the marble (glass) and the paper clips ?

From the fourth activity, how similar are your values for the density of water to the accepted value for the density of water? Did you try other Solutions, such as a Sugar Solution or a Salt Solution – how did their density values turn out and compare to water (they should be higher values)?

A different follow up is to use cooking oil after using water, but note that it can take some effort to clean out of the graduated cylinder – be patient.

EXTENSION ACTIVITY :

Why not choose a material that can be considered a solid or a liquid, namely the product Silly Putty!

In the case of Density, one first measures the Mass of the Putty on a scale and records this value.

Next, what of Volume?

The first and fastest way is to have a graduated cylinder of sufficient size and to shape the putty to put in it after some water it placed in it. This is using the Archimedes method of volumetric displacement to find the volume of the putty. – Now simply take Mass divided by Volume to find its Density.

Other ways of Volume are interesting for a challenge and learning experience : Shape the putty to be a cube or rectangular solid form and then measure the dimensions (Length, Width, and Thickness) like you did with a wooden block and calculate Volume from this. Note : Should it be similar to the Archimedes volumetric displacement method value? (Hint : yes).

Knowing that Volume can be affected, one can try sample pieces (say cutting the putty in half, for example) and test each one for mass and volume and see how similar the results are for density. – A fun method can be using the Archimedes volumetric displacement method and break off pieces that have measurable volumetric change in a graduated cylinder with water and test each piece independently and compare the results.

Have fun – Be curious – Explore & Enjoy – Be Brilliant!

Activity #14
Computing the Number of Moles in a Substance

Grade Level : High School
Math Level : Calculating

The Number of Moles Activity

Simple fun yet interesting – it is much like contemplating how many grains of salt are there in a teaspoon yet here we are in this Activity engaged in using actual ideas found in chemistry as we shall soon see. Read on. (Be sure to brush up on use of exponents – see the chapters and early Activities for summary of rules in their use).

One of the important ideas of understanding Atoms is to determine just how many there are. To approach this, we need to recall common items that we group and classify by a name. Take, for example, when we buy eggs in the store. Typically there are 12 in a group and this is clearly called a dozen. Of course many of us like a 'baker's dozen' – say 13 donuts for example – yum. There is a term – gross (not eww yucky) but here meaning 12 dozen or 144. A ream of paper has 500 sheets is another example of a term that represents a specific number of items.

In chemistry there is also just such a term. It is the Mole. It is the SI base unit to measure the amount of a substance. In 1 Mole there are 6.02×10^{23} items. This is also called Avogadro's Number and is named for the Italian physicist and lawyer who in 1811 had determined the volume occupied by 1 mole of a gas. It turns out that the idea of the mole is very useful in the Gas Laws as well, but can be applied to other states of matter too.

The sheer number represented by a Mole is larger than most of us can first consider. If we had this number of marbles (standard size) they would cover the entirety of the Earth to a depth of more than 6 kilometers. But because Atoms are so small, hence so too are Molecules, it does not take large amounts of materials to represent one mole of a given substance. There are 100 trillion atoms in one cell for example in our bodies – the average body could be around 10^{27} atoms – wow!

For any given substance, the mass of a Mole is unique to that substance, yet if you had a mole of substance A and 1 mole of substance B, they would have the same number of representative particles.

For example, if you have 18.0 g of water, which is 18.0 mL you would have 1 mole of molecules of water. By the way, an average soda pop has 500 mL (16.9 fl oz). This illustrates that it is a small amount of water. For comparison, To have one mole of iron, you would need 55.8 g of iron, however.

The mass 1 mole of any pure substance in grams is called its Molar Mass. It turns out that the Molar Mass of an Element or a Molecular Compound is numerically equal to its Atomic Mass. From the above example, Water is H_2O, and Hydrogen has an Atomic Mass of 1.0 and there are two of them in a water molecule and it is added to Oxygen's Atomic Mass of 16.0. So we have 1.0 + 1.0 + 16.0 = 18.0 g. In the case of Iron, its Atomic Mass is found on the Periodic Table, and recall for 1 mole of this it is equal to its Molar Mass in grams, so it is 55.8 g.

Note that the subscripts in a Chemical Formula indicate how many moles of each element are present in 1 mole (also written 1 mol) of the compound.

Using the various factors, one can convert readily between Molar Mass to Moles to the Number of Particles in a given substance. The conversions are found in the Calculation section of the Activity and an example is found in the Procedure.

This Activity uses common household items, but one still needs parental permission and supervision. In doing the Activity, exercise proper lab procedure and safety, such as wearing goggles and gloves. Keep equipment clean. Do not taste, touch, or mix materials. Practicing safety is a good thing.

Purpose : To determine the number of Moles of a substance in a given quantity.

Materials :

- Mass Scale,
- Periodic Table (Note Table of Information provided),
- ¼ cup Measuring Cup,
- 1 Tablespoon Measuring spoon,
- Small Plastic Cup (disposable is best),
- Packets of Sugar ($C_6H_{12}O_6$),
- Salt (NaCl),
- Water (H_2O),
- Baking Soda ($NaHCO_3$),
- Cooking Corn Oil ($C_6H_{12}O_6$),
- Vinegar (CH_3COOH),
- Aspirin ($C_9H_8O_4$),
- Wax Paper,
- Goggles,
- Dishwashing Gloves (appropriate size to allow materials handling),
- Slide Rule

Note : The list of items used will not only depend on materials on hand, but also you must have parent permission and supervision. In no instance are the items to be consumed. Do not mix items. Exercise proper cleaning of equipment used (measuring spoons, measuring cups) between items used. Avoid getting materials on your hands – you should consider using dishwashing gloves and wearing goggles to act in manner that exercises proper lab safety.

Note : We are assuming in the case of aspirin (as well as all other items in the list) that they are pure substances and there are no other fillers – clearly with aspirin this is not the case.

Procedure :

1. Be sure to zero out your scale in measurements. Be sure to wear goggles and gloves in handling materials.
2. Have your materials at hand.
3. In all cases of measuring mass, use sufficient amounts to be measurable according to the level of sensitivity of your scale being used. Measure all amounts used to the nearest 0.1 g if possible.
4. The following items can be measured on the scale by themselves on wax paper (or in the zeroed out small plastic cup on the scale) :
5. (Note : Be sure to zero out the scale with the wax paper on it first before any measurement)
6. (Note : For each item in the list below, use different wax paper pieces for each and dispose of the items once done)
7. (Note : Each in the list is done separately – there is no mixing).
8. List : Aspirin, Salt, Sugar, Baking Soda
9.
10. The following list of items can be measured with the small plastic cup on the scale (Tablespoon size preferred).
11. (Note : Be sure to first zero out the mass scale with the plastic cup on it, but it is empty – then put the sample in it)
12. (Note : Be sure to dispose of the measured materials after a given trial and to wash and dry the measuring spoon and the plastic cup between each trial – be sure to re-zero the cup out)
13. In putting mass of the material in the cup use the plastic measuring spoon (Tablespoon size is best).
14. (Note : In some, if not all cases, you may need more than one spoonful to have sufficient mass to be measured. In any case keep track of the number of spoonfuls you use in a given trial).
15. List : Vinegar, Cooking Oil, Water
16.
17. The following list of items can be measured with a measuring cup zeroed out on the scale (¼ c preferred)
18. (Note : Some of the items on this list are here only if one wants to have more for comparison to the Tablespoon measurement previously mentioned, but you need parent permission and supervision to use so much material and it is best to have a use for it following its measurement or a means to return it to its container)
19. The small plastic cup is used to act as the item to fill the measuring cup on the scale.
20. (Note : Between any of the items on this list, be sure to clean and dry, as well as re-zero out the measuring cup and the small plastic cup used to fill the measuring cup)
21. List (liquids) : Water, Cooking Oil, Vinegar,
22. List (solids) : Salt, Sugar, Baking Soda
23. Note : If approved and available, one might consider solid Iron bolts
24. Be sure to put all things away, clean up the area before continuing.
25.
26. Calculations :

27. For each material used, first determine the Gram Molecular Weight of one mole of the item by using the Chemical Formula and the Atomic Mass of each of the Atoms in the given Molecule.
28. (For example : In the case of Salt, the chemical formula is NaCl. It is composed of one atom of sodium and one of chlorine in each molecule. From the Table provided, sodium's atomic mass is 23.0 while chlorine's atomic mass is 35.5 – We add these together and recall that 1 mole of a given element will have the gram mass value of its atomic mass (23.0 g/mole + 35.5 g/mole = 58.5 g/mole)).
29. (Continuing the example, let's say we had 16.0 g of salt. We would have 16.0 / 58.5 or 0.273 moles of salt. Converting this to the number of particles, we would take that result and multiply it by Avogadro's Number to find an answer of 1.64×10^{23} – Note this is a rounded answer taken from a slide rule).
30. Place the Gram Molecular Weight on the Table.
31. For each material used in the exercise determine the number of Moles present by using the formula. Place this value on the Data Table.
32. For further calculations, you can determine the number of items (molecules) present in your sample by using the relation of how many items per mole conversion factor.
33. Also for those items you have volumetric measures of use (such as a Tablespoon or ¼ c) you can determine how many of these are needed to reach 1 (one) mole.

Data :

Trial	Substance	Chemical Formula	Gram Molecular Weight of Substance (g)	Mass Measurement (g)	Number of Moles of Substance
1	Salt	NaCl	58.5		
2	Sugar				
3	Water	H_2O			
4	Baking Soda				
5	Cooking Oil				
...					
Last one					

Calculations :

Be sure to use your Slide Rule!

Number of Moles :

$$\#m = \frac{\#\ \text{grams of substance present}}{\text{gram molecular weight of substance}}$$

Number of Items (Molecules, Particles) in the # Moles Present for a Material :

$$\#n = (\ \#\text{moles}\ ) * (\ 6.02 \times 10^{23}\ \frac{\text{items}}{\text{mole}}\ )$$

Basic Useful Information :

Density of Water : 1.0 g/cc
1 cc = 1 mL
1 kg = 1,000 g
1 mole = 6.02×10^{23} items
1 oz = 28.3 g
1 oz = 29.6 mL

Table of Information :

3-significant figure values of Atomic Mass of these given Elements :

Atomic Number	Element	Element Symbol	Atomic Mass
1	Hydrogen	H	1.01
6	Carbon	C	12.0
7	Nitrogen	N	14.0
8	Oxygen	O	16.0
11	Sodium	Na	23.0
14	Silicon	Si	28.1
15	Phosphorus	P	31.0
16	Sulfur	S	32.1
17	Chlorine	Cl	35.5
19	Potassium	K	39.1
20	Calcium	Ca	40.1
26	Iron	Fe	55.8
29	Copper	Cu	63.5

Conclusion :

Simple fun yet interesting – it is much like contemplating how many grains of salt are there in a teaspoon yet here we are engaged in using actual ideas found in chemistry.

Did any of your results surprise you as to how much of a mole or number of moles you had in your sample? What of your calculation to determine the amount of material needed to reach one mole?

Activity #15
Determining Weighted Average to explore Atomic Mass

Grade Level : Middle School
Math Level : Calculating

Weighted Averages and Atomic Mass Activity

Well Inquisitive Pioneer Adventurers – here we are with another Activity based on ideas found in Chemistry. Say you had a group of marbles that all appear the same in size, perhaps very similar in color – are they all the exact same mass hence weight? Probably not. So in taking their average mass we would not have a nice whole number answer therefore. The same is true for Atoms as this Activity illustrates by letting nuts and bolts represent Atoms of the same Atomic Number yet different Atomic Masses to find an average mass or weight of them. Read on.

The Atoms for any given Element all have one thing for certain : their Atomic Number which is the Number of Protons in the nucleus. This value determines the species of the atom and hence, most of its chemical and physical properties. For example, Carbon has 6 protons, Oxygen has 8 protons, and Hydrogen has 1 proton, while Iron has 26 protons.

It is noted that the previous statement said 'most' of its properties. In terms of chemical properties this is always true – that all of the properties are the same for any member of a given element species.

However, one physical property will be different for some members of a given species, namely mass. These species members have the same number of protons, yet can have the different numbers of Neutrons, and hence have different mass from their other species members.

For example, normal Hydrogen has 1 proton, but 0 neutrons. Another species of Hydrogen (given the name of Deuterium) has 1 proton (as do all Hydrogen members) but also 1 neutron. Yet another species of Hydrogen has 1 proton and 2 neutrons (called Tritium).

In nature, virtually all Elements are found as mixtures of their Isotopes. Any given sample will have a similar relative abundance of each of the Isotopes.

The Atomic Mass for a given Element is found from the Weighted Average of the Isotopes of a given element. The idea of a Weighted Average is simple enough to understand. Imagine you have 3 grocery bags. Two of them each weigh 1 pound and the third weighs 2 pounds. The total weight for the 3 bags is 4 pounds. The average weight for them is 1.33 lbs/bag (4 lbs / 3 bags). Notice that none of the bags has the same weight of the average. Also the average is not exactly in the middle of the highest and the lowest values. This is because there are more values at the lower end of the weights (two 1 pound bags) than at the higher end of weights (one 2 pound). Though the example uses weight here, the concept applies to mass as well. In fact, the expression atomic mass and atomic weight on such small scales is used oftentimes interchangeably.

Returning to our example of Hydrogen, it has an Average Atomic Mass of 1.008. Though the regular type has 1 proton and an Atomic Mass defined as '1' (more or less), Hydrogen is a bit more than one because some of the others in the group have greater mass (such as some of it being deuterium and some of it is tritium).

Isotopes can have importance too, since some are Radioactive. That is to say they will be unstable and will decay (give off various forms of Radiation – Alpha, Beta, Gamma Radiation) until they become stable. Some of these can be used to date materials. For example, Carbon-14 is used for organic materials up to about 40,000 years in age. This is because C14 decays into stable nitrogen 14 and gives off a Beta particle (electron).

Notice that the Atomic Mass is Not the Mass Number for a given Element. The Mass Number is the sum of the Atomic Number (the number of Protons for a given element) and the Number of Neutrons. The Mass Number for a given Isotope is unique since the number of Neutrons are different for a given member of a chemical species.

We are not going to explore isotopes in our Activity. We will use items to act in a representational way to illustrate how similar items can vary in mass. In our Activity we will take a Set of Nuts (metal and/or nylon ones that are normally used with Bolts) that was either set up by someone else (parent, teacher) or by ourselves. The set will represent a given Element which has members of different masses (the set is composed of nuts made of various metals since as brass, aluminum, steel, et al), hence act like different Isotopes. From the Set we will find the Weighted Average of the Set and hence the Average Mass of the Set.

Purpose : Using Nuts of varying mass in a given set these will be used to represent and illustrate the determination of Average Atomic Mass and the fact that elements of a given species all have the same number of Protons but may have different numbers of Neutrons, hence are Isotopes.

Materials :

- Best to use 3/8" Nuts,
- Nut types : galvanized, brass, zinc-coated steel, nylon,
- Depending on the number of sets you wish to construct will determine the numbers you need of each type,
- The level of complexity is determined by the numbers of nuts in a given sample set to evaluate,
- Mass Scale,
- Spray Paint (to paint the set so as to make them indistinguishable),
- Small Plastic Bags (Snack size is sufficient),
- Slide Rule

Notes :

1) The first choice is whether to paint the set of nuts to be used or not. This is put into this Activity to illustrate the fact that for any element under consideration all members of the species have the same number of protons and the exact same chemical properties. The isotopes only differ in the number of neutrons, hence differ in mass from each other.
2) If you choose to paint them, be sure to have parental permission and supervision. Be sure to let them dry thoroughly before use. Best to do this outside. Be sure to follow directions on the given paint can.
3) The second choice is the number in the set of nuts. Note that the set is not just the total number, but also the number of subgroup members too. Each group should be composed of no less than 2 subgroups to make a set. Each subgroup should have at least one member (> 1 is best). (see example set description)

Example Set Description
Recommended Set : (Assumes 3 types of Nuts)

9 of first nut type, 6 of the second nut type, 3 of the third nut type
Total number of nuts in this set : 18

Set Up :

- The only steps in set up are noted above in the notes in the two decisions that need to be made :
- To paint the nuts or not and
- The composition of a set of nuts composed of subgroups in the same set
- The definite thing that must be done in the set up is the creation of the Set to be analyzed in the exercise itself

Procedure :

1) The following directions presented here assumes all nuts are painted :
2) Record the Number of nuts in the Set (N).
3) With the Set established (see Set Up and Notes) use a piece of paper and create a 'checkerboard' pattern with squares large enough to place each member of the Set into it. Be sure to number each of the squares. In this step you are numbering each member of the Set. – This is done to create an 'identity' for each nut that is under investigation.
4) Take each nut in turn and determine its mass to the nearest $1/10^{th}$ of a gram on the Mass Scale. (Note : It is recommended to use a triple beam balance, if available since this allows for the practice of reading scales and estimating the final significant digit).
5) Calculations :
6) Sum up all of the masses (M).
7) Use the Slide Rule and determine the Average Mass (A) by dividing the Total Mass by the Number of Nuts in the Set (N).
8)
9) The following directions presented here assumes none of the nuts are painted and they are visually distinguishable so they can be classified into subgroups.

10) If you did not paint the members of the Set and can visually distinguish them, you can reduce the number of squares to match the numbers of Subgroups instead.
11) You can measure the mass of each member of the subgroup (b_X), sum these up and divide by the number of members in a subgroup (X) to determine the average mass of a subgroup member (m_X).
12) The Average Mass (A) is then found by the sum of the number of each subgroup (X) multiplied by the average mass of the given subgroup (m_X) divided by the number of members in the set (N).

Data :

Total Number of Nuts in Set (N) : ______

Nut Number	Mass (g)
1	
2	
...	
Last one in set	

Note : Sum up the total Mass of the Collection

Note : If able to create subgroups of nuts which have a 'common' mass, instead of Nut Number, this would be Subgroup Number and the number of these represent the number of subgroups. In this case an additional column is needed titled : Number of Nuts in Subgroup.

Calculations :

Be sure to use your Slide Rule!

Average 'Atomic' Nut Mass :

$$\mathbf{A = \frac{Total\ Mass}{Number\ of\ Nuts}}$$

$$\mathbf{A = \frac{M}{N}}$$

If able to separate nuts into subgroups of 'similar' mass :

$$\mathbf{A = \Sigma\left(\frac{X}{N} * m_X\right)}$$

(X = the number of nuts in subgroup, N = number of nuts in group, m_X = mass of typical member of subgroup X)

The equation is the sum of the ratio of number of members in a subgroup to the total members in a group times the common mass of each of the subgroup members.

Conclusion :

Looking at the Atomic Mass values in the Periodic Table, does this exercise help to illustrate the concept of Isotopes for a given Element species and the idea of Atomic Mass in its determination for you?

In examining the exercise, what happens to the average 'mass' as the set has more 'heavy' subgroup members included? In contrast, what happens to the average when less 'massive' members are in the subgroup?

How does this idea relate to the idea of a Weighted Average as found in Algebra in Math? Do some research on this topic and think on it.

Be Brilliant!

Activity #16
Measures and Calculations approximating Atmospheric Pressure

Grade Level : High School
Math Level : Calculating

Measuring Pressure with a Suction Cup Activity

Living on the surface of the Earth as we do, we live at the bottom of an ocean of air. All of this air atop of us is gravitationally held to the Earth just as we are and is exerting a pressure on us. Air has mass (at sea level, dry air at 20°C has a mass of 1.2 kg per cubic meter volume). With increasing altitude there is less air pressing down and the overall pressure therefore decreases with altitude (hence the need for pressurized cabins in airplanes). The weight of air from the surface of the Earth to what is considered outer space turns out to be about 101,000 N of weight on an area of 1 square meter. This converts to the more well-known 14.7 lbs/sq.in^2.

It is not this pressure that does something on its own, though, such as wind and the like. It is the differences in pressure from one region to the next that generates wind (primarily) (the pressure difference relates to uneven heating of the Earth's surface due to its make-up (water or land), reflectivity, opacity, and angle of incident sun rays).

The effects of pressure difference often leads to one of the chief misconceptions in science – many of us say, for example, that a vacuum 'sucks stuff up'. No, it does not. There is a pressure difference where the vacuum's engine whirling a fan in a specially-designed chamber creates a low pressure zone which is immersed in the higher pressure (really normal pressure) surrounding air. Pressure differences result in a net force that naturally moves from high pressure regions to areas of lower pressure regions. This comes naturally from the formula itself. Hence it is often said in jest, Science does not suck.

Pressure is the ratio of Force to unit Area. If the Areas in the two regions (low and high) are the same, then they can be dismissed. For one area to be a low region means that it therefore has a smaller force value than the high pressure region. Forces are vector quantities. The difference of the larger force value of the high pressure region minus the force value of the lower pressure region results in not only a nonzero value, but also in a direction for that net force moving from the high pressure region toward the lower region. (For more on Forces, see Newton's 2nd Law Activity Prelude #25). (For more on ideas involving Net Force and Pressure concepts see the Weight of a Car from Air Pressure Activity #18).

Needless to say, the vacuum does not suck (much like many of your science teachers may have told you – science does not suck), it is high pressure regions pushing into the lower pressure regions that does the action seen. To illustrate this idea further, consider a soft drink in a cup you may purchase. Though the cup appears rather enclosed there are small holes that allow air in, so that when you draw the air out of the straw, you create a lower pressure region, so then the air pressure acting on the fluid (the pop) can push it up the straw and into your mouth. If the cup were totally sealed, it becomes impossible to draw the liquid into the straw.

In our Activity, we are using common suction cups (see photo) that are often used to attach small items to the wall, window or refrigerators. We will press them onto a surface (such as a tile floor or a table) and pull up on them with a tension scale (all the while wearing goggles and blocking any sudden object movement with a oven mitt – see NOTES on safety for this Activity – Note that you must have parent permission and supervision for this Activity too).

One might think that our measured values of Force when divided by the Area of the suction cup should result in the Air's Pressure. Notice what this assumes – that there is a perfect vacuum under the suction cup, which is never the case. This means we can approximate Standard Air Pressure for the Earth, but there will be some margin of error.

Purpose : To measure the force needed to separate a suction cup held in place by pressure differences and calculate the pressure acting on it

Materials :

- Suction Cup with hook (about 1" dia is best – see Note),
- Tension Scale (measures 50 N for given suction cup size),
- Ruler or Caliper,
- Duct Tape,
- Goggles,
- Oven Mitt,
- Flat Surface – Table or Tile Floor (best),
- Slide Rule

Photo of Experiment Materials :

Note : Not in photo are your goggles nor the oven mitt for protection which are absolutely needed!

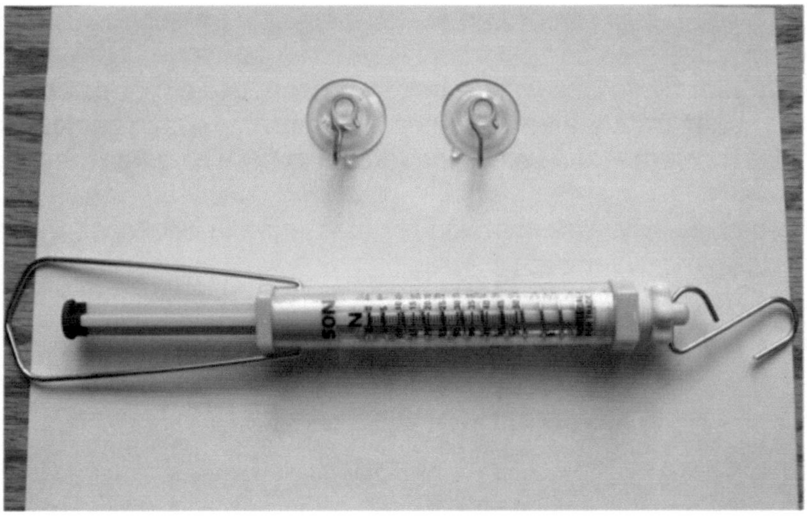

NOTE : You must have parental permission and supervision in this Activity. At all times in the Activity all participants are wearing goggles. Always act in a safe manner.

NOTE : The key to safety in this Activity is protecting your eyes and yourself. The best set up attaches the suction cup to the intended surface (such as a tile floor), then use some tape to attach the tension scale hook to the suction cup hook. When pulling up (read the Procedure), do so slowly, watching the scale reading. Use the oven mitt on your opposite hand to act as a shield covering the suction cup. When pulling up, pull the tension scale away from your direction as well as shielding with the mitt.

Procedure :

1. The first thing to do is to read and follow the NOTES proceeding this Procedure. Safety is always the first order of business. You must have parent permission and supervision in this Activity. All involved must wear goggles.
2. Measure the diameter to the nearest $1/10^{th}$ cm and record this measure in the data table (d).
3. With the materials ready, you have also therefore selected the best spot to attach the suction cup. As noted in the list, a tile floor works best, though a table is very good too. The goal is to have a flat (as few irregularities or bumps as possible) clean and clear surface.
4. Attach the suction cup. Pull at it slightly to insure it is adhered to the surface well.
5. Before attaching the Tension Scale, be sure to examine it, zero it out, and understand how to read it.
6. Attach the Tension Scale to the suction cup so that the scale can be read as you pull it as it pulls the suction cup. Note it is best to attach with some strong tape (duct is best).
7. Before the slow and gradual pull of the tension scale, be sure to have your oven mitt on and place this barrier between you and the suction cup (still be able to read the scale).
8. Gradually pull the tension scale at as small of an angle as possible (recognize that this is one of the errors in the process).
9. Alternative Measure Note : for those who want a better measure use a protractor to measure the angle of the scale from the vertical since it is the same as the angle it is pulled at – interior angles theorem – and use as the force $F*\cos\Theta$ where F is the tension scale reading.
10. Pull until the suction cup gives way. Note that you are measuring in Newtons (if not be sure to convert) Record this measurement (F).
11. Do this at least 3 times (more is always better in this case) and determine the Average of the results (F_{ave}).
12. Calculations :

13. Though the Slide Rule is a recommended tool, all of these calculations can be done with a regular or scientific calculator. Some scientific ones even have built-in averaging formulae. For those who like spreadsheets, the data can be typed in and the formulae then also be typed in its own cell where the formula references each of the measured variables in their respective cells (for example in cell B1 the force measured (F) is typed and a later cell, say B3 has the formula where it occurs).
14. Start with determining the Area of the suction cup. Recognize that you must convert the centimeters to meters (1 m = 100 cm).
15. Determine the Pressure (P) from your measures of Force (F_{ave}) and Area (A).
16. Compare this result to the normal Atmospheric Pressure value by dividing the regular Pressure by your determined pressure to see how many times it is as compared to your outcome.
17. Using the conversion information provided, convert your value into pounds per square inch.
18. Try other surfaces if available. If you have other suction cups, you might consider trying these. Be sure to note that in all cases, parental permission and supervision is needed. Also in using suction cups over 1 in. diameter, the amount of force goes up quite a bit, so you should consider using small suction cups only. (Consider by thinking of the formula and the regular value of the atmospheric pressure why this is so).

Data :

Surface the suction cup is in contact with : ____________

Diameter of Suction Cup (d) : __________ cm

Trial	Force [F] in (N) on Tension Scale
1	
2	
...	
Average	

Calculations :

Be sure to use your Slide Rule!

Average :

$$\mathbf{X_{ave}} = \frac{\Sigma \mathbf{x}}{\mathbf{n}}$$

(x = a given value in the set, n = the number of elements in the set, Σ means the sum of the items)

Area of Circle :

$$A = \frac{\pi * d^2}{4}$$

(A = Area, d = diameter)

Pressure :

$$P = \frac{F}{A}$$

(P = Pressure, F = Force, A = Area)

Assumed Constants for Comparison :

Atmospheric Pressure : 1 ATM = 101 kPA = 14.7 lbs/sq.in.

Possibly Needed Conversions :

1m = 100 cm
1 N = 0.224 lbs
1 kg = 2.2 lbs
1 in. = 2.54 cm

Conclusion :

How do your results for Pressure compare to the generally accepted value for the Atmospheric Pressure (more or less) and why did these results come out this way do you think ?

Activity #17
Personal Pressure Determination
Grade Level : Middle School
Math Level : Calculating

Basic Pressure Calculations with a Slide Rule

All weights are forces and they are due to the force of gravity acting on a mass. A rectangular brick or block of wood on the table has a weight. When lying flat or standing up the weight is the same and this is independent of the amount of surface area in contact with the table. The best example for this is a book being rotated from a flat position and then the same book standing on its edges.

If asked about the amount of force each exerts, the answer is the same. Yet intuitively we know something is different. It is the amount of pressure each case exerts. The book standing up with less surface area in contact with the table has a greater pressure pressing down on the table. This is because Pressure and Force are not the same thing. Pressure is Force per unit Area.

$$\mathbf{P = \frac{F}{A}}$$

We hear of pressure all the time. It is often associated with daily life stresses, but this is not what concerns physics. We live at the bottom of an ocean – not water, but instead air. It exerts a pressure on us. The average air pressure is 14.7 lbs/in^2. In other measuring systems this is called 1 atmosphere (1 atm), 1.013×10^5 Pascals or 760 mm or 30 inches of mercury. We measure this with a barometer and it is given to us in the weather. The changes in pressure indicate wind and the likelihood of storms.

Though 14.7 lbs/in^2 is an average at sea level it is not constant at different altitudes. In fact many foods have high altitude directions because with higher elevations there is a decrease in pressure. To illustrate this imagine or have available a few pillows. Place the first lying flat on a table and then stack one upon another on this pillow. Notice that the first is flattened a bit while with each layer there is less force, hence less pressure as one goes up. This is similar to the Earth's atmosphere where pressure decreases with altitude, hence a need for oxygen for high altitude mountain climbers and pressurized cabins in planes.

Pressure is also often found in discussion about going under water and the pressure that increases with depth. In any body of water, with every 33 feet of depth the pressure increases on us by 1 atm or 14.7 lbs/in^2. Imagine the hundreds to thousands of pounds per square inch miles below in the ocean!

To illustrate how powerful this pressure is in history scientists would take two hollow half-spheres of metal and join them not with bolts but letting the air pressure alone outside of the sphere hold them together. The two halves were held together and heated with a small amount of water inside to drive as much of the air out as possible. Then it was sealed. It could never reach zero pressure, but it was much less than the outside pressure. For every square inch of the sphere there might be 7, 9, even up to 12 pounds of force. If it were 10 pounds of force (due to the difference of pressure outside and inside the sphere) and the area were 200 square inches, the amount of force holding the two halves together would be 2000 pounds of net force!

When we measure pressure with a common gauge a rod pops out when we connect it to a basketball or a tire from a car or bicycle. This is measuring gauge pressure. Gauge pressure is the difference of the actual pressure of what is inside the item and the outside air pressure. For example, if the gauge read 30 lbs/in^2, then the actual pressure inside would be 44.7 lbs/in^2 (14.7 lbs/in^2 more). If there were no pressure at all outside (of course you would need a pressurized air suit like an astronaut – remember there is almost zero atmosphere in space, quite the opposite of problems of a diver deep in the ocean) then the gauge pressure would read what is actually the pressure of the item.

Pressure occurs when any force happens and there are surfaces in contact with each other that have a surface area to them. For example, if a person lies down or stand up, that person's weight remains the same yet the pressure that is exerted changes drastically. The amount of surface area lying down is much greater than the surface area of one's shoes when standing. This is why rescue workers when going across thin ice to save someone lie down. Also this explains how someone who weighs less yet may be wearing smaller surface area shoes can exert a much greater pressure on her feet such as when a girl wears high heels as compared to someone else in flat shoes.

We are going to do just that – compare the pressure of wearing one pair of shoes versus another for a given person. It helps to have a few pairs handy – such as sneakers, wedges, and heels - (see photos of my shoes I used on the next page).

With any and all Activities read through, then set up the process for the Activity and run it without measurement and watch where you may have bumps in the road so as to anticipate these. Be patient – Repeat processes. Once all is well when measuring see to the values – do you notice patterns or trends – what do you predict will happen next as you change one thing or another from one trial in an experiment to the next. Always Explore & Enjoy. Be Brilliant!

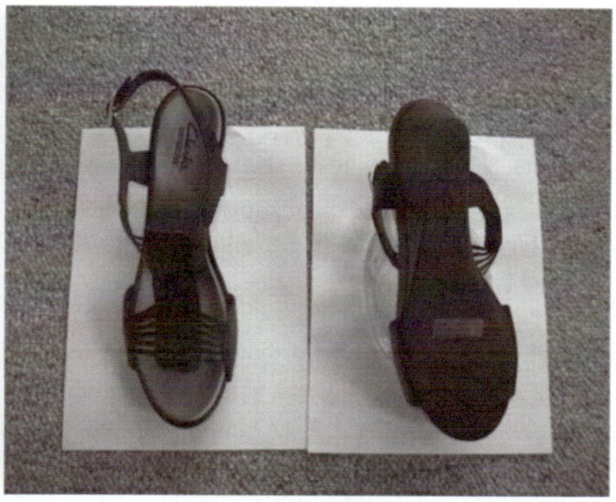

All one has to do is weigh oneself on a scale, then place one of the shoes on graph paper and trace the area of the sole in contact with the graph paper.

Next you can determine the area of the sole of the shoe in one of two ways. In a slower more sequential manner of area determination by counting all of the whole blocks in the area of the sole on the paper and then estimating how many of the fringe pieces it takes to make a whole block and determine this number. This is the total number of blocks a shoe takes up (be sure to double it, since there are two). For both methods : Be sure to find how many whole blocks per inch as well as per square inch area on the paper to find the total area from a simple multiplication. This will be used to find the area. In finding the area by our first method merely divide the total number of blocks by the number of blocks per square inch – this yields the number of square inches to be used as the area in our pressure determination.

A faster way other than counting the blocks is the second method, is to draw rectangles and/or squares in the area of the drawn sole of the shoe when and where possible in the shoe sole print. For each square or rectangle calculate its area from its length and width measures (by block numbers of each). Add together each of these values. This method only hastens the counting. Now add the bits and pieces to reach a grand total, which should be the same as the first method total. In both methods you have an area in blocks squared determined. Now divide by the number of blocks per square inch to find the area of a shoe – then double it to have both combined.

With the known weight and the surface area now treated as a ratio we can find the pounds per square inch for a given weight and a given pair of shoes. Doing different pairs, such as sneakers and heels can be fun! :)

Purpose : To determine the pressure of a person in shoes standing (P = F/A).

Materials :

- ruler,
- weight scale,
- pencil,
- graph paper,
- person,
- different types of shoes with different sole areas,
- Slide Rule

Procedure :

1. With any and all Activities read through, then set up the process for the Activity and run it without measurement and watch where you may have bumps in the road so as to anticipate these. Be patient – Repeat processes. Once all is well when measuring see to the values – do you notice patterns or trends – what do you predict will happen next as you change one thing or another from one trial in an experiment to the next. Always Explore & Enjoy. Be Brilliant!
2. All measurements can be done in English units, such as inches for the length and widths called for if you like or you can use metric, such as centimeters – note you will then have to convert your weight to Newtons (see conversion chart).
3. If you do not know how many squares per inch on your graph paper, then you will have to either determine this or use the rule to take measurements. In that case : Be sure to measure the inches properly and not round off to a whole number. Measure the fractions and turn into a decimal value with the slide rule!
4. For example, ¼ easily becomes 0.25, while 1/8 can readily be read as 0.125 on the slide rule. If uncertain as to how to do this, look at the Activity for ratios, fractions, and proportions with the slide rule. All ratios are readily turned into decimals through division on a slide rule.
5. It is best to first weigh yourself, record this value in the table (Q). This is our force.
6. Next using one of the shoes trace its pattern on the graph paper. Note that you may have to turn it sideways (as seen in the sneakers photo) or a better solution is to

join two overlapping pieces of graph paper (this prevents the squares of the graph paper from being angled and hence harder to count). Be sure to only trace the part of the sole in contact with the graph paper – for example, look at shoes that narrow or have heels.

7. Once you have your tracing, you need to determine the number of squares in the image of the sole of the shoe either by a process of counting directly or drawing rectangles and/or squares and determining their area and summing these along with all of the pieces counted as estimated whole squares. This becomes your judgment of just how many of these there are. Nonetheless your goal is the final count of whole number of squares – no need for fractional numbers – use a whole number estimate. Use one of the two methods mentioned in the opening discussion of Pressure.
8. Be sure to have doubled the number of squares from one sole, since there are two for the final Pressure calculation. You can always use your original value if you want to determine the pressure found by standing on one foot, however.
9. Recognize in using either method you have two goals, first know the number of blocks per inch to find the number of blocks per inch squared.
10. Second to find the total number of blocks per shoe (either by counting directly or by dividing up the area of the sole into rectangles and squares and summing up the number of areas of blocks and to each method next adding the bits and pieces through estimation), then doubling the number. Finally from this Area of the shoe in blocks total divide this by the number of blocks per square inch to find the area of the shoes.
11. Now you can determine the Pressure from your weight divided by the area in inches squared (or whatever chosen units)– if done in English, your answer is in the units of lbs/in^2.
12. Try other shoes for comparison! Have others join in for other comparisons. Have fun :)

Data :

Weight (Q) : ___________ (lbs or N)

Number of squares per inch on graph paper (G) : ________

Number of squares per square inch (N) : ___________

Note : Either use the Number of Squares or Length And Width if using rectangles to find the number of squares. Realize that there may be multiple lengths and widths if more than one rectangle and the Area (R) is for them while the Number of Squares is the first method mentioned.

Shoes	Number of	Length	Width	Area	Pressure
	Squares (C)	L (in)	W (in)	R (in^2)	P (lbs/in^2)
Sole					

Calculations

Be sure to use your Slide Rule for All Calculations!!

The best choice here is the C & D scales.
Remember that ratios have one value over the other (for example - the numerator on the C scale and the denominator on the D scale for example) and the answer, the quotient, is read opposite the denominator's index (from our example it is found opposite the D index and read on the C scale).
For multiplication, place the multiplicand on one scale (for example the value read on the D scale) opposite the index of the other scale (C scale). Now read along this latter scale (C scale) to the other operand and find the answer, or product, opposite it on the initial scale (D scale).

If using rectangles and/or squares
Area of Shoe in squared squares = Length of Rectangle in squares x Width of Rectangle in squares

$$\Sigma R = L*W$$

The Sum of R is used since there may be more than one rectangle drawn. All of these areas are first determined then added up!

Number of squares per square inch of graph paper = (Number of squares per inch)^2

$$N = G^2$$

Use one of the following Area formulae :

If using Rectangles, then use :
Area of Shoe (in sq in) = Area of Shoe in squares / Number of Squares per square inch

$$A(\text{sq in}) = R/N$$

If using the Number of squares then use :
Area of Shoe (in sq in) = Counted Number of Squares in sole area / Number of Squares per square inch

$$A(\text{sq in}) = C/N$$

Pressure = Force per unit Area

$$P = \frac{\text{Weight}(Q)}{A(\text{sq in})}$$

Conclusion :

Any surprises – how does Pressure change with changing area of the shoes? Does the area of the shoe make a difference in the pressure that is exerted regardless of the weight of the person? Think of it this way – Your weight does not change whether or not you stand on both feet or one foot, but the area you are standing on one foot is now half of the area you originally were standing on – what should happen to the pressure here?

Explore & Enjoy
Be Brilliant

Extension Activity :

If one is so inclined they could convert the English units to metric. It may be that the gauge one has already has metric readings, say Pascals along with or instead of lbs/in^2. If so the measurements would best be made in centimeters and converted to meters for the calculation.

1 ATM = 101.3 kPa = 14.7 lbs/in^2
1 inch = 2.54 cm
1 lb = 2.2 kg

Activity #18
Finding the Weight of a Car from Tire Pressure
Grade Level : Middle School
Math Level : Calculating

Weight from Pressure Calculations with a Slide Rule

This Activity is a good follow up to the Basic Pressure Activity and explores Pressure in a rather unique application of the idea of pressure :

Okay Inquisitive Pioneers – Here is the challenge - The best way to approach this activity is to consider it a puzzling question or challenge. Say someone hands you a ruler and a tire pressure gauge and asks you to measure the weight of your car or a bike. Can this be done?

Put your thinking caps on before reading on – think of the car, the tools you have and which direction is weight pointing (up, down, sideways) ? What part of the car is in touch with the ground? This is a good question clue.

This activity takes advantage of pressurized items like tires which exert a supportive force translated from the axles through the contained air to the tires themselves in a car or the wheels of a bike for the masses they hold up. All that needs to be known is the reading of the gauge pressure of the tires for either type of transport and a determination of the surface area of the tires in contact with the ground and from rearranging the equation for pressure, the force exerted on the ground, which is the weight of the object here, can be determined.
Wow – Imagine that!

Purpose : To determine the weight of a massive object as obtained from the points the object has in contact with the ground by means of the pressure of the inflated tires and their surface area in contact with the ground where the weight is found from the definition of pressure ($P = F/A$).

Materials :

- air pressure gauge,
- ruler,
- weight scale,
- carbon paper,
- large sketch pencil,
- graph paper,
- car or bicycle,
- Slide Rule

Procedure :

1. NOTE : This is to be done with parental supervision, especially in the area of car weight determination.
2. All measurements can be done in English units, such as inches for the length and widths called for and lbs/in^2 for the Pressure on the pressure gauge.
3. Be sure to measure the inches properly and not round off to a whole number. Measure the fractions and turn into a decimal value with the slide rule! Note all measurements need only be accurate to 0.1 inch or centimeter.
4. For example, ¼ easily becomes 0.25, while 1/8 can readily be read as 0.125 on the slide rule. If uncertain as to how to do this, look at the Activity for ratios, fractions, and proportions with the slide rule. All ratios are readily turned into decimals through division on a slide rule.
5. Note : Though the Slide Rule is a recommended tool, all of these calculations can be done with a regular or scientific calculator. Some scientific ones even have built-in averaging formulae. For those who like spreadsheets, the data can be typed in and the formulae then also be typed in its own cell where the formula references each of the measured variables in their respective cells (for example in cell B1 the pressure on the gauge (P) is typed and a later cell, say B3 has the formula where it occurs).
6. With any and all Activities read through, then set up the process for the Activity and run it without measurement and watch where you may have bumps in the road so as to anticipate these. Be patient – Repeat processes. Once all is well when measuring see to the values – do you notice patterns or trends – what do you predict will happen next as you change one thing or another from one trial in an experiment to the next. Always Explore & Enjoy. Be Brilliant!

Procedure for Activity I : Determining the Weight of an ordinary Bicycle :

1) Use a bike that can be held. Have the 'rider' stand on a scale with the bike so that the scale measures the total weight of both. Record this value.
2) Next weigh the 'rider' independently and record this value. – These two values are for comparison and considered true.
3) Place graph paper face up on the floor, spaced so that when standing the bikes tires are on it.
4) On top of the paper place the carbon paper face down so that it can make a mark.
5) Place the bike on the carbon paper.
6) Have the 'rider' sit on the bike, feet on the peddles and kick-stand up. Have a third party balance the bike and rider from the side but adding no pressure down.
7) Measure the air pressure using the tire pressure gauge and record the values on the table.
8) Carefully have the 'rider' dismount the bike and take up the graph paper for measurement.
9) Treat the impression by the tires on the graph paper as rectangles and compute the area of each impression.
10) Use the Area calculations and Pressure measurements to Determine the Weight on each wheel of the bike and 'rider'.
11) Subtract the 'rider's' weight and find the weight of the bike.

12) One could avoid the rider altogether and see if the bike leaves impressions through the carbon paper on the graph paper – note that it probably does not work well.
13) An alternate method could avoid the carbon paper and simply trace the tire area in contact with the graph paper by carefully drawing closely where the bike tire makes contact with the graph paper beneath it. - Note that the area used here is probably a better one than with a rider on it.
14) One last alternative is found in step #7 for the car where papers are places around the tire area in contact with the ground and the open space template is then measured and its area is found.
15) Using percent error, determine how close this value is to the actual weight of the bike.

<u>Procedure for Activity II : Determining the Weight of a Car :</u>

1) Have a car parked (parking brake on) on level cement.
2) Note the following procedure applies to all of the tires individually.
3) Around each of the tires take blank paper up to the tire where it contacts the ground.
4) Place enough paper around the area of the tire in contact with the ground.
5) Secure the paper with tape, leaving one edge open so that the template can be removed.
6) Once removed measure its length and width of the tire tread in contact space and then calculate its area.
7) An alternative is to place the pieces of paper in front and behind the tire up to the tire-ground contact point and simply measure with an available ruler (instead of the template method).
8) Another consideration : Ask yourself, does the percentage of the tire in contact with the ground matter? If you think so you need to estimate the percent amount of tire in contact with the ground.
9) To determine the percentage of tire tread in contact with the ground, place carbon paper on the tire facing outward.
10) Place over the carbon paper a piece of graph paper and rub with a pencil to leave only a surface impression of the outer tread. Do not press into the groves.
11) From this impression, determine roughly (by counting boxes totally and those that leave an impression) the percentage of tire tread in contact with the ground.
12) Measure the air pressure for each of the tires and record the readings on the table.
13) Calculate the weight force on each tire from the relation for pressure. This is Weight Force = (% in contact as a decimal)*(Pressure)*(Area)
14) Note that if you did not consider the % in contact, then that is 1.
15) Sum up the weights from all the tires and find the weight of the car.
16) Read the weight of the car on a label typically found on the driver's door noted as GVW – Gross Vehicle Weight (often in pounds).
17) Compute percent error from the known value as compared to the calculated value.

Data :

Experiment I :
Determining the Weight of an ordinary Bicycle :

Tire	Length	Width	Area	Pressure
	L (in)	W (in)	A (in^2)	P (lbs/in^2)
Front				
Back				

Actual Values for comparison :

Weight Table of Data	
Weight of Bike and Person : B_A	
Weight of Person : P	
Net Weight of Bike : TB_W	

Experiment II :
Determining the Weight of a Car :

Total number of squares in rectangle (shaded & not): __________
Total number of squares in rectangle shaded : ____________
Percentage (as decimal) of shaded squares : _____________

The Length and Width are the measures of the tire tread in question in contact with the ground (see procedure).

Tire	Length	Width	Area	Pressure
	L (in)	W (in)	A (in^2)	P (lbs/in^2)
Left Front				
Right Front				
Left Rear				
Right Rear				

Weight of Car from Information Provided : _________________

Calculations (Formulae needed listed here) :

Tire	Calculated Weight on Tire Wt_C (lbs)
Total Weight (lbs)	

Be sure to use your Slide Rule for All Calculations!!

The best choice here is the C & D scales.
Remember that ratios have one value over the other (the numerator on the C scale and the denominator on the D scale for example) and the answer, the quotient, is read opposite the denominator's index (found on the C scale). For multiplication, place the multiplicand on one scale (D scale) opposite the index of the other scale (C scale) and read along this latter scale to the other operand and find the answer, or product, opposite it on the initial scale (D scale).

Percent of Car Tire in contact with ground :

$$\%\text{ Contact} = \frac{\text{Total number of shaded squares in rectangle}}{\text{Total nubmer of squares in rectangle}}$$

Surface Area of Tire = Tread Length x Tread Width

$$\mathbf{A = L*W}$$

True Weight of Bike = Total Weight of Bike & Rider – Weight of Rider

$$\mathbf{TB_W = B_A - P}$$

B_A = Bike Weight + Person Weight

Weight on Tire = Tire Surface Area x Pressure of Tire

$$\mathbf{Wt_C = A*P}$$

Percent Error : Accepted Value is given (car) or found on scale (bike):

$$\%E = \frac{[\text{Experimental Value-Accepted Value}]}{\text{Accepted Value}}*100\%$$

Conclusion :

Wow – what a fun Activity – the weight of a car with a ruler and a tire pressure gauge! Would you believe it?! Quite amazing I feel.

The basic questions here are these: How close is the experimental value to the actual? If they are considerably far apart, what may cause this? (Consider in the case of the car whether the weight for the car considers anything that may be in the car at present that the tire's must now support including gas in the tank).

Extension Activity :

If one is so inclined they could convert the English units to metric. It may be that the gauge one has already has metric readings, say Pascals along with or instead of lbs/in^2. If so the measurements would best be made in centimeters and converted to meters for the calculation.

1 ATM = 101.3 kPa = 14.7 lbs/in^2

1 inch = 2.54 cm

1 lb = 2.2 kg

Another consideration : Do the pressure readings, hence the amount of tire in contact with the ground change as the day does? Consider how much cooler it is in the morning as compared to the afternoon temperatures – perhaps this is something to examine as well?

Activity #19

Buoyancy – Archimedes Principle

Grade Level : Middle School

Math Level : Calculating

Do you ever ponder – why can a boat made of metal float? Such a basic question is at the heart of all of science which at its core is the nature of curiosity. Always be curious and ask questions. The real exploration of this idea begins with the thoughts and writings of Archimedes. The story of Archimedes and Archimedes Principle can be found in the Activity concerning the Density Activity (p. 195 Activity #13).

In a short summary, Archimedes found that the weight of an object in air and when placed in water are different. It is different by what we refer to as the Buoyant Force and this in turn is equal to the weight of the displaced water, which equals the volume of the object. He famously used this idea to solve the question of whether a king's crown was made of pure gold or not.

Using this idea, it is easy to find the volume of an irregularly shaped object and with a good mass scale, these two quantities, mass and volume, allows the determination of density of the object, which can be a useful physical property of a given material.

Further, Archimedes Principle allows for a simple conclusion to be reached : If the density of the object placed into any fluid (liquid or gas) is greater than the density of the surrounding fluid (or medium) it will sink. If it is less, then it will float. This is the reason why large steel-hulled ships float – they are less dense than the water they are in. They are able to displace their weight. As the ship is loaded, it sits further down in the water.

This is also the reason hot-air and helium-filled balloons float. They too are less dense than the surrounding air. In fact as long as there is a force acting on the rising balloon, the buoyant force in this case, the balloon will rise. When the forces of weight and buoyant force balance out, then the balloon stops rising. This is all in accord with Newton's Laws of Motion, where a Net Force results in a change of motion (the rising of the balloon) and when the forces balance it comes to rest, since now when rising further there is an exerted force downward (the weight force) and hence the balloon reaches equilibrium.

In this Activity, we directly and indirectly measure the Buoyant Force for comparative purposes as well as determine density for a given stone. We will directly measure the Buoyant Force which is quite easy – also we will look at

Note that this Activity unlike the others in the book is two parts – here in the first one we are looking at an object that sinks and finding and calculating its Buoyant Force and comparing these two values. In the second Archimedes Activity we are investigating Buoyant Force yet again but now are looking at floating objects instead. Fantastic and Brilliant Fun!

Purpose : To take measurements of mass, weight in air, weight in water, and water displaced to determine the buoyant force acting on a mass of density greater than that of water.

Note : There are several distinct purposes here :
1) Determining the Buoyant Force through Scale Measurements,
2) Calculate the Buoyant Force through Volume of water displaced and its Weight determination, and
3) Determination of the Density of an Object through use of the ideas in Archimedes Principle

Materials :

- Small Bucket with Lip,
- Pan large enough for bucket to sit in and water over-spilling the bucket with immersed object will collect,
- Measuring Cup with markings (metric and English best),
- Note : If possible – use a Graduated Cylinder instead of measuring cup,
- Spring Mass/Weight Scale (as used in fishing, pounds or kilograms),
- Kitchen Mass Scale (measurements in grams),
- Water,
- Small palm-sized Stones,
- Dental Floss or Fishing Line String,
- Scissors,
- Slide Rule

Procedure :

1) With any and all Activities read through, then set up the process for the Activity and run it without measurement and watch where you may have bumps in the road so as to anticipate these. Be patient – Repeat processes. Once all is well when measuring see to the values – do you notice patterns or trends – what do you predict will happen next as you change one thing or another from one trial in an experiment to the next. Always Explore & Enjoy. Be Brilliant!
2) To determine if the bucket chosen is best, first try to put the stones to be used in it to see that they fit.
3) Next fill it to the top with water (with the bucket in the collection pan for spilled out water). Place a stone in. Be sure to have enough water so that the water spills out into the pan (this can be tricky due to the surface tension of water).
4) To set up all things – have the bucket in the pan and the bucket filled with water.
5) Have a small set of stones (3 – 6) all about palm-sized to be used.
6) First tie the string being used to the stone to be measured. (Be careful with fish line as this can cut – have supervision or help).
7) Measure the mass (in grams) of the stone on the kitchen scale and record this value (M).
8) Measure the weight of the stone (in pounds) with the spring scale in air and record this value (W_{air}).

9) While the stone is on the string and on the spring scale, lower the stone into the water. Once underneath the water, measure and record the value of the weight reading on the spring scale (W_{water}).
10) Be sure to have collected the water spilled out of the bucket when the stone was placed into the water. Carefully remove the stone, then the bucket and then pour the contents of water in the pan into the measuring cup and record this volume value (ΔV).
11) If possible, use a graduated cylinder to have more precise values of volume.
12) Perform all of these measurements aforementioned for each of the stones separately and place their values on a table of information.
13)
14) Calculations to do for each of the stone's data :
15)
16) First calculate the Buoyant Force (F_B) (in pounds) directly from the difference of the measures of Weight in Air and the Weight in Water of the stone. This is the Direct Buoyant Force.
17) Now we will use other measurements to find the Indirect Buoyant Force measurement for comparison.
18) Convert the Volume measure (ΔV) from cups into cm^3 if using cups.
19) Determine the Density (D_{stone}) of the Stone by using the ratio of the kitchen mass scale reading and the converted volume measure.
20) Calculate the Indirect Buoyant Force (F_{B2}) from the displaced volume of water, the density of water, and the acceleration due to gravity.
21) Be sure to be using the proper units for the Force. Your initial calculation may have units of : $\frac{*}{\quad}$. this means you will have to convert the calculated buoyant force into Newtons and then Pounds. This will be the second buoyant force measure, called the Indirect Buoyant Force (F_{B2}).
22) Compare this new predicted value to the originally directly determined Buoyant Force (F_B).
23)
24) Also for comparison, calculate the predicted weight in water of the stone using the indirect formula involving the use of densities of the stone and water.
25) Assuming that the direct measure is the correct value, what is the percent error of your calculated value and why would this occur? Can you find ways to eliminate the error involved?
26) Though the Slide Rule is a recommended tool, all of these calculations can be done with a regular or scientific calculator. Some scientific ones even have built-in averaging formulae. For those who like spreadsheets, the data can be typed in and the formulae then also be typed in its own cell where the formula references each of the measured variables in their respective cells, for example B1..BN has the measurements and values used in the equation while BN+1 has the formula for all of these variables (why not A? Simple – use it to label you variables)

Data :

Note : Give each stone an ID, such as a Number or Letter.

Stone ID	W_{air} (lbs)	W_{water} (lbs)	ΔV (cups or mL)	M (g)

Stone ID	Direct Buoyant Force F_B (lbs)	Indirect Buoyant Force F_{B2} (lbs)

Calculations :

Be sure to use your Slide Rule!

Direct Determination of Buoyant Force formula :

$$F_B = W_{air} - W_{water}$$

Indirect Determination of Weight in Water formula :

$$W_{Water} = g*\Delta V*(D_{stone} - D_{water})$$

Indirect Determination of Buoyant Force formula :

$$F_{B2} = g*\Delta V*D_{water}$$

Density : Density is mass per unit volume

$$D = -$$

Weight Force formula :

$$W = M*g$$

Table of some values you may need :

Acceleration due to gravity for Earth :
$g = 9.8\ m/s^2 = 980\ cm/s^2$

Density of Water : $D_{water} = 1\ g/cc = 1000\ kg/m^3$

1 m = 100 cm
1 mL = 1cc = $1cm^3$
1kg = 2.2 lbs.
$1\ N = \frac{*}{\quad} = 0.224$ lbs.
1 oz. = 28.3 g
1 oz. = 29.6 mL
1 qt. = 4 cups
1 cup = 8 ozs.
1L = 1000 mL = 1.06 qts.

Conclusion :

The basic idea here is to compare and reflect on results and consider options to refine the measurements. We are using math, science and reasoning to find and compare the ideas of Buoyant Force both directly and indirectly and seeing how close they come to each other. In general they should be the same, but there may be small errors in our measurements and the like to affect our outcomes.

Archimedes Activity Part II (Floating)

Floating Objects and Archimedes Principle Activity

More Science and Math involving ideas from the Ancient mathematician Archimedes – here involving Buoyance and Floating!

Look to the other Activities involving Archimedes Principle (with regards to sinking objects totally immersed in water and density through water displacement) for information on Archimedes, his life, ideas, and most importantly Archimedes Principle.

This Activity focuses on the concept of floating objects. To float an object must have an upward force to balance its weight force which points down. These forces must be balanced (a net force of zero) otherwise (according to Newton's 1st Law) the object would be accelerating up or down, but it is stationary (one of two possible states when forces are balanced).

This upward force is called the Buoyant Force and comes to us from Archimedes who noted this idea long ago. The weight of the displaced liquid when an object is placed in it is equal to the weight of the object. Hence a boat that weights 2 tonnes, even if it is made of metal, when it displaces 2 tonnes of water will float. The key in boats is to shape the material so that it spans a large volume so as to displace as much water as possible (yet at the same time having a streamlined shape for mobility in the water, since water will act as a drag as the motor powers it through the water).

Note that the volume of the water for the floating object is NOT equal to the volume of the boat itself! These values are only the same when the object is completely immersed in the water (hence sunk or submerged).

An interesting consequence of the fact that a floating object requires less volume than itself in the volume of displaced water means that water itself is more dense than the boat! How odd this must seem?! A metal boat having less density than that of water itself. Note that this is the average density and not the density of any particular piece of the boat. The density of the metal of the body does not change in the water – it remains what it is in air (for example aluminum is 2.7 g/cc). But when you take the total mass of the boat and divide it by its entire volume that it encompasses (all the air space too) then it turns out that its average density will be less than the density of water (1.0 g/cc).

The goal of the Activity is to try various 'boats' of differing masses, to measure the amount of displaced water by volume and calculate its mass and compare this to the mass of the boat itself.

Purpose : Compare the Mass of Displaced Water by a Floating Object Experimentally to the Calculated Mass of Expected Displaced Water by a Floating Object using Archimedes Principle.

Materials :

- 16 oz. Plastic Cups (at least 2),
- Mass Scale,
- Marbles,
- Water,
- Marker,
- 4 c. Measuring Cup,
- Optional Equipment : Tension Scale, Graduated Cylinder,
- If a 4 c. Measuring Cup not available use a square pan and a smaller measuring cup (2 cup size),
- Slide Rule

Pictures :

Procedure :

1. The overall cups system will be placed in a container to catch the displaced water (either a square pan or a 4 cup measuring cup). Place the cup for water in this container.
2. Fill this first cup up to the very top carefully.
3. In the other cup now decide to use a set of marbles (10 minimally for each activity – up to around 40 total) in an incremental manner for each data set (for example start with 10, then 20 in the next set, et al). Place this chosen number of marbles in the second cup and take its mass (B).
4. Now slowly and carefully let the boat settle into the first cup which has water in it. The water should displace into another container (see no. 1).
5. Measure the displaced volume of water (DV). Do this 3 times, recording the displaced volume each time.

6. Note : Measured displaced water in the measuring cup is easiest. In the absence of such an item you need to pour the water from the square pan into a measuring system (another smaller measuring cup or the use of a graduated cylinder).
7. Note : Be certain to not spill the water. Also be clear as to the degree of precision of the tools you are using when recording your results.
8. Do not quickly discard this water since you must measure its mass too for comparative purposes (M) and record it.
9. Reset the system and do the Activity at least 3 times (3 different sets of marbles making up the bulk of the mass in your boat).
10. Calculations :
11. Though the Slide Rule is a recommended tool, all of these calculations can be done with a regular or scientific calculator. Some scientific ones even have built-in averaging formulae. For those who like spreadsheets, the data can be typed in and the formulae then also be typed in its own cell where the formula references each of the measured variables in their respective cells (for example in cell B1 the mass of the boat (B) is typed and a later cell, say B3 has the formula where it occurs).
12. Determine the Average (V_{ave}) amount of Displaced Water for each 'Boat' in your Activity.
13. Using the Average Displaced Volume of Water, calculate the Weight of the Water Displaced (F_B).
14. Convert both the mass of the 'Boat' (B) and the mass of the measured displaced water (M) into weight.
15. Compare all of the weight values (they should ideally be identical).

Data :

Note : Recreate the Table for each 'Boat' you test & for each 'Boat' run 3 trials and average the volume displaced :

Mass of the 'Boat' [B] : _______ (kg)

Trial	Measured Volume of Water Displaced (ΔV)
1	
2	
3	

Mass of Displaced Water for Comparative Purposes [M] : ______ (kg)

Calculations :

Be sure to use your Slide Rule!

Average :

$$V_{ave} = \frac{\Sigma}{\quad}$$

Calculated Weight of Displaced Water :

$F_B = D * g * V_{ave}$

Converting Mass to Weight :

W = B*g

Conversions & Constants that may be needed :
Acceleration due to gravity for Earth :
$g = 9.8\ m/s^2 = 980\ cm/s^2$

Density of Water : $D_{water} = 1\ g/cc = 1000\ kg/m^3$

1 m = 100 cm
1 mL = 1cc = $1cm^3$
1kg = 2.2 lbs.
1 N = $\frac{*}{\quad}$ = 0.224 lbs.
1 oz. = 28.3 g
1 oz. = 29.6 mL
1 qt. = 4 cups
1 cup = 8 ozs.
1L = 1000 mL = 1.06 qts.

Conclusion :

How did your comparison Activity go? Did your results show that the weight of the 'boat' equals the weight of the displaced water? What does this mean in terms of a net force acting on the 'boat'? If each weight is a force and they are equal then the Net Force should be zero!

As an alternative in measuring, use the marker to mark the side of the 'boat' cup when it is floating in the other cup of water and determine a way to measure this amount of displaced water up to this volume of the cup (besides measuring the displaced water itself).

Activity # 20

A Marble's Small Size Determination to relate to Atom's Size

Grade Level : High School

Math Level : Calculating

The Size of an 'Atom' (Marble, Coin) and the Slide Rule

Though there seems to be an unusual title to this Activity – Marbles / Coins and Atoms we are not going to find direct information about Atoms here but are going to look at collisions of Marbles and using Probability to determine the size of a Marble much like in the history of the Atom the size and general structure of the Atom was realized through collisions of particles. Also this Activity is like our determination of Pi thorough Probability and there is another one too in this book, the Estimated Number of Leaves on a Tree – which too uses Probability rules from Math to find answers to questions we might have.

This Activity as noted explores Not the actual size of an atom, but instead a simple idea using basic materials to relate the idea of the size of an atom by determining the size of a marble in one case and a coin in another case. These sizes are determined indirectly from collisions, much like atoms size were long ago in the famed Gold Foil Experiment and Alpha Particle scattering. Before the actual Activities, though, here is a brief reading history of the atom as it is known is given here for historical reference and for those of you students of life who may have an interest in these fascinating topics (the book has some tales such as these) :

The history of the atom and the atomic theory is both ancient and yet its most significant work is very modern. Democritus of Ionia in Greece is credited with thinking that all things were composed of atoms (taken from the ancient Greek atomos meaning 'undivisible'). The basic atomic theory is a general theory of the nature of matter that states that matter is composed of discrete units called atoms (that is units that have their own unique properties).

Since atoms are so small, it took many centuries before significant work was done to uncover their nature. With these discoveries, the basis of modern chemistry and the wave-particle mental picture of reality of physics were born. Not only this, it was found that the atoms are composed of what were once considered fundamental particles and these too were found to be made of still smaller pieces.

The first primary idea comes from Antoine Lavoisier in 1789 who wrote about the Law of Conservation of Mass. This law states that the total mass in any chemical reaction remains constant. This is to say that the mass of the reactants equals the mass of the products and acts as the basis of balancing chemical equations. This along with the Law of Definite Proportions (states that if a compound is broken down into its constituent elements then the masses of these parts will always have the same proportions) by Joseph Louis Proust in 1799 was initially extended by John Dalton in the Law of Multiple Proportions (for combined elements into compounds) and later the basis of the Atomic Theory of Matter.

The Atomic Theory of Matter by Dalton notes that each chemical element is composed of atoms of a unique type that cannot be altered nor destroyed by ordinary chemical means. These atoms can combine to form more complex structures (today called chemical compounds).

Dalton's original idea was modified and corrected in time. Amedeo Avogadro in 1811 set up Avogadro's Law (Equal volumes of any two gases, at equal temperature and pressure, contain equal numbers of molecules). These ideas were combined with Boyle's and Charles' Laws to establish the Universal Gas Law. Also, in 1827 the British botanist Robert Brown noted moving pollen grains floating in water under a microscope. In 1905 Einstein mathematically describes the Brownian movement which is attributed to water molecules bouncing off the grains.

It wasn't until 1897 when J.J. Thomson discovered the first primary particle of the common atom, the electron. He was working on what are referred to cathode rays (electrons) which he found could be deflected by electrical or magnetic fields. He concluded that these 'corpuscles' as he referred to them were negatively charged. Later in 1911 Milliken determined the charge to mass ratio which led to determining the charge on an electron with his now famous Oil Drop Experiment. (Note that it was Ben Franklin who gave us the notion for describing charges as either positive or negative).

Initially Thomson's work led to the 'Plum Pudding Model' of the atom where the pudding is a mass of positive charge with embedded negatively-charged electrons (acting as plums or raisins in the pudding).

Ernest Rutherford, a former student of Thomson, disagreed with the model and proposed an experiment to shoot alpha particles (essentially helium nuclei) at more massive atoms (such as gold). This is now referred to as the Gold Foil Experiment. It was actually conducted by Hans Geiger and Ernest Marsden (hence is also named after them as well) where the impacting and deflecting particles were detected by a fluorescent screen. The original anticipation from the prior model expected the particles to pass through. To the surprise of the scientists, there were a small fraction deflected by significant angles, where some almost bounced back towards the emitter. Rutherford used Newtonian Laws of Motion and Mechanics to analyze these motions (which illustrates the range of application of Newton's ideas) which turned out to be the same equation derived under the later Quantum Mechanics laws for scattering.

The only model that explains this outcome was the 'planetary model' as proposed by Rutherford. Here the atom has most of its mass concentrated at the center of the atom in the form of positively charged particles which are orbited by the electrons.

Niels Bohr modifies the model in 1913 so that the electrons have fixed orbits related to discrete amounts of energy. He arrived at this from an idea proposed by Max Planck and Albert Einstein concerning light and its absorption and emission in discrete amounts called quanta. This idea is further modified for atoms of heavier masses.

The proton became the central identity for the element with the work of Mendeleev. Though his initial work was based on atomic mass, later it is found that a natural organizational pattern emerges when atoms are placed in order of their increasing atomic number, which is the number of protons in the atom. This pattern is commonly called the Periodic Table.

Later in 1918 Rutherford is using alpha particles again on neon and discovers the proton. This is combined with the work of others who found that atoms of a given element can have different masses from each other. This meant there was minimally a third unknown particle in the nucleus. This is the neutron, found by James Chadwick in 1932.

The atom today views the electrons as acting as either waves or particles in prescribed regions of the nucleus (from Louis de Broglie in 1924 and Erwin Schrodinger in 1926). To envision the atom think of a soccer ball on the midfield line in a stadium. The ball is the atom's nucleus and is 99+% of the mass of the atom. The stadium is the realm of the atom itself where the electrons can be found. There is no prescribed size for them as it has never been determined (the electron is some $1/1835^{th}$ the mass of the proton). These electrons have prescribed properties of where they orbit and how they orbit (such as spin orientation, etc). Unfortunately one cannot know the position and momentum of a given electron, which is the Uncertainty Principle.

Our Activity set here will use large numbers of 'atoms', in the form of disc-shaped objects (i.e. coins) and/or marbles (there are two activities and your choice of materials) to illustrate the idea behind how atoms and their size is investigated. Here, we let objects either collide or not in a random manner and use the laws of probability to help us determine the size of the objects in our experiment. This is somewhat indirectly parallel to Rutherford's Gold Foil Experiment where alpha particles collide with or pass through the atoms of gold in the thin foil leaf. Study of the data revealed the overall structure of the atom and can be used to discern size too. This type of activity clearly demonstrates three of the most important aspects of science : 1) Models are developed from the behavior of events and their measurements to explain how things are structured and operate, 2) Models change with information that is added or contradicts the original model, and 3) Indirect measurements are a critical part of science since the vast majority of reality is at scales too large or in this case too small to be seen literally.

Practical Application : This activity relates to the study of the atom, such as the gold foil experiment that determined the structure of the atom.

Activity Goal : To determine the size of a small diameter object (penny, other coin, washer or marble) through elastic collisions and probability math models in 2 activities (horizontal and vertical)

Activity I : Determination of the diameter of a marble through probability of an elastic collision (on a horizontal planet)

Purpose : To discern the size of a marble from collisions with other marbles in an indirect elastic collision study on a horizontal plane using probability of collision.

Materials :

- 3 Meter Sticks,
- 11 Marbles (all the same size) (1 of different color),
- Flat Area (tiled floor best, but can use table top),
- Caliper,
- Slide Rule

Note : To promote randomness a poster board or box can be used where the side meter sticks are stood on edge and a folded piece of poster board taped to it acts as a roof to the atom (marble) chamber

Procedure :

1) With any and all Activities read through, then set up the process for the Activity and run it without measurement and watch where you may have bumps in the road so as to anticipate these. Be patient – Repeat processes. Once all is well when measuring see to the values – do you notice patterns or trends – what do you predict will happen next as you change one thing or another from one trial in an experiment to the next. Always Explore & Enjoy. Be Brilliant!
2) Before the activity, use the caliper to measure the diameter of the marbles in question. Record this number for comparison later.
3) On a tile floor place a meter stick to act as the far target area wall.
4) Perpendicular to it place the other 2 meter sticks so that there is a well-defined width to the area. This can be any choice from 40 cm to 80 cm, though 50 cm is recommended (see photo below)
5) Whatever the choice of width, this is also the length of area as well so when placing the Marbles (Atoms) in the confined area place them randomly within the width and no further than the length. Place 10 of the marbles here (you can choose to have different numbers : 5,6,9,12 etc). Record the number used (N).

6) The last marble is to be used as the projectile marble (shooter) and must be shot from minimally the full 1 m length away (best range 1 m to 1.5 m)
7) The shooter is best to be a different color.
8) Know the number of trials you plan to use. Recognize that if in any case that if the shot marble goes outside the confined area during its initial path it is not counted. Nor is it counted if it hits the back wall and returns to strike a marble.
9) To help randomness in the collision possibility you might consider the note on making a roof for the confined area below. If not, be sure that when shooting two things are done – first pick a random starting point within the width of the track to shoot from and then close your eyes before firing.
10) It may be best to have more than one person (one to watch) for the collisions if they are hard to hear. Listen carefully for collisions. Regardless of the number of collisions the question to consider is only did it or did it not collide. If yes, then increase the count of collisions (C) by one.
11) From the data, determine the calculated size of the marble (d) and compare it through percent error to the actual size. Repeat the experiment as needed to examine other results for cases of differing numbers of trials.

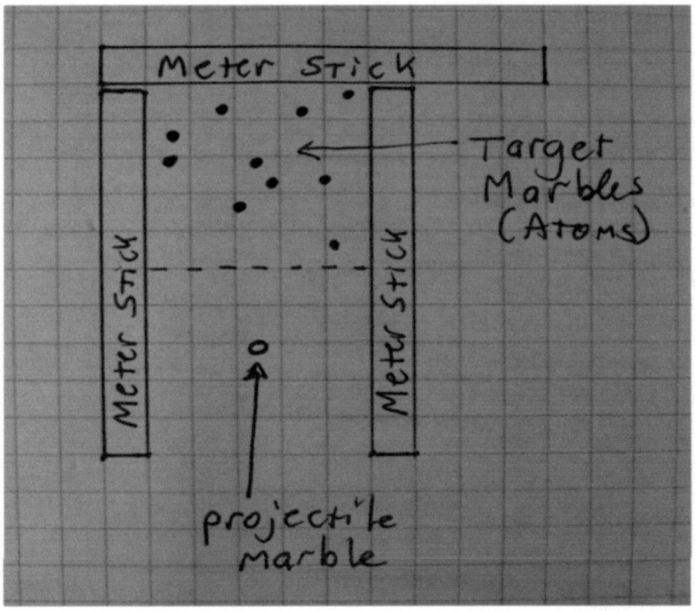

Data :

Width of Target Area [L] : ________ (cm)

Number of Trials [T] : ________

Number of Collisions [C] : ________

Number of Target Marbles used [N] : ________

Comparison Data :

Actual Diameter of Marble : _________ (cm)

Note : Both the Target Marble (Atom) and the Projectile Marble (Atom) are taken to be the same size, though the lab can be done with different size marbles!

Calculations :

Be sure to use your Slide Rule!

Diameter of Marble : $\mathbf{d = 2r}$

r = radius of target marble

R = radius of projectile colliding marble (note here using R=r)

Probability of Collision (as measured in experiment) : p

$$\mathbf{p = \frac{Collisons}{Total\ Number\ of\ Trials}} \qquad \mathbf{p = \frac{C}{T}}$$

$$\mathbf{p = \frac{Needed\ path\text{-}width\ for\ Collision}{Width\ of\ Target\ Area}}$$

$$\mathbf{p = \frac{2*diameter}{L} = \frac{2*d}{L}}$$

Note : The above equation is the probability of collision with 1 Target Marble

Theoretical Probability strictly based on area and marble numbers (no reference to the number of collisions which is the experimental value above)

$$\mathbf{p_{Th} = \frac{2*N*(R+r)}{L} =}$$

$$\mathbf{p_{Th} = \frac{4*r*N}{L}}$$

Note : In the above equation, the number of marbles that can be collided with is factored in too, N

Diameter of Target Marble Formula :

Note : The following is the derived equation for the diameter of the Target Marble (Atom) – combining our experimental and theoretical equations :

$$d = \frac{C*L}{2*N*T}$$

Percent Error Formula

$$\%E = \frac{[\text{Experimental Value-Accepted Value}]}{\text{Accepted Value}} * 100\%$$

Conclusion :

This experiment can be ran a number of times and in each increase the number of trials, much like the Pi Activity determined through Probability and dart throws – the more in the trial the better the results usually. Typically with more trials, there is a greater probability of determining a value for the size of the marble (atom) that is reasonably close to the actual value.

The measure of this experiment is the comparison of the calculated value for the marble's diameter to its actual value through the percent error formula.

Activity II : Determination of the diameter of a penny

Purpose : To use the experimental probability of a collision to determine indirectly the size of a coin from the number of collisions versus total attempts with a projectile (marker or art pencil).

Materials :

- Piece of regular paper,
- Caliper,
- Meter Stick,
- Ruler,
- 10-20 pennies (other coins, poker chips or washers),
- Pencil with rounded tip (art pencil is best, can use marker),
- White Nail Polish,
- Slide Rule

Set Up Procedure :

1) If using either washers or coins, paint one side white with nail polish, if using poker chips white is best
2) Measure the diameter (D) in centimeters of the disc you decide to use and record this value using the caliper

Procedure :

1) Place the blank piece of paper on the ground after measuring its length and width in centimeters. Record these values. (L,W)
2) Place the decided number of discs (coin, washer, chips) on the paper in a random pattern. Hereafter the term disc will be used for whichever one is chosen.
3) Us the meter stick to determine approximately one meter above the paper. This will be the consistent drop height of the pencil.
4) Hold the pencil point down at the noted height, close your eyes, move your hand with the pencil slightly so as not to keep track of its location and let the pencil go.
5) Check where it makes a mark on the paper or the disc. Every attempt is a trial (T) and every hit is recorded as an impact (R).
6) Do not move the marked disc. If possible wipe it with a damp paper towel to erase the mark (or use an eraser). Be careful to not let any of the discs move.
7) Note that if the paper is missed this is not counted as a trial.
8) When the number of trials is complete, calculate the area of the paper (A). (A = L x W)
9) From the data, determine the calculated area of the disc (Q).
10) Use the actual measurement of the disc to compute the percent error.
11) If interested, do the experiment more times with a different number of trials to see if this is effective. Other considerations can be done using other discs (other size coins or washers).

Data :

Paper : Length (L) : ________ cm

: Width (W) : ________ cm

Trials (T) : __________

Impacts (R) : _________

Comparative Data :

Diameter of Disc (Coin) : ________ cm

Calculations :

Be sure to use your Slide Rule!

$$\frac{\textbf{Impacts}}{\textbf{Trials}} = \frac{\textbf{Area of Penny}}{\textbf{Area of Paper}}$$

$$\frac{R}{T} = \frac{Q}{A}$$

Note : Your goal is to solve for Q, the area of the coin or washer

Conclusion :

From the experiment, find how close your calculated estimate for disc size comes to the actual size of the item used through percent error formula (see activity 1 for this activity for that formula).

Also test the idea of increasing the number of attempts. Do more trials affect the accuracy of your outcome?

Amazing how math and science blend when we try to indirectly measure something and try to find out about it!

Explore & Enjoy

Activity #21
Charles' Law exploration with a Balloon and Temperature

Volumetric Changes with Temperature (Charles' Law)

Ever watch a balloon on a cold day versus a warm day? What happens?! When colder the once rather robust filled balloon seems to have lost some of its internal gas – but it hasn't really. The molecules are just moving slower since they now have less energy to move about and bounce off the walls and each other when in an environment of less energy (ie lower temperatures) – so less pressure and the balloon seems to shrink some, yet taken back into a warmer room or environment and viola it is back to its regular self. Wow!

Temperature is by definition a measure of the average kinetic energy of the particles making up a material. Kinetic energy is energy due to motion, so it stands to reason that as temperature increases, so too does the average energy of the particles in a material. (There are many Activities that explore the idea of Energy later on in the book Galaxy Girl's Inquisitive Pioneer Adventures – Mechanical Energy (Activity # 27) is a good one to look at, as is the Efficiency Project (Activity # 52) plus many more)

If the substance is a gas, particles of a lower temperature will move more slowly and bounce off each other less actively. On average, they will remain closer together than in the case of a higher temperature gas, where the particles will have greater speed and motion.

What is generally found in nature is that as an object is heated, it expands and when cooled it contracts in volume, much like my opening question and observation to this Activity.

This behavior of gases was first described by Joseph Louis Gay-Lussac in 1802 but credits it to Jacques Charles work from the 1780s. Charles' Law is paraphrased thus :

When an ideal gas is at constant pressure, the volume of a known mass of this gas will increase or decrease by the same factor as its absolute-temperature-scale temperature is changed.

To illustrate this, say we take a given volume of gas at an initial given temperature and then double the temperature, what then happens to its volume? Since we doubled the temperature, the volume will double as well.

Charles' Law is also known as the Law of Volumes. As the temperature increases, the gas volume increases, and vice versa. Note that there are two critical ideas here minimally. First, this applies to an ideal gas, which is never the case, but many gases can exhibit this property to a certain degree. Also for all of these gases, the Pressure is held constant as well as the number of molecules under consideration. (As an aside, the relation of Pressure and Volume is Boyle's Law and when are all combined it is the Ideal Gas Law where $\frac{P*V}{T}$ is a constant)

Like other ideal gas laws, this law is seen as an extension of the Ideal Gas Law and its equation itself. This equation derives from the Kinetic Theory of Gases that basically presumes that gases are composed of point-like spherically-shaped atoms/molecules that occupy negligible volume, do not attract each other, and undergo perfectly elastic collisions (i.e. no loss of kinetic energy).

An interesting consequence of this law is that at absolute zero, 0 K, a gas would occupy zero volume. This is impossible, since all gases would turn into liquids at some given temperature point. It even is known from Quantum Theory that as the temperature approaches absolute zero the Uncertainty Principle states that the gas could not occupy zero volume.

The idea of the zero volume concept was being developed by Gay-Lussac, but first mentioned by Lord Kelvin (William Thomson) in 1848. His work defined the temperature in terms of the 2nd Law of Thermodynamics.

Our activity uses simple things like a balloon, a thermometer, a pair of tongs, and a measuring tape to investigate this idea. We will have a given volume of gas in a balloon and will cause its volume to change as we place it in environments of changing temperature. We will measure both the temperature and the volume of the balloon in each of these situations.

Note that one needs to be safe, so have adult supervision and permission in this activity. It is a good idea to wear goggles. Do not put one's hands in very hot or cold water.

Purpose : To measure changes in volume of atmospheric gas (mixture of nitrogen, oxygen, and carbon dioxide) with changes in temperature and compute the rate of change of volume with temperature.

Materials :

- Small balloons (as round as possible),
- Marker,
- timer or clock,
- thermometers (can be up to 3 lab quality -10°C to 110°C),
- flexible measuring tape,
- tongs (like for a salad),
- clamp to hold balloon in place,
- refrigerator (or cooler with ice),
- pot of cold temperature water from tap,
- pot of warm water as heated by heating source,
- goggles,
- Slide Rule

Procedure :

1) Be sure to follow all safety procedures. Have permission and supervision from an adult. Do not mishandle hot or cold water. For all measurements wear goggles and use the tongs to handle the balloon and efficiently (i.e. quickly) measure circumference when needed in the trials.
2) The directions given in the first few steps are the abbreviated version of the lab with only 3 data points for the graph. The steps after, note where other steps can be inserted if longer periods of time are available.
3) Inflate a balloon and let it sit on a table. Have a thermometer out to measure room temperature. Mark with a permanent marker a dot in 3 places around the balloon after wrapping firm but not squeezing with the flexible measuring tape. These points are used for each measurement for consistency.
4) After about 12 minutes (this will be treated as the time frame for each location of the balloon), record the room temperature and measure the circumference of the balloon.
5) For the second data point create a pot of cold water from the tap and place the balloon in it. To practice good technique, use the tongs to hold it 9 since this will be done with the hot water later). After 12 minutes, measure the water temperature and the circumference of the balloon.
6) Next place the balloon in the freezer of the refrigerator with a thermometer for 12 minutes. At that time take the same measurements of temperature and circumference of the balloon. – Note, if not using a freezer, put ice in a cooler and place the balloon in there.
7) For the hot water, first heat a pot of water to near boiling, and then turn it off and let it sit. It will remain hot for some time, so exercise caution near it. Use tongs to hold the balloon (since steam can burn). Keep your hands away from the steam.

You can even touch the balloon to the water surface. Use the time of 12 minutes, then remove the balloon.

8) Remove the balloon and measure the temperature of the water and the circumference of the balloon. (Be sure to wear protective eye-wear and be cautious with your hands as well).
9) Now with all of this needed measured data, At this point, the calculations can be done for radius, volume. Note, assume that the balloon is a perfect sphere.
10) Create a plot of Volume (Y-axis) versus Temperature (X-axis) and calculate the slope for the best fit line.
11) For a challenge compute the line of the graph and find the Temperature-axis (x-axis) intercept. See how close it is to -273° C, if using Celsius (note : it is best to use Kelvin, so convert all temperatures to this).
12) You can also use Charles' Law for each set of data (V & T) to see if their ratio is a constant value and how close they are to each other.
13) Note :
14) Though the Slide Rule is a recommended tool, all of the calculations can be done with a graphing scientific calculator or the use of a spreadsheet program. In these calculations you have to generate a table of data, graph it, and then find the slope and/or equation of the best fit line for the data. Other formula calculations can be done with these tools as well.

Data :

Trial	Location	Circumference (cm)	Radius (cm)	Calculated Volume (cm^3)	Temperature (°C)
1					
2					
3					
4					

Calculations :

Be sure to Use a Slide Rule for any and all calculations !

Formula for determining radius 'r' from circumference 'C' :

$$C = 2*\pi*r$$

Volume of balloon V :

$$V = \frac{4}{3}*\pi*r^3$$

$$N = \frac{\Delta V}{\Delta T}$$ (useful for slope and finding intercepts !)

$$\frac{V_1}{T_1} = \frac{V_2}{T_2} \text{ (Charles' Law)}$$

$$K = {}^{o}C + 273$$

For Fahrenheit to Celsius conversions :

$${}^{0}C = \frac{({}^{o}F-32)}{1.8}$$

Conclusion :

The first conclusion to consider is the conceptual behavior as expressed by the outcome in your numbers. How does volume change with temperature? Did our expectation in terms of behavior hold true? What happens when the temperature is changed and did the volume change in the same manner basically?

Depending on the calculations done, there are several ways to examine the outcome here. If you used Charles' Law, you can compare ratios to see how similar they are. With a graph, the Temperature-intercept can be found and examine how close it is to the expected value of -273°C.

Activity #22
Acceleration on an Incline calculation
Prelude : Position-Time Data Graphing

This next Activity begins with an essay that deals with the notion of a simple lab design using marbles, stopwatches, meter sticks and a stack of books to make tables of data with measurements of distance traveled and time elapsed so as to generate position-time graphs – one of the most useful considerations in both the science and math classroom today. Further I then discuss why the slide rule is a great tool to analyze this data as well.

Prelude to Acceleration Activities
Graphing Data and the Slide Rule :

Position-Time Data Analysis Considerations

One of the most important things about science is the study of how things change since that is the heart of science itself! So what better way than the study of motion. Being able to set up a basic experiment, take some essential measurements, and examine or analyze this data to uncover relations of time and distance covered whether it is constant speed or acceleration is essential in all of science, physics in particular, and often called the area of the study of motion or kinematics.

One of the most common data table-driven exercises examined in physics is masses undergoing constant acceleration and its' motion being presented on a position-time graph. The regular means of finding the displacement and time-squared relation is useful, but there may be fun ways to do this with a hint of yesteryear as well. Also, in the process, we become the data analyzers.

Instead, why not have a simplified lab using a measured incline, a marble, and a stopwatch to construct a personal data table of displacement and time values? All we need to do is construct tables of predetermined displacements and time them. From these basic materials and using a slide rule, students not only can graph the data, they can come to find the relation of displacement and time as well as the rate of acceleration for themselves without being told the answer. The strength of this exercise is that various angles of the incline can be used (hence different acceleration rates). One of many of the further questions to explore is just that, how does the angle of inclination affect the rate of acceleration.

What is best here is that the analysis method followed here will work for any and all data situations, regardless of the relation of the variables.

The data analysis tool here is a common 9-scale slide rule. Note that this slide rule is as powerful as a common scientific calculator today. In this activity and not even the full power of such a tool is needed. All that is essentially needed here are the C, D, and L scales to accomplish the task at hand.

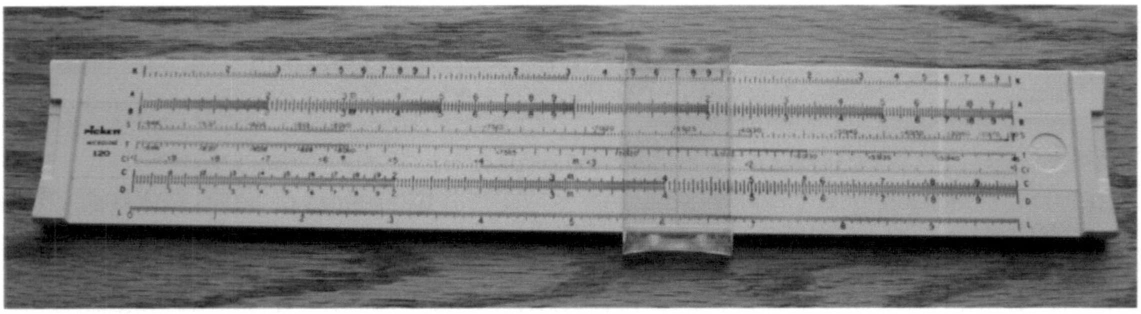

Why this tool? The **slide rule** is a tangible and visual bridge connecting numbers to the measured real world. It can act as a motivation for reasoning. To use it, one must first estimate answers, know what and why the measured values used are, their place values, and then how they are placed in relation to each other on the slide rule and how to find and read the answer. Hence, we can become the 'computer', once again.

The chief problem, then, is to find such a tool. The references at the end of the article note the International Slide Rule Museum web site, where there is a student-loaner program. For the cost of about $11 per semester, a teacher can be loaned a classroom set of slide rules. There is a power point on how to use a slide rule, along with history, and a way to have medals for slide rule competitions as well. Also in the references is a web site, Cosmic Quest Thinker, a place for many science and math activities using slide rules.

With this **incline activity** examining constant acceleration, we take the slide rule in hand, and then we can take the data tables (either given or generated by activity) and first graph them. Next, we take the log values of each of the data points and graph log (displacement) on the y-axis and log (time) on the x-axis. Draw a best fit line and determine slope. Given the precision of the tool and the values being used, they should find a slope of 2. This implies, from cross-multiplication, that displacement is proportional to time-squared ($d \sim t^2$). From *this, it is easy to conclude that the shape of the graph is, indeed, a parabola.*

With this knowledge in hand, we can then use the slide rule to square all of the time values. This can be done one the C & D scales or the A & B scales for further slide rule use exploration. Next we then graphs displacement (y-axis) vs. time-squared (x-axis) and again draws a best fit line, then determines the slope. The slope of this line will reveal 0.5*acceleration. Now the constant acceleration can be found for this case as well. They can then compare this slope value to calculated acceleration values for each of the displacements through the equation ($a = \frac{2*d}{t^2}$) and this can be compared to the theoretical ideal acceleration. Using a common marble, the 'ideal' or expected rate of acceleration is best approximated to be $a = \frac{5}{7}*g*\sin(\Theta)$. Note, one might ask why use the S scale for sine values? Sure you can but all that is needed is the ratio of height of the incline used to its hypotenuse length using the C & D scales which is the sine value. Most importantly, note that all the calculations needed here can be done with this math tool, the slide rule.

With this approach, our goal is to do the work and discover the outcome will be achieved here. We engage in the art and act of discovery through actually doing it. The students come to find the displacement-time relation. This way is just an alternative and a way to inspire a path to understanding. This is the primary goal of science – to not only construct a logical system to examine a measurable question, not only take measurements, but to analyze the data to find relations. With analysis, we now have a firm foundation to explain their conclusions, mentally 'see' the relations, and then explore further questions.

This approach does not exclude the presented data tables in texts (which are good follow-up reading to reinforce the idea) nor the use of other lab materials. *In fact, the calculator and the computer can act now as a follow-up to check the answers. We can find the equation for the line, graph the results, and compare these answers to their measurements and calculations done with the slide rule. They check their own work!*

What of the use of logarithms and the need to explore them? This can be done here or in math class, if the students are at that level for understanding. Otherwise, letting them know that logarithms are a tool to uncover such relations may be sufficient at this time. *Also note that this use of logarithms idea can be extended to any and all other variable relations they encounter in various science classes as well as math classes.*

This can be a cool math tool adventure. We can and will examine data for itself and use their minds to find the answers. We use tools that help them visualize the concepts and make finding answers a personal responsibility and journey. We can also be connected to history as they read about the slide rule. This is because the slide rule has a 350 year history (1620-1970). It has a history of being part of the making of the Panama Canal, the Empire State Building, the Golden Gate Bridge, along with development of the steam engine, the discovery of oxygen, and the determination of the density of the Earth. Both Einstein and von Braun used the same 9-scale model themselves – one from the realm of theoretical physics while the other in the practical realm of applied physics to rocket engineering where he built the Saturn V, the largest human-made device to leave the Earth carrying aloft astronauts to the Moon, each carrying a Pickett 600 slide rule.

Plus realize here in Galaxy Girl's Inquisitive Pioneer Adventures with Ms Briana Andromeda Purcell there is a 9 scale slide rule in paper form at the end of the book you can make for yourself and do any and all calculations in the book in these science and math adventures too. As well with this essay and the following exercise that opens the world of kinematics in Physics there are many references to help you in your learning the slide rule and its many uses – this book opens with chapters on how to use one and the first two Activities give step by step discussion in everyday uses of the slide rule and many of the Activities walk you through its use. Take time to learn it as well as the study of motion which is the beginning of not only kinematics, but also Newton's Laws, and the study of Energy – Kinetic & Potential and other ideas that follow.

Resources :

Slide Rule Loaner Program, Directions for Slide Rule Use, Make your own slide rule :
http://sliderulemuseum.com
Other Classroom Ideas for the use of the Slide Rule :
http://galaxygirl-brianaandromedapurcell.com/index.html

Activity #22 : Activity

Acceleration on an Inclined Plane and the Slide Rule Activity –

Galileo is famous for many things, like observations of the Moon and finding Jupiter's moons, as well as experiments, ideas, and devices, such as studying the motion of pendulums (my book here Galaxy Girl's Inquisitive Pioneer Adventures has a Pendulum lab set), creating a sector and a thermometer, and is most well-known for his studies of motion and the quantifying of the ideas we have come to call kinematics, the study of motion (besides this Activity which is Galileo inspired we have also in my book studies of the Acceleration due to Gravity as well). His primary work was with a long, smooth wooden ramp down which he rolled spherical orbs and timed them with water-drop clocks. These studies lead to conclusions about inertia, friction, acceleration due to gravity, and laid the foundations for conservation of mechanical energy (since he noticed that two inclines facing one another resulted in an orb only returning to its original starting height, despite the angle of each of the inclines).

In this Incline Motion Activity, we use a marble rolling down a enumerated scale (two parallel meter sticks) at a pre-determined angle with known distance and it is timed. From this, acceleration is determined from basic kinematic formulas of motion. In essence we are re-performing Galileo's experiments and trying to uncover his findings concerning acceleration on an incline for ourselves. Wow we are parallel to Galileo's sort of experimental design and study!

His findings were that the ball would roll a distance proportional to the square of the time traveled by the rolling sphere. Basically for each moment in time – one click of the clock – the distance covered would be the square of the prior distance covered (for example say in one unit of time 1 unit of distance is covered, so in the second unit of time, 4 units of distance would be covered in total (2x2 or 2^2). The acceleration rate would depend on the angle, where the steeper the angle the greater the acceleration rate. Galileo was trying to come close to approximating the acceleration due to gravity, since in his time accurate measures of parts of time where technologically impossible and the acceleration rate due to gravity for the Earth is rather large (980 cm/s^2).

Note Galileo correctly assumed that the horizontal and vertical motions were and are independent of each other. So his use of the ramp did not directly affect masses in their inclined path. He recognized that acceleration occurs on a straight-line path between the centers of mass of the objects, the item and the Earth here. He sees the ramp as a dilution of the Earth's acceleration and he needs this dilution since in these times the ability to measure small increments of time and distance are far too hard to do.

From our Activity, Though we can measure the outcome of acceleration we have witnessed, the question remains, however, what value of acceleration do the laws of physics predict. Then from this, we can ask the question, why are they different?

Which laws make such predictions of acceleration? These came not only after Galileo but even after Newton as well, but extend from both of them. These are found in the Laws of the Conservation of Energy. How and Why is this? The energy here of the marble comes only from gravitational potential energy. When at its highest point (the start of its motion) is the place it has all of the energy that it can have through the entire trip.

There is even more than potential energy and basic kinematic kinetic here, though. There is also kinetic energy of angular motion of the marble that must be taken into consideration. Any rotating object has its own formula based on the distribution of mass, for a sphere it is :

Moment Inertia for a Sphere $I = \frac{2}{5} * M * R^2$. Instead of straightforward velocity, angular velocity is considered here, but there is a connection, where : $\omega = \frac{v}{R}$.

All of this results in this derivation and final conclusion :

$$P.E._{grav} = K.E._{linear} + K.E._{rotational}$$

$$m*g*h = 0.5*m*v^2 + 0.5*I*w^2$$

$$2*m*g*d*\sin\Theta = m*v^2 + (\tfrac{2}{5}*m*R^2)(\tfrac{v}{R})^2$$

$$v^2 = 2*d*(\tfrac{5}{7}*g*\sin\Theta)$$

$$a= \tfrac{5}{7}*g*\sin\Theta$$

This formula will be considered the ideal acceleration and will be used for comparison purposes to our determination of acceleration of the marble on the incline from the data of distance and time measures. For further ideas on this investigation see the preceding essay on "Position-Time Data Analysis Considerations".

Here we retrace the path of Galileo in trying to exclude all that does not matter from an experiment and focus and the cause and effect – to see how motion changes with time, to see how elevation effects motion and the like.

Studying and understanding acceleration is essential in science since acceleration only occurs due to a net Force as Newton later tells us in his Laws of Motion.

Explore and Enjoy!
Be Brilliant! :)

Purpose : To calculate the acceleration from the graphical investigation of the relationship between distance and time for a marble rolling down an incline.

Purpose : To compare measured acceleration rates for a marble rolling down an incline as compared to the predicted outcome from mathematical analysis using Newtonian Mechanics and Conservation of Energy.

Extended Activity : If you try different angles

Purpose : To determine the relation of acceleration of a marble down an incline with respect to the angle of the incline.

Materials :

- Timer,
- Marble,
- 2 meter sticks,
- stack of Books,
- Protractor,
- Ruler,
- Measuring Tape,
- Slide Rule

Procedure :

1) As with all Activities be sure to test and retest your set up and watch how it operates – be used to it. Anticipate and understand when things don't work well and redo things as needed. I always have to as well.
2) Use a Table where the meter sticks can act as a track for a marble to roll on between the 2 elevated meter sticks which sit atop a stack of books.
3) For this Activity choose a height that approximates about a 30° angle or less of the meter sticks-track to the table.
4) Be sure to test the track with the marble so that there are no problems. Choose the edge of the meter stick that appears smoothest. Keep separation distance consistent. Be able to read the numbers on the meter stick. It is best to place '0' at the bottom of the track.
5) For a given distance in the Activity always do 1 or 2 trial runs and practice with the timer.
6) There are two ways for the end of the run for the marble. Either place a book there that it hits (be sure to note that each distance will be about 1 cm off as it does not fall off the track) OR let the marble roll off the track and the sound of it hitting the table acts as the stopping moment.
7) For each run, have 3 timed runs and average this time for the second data table. This value will be used for determination of acceleration.
8) In this Activity start at either the smallest or largest distance and then continue through the values in the trials.

9) Find all formulae for calculations in the calculations section.
10) The first calculation is the angle of your track. Measure the total length of the track as well as the height of the track at the topmost point. Record these two values and use them as a ratio for the sine of the angle to be used.
11) For the calculations compute the measured acceleration from the data using only distance (d) and time (t).
12) Next compute the theoretical acceleration and use this value for comparison to the measured acceleration. (Here is where the sine of the angle is used)
13) Graph distance versus time (though we have used distance as the independent variable, it is traditionally placed on the 'y' axis). A best fit line should approximate a parabola. Ideally it is a parabola.
14) To see if it is a parabola, next compute the log values for both the distance and the time.
15) Then graph log(distance) and log(time) and draw a best fit line through the origin and determine the slope (it should be 2 ideally).
16) Compute the percent error (assuming the theoretical acceleration is correct or accepted value and the measures value is experimental value).
17) Now graph distance vs. time-squared. (first square time, remember)
18) Draw a best fit line and find the slope. This slope is half the acceleration rate. Multiply the slope by two and find the acceleration rate and compare it to the previously calculated value.
19) Note : For other comparisons consider trying other heights and following all of the steps. What pattern do you expect for acceleration with increasing angle of inclination?
20) Note :
21) Though the Slide Rule is a recommended tool, all of the calculations can be done with a graphing scientific calculator or the use of a spreadsheet program. In these calculations you have to generate a table of data, graph it, and then find the slope and/or equation of the best fit line for the data. Other formula calculations can be done with these tools as well.

Data : Acceleration with distance at a given angle

Height of track : ____________ (cm)
Total Length of track : _______ (cm)

Trial	d (cm)	t_1 (s)	t_2 (s)	t_3 (s)
1	100			
2	80			
3	75			
4	60			
5	50			
6	40			

Trial	d (cm)	t_{ave} (s)
1	100	
2	80	
3	75	
4	60	
5	50	
6	40	

Calculations :

Be sure to use your Slide Rule!

Angle Determination and sine value for theoretical acceleration :

$$\sin\Theta = \frac{\text{Height of Track}}{\text{Length of Track}}$$

Average Time :

$$t_{ave} = \frac{\Sigma t}{\text{Number of Trials}}$$

Measured Acceleration :

$$a = \frac{2*d}{t^2}$$

Theoretical Acceleration :

$$a = \frac{5}{7}*g*\sin(\Theta)$$

let g= 980 cm/s/s

Slope :

$$m = \frac{\Delta y}{\Delta x}$$

Percent Error :

$$\%E = (100\%)*\frac{[\text{accepted value-experimental value}]}{\text{accepted value}}$$

Conclusion :

Realize we were engaged in modern parallel ideas that Galileo explored. Can you generalize the behavior of the marble and its motion on the incline? As the height of the marble at its starting point was increased what happened to its overall speed for example. Observation and description are the beginnings of real science. Next is the measurement and looking at the data too.

The goal here was to find that the acceleration rate is constant. Next one should find that the theoretical and measured values should be reasonably close (factors such as friction and other errors though may be present). One should find that in time, the speed increases at a steady rate with time, while the distance travelled is related to the time-squared.

Examine your data to see how the rates of acceleration at a given distance traveled for a given angle turned out. They should be ideally identical. What were the results of the graphs, did they match the expectations noted in the description? (Parabolic for d vs. t and linear with slope of 2 for the log-log graph). If you were to graph d vs. t^2, what sort of line would you get and what is the slope of this line?

Other Activities :

Another graphing exercise can be final speed vs. time. Final speed is found to be 2 times the distance divided by time. The slope of the best fit line here would be our acceleration rate for the given incline. Also this is a good illustration of what one expects for a system undergoing constant acceleration (it will be a line of constant slope).

In this Activity, we only tried distance at a given angle. First try the directions at different angles (one small, one large) for comparison.

Try other sized marbles to see how the results compare. Look at the derived equations – the marble radius cancels out, so all solid spheres of uniform material and density should behave this way. Also try to place inclines in opposition to your initial incline like Galileo did and determine the height to which the marble rises to on it.

Activity #23
Acceleration due to Gravity determination

Grade Level : High School
Math Level : Challenging

Acceleration due to Gravity calculated with a Slide Rule Activity

The acceleration due to gravity is considered one of the most important ideas written about, discussed, and mathematically analyzed in the history of science and many of us know the chief architect of this notion as Galileo. What is first important to realize is that he did not discover that objects fall faster once they are let go, this has been a known fact by everyone before the time of writing.

Galileo concluded that independent of mass, all objects fall at the same rate in the absence of air resistance. He correctly determines that air resistance basically acts like friction, the opposing force of motion for two solids moving against each other. The value of it is approximately 9.8 m/s^2 or 32 ft/s^2. Here we will use two methods to see how close to this value we come.

Many of us have seen the classic illustration of this in a classroom. A teacher holds up a piece of paper and a textbook which is many sheets of paper bound to each other. When let go, clearly the book falls much faster than the single sheet of paper. What then happens if the shape of the paper is changed and not its mass? Crumple up the paper and now let the two go. They fall at the same rate!

Galileo is purported to have dropped two different masses from the top of the leaning tower of Pisa in Italy. There are only small reports of this. Also from one report, the objects did not hit the ground at the same time, there was a small difference between them. This would seem, at first, to contradict even Galileo's conclusion. Keep in mind the presence of air resistance, whether or not both objects were indeed let go simultaneously and exactly at the same height. More importantly, however, is the fact that the masses were markedly different, yet the time between them was incredibly small. This stood in sharp contrast to older thoughts on the matter and heavily favored Galileo's idea that ran in the opposite direction to them.

More information is contained in the Acceleration on an Incline Activity, another of Galileo's great experiments. There he shows the independence of horizontal and vertical motions. As noted there, Acceleration is a central concept in all of Physics and is central to both Kinematics and Newton's Laws, which are the foundations of all of Newtonian Physics.

This Activity has a number of connections to other Activities, such as Newton's 2nd Law Activity, where the Atwood system is reconfigured to a horizontal-vertical system from the double-vertical system as employed here. There is also a separate Pendulum Activity (another study done by Galileo by the way) studying the question as to what effects a pendulum's period. In several other of our Activities, such as the Efficiency Project (#52) and Mechanical Energy (#27) ones we must rely upon the Acceleration due to Gravity in doing the math analysis of our work there. This is the heart of science right there in these ideas – where ideas and conclusions reached in one experiment are found to be valid and reliable – this then becomes a tool that will be employed elsewhere.

A final point is this – the Acceleration due to Gravity as noted here is an 'accepted value' to 2 significant digits, but does depend on some other variables, such as latitude, the concentration of mass in the Earth beneath the area in question, and to a much lesser extent, the altitude.

From Newton's Universal Law of Gravity and from Newton's general expression of Force in his 2nd Law we can find Acceleration due to Gravity :

$$F = \frac{G*m*m_e}{d^2}$$

$$F = m*a = m*g$$

Set these equal to each other and cancel out where needed to find :

$$g = \frac{G*m_e}{d^2}$$

'g' is the Acceleration due to Gravity, 'm_e' is the Mass of the Earth itself (excluding the object in question, technically), and 'd' is the distance between the centers of mass of the Earth and the object in question – squared. What this means is what Galileo determined before Newton : All objects, regardless of their mass, fall to the Earth at the same Rate of Acceleration when dropped from the same height. We know that differences are due to other forces acting on the objects, such as air resistance.

An important conclusion comes from this – the Acceleration due to Gravity mostly depends on the main body in question (here the Earth) and the object's distance from its center (more or less). Think on this when considering other objects, such as the Moon, Mars, and other worlds. They have different masses and different radii, hence the Acceleration due to Gravity for each of them will be different.

The Activity here is to determine the acceleration due to gravity for the Earth. Unlike Galileo, one of the key tools we will have is a timer. This is the main reason it was difficult for Galileo to even study this phenomena directly. The experiments will try 3 different approaches to this method, where one employs pendulums of varying length (since it turns out length is the key factor for the period of a pendulum) and another employs the use of net forces so as to decrease the falling rate by using a falling mass attached to a mass that rises (much like an elevator) in a device called an Atwood machine. Using Newton's Laws we uncover the acceleration due to gravity here, while the 3rd merely uses a bouncing super ball. Note that there can be variations to the materials noted – for example – it is noted to use round rods – either wooden or knitting needles in the Atwood Machine approach – but perhaps you want to try small pulleys as they might be available. Always consider what may work best, but always employ safety in the approach.

So set forth and be the Inquisitive Pioneer and join Galileo in examining the Acceleration due to Gravity for the Earth itself! Explore & Enjoy! Be Brilliant!

Activity Goal : Determination of the Acceleration due to Gravity by --

Activity 1: The Atwood Machine Method

Purpose : To determine the acceleration due to gravity using an Atwood Machine System so that in an elevator fashion masses are falling and rising so as to affect the rate of acceleration.

Activity 2: The Pendulum Method

Purpose : To determine the acceleration due to gravity by using a Pendulum of known length and measured period.

Activity 3 : The Bouncing Superball Method

Purpose : To determine the acceleration due to gravity from the distance and time measures of a falling and bouncing mass.

Materials (for all Activities) :

- Meter Stick,
- Timer,
-
- **Activities 1 & 2 :**
- Mass Scale (electronic kitchen one is good),
- String, Dental Floss or Fishing Line,
- Masses (2 Large Eye Hook Bolts with Multiple Nuts),
-
- **Activity 1 (Atwood) only :**
- Knitting Needles or Round Wooden Dowel Rods (2),
- Metal Sewing Machine Spools (2) (not necessary),
- 2 Chairs (wood or plastic with level backs),
-
- **Activity 2 (pendulum) only :**
- Protractor,
- Wooden square dowel rod (1 m length),
- Small Eye Hook,
-
- **Activity 3 (superball) :**
- Superball,
-
- **For All Activities :**
- Slide Rule

Process Set Up before Procedure :

2) With any and all Activities read through, then set up the process for the Activity and run it without measurement and watch where you may have bumps in the road so as to anticipate these. Be patient – Repeat processes. Once all is well when measuring see to the values – do you notice patterns or trends – what do you predict will happen next as you change one thing or another from one trial in an experiment to the next. Always Explore & Enjoy. Be Brilliant!
3) Each Activity has its own set of materials – choose the activity(ies) that best suit your materials and time.
4) Each Activity uses a measure of distance and time, so each Activity employs the meter stick and the timer.
5) In most cases, set up the activity and perform it without recording the data initially to understand the workings. Take sample data on one or two trial runs and then start the regular measurements to be used.

Procedure :

Activity 1 : The Atwood Machine Method

1) Use 2 chairs with level backs and have them face each other.
2) Place across the backs two rods (can be knitting needles with metal sewing spools on them to act as pulleys).
3) Use 2 eye hook bolts and place no nuts on one while the other can have 1,2, or 3 depending on the speed of falling desired.
4) Use a scale to measure the mass of each mass (m_1 & m_2) and record it.
5) Note m_1 is the falling mass and m_2 is the rising mass.
6) Cut a piece of string (or dental floss or fishing line) [note: the lower the mass the better] so that it goes from the floor over both rods and extends no further than 1/3 of the way back down to the floor over the second rod.
7) Attach the masses to the ends of the string.
8) Have someone or something hold the more massive piece at the floor and measure the distance to the bottom of the less massive piece in the air. Record this distance (d).
9) Test the apparatus by having the larger mass in the air and the smaller mass start at the ground. Let it go and measure how much time it takes for the in-air mass to reach the ground. If too fast, make the masses more similar.
10) With testing done, now measure for each trial the amount of time (t) it takes for the elevated mass to reach the ground. Be sure to have recorded the masses used and the distance traveled.
11) Perform this at least 3 times for a given mass and perhaps for different mass combinations.
12) For a given mass system, this is a set of data. For each set, conduct 3 trials and determine the average amount of time. Be sure to use the same distance (d) as the starting height.
13) Note in measurements to convert centimeters to meters. Measure time in seconds.
14) From these sets of data, separately determine the calculated acceleration due to gravity and compare by percent error calculation how close it comes to the accepted value of 9.8 m/s^2.

Activity 2 : The Pendulum Method

1) Cut a piece of string approximately 60 cm in length and attach it to a firm rod (thick square dowel rod or metal rod) to wind it to a length no more than 20-30 cm long.
2) At the end of the string attach an eyehook bolt and connect 2 or 3 nuts to it.
3) Place the rod on the backs of chairs with level backs facing each other. Find a way to secure it so that the rod does not move or vibrate significantly when the pendulum is in operation.
4) Measure and record the length of the pendulum with the mass from the rod to the end of the mass (L in meters).
5) Use the protractor to determine the maximum displacement to be used in all trials (no more than 15°). Look down on the rod when drawing back the pendulum and find or make a mark with tape so that each trial begins at the same position.
6) Perform a practice run for timing and testing the operation of the pendulum.

7) For each trial have the length recorded for the pendulum. Let the pendulum have the same displacement and swing 10 times.
8) Record the amount of time (t) for the 10 oscillations.
9) Determine the Period from this (T).
10) For each subsequent trial unroll the wound string about the rod for a longer pendulum length. Have at least 4 data points (4 lengths) that are treated independently.
11) Calculations :
12) Your will construct 2 graphs for each set of data you have measured.
13) From each set of the data, now construct a table of log(T) and log(L) values by using the L scale of your slide rule.
14) Graph on the y-axis the log(T) values and on the x-axis log(L) values. Draw a best fit line and determine slope. If the period and length are done well in your activity, then the slope should be 2.
15) A further test of the period-squared to length relation can be found by graphing the length of Pendulum (L) on the x-axis versus the Period-squared (T^2) on the y-axis. Again, Draw a best fit line and determine slope.
16) Multiply the slope by $4\pi^2$ to compare to the accepted value of acceleration due to gravity (9.8 m/s^2)
17) Further calculations include for each trial, use the Pendulum Period Equation in the table of Calculations to find the calculated acceleration due to gravity for comparison directly.
18) For each comparison use the percent error formula, where g = 9.8 m/s^2 is the accepted value.

Activity 3 : The Bouncing Superball Method

1) Either on a Table or on level hard ground against a wall, first tape a meter stick with 0 (zero) at the bottom and 100 at the top.
2) The ball will bounce in front of the meter stick so you need to be able to watch both.
3) Use a superball and drop it from 100 cm mark and try to determine to what height it bounces (typical values from 40 cm to 80 cm)
4) This may take some practice – so watch carefully. You should be able to estimate to the nearest centimeter in height (s).
5) Note : If you can use a measuring tape, starting distances of greater than 1 m can help in that you will have greater time to start and stop the timer due to the longer time the ball is in the air.
6) At the same time, when the ball first reaches the ground, start the timer and stop it when it returns to the ground.
7) In essence you are measuring 2 things :
8) The maximum height achieved by the bouncing superball and
9) The amount of time it takes for the ball to go from the ground to its maximum height and return once again to the table (or ground)
10) After enough practice, for each trial write down the height to which the ball returns and the total time (t) for the trip of the first bounce to the second impact on the table (or ground).
11) Use the data in the kinematics formulae to determine the acceleration due to gravity for each trial. Perform this no less than 5 times.

12) Use your data values for time and maximum height to Determine the average of these values (t_{ave}) and the average maximum height (s_{ave}).
13) Now be sure to use 1/2 the average time for your time – the ball took half the time to go up and half the time to return to the ground.
14) With the calculated average time and average distance traveled values determine the experimental Acceleration due to Gravity value using your Slide Rule.
15) Calculate percent error from calculated values for acceleration due to gravity (the accepted value is 9.8 m/s/s).

Data :

Activity 1 : The Atwood Machine Method

Trial	m_1 (g)	m_2 (g)	d (cm)	t (s)

Activity 2 : The Pendulum Method

Trial	Length [L](cm)	Time for 10 oscillations (s)	Period [T](s)

Activity 3 : The Bouncing Superball Method

Trial	s (cm)	t (s)	½ t (s)

Calculations :

(g = acceleration due to gravity, a = acceleration)

For any and all Activities, be sure to use your Slide Rule!

Average :

$$X_{ave} = \frac{\text{Total of all trial values}}{\text{Number of Trials}}$$

$$X_{ave} = \frac{\Sigma x_i}{n}$$

Activity 1 :

$$a = \frac{2*d}{t^2}$$

$$g = \frac{(m_1+m_2)}{(m_1-m_2)}*a$$

$$g = a*(m_1 + m_2) / (m_1 - m_2)$$

The above equation is derived from :

$$\Sigma F = m*a_{net}$$

(Note : the ascending mass is m_2, the descending mass is m_1)

Activity 2 :

Period-Time Equation :

$$T = \frac{t}{10}$$

Pendulum Period Equation to derive from (for those interested):

$$T = 2*\pi*(L/g)^{1/2}$$

Acceleration due to Gravity equation from the Pendulum Period Equation :

$$g = \frac{4*\pi^2*L}{T^2}$$

Activity 3 :

Original Equation to derive (if desired)

$d = d_o + v_i*t + 0.5*g*t^2$
(d_o and v_i = 0)

Equation to Use for Activity 3 :

$$g = \frac{2*d}{t^2}$$

Note : the time used is half the time since it took this amount to travel the measured distance from the ground to the maximum height!

For All Activities, for comparative purposes let the Acceleration due to Gravity for the Earth be 9.8 m/s^2

$$\%E = \frac{[\text{Experimental Value-Accepted Value}]}{\text{Accepted Value}} * 100\%$$

Conclusion :

Examine your data and compare your results to the anticipated outcome.
Look for possible sources of error in the procedure.
If you tried other procedures, how do they compare? Which resulted in the most accurate results?
Amazing it is only about measuring distance and time when you think about it.

Activity #24
Graphical Analysis of a Bouncing Ball

Grade Level : High School
Math Level : Challenging

Graphical Analysis of a Bouncing Ball Activity –

The first thought about a bouncing ball might be : how much can one do with it?

Here in this awesome Activity we have to put our thinking caps on and will look with great mathematical, yet simple models, detail at the wonders of a superball and it bouncing before us as we measure distance and at most time yet will find all sorts of questions we can explore and answer with this childhood toy perhaps still in a junk drawer somewhere.

When it comes to solids, there are a number of factors one could characterize them by : such as density, melting point, and the like, but there is also the physical property of elasticity. Elasticity is the tendency of a material to return to its original shape when stressed through some sort of deformation or bending. The classic example is the bow in archery. One draws upon the bowstring, pulling it back, which in turn, bends the bow itself. This bending stores elastic potential energy (see Elastic Potential Energy Activity), that when the string is released turns into the kinetic energy of the bowstring which is transferred to the arrow putting it into flight.

When it comes to a ball made of rubber, such as a basketball – tennis ball – super ball, the property of elasticity is very important to its functioning. The same is true for shock absorbers in vehicles.

What is happening when a ball bounces. A ball lifted from the surface of the Earth has increased gravitational Potential Energy. Once released, due to the attraction of masses towards each other, both the ball and the Earth move towards each other. (Not to worry, the Earth moves an immeasurable amount hence it is the ball's distance we measure to the Earth when it falls). From Newton's 2nd Law, since a force is acting on the mass of the ball, it is accelerated. This is our measure of the acceleration due to gravity. In the process of falling potential energy is being turned into kinetic energy, the energy of motion. We know it must be increasing, not only since the distance between the Earth and the ball is decreasing (hence a decrease in gravitational potential energy since it is due to height) but there is a change in velocity (the acceleration due to gravity) hence an increase in kinetic energy. At the midway point in the fall, both the potential energy and the kinetic energy values are the same, by the way. Just as it strikes the Earth, all of the potential energy has been turned into kinetic energy. Since energy cannot be created nor destroyed, it goes somewhere. In striking the Earth, the ball deforms, which in turn increases its Elastic Potential Energy. Depending on how elastic the material is, a small to large fraction of the kinetic energy can turn into this. The amount that does so is illustrated when the object returns to its original shape and its bounces in the opposite direction from its original drop (assuming no redirection due to shape or obstacles). The ball will not return to its original height, because no matter how elastic the material, some of the kinetic energy will go into sound, heat, and snapping of some of the molecular bonds in the ball.

The opening statement about this Activity is simply this – We are going to bounce a ball of one type or another, record some data on its starting and bounce back height and the amount of time between bounces and with such scant information, we are going to be able to : 1) Compare the 'elastic response' or 'bounciness' of any two balls in question, 2) Determine a measure for the Acceleration due to Gravity of the Earth, 3) Examine the data to test if it has an Inverse-square law relation between Acceleration and Time, and 4) Construct and Mathematically determine a Graph that resembles Radioactive Decay. Some of the other mathematical ideas explored are : Average & Slope, Graphing Points, and using Logs as well.

That being said, there are a number of concepts that are being addressed in this Activity, that are discussed at great length in other Activities. For example, the reason a ball does not return to its original height once let go is examined and noted in the Mechanical Energy Activity where gravitational Potential Energy, Kinetic Energy, and most importantly, the concept of Conservation of Energy is considered. Also recognize the simple fact that elasticity is addressed in the Elastic Potential Energy Activity. For more ideas on the Acceleration due to Gravity, see that very Activity of the same name to look at Galileo and some other ways to measure such a value. Also we explore an Inverse-square relation here as well, as can be found in the Inverse Square Law of Light Activity. When it comes to negative exponential graphs, such as for radioactive decay, the idea of these types of functions is addressed in the Rate of Cooling Activity.

So find a super ball, a tennis ball, and/or a ping pong ball and have some fun.

Enjoy & Explore
Be Brilliant

Overall Activity Goal : Examine a Bouncing Ball through Data Collection and Graphical & Mathematical Analysis

Activity : Comparison of Bounce
Purpose : To determine the 'elastic response' (the relationship of the height of drop to the height of return after the first bounce) of different types of balls that can bounce

Activity : The Bouncing Ball Method to determine Acceleration due to Gravity
Purpose : To determine the acceleration due to gravity from the distance and time measures of a falling and bouncing mass.
Note : This is done two different ways

Activity : Bouncing Ball to Approximate Radioactive Decay
Purpose : Using data from a bouncing ball in a series of bounces, measuring both maximum height and time, to generate and mathematically formulate an negatively exponential graph that approximates the phenomena of radioactive decay

Materials :

- All Activities Require These Items :
-
- It is best to have a selection of Ball Types to use :
- Ball choices include : ping pong ball, tennis ball, super ball, golf ball,
- Meter Stick,
- Graph Paper,
- Slide Rule
-
- Activity 1 : Comparison of Bounce
- (See list for all),
- 2 or more choices from ball choices,
- Ruler,
-
- Activity 2 : Acceleration due to Gravity (basic approach) :
- (See list for all),
- Need only 1 ball selection,
- Timer,
-
- Activity 3 : Acceleration due to Gravity (advanced approach) :
- (See list for all),
- Need only 1 ball selection,
- Timer,
-
- Activity 4 : Bouncing Ball Graphical Analysis (Decay Model) :
- (See list for all),
- Need only 1 ball selection

Procedure :

For All Activities :

1. With any and all Activities read through, then set up the process for the Activity and run it without measurement and watch where you may have bumps in the road so as to anticipate these. Be patient – Repeat processes. Once all is well when measuring see to the values – do you notice patterns or trends – what do you predict will happen next as you change one thing or another from one trial in an experiment to the next. Always Explore & Enjoy. Be Brilliant!
2. Either on a Table or on level hard ground against a wall, first tape a meter stick with 0 (zero) at the bottom and 100 at the top. This is done for all of the Activities.

Activity 1 : Comparison of Bounce

1) See Step 1 for all activities, which means choosing a surface to bounce on and the set up for measurements.
2) Choose at least two different ball types for this Activity. (See Conclusion for Alternatives, such as examining different surfaces or temperature for further ideas).
3) It is best to test the balls you have chosen. Hold each at 100 cm mark and note where they bounce back to (most are between 40 cm and 80 cm). This will give you an idea where to look as straight on as possible to determine the height of bounce in each of the trials.
4) Note that it is best to always look to the same part of a ball to use as the mark to determine height (the bottom or the top). The bottom is recommended.
5) To help the process, you might consider having someone help by holding the ball consistently at the same height for each of the trials and let it go, while you look to determine its bounce back height.
6) Choose the starting heights (H) you plan to use (which means the number of trials as well). A good idea is 100 cm, 70 cm, 50 cm, 40 cm for example. You can increase this set and make it every 10 cm from 40 cm to 100 cm, or go by 20 cm increments.
7) For each Starting Drop Heights (H) you are to conduct 3 trials for each ball. So for example, at 100 cm, the first ball is dropped 3 times and each of these has its Bounce Back Height recorded (b_1, b_2, b_3).
8) For each given set of Bounce Back Heights (b_1, b_2, b_3) for a given height for a given ball the average is calculated (b_{ave}). This value is used in the graphing process.
9) Each ball is dropped in the same process in steps 7 & 8 and the information is recorded.

10) For each ball, there is now a set of data points (starting height, average bounce back height). Each of these pairs for a given ball is plotted on a graph of Bounce Back Height on the y-axis and the Drop Height on the x-axis.
11) For each set of data, draw a best fit line through the points. This means that there are as many lines as there are balls tested.
12) For each line determine the slope of the line (m).
13) Compare the results the balls used. Consider what do the slopes represent? The larger the slope, what does this indicate as compared to one's of smaller values?

Activity 2 : Acceleration due to Gravity via The Bouncing Ball Method

1) See step 1 for all activities and set up the meter stick for the observations. The ball will bounce in front of the meter stick so you need to be able to watch both.
2) Use a super ball (best choice) and drop it from 100 cm mark and try to determine to what height it bounces (for most the typical values from 40 cm to 80 cm).
3) This may take some practice – so watch carefully. You should be able to estimate to the nearest centimeter in height (s).
4) At the same time, when the ball first reaches the ground, start the timer and stop it when it returns to the ground.
5) In essence you are measuring 2 things :
6) The maximum height achieved by the bouncing super ball (s) and
7) The amount of time it takes for the ball to go from the ground to its maximum height and return once again to the table (or ground)
8) After enough practice, for each trial write down the height to which the ball returns and the total time (t) for the trip of the first bounce to the second impact on the table (or ground).
9) Use the data in the kinematics formulae to determine the acceleration due to gravity for each trial. Perform this no less than 5 times.
10) The first calculation to do is to take half of the time value. Realize that the ball started from the ground, reached its maximum and measured height, and then returned to the Earth – it spent half the time going up and half the time going down, so we need half of the time value.
11) Use your data values for time and maximum height to Determine the average of these values (t_{ave}) and the average maximum height (S_{ave}). Recall that time is now half the time it originally was.
12) With the average time and average distance traveled values determine the experimental Acceleration due to Gravity value using your Slide Rule. The formula is in the Calculation table.
13) Note that the number of values for the Acceleration due to Gravity will be the number of Trials you have. You can average these amounts and then determine percent error from this.
14) Calculate Percent Error from the average of the calculated values for Acceleration due to Gravity (the accepted value is 9.8 m/s/s).

Activity 3 : Acceleration due to Gravity – Advanced Approach to the Data

1) This next activity has some of the same properties of Activity 2, but employs the techniques of Activity 1 in data gathering.
2) Look to the Data section below and note which values are needed. Realize that the more points of data you have the better for the process.
3) As in Activity 1, you need to have a variety of starting drop heights. It is best to have a sequence, say for example : 100 cm, 90 cm, 80 cm... This should be done for no less than 4 starting drop heights. This is denoted as 'D' in the table under Trial.
4) It is best to have a decreasing table of heights from 100 cm to the smallest one used and have their corresponding times (which should also decrease).
5) For each trial (a given starting drop height), drop the ball and let it bounce. Once it strikes the ground on the first bounce start the timer. When it rises and falls to strike a second time then stop the timer and record this value (t).
6) In the process of timing the bounce, also look straight on at the ball and note the bounce height to which it rises (s) and record it.
7) For a given drop height (D) do the prior two steps 3 times and take the average of the bounce height (s_{ave}). Also take the average of the time as well (t_{ave}).
8) Note you will have to take one-half of the time value, since we only want the amount of time it takes to fall to the ground from the maximum height to which it bounced to in the first place, so divide the average time value in half and record this value (T).
9) The first two tables are merely the data, the 3rd table, the averages of bounce height (s_{ave}) and half the average time (T) will be the data used for most calculations.
10) Also as noted in the Data section, you need to construct a $\log(s_{ave})$ and $\log(T)$ table.
11) Graph the $\log(s_{ave})$ on the y-axis and the $\log(T)$ on the x-axis. Draw a best fit line and determine slope. It should be 2 ideally.
12) To test the squared relationship (as noted in the prior step) now construct a table of average bounce height (s_{ave}) and average time to fall squared (T^2).
13) Graph the average bounce height (s_{ave}) on the y-axis and the average time to fall (T^2) on the x-axis. Draw a best fit line through the data points and determine slope.
14) The slope to the line is important. It is equal to ½ of your acceleration due to gravity value! Hence multiplying the slope by 2 yields your estimated value for the Acceleration due to Gravity to compare to the accepted value of 980 cm/s^2 in the percent error formula.

Activity 4 : Bouncing Ball Graphical Analysis (Decay Model)

1) This Activity is similar to Activity 1, but there is a critical difference in accumulating the data. Instead of choosing the starting drop heights, you have only the initial height of choice, 100 cm.
2) You let the ball drop 3 times and bounce back, recording the bounce height. Then you average these 3 values (b_{ave}).
3) In the next trial, you now use the average bounce back height as your new starting height and perform step 2 again. It will have a new bounce back height that is recorded and averaged as well.
4) For a third trial you use the bounce back height average from the second trial, and so on. You should try to have at least 4 average bounce back heights to use for your graph.
5) At this stage, you have 4 or more average bounce back heights in succession (largest to smallest). We can call the first one 1, the next 2, and so on. (This is called N)
6) On a Graph, plot the average bounce back height on the y-axis and on the x-axis, simply the number of the bounce (N)(where N= 1, 2, 3). Note that at the x-axis, x=0, the first data point is 100 cm, since this is the first height.
7) Draw a curve through these data points – it should have a decreasing nature to it.
8) Construct a new table of data. Use the average bounce back heights value and take the log of it.
9) Graph the log(b_{ave}) vs. the number of the bounce (N). Note that at the x-axis, x=0, the first data point is 2, since this is the first height and the log of 100 is 2.
10) Draw a best fit line through these points. Determine slope (it should be negative and a small number).
11) Use the formula for converting this slope into the exponent coefficient (A = $\frac{\qquad}{(\ .\ )}$) for the natural log, e, in the formula $b = 100e^{-AN}$. Note that the coefficient of 100 is from the initial starting height of 100 cm.
12) You have constructed a negative natural logarithmic formula that can be used to predict the height of the ball on a given bounce!
13) With the 'A' coefficient factor you can now examine your formula for predicting the height of a ball on a given bounce. This will take some effort with a slide rule, but can be done.
14) Testing your formula :
15) Recognize that you already have the log(e), so you merely multiply this value by the factor A*N, with N being a given bounce number, and recognizing that this product is a negative value. Write this down and set it aside for the moment.
16) Realize we have not examined the 100. The log of it is 2. Now take the aforementioned value, which is a negative number, and add its value to 2. The answer should be of the form : 1.XYZ
17) The 1 in the answer places the decimal point (that is multiply the answer by 10). The XYZ is found on the L scale and read back to the D (or C scale) to find the corresponding value and read as a whole number. (Be sure to multiply it by 10). Your answer should closely match the given bounce value sought!

Data :

Activity 1 : Comparison of Bounce

Note : The following table is created for each ball used

Chosen Ball : ______________

Starting Drop Height [D] (cm)	Bounce Back Height b_1 (cm)	Bounce Back Height b_2 (cm)	Bounce Back Height b_3 (cm)	Bounce Back Average Height b_{ave} (cm)

Activity 2 : The Acceleration due to Gravity (Basic Approach)

Trial	s (cm)	t (s)	½ t (s)

Activity 3 : Acceleration due to Gravity – (Advanced Approach)

Starting Drop Height [D] (cm)	Bounce Back Height s_1 (cm)	Bounce Back Height s_2 (cm)	Bounce Back Height s_3 (cm)	Bounce Back Average Height s_{ave} (cm)

Starting Drop Height [D] (cm)	Time for bounce trip t_1 (s)	Time for bounce trip t_2 (s)	Time for bounce trip t_3 (s)	Time for bounce trip t_{ave} (s)	T [½*t_{ave}] (s)

Trial	s_{ave} (cm)	T (s)

Note : Need to construct : 'log(s_{ave})' and 'log(T)' table as well as a table of 's_{ave}' and ' T^2'

Activity 4 : Bouncing Ball Graphical Analysis (Decay Model)

Starting Drop Height [D] (cm)	Bounce Back Height b_1 (cm)	Bounce Back Height b_2 (cm)	Bounce Back Height b_3 (cm)	Bounce Back Average Height b_{ave} (cm)

Calculations :

Be sure to use your Slide Rule!

Though the Slide Rule is a recommended tool, all of the calculations can be done with a graphing scientific calculator or the use of a spreadsheet program. In these calculations you have to generate a table of data, graph it, and then find the slope and/or equation of the best fit line for the data. Other formula calculations can be done with these tools as well.

(g = acceleration due to gravity, a = acceleration)

Conversions (if needed) :

1.0 m = 100 cm
1 in. = 2.54 cm

Note : The following formula will be used in a number of the Activities

Average formula (used in all Activities) :

$$\mathbf{x_{ave}} = \frac{\textbf{Total of all trial values}}{\textbf{Number of Trials}}$$

$$\mathbf{x_{ave}} = \frac{\Sigma x_i}{n}$$

To be used in Activity 1, 3, 4 :

Slope :

$$m = \frac{\Delta Y}{\Delta X}$$

Activity 2 :

Acceleration due to Gravity formula :

Original Equation to derive (if desired)

$d = d_o + v_i * t + 0.5 * g * t^2$
(d_o and $v_i = 0$)

Equation to Use for Activity 2 & 3 :

$$\mathbf{g = \frac{2*d}{t^2}}$$

Note : Be sure to use the half-time average value since this is the amount of time to travel the noted distance.

Acceleration due to Gravity for the Earth be 9.8 m/s^2 (980 cm/s^2)

$$\mathbf{\%E = \frac{[Experimental\ Value\text{-}Accepted\ Value]}{Accepted\ Value} * 100\%}$$

Activity 4 :

Conversion of Slope into Coefficient 'A' :

$$\mathbf{A = \frac{slope}{\log(2.72)}}$$

(Note : e is taken as 2.72 (from 2.71828...))

Formula for determining the height of the ball on a particular bounce : b

$$\mathbf{b = 100e^{-AN}}$$

Conclusion :

Wow – what an adventure – so many ideas from a super ball, a measuring tape and at most the addition of a timer. This Science adventure shows how we need to spark our creativity, our imaginations and think outside the box on different things. How is this similar to that over there? What information can this provide for me and how can I use it to uncover other parts of nature itself?! This is the heart of being the Inquisitive Pioneer, this Galaxy Girl, Ms Briana Andromeda Purcell knows this for sure. I love these sort of adventures. As I note regularly : Enjoy & Explore as well as Be Brilliant. I am.

In Activity 1, which type of ball had the greatest bounce back? How does this ability to bounce back relate to the slope on your graph? For the given game that the ball is used in, how do you think this affects the outcome in the game?

From Activity 1 : Are you surprised that no ball bounces back to its starting point? Consider this idea from the perspective of Energy and the Conservation of Energy concept (see this in the Mechanical Energy Activity prelude). Where does this 'lost' energy go?

In the case of Activity 1, consider trying different surfaces : Table Top, Cement, Tiled Floor, Wood Tiled Floor, Hard Ground, Thin Carpeting on the surface. How do you think these will affect the outcome of the slope of your lines?

In the case of Activity 1, what if you used two of the same balls, only one is at a different temperature? For example, use 2 tennis balls. One is just left at room temperature while the other is placed in a freezer overnight and during the experiment sits in a cooler with ice between trials. Note : Be sure if doing this, you wear a winter glove so as to protect your fingers and also to not warm the ball too much. How does the temperature affect the bounce of the ball?

In the case of the Acceleration due to Gravity values in Activity 2 & 3, Examine your data and compare your results to the anticipated outcome. How close are they? How do the two methods compare?

Also realize that the accepted value for Acceleration due to Gravity is an average for the Earth and can vary slightly from place to place, but in the case of two significant figures the value used here is sufficient.

How does the data from Activity 4 look? Note that in Radioactive Decay, each of our points on the graph would have equal units of time separating them, whereas ours do not (we are using bounces). If you have had the opportunity to do the Rate of Cooling Activity, how do the shapes of these graphs compare?

In all Activities : Look for possible sources of error in the procedure.
But above all – Be Brilliant!

Activity # 25

Newton's 2nd Law relation examination

Grade Level : High School

Math Level : Challenging

Newton's 2nd Law and the Slide Rule

Newton's Laws of Motion (there are 3) are the basis of all of Physics in the Mechanical sense. This is because it is an explanation of what causes any change in motion that is measured. These laws can be considered such though written in the 1600s even today because in the range of proper application they are accurate.

Newton summarized the behavior of all masses thus :

> 1st Law : Every mass either remains in a state of rest or at constant velocity (constant speed in a straight line) unless acted upon by an external net force compelling it to change that state of motion.

> 2nd Law : The Acceleration (change of velocity) of an object is directly related to the net force acting upon a given mass and inversely proportional to the mass of the object. **($a = \frac{F_{net}}{m}$)**

> 3rd Law : To every action force exerted on a mass, there is a reaction force of the mass equal in magnitude and opposite in direction to it exerted on the initiator of the force onto the mass.

The first law is very important for several reasons. First, it does not separate the universe into the Earth realm and the rest of the universe, which many philosophies of the past had done. Newton recognizes that these rules apply to any and all independent of location or even composition. For example, even on Earth, we often think of Newton's Laws as they apply to solids, yet the same goes for liquids and gases too – they are, of course, made of matter just in a different state. Second, the first law notes that all objects from the perspective of a frame of reference are either going to be at rest or moving in a straight line motion at a constant speed (best to call this latter phrase a constant velocity). If it does not follow this simple guideline, say it speeds up or slows down, or moves off a straight line even at a constant speed or not, then it must be affected by an outside net force. Even without Newton's description, we know what characteristic this is. This is acceleration.

In essence either a mass has a net force acting on it or it does not. If so, then there is an associated acceleration in the direction of that net force (see 2nd Law below). If there is no net force, then the mass is either at rest or moving at constant velocity.

Newton ascribes the acceleration of any mass to a 'net force' acting on the mass and the fact that the object has mass which in turn affects the degree to which the object accelerates. The power of this law operates on several levels. First, it is not any particular force. This means those investigating the cause of an acceleration look to identify the force. As a side note, the net force is merely the vector sum of all the forces acting on a mass. Second, it now gives a quantifiable way to measure this force from the acceleration of a given mass. Third, Newton's ideas apply to any and all situations, whether celestial (space-based) or terrestrial (earth-based) within regular measures. The force could be due to gravity, electric fields, or some other mechanical force such as elasticity. This is why it acts as the foundations upon which all of Physics is built. His conceptual frame necessitated a search for cause from the observed effects and a means to measure, calculate, and predict them.

Simply put in **the 2nd Law, acceleration and force are proportional**. This means as the net force increases or decreases so to by the same factor does acceleration. This makes perfect sense – if one pushes on a cart with greater effort, its speed increases. If the net force of pushing or pulling is doubled on a given mass, then so too is the rate of acceleration. The exact opposite is true as well - the smaller the force acting on a mass, the slower its rate of acceleration.

Also from the **2nd Law, acceleration and mass are inversely proportional**. That is to say, as mass increases, acceleration decreases by the inverse of the mass increase. For example, imagine two carts where each receives the same amount of effort force, but one of them is twice the mass of the other (maybe because it has twice the amount of items in it). What happens? The more massive does not accelerate nor move as fast as the lighter cart. By how much? The cart that is 2x as massive accelerates at ½ the rate of the other one. If the cart were 3x as massive, the acceleration from the same amount of force would be 1/3 as much.

The **3rd Law basically notes that no force in nature acts alone and all act in pairs**. If something pushes or pulls on something else, then that something else pushes or pulls in the opposite direction with the same amount of force on that first object. Though apparently simple, this notion is best examined through extension as a conservation law.

Before the activity, it is best to summarize this prelude with these thoughts – first, Newton changed the entire frame of thinking of the motion of objects in space and those on the Earth, uniting them instead of separating them. This meant that the Earth is part of the universe and not alone and adrift from it where there are different rules for each. Due to that, the exploration of all things both here and there is now inevitable due to this notion. Plus the frame to examine things is provided by these basic laws. With the basis of what can be called force established, this idea can be extended to any and all investigations – magnetism, electricity, gravity, and the like. At the heart of all physics books it can be said that Newtonian Mechanics is the basis of all the subsequent chapters and all of their ideas.

To illustrate, Newton's Laws give an explanation to the kinematic relations formulated by Galileo. It describes why all things undergo constant acceleration despite having different masses hence different amounts of gravitational force acting on them (see Acceleration due to Gravity Activity #20). Further, Force is the foundation of the definition of Work, which is described as the change in Energy of a System. Later scientists working in electricity, magnetism, even the nature of the atom used Newtonian Mechanics to describe the behavior of matter in these situations.
This discussion of the 2nd Law is the basis of the exploration in this activity – to examine and demonstrate the effects of force and mass on acceleration.

Activity I : A Constant Force acting on a Changing Mass
Purpose : To graphically analyze the effects of increasing mass on an accelerating system undergoing a constant force.

Activity II : A Changing Force acting on a Constant Mass
Purpose : To graphically analyze the effects of increasing force on an accelerating system of constant mass.

Materials :

- Timer,
- Meter Sticks or Measuring Tape,
- Skateboard or Regular Roller Skate,
- Small Plastic Box or Bag (acts as mass container),
- String,
- Rounded or Smooth Edge Board,
- C-Clamp (to clamp Board to edge of Table),
- Table,
- Nuts & Bolts in boxes, Marbles in boxes or Cups,
- Mass Scale,
- Slide Rule

Pre-Activity Notes :

1) With any and all Activities read through, then set up the process for the Activity and run it without measurement and watch where you may have bumps in the road so as to anticipate these. Be patient – Repeat processes. Once all is well when measuring see to the values – do you notice patterns or trends – what do you predict will happen next as you change one thing or another from one trial in an experiment to the next. Always Explore & Enjoy. Be Brilliant!
2) The materials used here depend on what you have available. The key is to have a rolling item (small skateboard or 4-wheeled basic or old-time roller skate). This will be referred to as the 'car' henceforth.
3) Depending on the 'car' used will affect the amount of mass needed to move it. For the literal small car or roller skate, marbles and/or metal nuts may be used. The larger the car or a skateboard may require masses that are larger (each item is 200 g + for example). – It is best to use a small mass system as possible for ease of use.
4) The key is to experiment to see what is needed in this case.
5) Set up the 'car' on a level surface where about 1.1 m is available and the drop from the table or counter is 0.7 m or taller.
6) Place the cart so that it can have a start and finish line (use tape, pencil or chalk where appropriate) on the table / countertop. A good track length is about 40 cm minimally. The decided upon distance is measured and recorded in the table below. (d)

7) Note that this distance must be accelerated by the falling mass for the entire time. This means if the mass has reached the floor, yet the 'car' has still not reached the finish line, then alterations need to be made.
8) Clamp the rounded-edge board to the edge of the table so that it juts out a bit so that a mass can hang over the edge.
9) Attach the string to the 'car' and pass it over the edge of the board to the mass system below. (This may be a box or bag for the materials to be put into).
10) Also do not seal the bag or box, since in one of the activities mass is added to this from the 'car'. By not sealing does not mean it cannot be closed. A resealing freezer bag works very well.
11) Also, there should be a corresponding bag or box attached to the 'car' for masses to travel in it.
12) This system is essentially an Atwood Machine, like the one used in Acceleration due to Gravity Activity #20. Here there is a falling mass and not one rising, however. In this case, one item (the 'car' - skateboard or roller skate moves horizontally while the mass falls vertically).
13) Test the system to have the minimum mass to make the 'car' move efficiently when the 'car' is at maximum mass. Be able to add mass as needed to the falling mass system.
14) Always start with the 'car' just at the start line and start the timer when it passes it. Stop it when the same part of the 'car' reaches the stop line. Be sure to have a means to stop the 'car' so that it does not fly off the table. Clamped down wood blocks either side of the pulley clamp are good or something similar to this.

Procedure :

Activity I : A Constant Force acting on a Changing Mass

1) In this activity, the 'car' M1 will receive mass from an external source (you are putting mass in/on the 'car'). The force causing the acceleration remains constant for the entire exercise (this is the falling mass).
2) Before doing the exercise make sure that when the 'car' is at its maximum mass that the accelerating force due to mass (M2) is sufficient to move the whole system.
3) You can either start with a max load for the 'car' and reduce mass with each completed trial or vice versa, where the 'car' begins with the least mass and it is added for each completed trial.
4) Note that the Force Causing Acceleration does not change here, yet the total mass M_T of the system does.
5) Realize that a completed trial is when the mass in the 'car' is ran 3 times and timed for the distance (d) measured off on the table/counter.
6) Conduct each trial, record the time (t).
7) Add or Remove Mass as needed for the next trial and do the same.
8) When completed, Average the times for a given trial.
9) Calculate the amount of force used to accelerate the system.
10) Calculate the measured Acceleration (distance and time)
11) Create a Graph of Measured Acceleration vs. Mass (M_T) [total mass]. This graph should appear as an inverse relation. Create a table of log(measured acceleration) and log(total mass). Graph these to test the idea of an inverse relation.

12) For the log-log plot, draw a best fit line and determine slope (ideally it should be -1 since $a \sim \frac{1}{m}$)

13) To further test the idea graph Acceleration vs. $\frac{1}{\text{mass}}$. Draw a best fit line that should be linear. (NOTE : It should be a straight line with slope equal to the Force being used here). [Use the C1 Scale to find the inverse values !]

14) For comparison values, compute theoretical acceleration (mass)

15) For the compared Acceleration values, compute the percent error

Activity II : A Changing Force acting on a Constant Mass

1) For this activity, there is a total mass for the entire system that does not change. First test when the maximum mass is on the accelerating (falling) mass M2 and the 'car' has the least. Do not make the system out of control.
2) Begin this activity with the majority of masses in the 'car' (M1) and the least to make it move in the falling mass (M2).
3) For each trial run it 3 times recording the time (t) from each run for the given distance (d).
4) To move on to the next trial, the force for acceleration is altered by moving mass from the 'car' to the falling mass. (hence M_T will remain the same for each calculation, yet M1 and M2 are changing, so that the accelerating force is changing)
5) When completed, compute the Measured Acceleration (distance, time)
6) Next calculate the Theoretical Acceleration for comparison.
7) Determine the percent error between the Measured and Theoretical values of acceleration.
8) Graph Force Causing Acceleration vs. Measured Acceleration (this is opposite the normal direction of things, but the hypothetical expectation shows that this slope should be the mass of the system)
9) For purists, graph Measured Acceleration vs. Force Causing Acceleration (note that slope of this line will be the inverse of mass).
10) In either case, draw the best fit line and determine the slope of the line.

Data :

Activity I : A Constant Force acting on a Changing Mass

Mass used for Acceleration : _________ kg (M2)

Distance traveled by 'car' (d) : __________ m

Mass of 'Car' (kg) (M1)	Time 1 (s)	Time 2 (s)	Time 3 (s)	Average Time (s)

Note : M2 remains constant here, hence constant force affecting the system, while M1 increases in mass from its smallest value to its largest. Recognize that the total mass of the system grows with each trial.

Activity II : A Changing Force acting on a Constant Mass

Total Mass of the Items, M1 & M2 : __________ kg

Distance Traveled by 'Car' (d) : _________ m

Mass of 'Car' (kg) (M1)	Mass of Falling System (kg) (M2)	Time 1 (s)	Time 2 (s)	Time 3 (s)	Average Time (s)

Note : M1 starts with the majority of mass, which is shifted to M2 as the system is measured for acceleration. The total mass of the system remains constant, while the force accelerating the system, M2 increases.

Calculations :

Be sure to use your Slide Rule!

Though the Slide Rule is a recommended tool, all of the calculations can be done with a graphing scientific calculator or the use of a spreadsheet program. In these calculations you have to generate a table of data, graph it, and then find the slope and/or equation of the best fit line for the data. Other formula calculations can be done with these tools as well.

Formula for what is measured :

$$\mathbf{a = \frac{2*d}{t^2}}$$

Accelerating Force

$F = m*g$

Total mass of the system

$$\mathbf{M_T = M1 + M2}$$

Theoretical Formula to compare to :

$$a = \frac{F_{net}}{m}$$

$$\mathbf{a = \frac{g*m_2}{(m_1+m_2)}}$$

Percent Error :

$$\mathbf{\%E = \frac{[\text{Experimental Value-Accepted Value}]}{\text{Accepted Value}}*100\%}$$

The accepted value here is the theoretical formula, while the experimental was what was measured in lab.

For 'g' use 9.8 m/s^2 for all calculations.

Note : Here it has been assumed that all measures are taken in the metric system, so if not, conversions are needed. Recall that the distances are noted in the Procedure in centimeters yet in the data table are noted in meters (1 m = 100 cm). Other useful conversions, if needed : 1 kg = 1000g and 1 in. = 2.54 cm

Conclusion :

Realize we have been participating in history in our own way – looking at Newton's Laws not just as words but measured items in our own labs of our design and exploration! These laws are fundamental to all of history as they shaped our world view and paved the way to the Mechanical Age, the Industrial Revolution with the Steam Engine, and the later Industrial Revolution involving Electricity & Magnetism since Newton's Laws naturally branched into looking at all things through the lenses of Energy (Kinetic and Potential for the most part) in terms of understanding not just things we make but how large and small things in the Universe operate.

The primary conclusion points in our Activities to reach are these : 1) What is predicted and what were the results when the force acting on the system to accelerate it had a changing mass itself? 2) What is predicted and what were the results if the total mass of a system is constant yet the amount of force affecting it is changed in an increasing manner? 3) What did the graphs show as the relation ? 4) How similar (or dissimilar) were the results of actual measurement versus theoretical measurement? (Hint : consider the theoretical and what assumptions is it making to reach that conclusion along with the other experimental areas that may have error in them). – Always be curious and Be Brilliant!

Activity #26
Coefficient of Sliding & Static Friction
Grade : Middle School
Math Level : Calculating

Friction was a term first noted and described by Galileo in his work concerning the motion of bodies. He realized that in the absence of friction a body in motion would stay in motion (law of inertia) [later described in more mathematical terms and relating to force by Newton]. His thoughts on this are what we consider obvious today, that friction is primarily caused by irregularities in the surfaces in contact. On a scale of a little more than a molecule, all surfaces seem like uneven, broken, and bumpy, undulating terrains. Here surfaces can 'catch' each other as one slides past another. More modern analysis includes electrostatic forces due to the molecules themselves as being part of the frictional forces affecting masses in contact with each other.

Friction is a force that always opposes motion and is a force due to the contact of surfaces (typically seen as solids) moving relative to each other. In common practice there is a sort of negative thought about friction. It seems to imply an inability or reduction in motion. In general this is true, but when it comes to objects, like cars or even people moving across the planet, it is a necessity. Imagine, however, being in the middle of a frictionless pond. If there were absolutely no friction whatsoever, no matter how much you moved your feet, you would be motionless. It is the friction of our shoes against the ground that allows us to impart a force to the ground, which in turn responds with a force that acts on us to move us. Of course, flying objects such as planes and rockets would be in better shape with little to no friction. Also friction means that no system can run at 100% since some of the energy is lost as heat (3rd Law Thermodynamics). Then again, air resistance is a form of friction and is greatly appreciated by parachutists and friction is the reason braking systems for cars operate. Though it opposes motion and if we assume motion in one direction is positive and the resistance is negative, it can have positive consequences. Science does not assign emotion, only direction and magnitude. – Though admittedly we dislike personal friction in our attempts to do things and wish for only a smooth go of it.

In Newton's Laws where Force is defined and by examining objects in free-body diagrams (pictures that detail the magnitude and direction of forces acting on a mass) it can be easily found a means to measure the force due to friction and to find a value useful in classifying different surfaces in contact with each other called the coefficient of friction.

The coefficient of friction has several different types : the Coefficients of Static Friction, Sliding Friction, and Rolling Friction. In this Activity we are going to explore only one the Coefficient of Sliding Friction, but at the end of the Activity in the Extension Section there is a means to determine another one called the Coefficient of Static Friction. In terms of values the list of them given is one of decreasing amount, that is to say :

Coefficients for each of the following tend to follow this pattern :
of Static Friction > of Sliding Friction > of Rolling Friction

Each of them is unit-less because it is the ratio of 2 forces acting on the surfaces in contact with each other. (see diagrams)

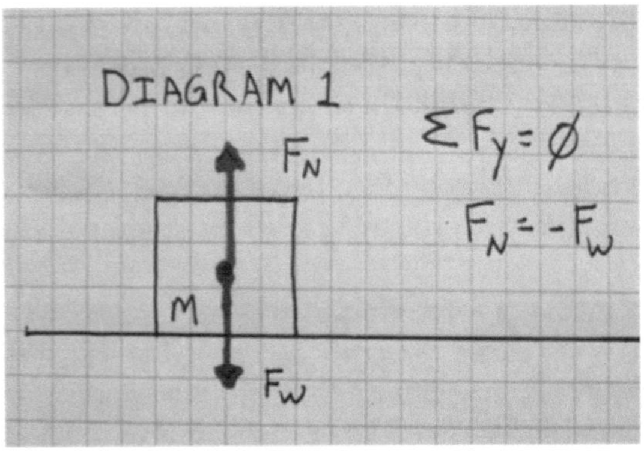

Diagram 1

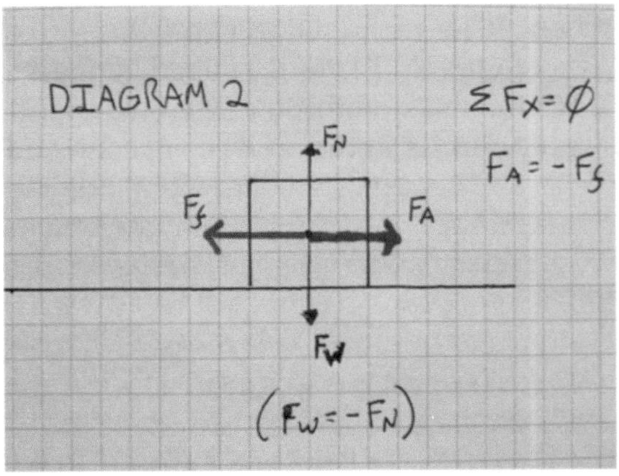

Diagram 2

In the first diagram we have a body at rest on a surface, therefore it has two equal forces acting on it, the force due to gravity we call weight acting down and the normal force (in this case electrostatic) acting up (F_W & F_N respectively) . Each are equal in magnitude and opposite in direction, hence there is no net force in the y-direction, therefore no acceleration for the object. (Note as an aside : These are NOT action-reaction pairs as they act on different objects. The reaction force to the action force of gravity on the mass is its own force due to gravity acting on the Earth itself).

In the second diagram we now have the same mass now in motion at a constant speed. To be at a constant speed in a straight line means that there is no net force in the x-direction according to Newton's 1st Law. If we are pulling this with an Applied Force, what is impeding it so that the net Force equals zero? It is the force due to friction between the 2 surfaces in contact. Our Applied Force just balances the Force due to Friction when the moving mass moves at a constant speed. The action force is F_A and the force due to friction is F_f. Notice in all of this writing of the description we are employing Newton's Laws of motion and his system of how we see the actions of forces in play here.

The coefficient of sliding friction is defined as the force due to friction divided by the normal force. We have chosen to have the mass rest on a level surface so that the normal force (F_N) balances and equals the weight force, so that we can substitute the weight (F_w) in our equation. In the case of the force due to friction, we are applying a force that equals it so that the mass moves at a constant velocity so that we can measure it with a tension scale as we pull the object. Again we can substitute the applied force (F_A) for the frictional force (F_f) in our equation. We end up with the following :

$$\mu = \frac{F_f}{F_N} = \frac{F_A}{F_W}$$

The following paragraphs take the aforementioned formula and now think in terms of thought experiments – what if we change this – what should happen?

Before determining this coefficient of sliding friction, we must further explore the nature of friction. For example, if we have a rectangular solid where the surface areas are different and then placed in contact with a given surface, will the coefficient of friction be the same or not? We can hypothesize and test this idea (For example : As surface area increases so too does the coefficient of friction). We could further consider the prior formulae and note there is no provision for surface area, so we might also assume that nothing at all happens.

Another question we explore is the impact of mass on our situation. Let's say we keep the surface areas in contact constant, but now place a second rectangular solid atop the other one effectively doubling its mass? Would this affect the outcome? Looking to the formula above, of course the force due to weight is now twice as much, but what happens to our applied force to make the object move at a constant speed? From Newton's 2nd Law we realize we must apply twice as much force to achieve this. In essence, the increase in mass should result in no changes, but again we must test the idea to prove this to be true.

Next we can then explore various surfaces in contact with each other and determine the coefficients of sliding friction. It would be easy to assume that we should find that as the surfaces in contact increase in roughness, the force due to friction would increase, hence so too would the coefficient of sliding friction values. Also the converse is expected to be true, the smoother the surfaces in contact, the smaller the value. In both cases they should normally range from 0.0 (ideally smooth) to 1.0+ (difficult to move).

In a separate investigation we will observe and measure that there is two types of friction. The first we measured was sliding friction and deals with an object in motion. So this evokes the question, what if the mass were initially at rest? Anyone who has pushed a large cabinet or table knows that once in motion it seems to take less effort than it took to get the object to start moving. That initial resistance to effort is due to what is called static friction (μ). [Each has its own subscript, whether for static, sliding, or rolling but all have the same variable term]. The amount of force needed to get the object moving is found by taking the coefficient of static friction and multiplying it by the normal force of the object on the surface it is on. Like the coefficient of sliding friction, this one depends primarily on the types of surfaces in contact.

Unlike the coefficient of sliding friction where we balanced the applied force to the force due to friction between surfaces, here in the case of the coefficient of static friction we need a way to just get the object moving, that initial push. From experience we know that this initial push is greater than the force needed to maintain motion. A little geometry here enables us to find the force just needed to get the object in question to move with the assistance of gravitational force. Examine the diagram (#3) and recall the definition of the force due to friction and its relation to the coefficient of friction.

$$\mu = \frac{F_f}{F_N} = \frac{F_w * \sin\theta}{F_w * \cos\theta} = \tan\Theta$$

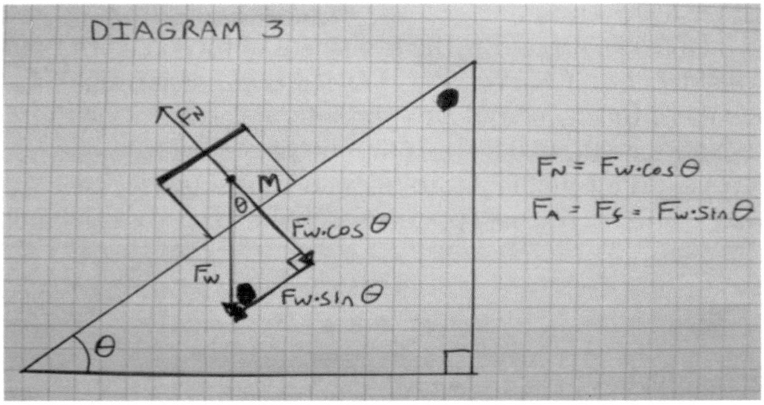

Diagram 3 for static friction (angle)

To find the coefficient of static friction, we need to only know the measure of the angle at which motion just begins. Apparently independent of mass or surface area.

Experiment : Determining the Coefficient of Sliding Friction and Examining Factors that May Effect It.

Purpose : To determine the coefficient of sliding friction for specific surfaces in contact with each other where the surfaces slide past one another with no net force (i.e. constant velocity).

Purpose : To determine if the coefficient of sliding friction for specific surfaces is affected by either the surface area or the amount of masses involved.

Materials :

- Rectangular Wood Blocks,
- Hooks,
- Hanging Tension Scales,
- Mass Scale (if needed),
- Materials for Surfaces : fine and coarse grain Sandpaper, Wax Paper, Plastic Wrap, Newspaper, Rubber Mat, et al
- surfaces to slide on such as Cement, Wood, Tile, or long cutting boards made of Glass or Plastic,
- Ruler,
- Slide Rule

Procedure :

1) With any and all Activities read through, then set up the process for the Activity and run it without measurement and watch where you may have bumps in the road so as to anticipate these. Be patient – Repeat processes. Once all is well when measuring see to the values – do you notice patterns or trends – what do you predict will happen next as you change one thing or another from one trial in an experiment to the next. Always Explore & Enjoy. Be Brilliant!
2) Here we become the force of investigation and come to understand friction!
3) The first experiment (Activity I) will involve only a table top or floor and a set of no more than 3 blocks of wood each with hooks and a tension scale. It is best if the blocks are rectangular so that the surface areas are different from each other.
4) Measure and record the surface areas of the blocks along with the mass of the blocks using the tension scale. (be sure to zero the scale out). Record the weights in the table.
5) Lay the block on the table (floor) and attach the tension scale. Pull it in a manner to be at a constant speed (the scale will not continue to increase in force measure). Pull it slow enough to read it.
6) Do this 2-3 times and mark down these values and average them before placing the recorded value in the table for applied tension force.
7) For Activity II we investigate surface area and the effect on coefficient of friction. Now place the block on its side and repeat steps 3 & 4 for a different surface area.
8) For Activity III we are looking to see if mass effects the coefficient of friction. Replace the original block used in Activity I with the largest surface area in contact with the table (floor) and place a second block of known weight on it. Record the total weight of the system in a new table.
9) Pull both blocks via the hook on the lower block and measure the amount of applied tension force as was done in steps 3 &4.
10) Repeat step 6 with a third block on the other two.
11) Steps 1-8 allow you to determine the coefficient of sliding friction and whether or not surface area or mass affects the outcome. The next set of directions are for different surfaces so as to determine a variety of coefficients of sliding friction.
12) First determine which surfaces are going to be used for the experiment (wooden, tiled, linoleum table top, tiled or cement or carpeted floor) and in the case of the wooden block what surfaces will cover it (none, wax paper, plastic wrap, cloth, aluminum foil, sandpaper).
13) For every choice this means a set of data for that determination of the coefficient of sliding friction. Follow steps 3 & 4 for recording data for a given set of surfaces in contact with each other.
14) Determine the coefficient of sliding friction from the data for each of the situations explored. Are they similar or different for a given set of surfaces in contact with each other?

Data :

For each case, do at least twice and take the average value of the tension scale reading and then determine the coefficient of sliding friction using this value.

Activity I : Determining the coefficient of friction for 2 surfaces

Trial	Weight Force	Tension Force

Activity II :

Surface Area and Coefficient of Sliding Friction :

Use the following table at least twice to see if there is any relation between the size of surface areas in contact and the coefficient of sliding friction for these surfaces.

Type of Surfaces in Contact :
A)__________________ B)__________________
Surface Area : __________
(from moving block)
(measure and calculate each surface area used)

Trial	Weight Force	Tension Force

Activity III :

Mass and Coefficient of Sliding Friction :

Use the following table at least twice to see if there is any relation between the mass for a given surface area in contact and the coefficient of sliding friction for these surfaces.

Type of Surfaces in Contact :
A)__________________ B)__________________

Trial	Weight Force	Tension Force

For all 3 Activities reproduce the Tables as needed for each set of surfaces explored.

Calculations :

Be sure to Use a Slide Rule in all of your calculations !

$$\mu = \frac{F_f}{F_N} = \frac{F_A}{F_w}$$

Average formula :

$$X_{ave} = \frac{\sum x}{n}$$

Conclusions :

Hopefully there was no resistance to your efforts – intentional pun there.

To analyze results consider the questions – did mass or surface area affect the coefficient of sliding friction? Examine the formula and see if whether or not these factors come into play. Next examine the various coefficients of sliding friction you have determined. Do they match your expectations? For example, when using two of the smoothest surfaces in contact with each other was the value of μ smaller or larger than the others? What should the value of μ do as a surface becomes more smooth or in opposite, more irregular?

Experiment : Determination of the Coefficient of Static Friction

Purpose : To determine the coefficient of static friction between two surfaces.

Materials :

- Blocks of wood
- Board (1.5 – 2.0 ft long,)
- Long cutting board made of glass and/or plastic
- Materials : tin foil, plastic wrap, newspaper, wax paper, other
- Protractor
- Slide rule

Procedure :

1) Determine which surfaces will be placed in contact with each other.
2) Choose one block that will be covered with the materials to slide down the board, which too will have various coverings.
3) Note, if the board is wood and the block is covered with aluminum foil, then this experiment is foil on wood. If the board had foil as well, then it is foil on foil.
4) Place the block about 2/3 of the way along the board from the end that will act as the pivot point.
5) Slowly raise the board with the block on it.
6) With each small increment just lightly tap the block and see if it starts moving continuously on its own.
7) Stop at the angle at which the block once lightly tap continues all the way down the board.
8) The best way to measure this is to have 2 people. If not, create a system where the ramp as it is elevated is held in place. Books or magazines can do this, or having an eyelet in a nearby wall and at the far end of the board and drawing the board up with a string is possible. – What is needed is an incrementally rising ramp so as to find the critical angle where motion just begins.
9) With the angle found, it is very easy to determine the coefficient of static friction by looking up the tangent value of this angle on your slide rule!

Data :

Types of Surfaces :

Block Surface : ________________

Board Surface : ________________

Trial	Angle at which block moves

Calculations :

Use a slide rule in all of your calculations !
Note here the only thing to do is look up the value of the tangent of the angle.

$$\mu = \frac{F_f}{F_N} = \frac{F_w * \sin\theta}{F_w * \cos\theta} = \tan\Theta$$

Conclusion :

From this experiment, consider the types of surfaces in contact and think of them as smooth or not smooth and examine the coefficients of static friction you have calculated. As the smoothness of surfaces increased, did the coefficient of static friction increase or decrease? Another consideration comes from comparing comparable surfaces from the first part of the lab and the determination of the coefficient of sliding friction. Were these corresponding values similar? Was one larger than the other?

Explore & Enjoy
Be Brilliant!

Activity #27
Mechanical Energy calculation exploration
Grade Level : High School
Math Level : Calculating

Mechanical Energy : PE & KE Considerations with the Slide Rule

The Story of Energy

Hey Inquisitive Pioneers! Ms Briana Andromeda Purcell, aka Galaxy Girl here and with a Note about this Activity : First, There is a lengthy story of Energy before the actual Activity which is several pages later, but for those interested in science, physics, or energy in general, this may be interesting to read about, otherwise you can turn to the Activity and engage in the measuring and calculating for yourself. Enjoy.

All of us have heard of the concept of Energy, but often when asked we have a hard time defining or describing it. We know we need it, and there seems to be a wide variety of types. It is in the paper and/or news most days. One of the key things is this – energy is essential. In matter of fact, Energy is probably the most central of all concepts to all sciences, since nothing is done without energy basically. The best road to Energy begins with Work. **Work**, by definition, is a Force acting through a Displacement.

W = F*d*cosΘ

*If Θ = 0º,(i.e. they are in the same direction) then W = F*d*

In all cases for Work being done, there must be at least two things :

1) There is a Force acting on an object and
2) there is a displacement or movement of that object (AND noting that the force is acting in the direction of the displacement as well).

For more on Force, look to the Activity on Newton's 2nd Law #22 which shows the connection of Force and Acceleration.

This definition of work leads to the famous question posed in Physics. If you exert a force, say 10 N on a book and the force is parallel to the table. You push the book a distance of 10 cm. Have you done any work and how much? From our formula, it would be :

W = F*d (Θ = 0º here, so cosΘ = 0)
W = 10 N *0.10 m = 1 N*m = 1 J

This introduces us to the unit of Work, namely the newton*meter, commonly referred to as the **Joule**, named after James Joule who was a physicist who worked in heat-energy relations. This, however, is not the 'trick' question in physics. If you hold the book in the palm of your hand and exert 10 N straight up to hold it in place (i.e. balanced forces) and now step 10 cm forward, have you done any work and how much?

Before you say that these are the same numbers, hence the same answer, read it carefully. The force is at an angle to the displacement. In fact it is at 90°. The cos(90°) = 0, hence W=0! We have done no work according to the rules here.

Does this mean that there is Work or Not only. Obviously not, since there are a wide range of angles between 0° and 90°. Each of these has some amount of work according to the formula. In many introductory physics classes the forces are considered to be in the direction of the displacement, hence the need for angles is not considered.

Returning for the moment to the **Joule** – how much is 1 Joule? Simply put, we would have to move a 1 N object 1 m and therefore we have done 1 J of work. (For English units : 1 N is approximately 0.22 lbs and 1m is about 3 ft. So lift a ¼ lb. apple one yard in one second and you have exerted the equivalent of 1 J of work). In essence, a Joule is a small amount.

How does this relate to Energy? So in this case here, looking at our first 'work' calculation, we pushed a book horizontally. If after that 10cm push we stopped pushing, what would happen? According to Newton's Laws, it should travel now at a constant speed unless affected by an external force. How fast depends on its mass. Push a large massive book, it would move much more slowly than a lightweight book. The question is, can we find a relation for how fast the book should be traveling?

Let's use the Equations :

$$W = F*\Delta d$$
$$W = m*a*\Delta d$$

Our goal is a relation having only mass (m) and speed (v)
We need two formula to substitute for 'a' and 'd'.
To further simplify, let $d_i = 0$ and $t_i = 0$

$$a = \frac{\Delta v}{\Delta t}$$

$$d = v_{av}*\Delta t$$

$$v_{av} = \frac{v_f+v_i}{2}$$

Substitute for 'a' and for 'd' and also put in for 'v_{av}' in 'd'. With some algebraic manipulation, you end up with :

$$\mathbf{W = \tfrac{1}{2}*m*(v_f^2 - v_i^2)}$$

We define '$\frac{1}{2}*m*v^2$' as Kinetic Energy (abbreviated KE)

$$W = KE_f - KE_i$$

From this we have the general equation :

$$K.E. = \frac{1}{2}*m*v^2$$

Before pressing on, how fast is our book moving? Recall that we had 1 J of work done to move it. Let the mass be approximately 1 kg (you'll see why in the potential energy discussion below). The speed then turns out to be about 1.4 m/s (try it for yourself on the slide rule).

Understanding both constant motion and acceleration can be found in the Activities on Acceleration on an Incline #22, and Acceleration due to Gravity #23.

Kinetic Energy is the energy due to motion of an object. Looking at the Work relation to it, we have what is now called the **Work-Energy Relation**.

Work is now defined as the change in energy of a system. If the change of energy is positive, then the system has taken in energy. If the value is negative, it has given off energy.

$$W = \Delta(KE)$$

Can we do work to something, yet not change its overall motion once we are done doing the work? For example, we take our book in the aforementioned example and now lift it to a book shelf that is 2 m above the floor. Recall we needed to exert 10 N of force to hold the book. From Newton's 2nd Law we know that the force due to gravity is what we call weight and is expressed as :

$F_w = m*g = 10\ N = m*9.8m/s/s$
$F_w \sim 1\ kg$

First note, this is where the approximate mass for the book in the prior example came from. Notice here we have a force and a displacement, the needed elements for work to be done. Second, note we are moving the force in the direction of displacement (unlike our prior 'trick' question).

$W = F*d = 10N*2m = 20\ J$

We also know that Work is the change of energy, and it is positive, so we have added energy to the book, yet it is not moving. Where, then, is the energy? Consider, if the book were to fall off the shelf, would its speed change? Certainly, due to the force of gravity, its velocity changes at the rate of the acceleration due to gravity. If we were to calculate the amount of energy it has as it reached the height from which we lifted (the 2 m) it would have 20 J and we could determine its speed at that point. The energy we put in has returned in the form of motion of the falling object.

This stored energy is what is called **Potential Energy**. Potential Energy, by definition, is the energy associated with an object because of the position, shape, or condition of the object. In this case, it is specifically **Gravitational Potential Energy**. Gravitational Potential Energy has to be measured from a **Reference Point** (a base or starting point). We measured ours from where we lifted the book 'from' and 'to'. Most often in books, the reference point is the ground, where the potential energy is defined as zero. (Note also unless specified, if there is no descriptor in front of potential energy, it usually implies gravitational).

$$\mathbf{P.E. = m*g*h}$$

$$\mathbf{W = \Delta(PE) = m*g*\Delta h}$$

Since there is an adjective describing the noun, there must be others. There is Elastic Potential Energy as found in stretched rubber bands, compressed and stretched springs, and any other matter that can snap back when tension forces are applied to it. By definition this is the energy available to use when a deformed elastic object returns to its original configuration. There is also Chemical Potential Energy that is similar yet deals with the arrangement of atoms and molecules. There are other common forms of energy too such as Heat, Sound, and Light all of which can be expressed and analyzed mathematically as Energy.

(Aside : There is an Activity on Elastic Potential Energy #25 and in it is an examination of Hooke's Law which deals with forces on elastic materials).

These first two energies (KE and PE) and their subsets constitute what are called Mechanical Energy (there are other potential energy forms too – these are the most commonly referred to here).
Mechanical Energy is the sum of the Kinetic Energy and all of the Potential Energy of a material.

$$\mathbf{M.E. = K.E. + \Sigma\ P.E.}$$

Mechanical Energy can be conserved. That is to say, what there was initially will be there at the end. Recall the discussion of lifting the book to the shelf and it falling. The amount of energy put in was all that was available when it fell to its starting height.

Initial Mechanical Energy = Final Mechanical Energy

$$ME_i = ME_f$$

In many cases, there is often only one potential energy form under consideration, such as gravitational potential energy, so we can write :

$$KE_i + PEi = KE_f + PE_f$$

There appears to be many variables in this equation. But we need to choose the best places for analysis to simplify the situation. At the highest point for the book on the shelf, it is not moving, so there is 0 KE here. When it has reached the bottom of the fall, the PE is zero and all of it is now KE only!

Of course the question then is how much of it is KE and how much is PE say halfway down? One-half the energy is KE and one-half is PE here. The further the book falls, the more of the energy is manifested as KE, since the speed is changing. The closer to the top of its path, the more PE it has, hence less KE.

This idea is best illustrated by the classic roller coaster. A motor driven system pulls the cars to the top of the hill, hence they have their maximum gravitational potential energy. As they travel down the hill, their potential energy decreases, and their kinetic energy increases. The wheels have low rolling coefficient of friction values so that all the successive hills are easily managed by the car train. Notice that the later hills are shorter than the first big hill. This is not only a good heart-racing effect, but the energy gained there is used to let the cars go through the whole of the track with little to no input of additional energy. In fact, there is enough energy so that the momentum of the system is still there at the end and a brake is applied to bring the cars to a safe stop.

Despite a description that seems very easy, this is the basis of all science and its analysis of energy flow through a system, whether it is in chemistry, biology, physics, astronomy, geology, or meteorology. The ideas of energy created the roller coaster idea mentioned – it is the imagination of the designer and the skills of the engineer all in one person to make such magic happen. In the case of energy production, we will dam up a river to create a lake behind the dam, divert some of the water down a tube to hit and spin turbine blades that are attached to a generator to produce electrical power. So the gravitational potential energy becomes the kinetic energy of the water, which in turn becomes the kinetic energy of the turbine, then this is the kinetic energy of the rotor in the generator to produce electricity, which generates the electric field in the wires which becomes the kinetic energy of the electrons in the wires to our homes and businesses.

This all leads to the **Law of Conservation of Energy**.

It states that energy cannot be created nor destroyed, it can only be transformed or transferred from one form to another. The total amount of energy in a system does not change. Like other conservation laws this applies to a closed system.

The energy of a dropped object turns into heat and sound once it strikes the floor, like if our book we placed on the shelf should fall. Some of the energy can also deform the book too. But most importantly, all of the gravitational potential energy turned initially into kinetic energy which then turned into these other forms of energy. The total at the beginning is the same as the total amount of energy at the end of this situation.

Are there ways to change the equation for conservation of energy ? What of **Simple Machines**, such as the lever, the ramp, the pulley, the screw, and the wheel & axle? All simple machines can either redirect force and/or multiply the amount of force (at the expense of displacement) but cannot increase the amount of work done! When using a lever, for example, it appears that we can lift incredibly heavy objects, such as a car with a car jack. Notice though that even using a hydraulic lift, we move the lever arm a great deal of distance, say 1 m, while the car only moves in height 1 cm. No matter the case : The work in can only equal the work out ideally and never actually does due to friction!

$$\mathbf{W_{in} = W_{out} + W_{friction}}$$

$$\mathbf{Efficiency = \frac{W_{out}}{W_{in}}}$$

Thus far, we have noted the amount of Work done, but not the Rate at which the Work is done – that is the definition of Power. **Power** is the rate at which work is done.

$$\mathbf{P = \frac{W}{t}}$$

The units are joule/sec which is called a **Watt**. This is the same unit we note in electrical devices, such as light bulbs. Two people can lift the same amount of weight when walking a flight of stairs, so they have done the same amount of work. Yet if one of them does it in half the time than the other then the one who takes less time uses more Power than the other.

Interesting to note is that the conservation of energy applies to power as well. The amount of energy per unit time going into a system, like a transformer which can alter the voltage, the power is the voltage times the current in the system, and this product going into the transformer

and coming out will be the same. In essence the power going into a system always equal the power coming out of a given system. This too cannot be altered. (Hence $Power_{in} = Power_{out}$)

The ideas of Energy, Work, the Work-Energy Theorem, and Power are central to All Sciences and how they are looked at. In fact, how much energy is available can be found from a given piece of matter. Einstein's famous equation : $E = m*c^2$. This is the equivalence of matter and energy itself.

Consider the Weather. What drives the winds, the waves, and causes the temperature of the environment around us? **The Sun is the primary source of all energy on this planet for the most part.** The Sun heats the matter on the planet, the solids, liquids, and gases, and since different materials heat at different rates, and because the sun strikes these materials for different amounts of time at different angles, it results in differentiated heating, hence regions of hither and lower temperatures on the land, in the air and in the water. Also there are areas of different pressure in the air, convection cells of higher and lower pressures. These pressure differences result in our winds, which drive the waves too. Evaporated water becomes clouds which again fall as precipitation.

Geologic processes are driven by the stored energy of the Earth internally from radioactively decaying materials which heat and melt the interior regions of the Earth. This results in convection currents which in turn drive and move the plates of the Earth.

Biologically, all life depends on the Sun as well. The Sunlight is absorbed by plants through photosynthesis, which uses the energy, water, and carbon dioxide in a chemical process to produce sugars. This acts as the basis of energy and food for all food chains. The food we eat powers our bodies, for example.

Not only does the Sun power today through the foods taken in, but it was obviously shining long ago on plants then. Some died in shallow swamps and were buried over. Years of sediment piled atop them and chemical processing in the Earth has led to oil, natural gas, and coal deposits. Hence their name or term for them, the fossil fuels. These, in turn, have been excavated, mined, and pumped to be are used as power sources today. We burn coal in an electrical plant to heat water which in turn turns turbine blades that spin a generator (a system of magnets and wires) to generate electricity. Transformers change the voltage (increasing it, but therefore decreasing the current since the Power cannot be changed) to send it along power lines to our homes. Here transformers step down the voltage to usable levels in the home. In the house we plug in our appliances to use the electricity to create heat (for the toast in the toaster), light (from the light bulbs) , and sound (from the radio or TV) energy from burned coal that heated water to generate electricity. We can only use the energy available to do things, such as move electrons in wires.

As noted in the conservation of energy idea, energy cannot be created nor destroyed, only transformed and/or transferred.

This is true even for the seemingly inexhaustible energy source for many things, the Sun. First, a word on **Renewable** vs. **non-renewable Energy forms**. Actually given enough time and the right circumstances, energy forms like coal and oil can renew, but this takes millions of years to occur. What makes them non-renewable is the fact we are taking out far faster than their recovery rate. On the renewable energy side are things like solar, wind, and wave forms of energy. The wind and wave ones are driven by the sun and solar can be used as passive heating systems or active systems which tap the sun using solar cells to generate an electric current.

In any case, the Sun, as noted is not infinite either.
But where does its energy stem from? Recall the aforementioned equation by Einstein equating mass and energy. In the core of the Sun some 5 billion tons of Hydrogen are turned into pure energy every second. The reaction actually has a byproduct too. The original amount in the reaction 500 million tons of Hydrogen is turned into 495 million tons of Helium. The difference, noted above, goes into energy and becomes the light and all other wavelengths given off by the Sun. Stars are fusion nuclear reactors with sufficient pressure (due to its mass and gravitational pull) and temperature (10s of millions of degrees) which turn low mass elements like Hydrogen into heavier ones like Helium and in the process give off energy.

BTW We are all Star Stuff! Be the spark of energy and engage in this and all the Activities in the book – have fun, be creative, think out loud – I do all of these myself and love the adventure – here in my book – Galaxy Girl's Inquisitive Pioneer Adventures by Ms Briana Andromeda Purcell – as I always note : Explore and Enjoy! Be Brilliant – I am!

The Mechanical Energy Activity

Our lab will deal with the ideal and overlook the friction and assume complete conversion from gravitational potential energy to kinetic energy.

In the process of the Activity, we will take measurements of distance and time and use these to calculate speed and acceleration which will be actual measures and calculated theoretical values for comparison. Next with the former measures and with measures of mass we will then determine the gravitational potential energy and the kinetic energy of a rolling marble down an incline. As an aside we can even use some 2-dimensional vectors of velocity to determine the impacting speed (the magnitude of the velocity) of the marble as it strikes the ground. We will also graph the relations in a couple of ways to analyze the information to make predictions and find the relation mathematically between gravitational potential energy and kinetic energy.

Why so much? Energy allows for the exploration of kinematics (the science of motion), Newton's Laws (the science of the causes of motion), and Energy itself. These equations are encapsulated combinations of all of the ideas in science.

Have fun with your Slide Rule in exploring energy!

Activity : Using the Ramp to illustrate Mechanical Energy

Purpose : To measure and calculate the gravitational potential energy of a marble rolling off a ramp on a table and compare it to the kinetic energy of the marble as it lands on the floor.

Purpose : To compare the actual time of a falling object to the predicted time of a falling object from theoretical calculation.

Purpose : To compare the measured speed of the marble exiting the ramp and table as compared to the estimated speed of the marble as it exits the ramp on the table.

Purpose : To compute the impacting speed of the marble from combining both the horizontal and vertical components of speed by using vector addition.

Materials :

- 2 Meter Sticks OR 2 long dowel rods,
- Measuring Tape,
- Timer,
- Stack of Books (4 books is good),
- Table,
- Transparent Tape,
- Marble (large is best),
- Kitchen Mass Scale,
- Blank Paper,
- Carbon Paper,
- Slide Rule

Pre-Activity Set Up :

1) With any and all Activities read through, then set up the process for the Activity and run it without measurement and watch where you may have bumps in the road so as to anticipate these. Be patient – Repeat processes. Once all is well when measuring see to the values – do you notice patterns or trends – what do you predict will happen next as you change one thing or another from one trial in an experiment to the next. Always Explore & Enjoy. Be Brilliant!
2) Create a track ramp on the Table with the long dowel rods or the meter sticks where you use for the first ramp 2 books and then for the second set of trials use 4 books to make a shallow angle ramp. In either case, the ramp is between 10 and 25 cm in height. The ramp track should be close to 1 m in length if possible. (Read step 2 for ramp placement)
3) In creating the ramp, be sure to leave a gap between the ramp and the edge of the table that it approximately 2-3 cm.
4) Record the Length of the track (L).
5) Use the Kitchen Mass Scale to measure the mass of the marble being used and record it (M).
6) Note : If the mass is too small for the precision of the scale, then take 10 or more marbles in order to obtain a reading. Divide the total mass of the marbles by the number of marbles used. Use the Average Mass as the mass of the marble.
7) From the center of the marble sitting atop the ramp to the table surface measure and record this starting height (H1).
8) Use the Height of the ramp with marble value (H1) and the Length of the Ramp (L) to determine the angle (Θ) that the ramp is at with respect to the table. (Note that the ratio $\frac{H1}{L}$ is the sinΘ which we will use)
9) Measure the vertical falling height from the edge of the table where the marble will fly off at to a point directly below on the floor and record this height (H2).
10) Also place a piece of tape on the floor directly beneath the end of the ramp. This will be the initial point to measure from for distance.
11) Measurement of the Fall Time of the Marble :
12) At the end of the Ramp at the Table's edge where it would fly off, hold the marble so as to be able to drop it.

13) Drop the marble 3 times and fall time ($\mathbf{T_x}$) each fall.
14) Calculate the average time ($\mathbf{T_f}$) for falling to the ground.
15) (Note the vertical fall will take the same amount of time as the marble would when flying off the table since gravitational force acts only vertically).

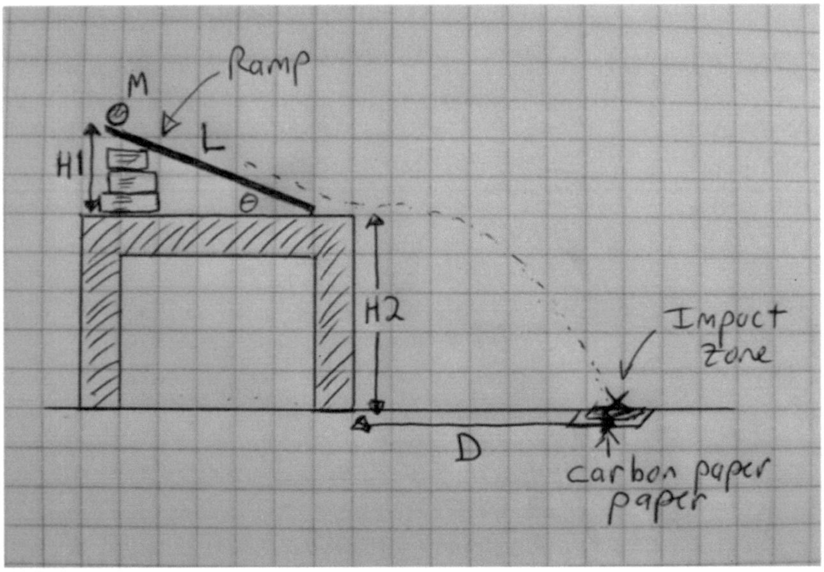

Procedure :

1) First let the Marble roll down the ramp and note where it lands.
2) In the general area of its landing tape down a piece of paper and atop of it place a carbon piece of paper so that when struck by the Marble it will leave a mark.
3) Test the system so that it operates correctly.
4) While the marble rolls down the ramp, time it and record these rolling times ($\mathbf{t_R}$).
5) Determine the average of the ramp time ($\mathbf{t_1}$)
6) With each run on the ramp, mark where the marble lands by lifting the carbon paper and mark the mark with a pen.
7) Take time to measure the distance from the marked point to the initial point already marked. This distance is the horizontal flight distance of the marble ($\mathbf{D_x}$).
8) Perform the run of the marble 3 times, recording the horizontal flight distances and averaging them (D)
9) At this point you have enough data to perform all of the necessary calculations.
10) <u>2nd Set of Data?</u>
11) If you want a second set of data, change the height of the ramp using a different number of books. Note : This, along with a 3rd set of data is recommended for graphical analysis.
12) Note that the height is changed (now a new H1) hence a new angle (Θ) from the new sine of the angle.
13) With a new height, redo steps 1-8 in the Procedure on a new set of tables of data you create.
14) Note that this data can be useful for other Calculations noted below, such as the graphing of data, so this is recommended

15) Note that only two sets of data are often not enough, so it is recommended to have 3 sets!
16) <u>Calculation Procedures :</u>
17) Use a Slide Rule for all Calculations!
18) As noted above in the directions : first calculate the angle from the height of the ramp (H1) and its length (L) using sine on your slide rule
19) Sum up the Falling Time ($\mathbf{T_x}$) and determine the average from your trials to find the average time ($\mathbf{T_f}$)
20) Sum up the Rolling Time ($\mathbf{T_R}$) and determine the average from your trials to find the average time ($\mathbf{t_1}$)
21) Sum up the horizontal flight distances ($\mathbf{D_x}$) and determine the average horizontal flight distance (D) for your calculations.
22) Note : For All Calculations, calculate the average of the values for the given variable for use in comparisons.
23) Calculate Predicted Time of Fall ($\mathbf{T_P}$) and compare to your actual average measured time of fall ($\mathbf{T_f}$)
24) Calculate Average Measured Acceleration ($\mathbf{A_M}$) and calculate Average Predicted Acceleration ($\mathbf{A_P}$) and compare these two values. How similar are these values?
25) Calculate the Average Horizontal Speed (V)
26) From your 3 sets of data, then you can : Graph Horizontal Flight Speed (y-axis) vs. the Height (H1) started from (on the x-axis). Note : You need more than one set of data to do this, and 3 is recommended.
27) With the graph from the prior step, now draw a best fit line and calculate slope from this line.
28) From the data set of points for graphing, create a new table of the log values of each of the data points (use the L scale for this)
29) Graph the log values of each of the new data points, so that it is the log (horizontal flight speed) on the y-axis and log (ramp height starting point) on the x-axis.
30) On this log graph, draw a best fit line and calculate slope from it.
31) If done correctly, the slope should be ½ since the ratio of KE/PE is this – solve it for yourself to find the relation of horizontal flight speed as derived from gravitational potential energy.
32)
33) Another Graph and Testing an idea :
34) Graph the Measured Range of Flight (D_x) on the y-axis vs. the Starting Height (H1) on the x-axis.
35) Draw a best fit line and calculate slope of this line.
36) Test this line by placing a cup at a given range that has not been done before and note from the graph what should be the starting height of the marble on the ramp. Test the idea and see if your estimate works or not?
37)
38) Other Calculations :
39) Calculate Average Measured Speed on Ramp ($\mathbf{V_R}$) and Average Predicted Speed from Acceleration ($\mathbf{V_P}$) and compare these values to the measured Horizontal Speed. Which is the best estimate of speed here?
40) Estimate by calculation the Vertical Speed component (**Vy**)
41) Calculate the Impacting Speed of the Marble ($\mathbf{V_I}$)

42) Calculate the Gravitational Potential Energy (PE) of the Marble at the top of the ramp.
43) Calculate the Kinetic Energy (KE) of the Marble at the bottom of the ramp using the Average Horizontal Speed as being the valid measure of its motion.
44) Compare the values for the Potential Energy and Kinetic Energy and determine why or why not they should be or are similar.

Data :

[M] mass of the marble (ball) : ________ (kg)

Height of Ramp Triangle (H1) : ________ (m)

Length of Ramp (L) : _________ (m)

[H2] Height from floor to Launch : __________ (m)

Data for Falling Time :

Trial	Time to Fall ($\mathbf{T_x}$) (sec)
1	
2	
3	

Compute : $\mathbf{T_f}$ = __________

Data for Rolling down the Ramp Time :

Trial	Time to Roll ($\mathbf{t_R}$) (sec)
1	
2	
3	

Compute : $\mathbf{t_1}$ = ___________

Data for Horizontal Flight Distance :

Trial	Time to Fly the Range ($\mathbf{D_x}$) (m)
1	
2	
3	

Compute : D = __________

Calculations :

Be sure to use your Slide Rule!

Be sure to convert as needed
1 m = 100 cm
1kg = 1000g
1 in. = 2.54 cm
1 kg = 2.2 lbs

Use for acceleration due to gravity, g = 9.8 m/s/s

Angle of Ramp to Table :

$$\mathbf{Sin(\Theta) = \frac{H1}{L}}$$

$$\mathbf{\Theta = sin^{-1} (\frac{H1}{L})}$$

Slope :

$$\mathbf{m = \frac{\Delta Y}{\Delta X}}$$

$$\mathbf{\Delta N = N_2 - N_1}$$

Formula for determining Averages :

$$\mathbf{Average = \frac{Sum\ of\ the\ Measured\ Values}{Number\ of\ Measured\ Values}}$$

Predicted Time of falling marble :

$$\mathbf{t =(\frac{2*H2}{g})^{1/2}}$$

Predicted Acceleration (A_P) :

$$\mathbf{a = \frac{5}{7}*g*sin\Theta}$$

Measured Acceleration (A_M) :

$$\mathbf{a = \frac{2*L}{t_1^2}}$$

Horizontal Flight Speed (V) :

$$V = \frac{D}{T_f}$$

Measured Speed on Ramp (V_R) :

$$V_R = \frac{2*L}{t_1}$$

Predicted Speed from Acceleration (V_P) :

$$V_P = a*t_1$$

Estimated Vertical Speed (Vy) :

$$Vy = \frac{2*(H2)}{T_f}$$

Impacting Speed (V_I) :

(Note use Horizontal Flight Speed & Est. Vertical Speed) :

$$V_I = (V^2 + V_y^2)^{1/2}$$

Energy Formulae :

PE = m*g*h

KE = ½*m*v²

Mechanical Energy :

M.E. = K.E. + P.E.

M.E. (before) = M.E. (after)

In this case :

P.E. = K.E.

Conclusion :

Wow – What adventures in Energy Exploration we have done here. There are many more of these ideas extended and expanded upon in the Efficiency Project which uses many of these ideas and takes it further and includes devices and a system we create ourselves! Too cool! Awesome Sauce!

The idea is to examine how close the predicted by formula to actual values are and determine how and why they are not the same. Consider what could be done to have the values of PE and KE closer to each other and what can be done in terms of your measurements. Next notice the range of topics that can be covered in this lab from the various realms of physics.

Explore & Enjoy!
Be Brilliant!

Activity #28
Elastic Potential Energy determination
& Hooke's Law Exploration
Grade Level : High School
Math Level : Calculating

Elastic Potential Energy Activity

At one time or another we have all bent a flexible solid material, such as plastic or metal, and let it go. As we pulled it, we notice there is a force pulling back on us. We find that it moves opposite to our original motion, from a restoring force. Typically the object will move quickly right past its original starting position and move almost as far in the opposite direction that it was originally moved. Once there it slows to a stop and moves back in the opposite direction again. The mass is vibrating back and forth past its starting point, called the equilibrium position. Given time it once again comes to rest at this point. Consider placing a ruler several inches off of a table letting one end stick out and flex it and let it go whilst holding it in place at the other end on the table – this vibrating ruler is what we just described in the opening paragraph.

Think about its motion in your mind as a sort of animation for a moment. The motion of the object has zero speeds at its extremes and has its maximum speed at the equilibrium point in the center. Since it is changing speed, it is therefore accelerating, recalling the idea we covered in Newton's 2nd Law Activity. This means a force is acting on the mass. This force is given a name. It is often referred to as the restoring force.

Also any change of speed means a change in the kinetic energy of the object. But energy cannot be created or destroyed – it had to be put into this system – the vibrating ruler for example. For any change in energy means that work is being done or by that system. First work had to be done to move the mass (can be a stretched or compressed spring, a stretched rubber band, or plucking a taught wire, or the slight bend of our ruler). This initial work is a change in elastic potential energy of the system. This in turn becomes the change in kinetic energy of the system. The reason it does not move indefinitely means that there must be a force acting on the system (according to Newton's Laws of Motion). This force is friction (essentially heat and sound), which brings the system to a stop and its energy, which cannot be destroyed is turned into heat and sound. The decrease in motion is called damping.

A key question arises, however. What does this restoring force depend on? This question is not new and even centuries ago there were some thinking about it and working to describe it. . The answer came from Robert Hooke in 1678 when he described the restoring force and its dependency on the displacement of the mass. That is to say, the more one stretches a spring, the greater the restoring speed. The force, however, is only proportional to the displacement. To make it an equality, a constant of proportionality is needed, here called the spring constant. The value of this is not universal for any and all springs, but instead depends on the spring. The harder it is to move the spring, the stiffer it is, hence the 'k' value is greater. It's units are N/m, which can be seen from the formula described here :

$$\mathbf{F = -k*\Delta x}$$

The spring constant is a characteristic or parameter of an elastic materials, such as a spring (for which it was originally named) to represent a measure of a spring's resistance to being stretched or compressed.

Note, it is important to realize that too much stretching can extend an object beyond its elastic limit, and will render the spring basically permanently sprung and in the case of a rubber band, typically broken.

Also in the case of the Activity, springs within a given range of stretching will be very consistent, while rubber bands have as many as 3 different ranges over which they have 'k' values, so in obtaining data be careful with rubber bands and keep the ranges small for a reliable figure for the calculation of its 'k' value.

The application of this motion can be seen by considering simple harmonic motion. Simply defined, simple harmonic motion is where there is a vibration about an equilibrium position in which a restoring force is proportional to the displacement from equilibrium. This type of motion needs to be understood and controlled in all spring-shock systems for vehicles or anything that needs to remain stable when in randomly-oscillating, or bumpy situations. It also is parallel to Pendulums (see Period of a Pendulum Activity #44) since both exhibit the same characteristic back and forth motion where the force is a max at the extremes and is zero at equilibrium. One key difference comes from the formulas for determining the period of a pendulum as compared to a mass-spring system. The simple pendulum does not depend on mass as the amount of restoring force (gravitational force) here increases as the mass, hence inertia, of the pendulum does. The mass-spring system on a level surface and only relying on the force due to the spring, has a situation where the mass's magnitude does not affect the restoring force. An increase of mass here results in a smaller acceleration, hence a greater time to complete one oscillatory cycle.

Other types of mathematical applications for restoring forces look at bow strings for archery, pendulums in clocks or when acrobats such as trapeze artists are in motion.

The activity here first examines the determination of the spring constant for a rubber band. This idea is from Hooke's Law. In the main activity that follows we then examine the energy of the rubber band when stretched. Here it has what we call Elastic Potential Energy.

Potential Energy can come in a variety of forms and is all due to the properties or conditions of the matter. Elastic Potential Energy can be stored in any solid material that can be deformed and it then returns to its original state when released, such as a stretched or compressed spring, a stretched rubber band, a bounced super ball, or any other rubber or elastic ball for that matter. It does not always have to be things that are made of rubber or plastic, however. Virtually all materials have a certain amount of response or return to its natural 'at-rest' state when distorted. Take, for example the fact that most items bounce some when they hit the ground after being dropped. This is due to their flexing and returning to their original form.

Stored elastic potential energy has a number of uses. Take into consideration the use of springs or bungee cords. The simplest, most common we encounter are spring-activated toys that pop back after being compressed. There are rubber elastic counterparts to this, such as the poppers, which are essentially half of a hollow tennis ball that one inverts and places on the ground and then it pops back into shape and in snapping against the ground launches into the air.

Like all energy forms, elastic potential energy can only be stored or converted into other forms of energy. Like all energy it cannot be created nor destroyed. (For more on the topic of Energy, its forms and the like, read the prelude to Mechanical Energy Activity #27)

The best way to analyze it is to determine how much elastic potential energy an object has, here we are using rubber bands, and then converting it into another known energy form we can measure, here gravitational potential energy (after first becoming the kinetic energy of the moving rubber band) and comparing the two values. Of course along the way we are assuming a complete conversion of one energy form into another, but we need to realize that energy can also manifest itself in terms of other things such as heat, light, and sound. Here no light is given off, but some of the elastic potential energy when converting to both kinetic and finally the gravitational potential energy will go into heating the rubber band slightly, cause some sound (the snapping of the rubber band), as well as breaking some of the bonds of the molecules of the rubber band. Note, that this is not a violation of the conservation of energy principle, but an affirmation of it. We will never get more out of the system than that which we put in.

Prelude Activity to Main Activity : Hooke's Law (aka Pre-Activity) :

Purpose : To determine the spring constant, k, for a rubber band through increasing force and measured displacement values.

Materials :

- Rubber Bands (buy bags of specific types for uniformity),
- known masses – here nuts for eyebolt or hook,
- eye-hook or hook bolt where nuts can attach – a very good suggestion for this item which is used to connect masses to the rubber band is to simply use a twist tie instead of items that possess so much mass,
- Kitchen mass scale (to 0.1 g),
- System to suspend the rubber band and be able to attach masses one at a time so that it stretches and the amount of stretch is measured – typically it is easiest to use a dowel rod sandwiched between two stacks of books that act as towers to suspend the rod over the table. The rubber band is looped on the rod,
- Ruler,
- Slide Rule

Note : If masses too small – place a measured number of them on scale, measure mass and divide by number of items and use average

Note : If comparative experiments are wanted then having 3 varieties of rubber bands is a good idea.
Note : measure all masses used separately – that is each eye hook (if used), each nut, et al – find a means to keep them labeled and known when adding masses to your spring to stretch it.

Procedure : Determining 'k' for a Rubber Band

1. With any and all Activities read through, then set up the process for the Activity and run it without measurement and watch where you may have bumps in the road so as to anticipate these. Be patient – Repeat processes. Once all is well when measuring see to the values – do you notice patterns or trends – what do you predict will happen next as you change one thing or another from one trial in an experiment to the next. Always Explore & Enjoy. Be Brilliant!
2. Know the masses being used by measuring them on the mass scale. Record these results in a table. – Note if using a twist tie, we will assume that its mass is negligible.
3. A convenient stand for the rubber bands is having to stacks of books near each other and bridged by a dowel rod. Make it level and tall enough so that when the rubber bands when elongated by the addition of masses allows for it to stretch. (The stack should be approximately 1 to 1.5 ft)
4. The rod can be wood and can be notched slightly to let the rubber band hang down from the rod.

5. Arrange a ruler nearby so as to stand vertically and close enough so that the hook on the rubber band will be readable.
6. First record the initial position of the bottom of the rubber band according to the scale
7. Attach the eyebolt hook to the rubber band – or as noted earlier use a twist tie which is more convenient (it merely has to be long enough to allow ease of attachment of the additional nuts – see step 8). Be sure that it does not touch the scale and is parallel to the table top.
8. Record both the mass of the eyebolt hook and the new position of the rubber band – note you first measured to the bottom of the rubber band, so do so again. This is your measuring point for each of the trials as mass is added.
9. Add nuts (known masses) sequentially. With each added mass record the total mass for the system as a separate trial.
10. For each trial - Read and Record the total displacement for each trial.
11. Minimally have 3 data points (trials) but the more you have the better.
12. With all measurements made, perform calculations using the slide rule :
13. First, Calculate the weight for each trial (this is the force acting on the rubber band in each case) (W)
14. Convert the Displacement from cm to m, then
15. Graph the points Force (W) on the y-axis vs Displacement (Δx) on the x-axis.
16. Draw a best fit line through the data points – it should be linear.
17. Next, Determine the Spring Constant from the slope of the graph from the best fit line
18. If doing more than one rubber band for comparisons, Follow all of these steps for each of the rubber bands used.
19. In the next phase of the Activity, use the spring constant values 'k' and the mass for each of the rubber bands involved. Be sure to keep track of them. Using a sequential ID for them is best, such as #1, #2, et al.

Data :

First generate a table for the known masses to be used. Number them sequentially. The 1st is the eyebolt or hook. Note that each rubber band has its own set of tables.

Trial	Total Mass (g)	Weight (N) Calculated	Displacement place (cm)
1			
2			
3			

Calculations :

Be sure to use your Slide Rule!

Note : Mass is taken in grams in this exercise.
Convert the mass to kilograms! 1kg = 1000g
Calculated Weight : W = m*g (let g = 9.8 m/s^2)
As the force from Hooke's Law is a response force acting in opposition to displacement, the formula reads :

$$\mathbf{F = -k*\Delta x}$$

Here F = W, Δx is converted to meters and will be written as 'x' since measurements for the initial 'x' will be called '0'. The absolute value will be used here for the 'k' or spring constant value and the formula is rearranged as :

$$\mathbf{k = \frac{W}{x}}$$

the best way to deal with the data is to graph the points on a graph, draw a best fit line and then determine the slope of the line to determine 'k'. Note that the y-axis is Force and the x-axis is displacement. Slope is rise over run or :

$$m = \frac{\Delta y}{\Delta x}$$

Main Activity :

Calculation and Comparison of Elastic Potential Energy to Gravitational Potential Energy

Purpose : To use calculated elastic potential energy values and compare them to gravitational potential energy outcomes (based on altitude acquired) from launched stretched rubber bands.

Materials :

- Rubber Bands (from prior pre-activity of 'k' determination),
- Mass Scale (note good to keep measures of mass from pre-activity for each of the rubber bands),
- Ruler,
- Measuring Tape,
- Wall,
- Duct Tape or Tack Pins that can be pushed into wall,
- Rubber Mat or Other mat-shaped material that rubber bands can cling to,
- Slide Rule

Note : It is assumed that the 'k' value from the pre-activity (Hooke's Law) has been done and determined for the rubber band(s) being used.

Set Up :

1) From the floor hold a ruler upright with the edge away (perpendicular) to the wall. Slightly angle it towards the wall. It is best to use a folded sheet of paper or index card to maintain a constant slight angle,
2) Take one of the rubber bands and stretch it slightly from the top edge of the ruler and pull back towards the floor. This is done so the rubber band can be launched essentially straight upwards.
3) Keeping your eyes out of the way (if uncertain – wear goggles), launch the rubber band to see how far it goes.
4) Do this for several practice runs to estimate where the rubber band will reach to. Stretching should be from 2.0 cm to 6.0 cm (the amount is determined by the stretch uses in the pre-activity and the amount of elasticity of your rubber bands).
5) With the idea of range in mind, either tape or pin the rubber mat (or other material) to the wall.
6) Try the experiment again and perfect the technique of launching the rubber band successfully and finding an angle so that the rubber band does not hit the wall too soon or too late. Plus it must stick to the wall long enough so that you can stop it from falling. Be sure not to affect its attained altitude in stopping it.

Procedure :

1) With any and all Activities read through, then set up the process for the Activity and run it without measurement and watch where you may have bumps in the road so as to anticipate these. Be patient – Repeat processes. Once all is well when measuring see to the values – do you notice patterns or trends – what do you predict will happen next as you change one thing or another from one trial in an experiment to the next. Always Explore & Enjoy. Be Brilliant!
2) With the pre-activity and the set up exercise done, you can now measure each of the rubber bands used in this Activity for comparison of Elastic Potential Energy to Gravitational Potential Energy.
3) Record the amount of mass of the rubber band (m), the stretched distance (d). Also have for it the 'spring' constant 'k' as determined for that rubber band placed into the table too.
4) With each rubber band, in turn, set up to launch it with a pre-determined amount of stretch. Do this 3 times for each rubber band at a given amount of stretch.
5) Launch and quickly stop the rubber band on the mat on the wall.
6) Measure the height from the starting point (the top of the ruler) for the maximum altitude attained by the rubber band and record this value (h_i). It is best to use the same method in measuring for each trial. For example, measure to the top of the rubber band on the wall mat.
7) Record the maximum height attained value.
8) Note there is an easier alternative – the prior procedure assumes one person – with someone else helping one can launch the rubber band while another looks straight on from the side watching to see how high does the rubber band go – be sure to wear goggles and not be too near. To help in measurement, simply tape a meter stick to the wall to look at as well.
9) Calculations :
10) From the accumulated data : Determine the Average Height that the rubber band achieves for a given amount of stretch. (h_{ave}).
11) Using your slide rule, calculate the Elastic Potential Energy (PE_E) of the rubber band.
12) Now determine the Gravitational Potential Energy (PE_G) and place it as well on the table.
13) Graph the gravitational PE (y-axis) versus the elastic PE (x-axis), draw a best fit line and determine the slope.
14) Graph Average Height achieved (h_{ave}) on the y-axis versus the stretched distance (d). Draw a best fit line through the curve (it should approximate a parabola).
15) Graph Average Height achieved (h_{ave}) on the y-axis versus the stretched distance-squared (d^2). Draw a best fit line through the line and determine slope. (If done well, the slope should be 2).
16) Another type of graph to use to see if there is a square relation between height attained and stretched distance is to graph the log values of each. Here $\log(h_{ave})$ is on the y-axis and $\log(d)$ is on the x-axis. Again draw a best fit line and determine slope (which should be 2).

Data :

Note : Each rubber band should be given an ID, such as a number : 1, 2, et al.

Rubber Band ID	Mass [m](g)	Stretch [d](cm)

Rubber Band ID	Height 1 [H_1] (cm)	Height 2 [H_2] (cm)	Height 3 [H_3] (cm)	Height [H_{ave}] (cm)

Calculations :

Be sure to use your Slide Rule!

Calculation Steps :

Convert all mass units from grams into kilograms (1 kg = 1000 g)
Convert all distance units from centimeters to meters (1m = 100cm)

Elastic Potential Energy :

$$PE_K = 0.5*k*d^2$$

Gravitational Potential Energy :

$$PE_G = m*g*h$$

Note : g is taken as 9.8 m/s^2 and 'h' here is h_{ave}

Average :

$$= \frac{\Sigma x_i}{n}$$

Slope :

$$m = \frac{\Delta y}{\Delta x}$$

Conclusion :

Certainly amazing Rubber Bands could be something we turn into a Science Experiment and Exploration – Wow! All things have a story to tell us – we have to be creative in thought, curious in nature, and passionate about our need to explore to learn and be Brilliant!

How well did your graphs turn out with regards to showing a square relationship between height attained by the launched rubber band and its initial stretching? How comparable are the results of the values of the types of potential energy? Why might you think that there are differences? (Consider conservation of energy, which is being employed here and yet does all of the elastic potential energy only go into the motion of the rubber band – can some of the bonds in the band break? When stretched, does the band become warm – why, where does this energy come from? And many more questions to think on).

Activity #29
Doll Bungee Jump
Grade Level : Middle School
Math Level : Calculating

Prelude :

At one point or another most of us have heard of and seen bungee jumping. Safety is paramount in such an endeavor. This Activity does not involve people, but creates a small-scale version where a doll, such as Barbie, can be a bungee jumper for math and science considerations. We get to explore on a small scale bungee jumping using one of our favorite dolls – too sweet and fun!

Here we are going to use rubber bands which stretch in a limited range in a manner akin to Hooke's Law (see that Activity #28 for more on this). In actuality rubber bands only exhibit this Hooke's Law behavior over a short range of stretch.

Here is a more technical aside to this activity for those considering other depth on an overly stretched rubber band - if they were stretched at longer distances it would show what is referred to as hysteresis (a non-linear relation). Here since the masses are small and the number of rubber bands relatively large the linear relation is reasonably accurate.

The basic goal is to obtain data in the lab (so to speak) on a small scale which will be used to create a line equation to determine the number of needed rubber bands connected to each other in a slip knot fashion so that when the doll is dropped from a higher chosen height there is a balance of safety versus the maximum thrill of the fall (that is to say the doll comes as close to the ground as possible but not hitting it and causing injury).

For those interested in the math of the Activity this one requires gathering data, data analysis by graphing data points, finding the slope and intercept of a line and use the line to find a reasonable estimate for the needed number of rubber bands to maximize thrill versus safety for the doll. It should be found that the distance the doll will fall to has a directly proportional relationship to the number of rubber bands employed in a given trial. So it can be said that this exercise (good for math and/or science) is an examination of a linear function.

Purpose :

To determine the length of bungee cord needed as determined from graphed data to reach certain maximum yet safe lengths of stretching potential

Materials :

- Rubber Bands (best to have uniform type bought in large amount – typically bagged or boxed and found in an office supply store),
- Measuring Tape,
- Stairway with railing (so as to provide the maximum vertical distance safely), - see note if using a triangular standard ladder instead
- Doll (Barbie works best, but can also be GI JOE – a standard type doll about 7" works well),
- Table,
- Stack of books or a plastic crate (best choice),
- Plastic cutting board or rod (acts as jumping off point from the crate atop the table (all held by C-clamps)),
- C-clamps (2-3),
- Goggles,
- Graph Paper,
- Slide Rule

Note : The testing of both the short falls with increasing numbers of rubber bands to generate data points and the final long fall testing the estimated number of rubber bands needed should be done with parental permission and supervision. Note – wear your goggles and be careful of the rebounding doll.

Note : In the case of the maximum fall, this is an agreed to distance ahead of the experiment and must have parental permission and supervision. It is also recommended to be conducted by the parent(s) as needed. Never engage in unsafe conduct involving vertical heights, stairways, railings, and the like.

Note : The number of rubber bands depends on the chosen maximum falling distance the bungee-jumping doll will fall. Note that for the trial run to gather data one will need initially up to 15-20 (probably less depending on size) rubber bands and then you will determine from your generated equation from the data points the estimated number needed to safely fall and return from the chosen maximum height (this could be a large number – 30 and up)

Note : Though only stairs are considered and described here – a regular triangular ladder can be used – but be sure to have adult help and supervision – never be up on the upper steps of the ladder at all – all you have to do there is let the doll fall between the legs of the ladder which can be reached rather safely.

Picture of possible Dolls (Barbie on left, GI Joe – Jane version on right – note the rubber bands besides them. Also important is the fact that other than a basic outfit, you should have no other loose items on the doll (shoes, gloves, necklaces, hairpieces, et al) :

Procedure :

1. With any and all Activities read through, then set up the process for the Activity and run it without measurement and watch where you may have bumps in the road so as to anticipate these. Be patient – Repeat processes. Once all is well when measuring see to the values – do you notice patterns or trends – what do you predict will happen next as you change one thing or another from one trial in an experiment to the next. Always Explore & Enjoy. Be Brilliant!
2. Set Up for Activity :
3.
4. First a Maximum Fall Distance place is chosen and known ahead of time. Its height is measuring for the maximum fall. Note that parental permission and supervision is needed here.
5. Read the Materials list and Notes to be sure you have the correct and needed materials. It is important to have parent permission and supervision as well as wearing goggles. Be sure that the doll used has no loose materials associated with it that can fly off when undergoing the bungee jump.
6. The initial set up involves a small scale version of the bungee jump so as to collect data points, generate a graph and then use the equation of the line formed to determine the number of rubber bands needed for a much larger scale bungee jump and then test that estimated value. The goal of the successful bungee jump is to have the doll avoid injury

yet at the same time maximize the thrill of the fall (be as close as possible without contact).

7. Essentially the following describes placing a table near a stairwell (usually with a railing whereby a small 'platform' for the doll to fall from is set up. Our goal is to maximize the falling distance. One could simply avoid the table, the clamps and tape and such and have the doll fall from the railing but it may be too close to the wall and have too much interference – so a small ramp one can reach from easily could be in place with little to no effort. An even better idea may be the use of a standard ladder where the doll falls between the legs.
8. The initial set up described here involves a stable table where a plastic crate is turned so as to be its tallest in the vertical direction and then it is secured to the table with one or more C-clamps (Note if it is a typical crate with holes and it is hard to find a place to clamp across, then place a thin book that fits into the crate to act as a flat plane). Note : In using C-clamps on a table or book first you need parental permission and supervision – next it may be a good idea to use a thin block of wood to be placed between the clamp and the table or the crate so that it prevents damage.
9. Next – at the highest position (the top of the crate) now attach the chosen rod or thin board to be used. C-clamps or tape may be used. It needs to be reasonably secure not only that it does not become unsecure, but also so that it does not vibrate too much when the doll is bungee jumping – the greater the vibration, the greater the error in the data.
10. The attached rod / board projects out over the edge of the table so as to act like a diving platform essentially.
11. Now set the measuring tape on the floor and extend it up to the platform and lock it in place there so that measurements can be taken against it.
12. A critical idea here is that even in executing the trial runs it will require help. It is best to have 2 people to do the experiment. One will hold and then let the doll fall while the other is positioned in such a manner to see the doll fall and be able to watch the maximum extension of the fall of the doll against the measuring tape – and then comes up to the tape and read the value (to the nearest cm reading) for each trial.
13.
14. How to Attach Rubber Bands :
15.
16. It is best to use what is referred to as a double-loop when not only adding rubber bands in sequence (two per each trial is recommended) but also when first securing the dolls feet.
17. The double-loop is made when there are two rubber bands and one loop is wrapped over the other stationary band and one edge of the looping band is drawn back through the band's other side after wrapping around the stationary band. This is what is called a slip knot. Research it if further information is needed.
18. With each trial add 2 (be consistent for each trial). Always attach at the end of the jump line that is attached to the diving platform.
19.
20. Activity : Trial Runs for Data
21.
22. At this point you should have all components set to do the trial runs of the bungee jumping doll. As a reminder it is best to have 2 people to do the exercise since one is letting the doll fall and rebound while the other is reading the scale.

23. With each trial (a set number of rubber bands) perform it at least 3 separate times and record each of the results.
24. Collect data for at least 3 (preferably 4 or 5) data points. Note, if needed do to running out of room for the fall distance of the doll and the floor then consider using only one rubber band for each trial run.
25.
26. Calculations :
27.
28. Note : do conversions when and where needed (see Calculation section).
29. With all the data collected the first calculation is to calculate the average jump distance for each value of rubber bands. Use the Average Formula. Place these values in the second table.
30. Now graph the Number of Rubber Bands (x-axis) vs. the Average Jump Distance (y-axis) as data points.
31. Using only the data points draw a best fit line. Notice the direction here – Only the data points – your line should NOT go through the origin of the axes but instead should have a y-intercept (which is important).
32. For this line determine the Slope by choosing two points on it and using the slope formula.
33. Now using the Line Equation, plug in the value of the slope and one of the points on the line and determine the y-intercept (b).
34. Compare your intercept value to the measured height of the doll. Use the percent error formula to determine the % Error.
35. Now With the line equation completed now let y equal your maximum distance of fall and determine the needed number of rubber bands (x). With parental permission and supervision test this hypothesis to see if it works.
36. Did your prediction work – if not, what things are you potentially overlooking. Try the experiment again after first going over your data and rechecking the calculations. Also consider such things as the use of the rubber bands – for example there are ones that are attached for some time to the doll while others are not – could they be too stretched out while the new ones are not as flexible – what can be done to change this (stretch them ahead of attachment).

Data :

Chosen Maximum Distance of Fall : ____________ m
Measured Height of Doll : ________ cm

Trial	Number of Rubber Bands in Jump (n) [x]	Distance of Jump (cm) [y]	Distance of Jump (cm) [y]	Distance of Jump (cm) [y]

Number of Rubber Bands in Jump (n) [x]	Average Distance of Jump (cm)[y_{ave}]

Calculations :

Be sure to use your Slide Rule! :)

Slope :

$$\mathbf{m} = \frac{\Delta}{\Delta}$$

Average :

$$\mathbf{y_{ave}} = \frac{\Sigma}{\quad}$$

Equation of Line :

$$\mathbf{y = mx + b}$$

Percent Error :

$$\%E = \frac{[\qquad - \qquad]}{\qquad} * 100\%$$

Conversions :
1 m = 100 cm
1 in. = 2.54 cm
1 ft = 12 in.

Conclusion :

Wow! All things have a story to tell us – we have to be creative in thought, curious in nature, and passionate about our need to explore to learn and be Brilliant!

What a fun adventure – our doll has a journey of excitement – we can supply the sound effects if you like and we not only partake in this in our minds but we are looking at it with scientific thoughts as well.

How does your estimated values for the needed number of rubber bands compare to the actual distance – was it safe or not ? What sort of factors could have affected the outcome ? How close was your predicted measurement for the height of the doll as compared to the actual measurement?

Activity #30
Finding Energy and Power from climbing stairs

Grade Level : Middle School
Math Level : Calculating

Individual Power on the Stairs and the Slide Rule Activity

Too often physics seems to connect only to the mechanical world and not the living, physical, or biological. In the case of the following activity, instead of racing marbles or measuring the mass of a piece of wood, you actually measure yourself and what you do physically. We will explore our expression of energy and power in our activity – wow!

Why? The units of energy are universal. Though some use Joules and others use Calories, these are merely two different scales that tell you the same relative information – how much energy is there, or needed, or was used in some given case. In our activity we will use both to illustrate the connection.

Energy is the central building block of all sciences – physics, astronomy, chemistry, biology, and any other science for that matter. We are as much a part of science as any other item in any other science is. Every action we do, walk, lift, even think, requires the use of energy. This, of course, is why we eat. The chemical bonds of the foods we consume contain energy in that their interactions in our bodies as we break them down allow the energy to transfer from those molecules to the molecules that make us up.

Look at how energy is oftentimes defined : "the capacity or capability to do work". The very equation for work is work is the change of energy of a system! $W = \Delta E$. What this means is that any and all actions and phenomena in nature require energy to happen. Energy cannot be created nor destroyed, it can only be transferred from one item to another and typically it changes from one form into another. We convert the chemical energy of our food into the kinetic energy of motion of our bodies.

This activity is a fun activity to measure the amount of energy we use in doing various common activities and we can compare these results to the food we eat (a good idea for a personal Activity in conjunction to this one is to keep a log of your dietary intake, add up the calories both for meals and the types of foods they are – look at various web sites for this information and create pie charts (see Activities where you are doing this)). Also in this Activity we can convert the energy we use into horsepower, a common unit of power still in use today for engines and other mechanical devices.

This is another reason for this activity – to illustrate the differences between energy and power. Energy is simply related to a value to illustrate how much there is or needed for a situation, but it does not tell you how long it took for that energy to allow the phenomena in question to take place. For example, does it take more energy to climb a flight of stairs slowly or quickly?

This is a trick question! The answer is that both take the same amount of energy! We will do calculations to illustrate this in the activity. You might say that in one case you took ½ the amount of time to climb the stairs as compared to the other effort at it. What then is different about them? Only the amount of time is different. Since Work is the Change of Energy, recall the formula for Work (see the Mechanical Energy Activity and the Story of Energy prelude p 308, p. 316 and Activity #27). Work is a Force acting through a Displacement. Note that there is no mention of time. You can exert a force on a box and push it a distance across the floor. Independent of the speed, the amount of work done is the same for all cases. Returning to the stair-climbing in half of the time : the factor that changed here is not the work, it is the amount of time it takes to do it. There is a relation used in physics to examine Work per unit Time; this is called Power. (Power = Work per unit Time, $P = \frac{W}{t}$).

So a better question would be does it take more power to climb a flight of stairs slowly or quickly ? Obviously though they take the same amount of energy, the faster it is done, the more power is needed to accomplish the task. The rate of energy consumption is what makes one tired and causes us to heat up and hence a need to sweat in order to cool down.

Clearly many precautionary notes need to come with this Activity – do not engage in any physical activity if you are not supposed to and not advised to do so and do not have proper supervision. Do not overly exert yourself – when able to engage in an activity – do so in moderation and safely with supervision and permission.

Purpose : To calculate the amount of energy needed to climb stairs and with time factored in, determine the amount of power used to do this task. These calculations can be compared to one's foods eaten as well as other comparisons.

Note : A child should be advised and watched by an adult when exercising so as to avoid injury. Also always consult with a doctor when necessary so as to avoid any problems.

Note : Do not do any of this activity without proper preparation, consultation with a doctor, and having permission and supervision from a parent. Also never overdo any activity. When engaging in any exercise program or physical activity, do so with safety.

Materials :

- Self,
- Stairway,
- Timer,
- Weight Scale,
- Measuring Tape,
- List of Foods with Nutritional Information,
- Slide Rule

Procedure :

Activity :

Power to Climb the Stairs

1) Note : Only do this in safety, with parent and doctor permission.
2) Measure your weight on a standard scale in pounds. Record this information.
3) If possible : Using the measuring tape measure from the bottom of the stairs to the top in a vertical fashion (not at an angle) – This, of course, assumes that you can stand alongside your stairway – however, most homes do not have this option. So, If there is no way to do this then do the following :
4) Measure the height of a given stair and multiply this by the number of stairs to determine the vertical height of the stairs from the bottom step to the top step.
5) If you do not consider the stairs to be of the same height, then pick three of them at random and average their heights that you measure and multiply this average height by the number of stairs.
6)
7) Doing the Physical Activity of climbing the stairs :
8)
9) Measure the amount of time with a Timer to reach the top of the stairs. With each trial, try to walk at the same pace. Even if they are not the same that is okay. We are going to take the average time.
10) In terms of pace, try a slow pace for the first 3 trials.
11) In the second set of 3 trials move at a faster pace up the stairs. Note : When moving faster be sure to be careful. Note that fast here merely means at a quicker pace than slow. This is not a race, and should only be done with permission from adults and doctor, focus on one's health, and with care.
12) If doing this with other participants, and it is recommended to have a partner who want to see their results, share and obtain the information with them.
13)
14) When all the measurements are taken, Calculate the following :
15)
16) First, Convert your weight into kilograms, then into Newtons. (formulae provided)
17) Determine the Average values for the times you went up the stairs. An Average time for Slow and one for Fast separately.
18) Now Calculate the amount of work each person has done in Joules from the gravitational potential energy formula.
19) Next, Calculate the amount of Power used for the Power relation.
20) If you have others to compare to or if you have created a list of foods you eat or have foods available with nutritional information on them, then do one or more of the following :
21) Compare your results to others who climbed the stairs, or
22) Compare the amount of energy needed to climb the stairs to various foods you eat in one day (for example, an apple, glass of milk, slice of bread, etc),

23) And Calculate as well, the amount of horsepower used to accomplish this stair climb.
24) Note of course you have to convert your Calories to kilojoules :
    1 Calorie = 4.186 kiloJoules, also convert your Joules to kilojoules (1kJ = 1000 J).
25) Be sure to use a slide rule in all calculation cases!

Data :

Stair Climb Activity :

Weight (Wt) : _________ (lbs)

Height of Stairs (Ht) : ____________ (m)

Trial	First time $[t_1]$(s)	Second time $[t_2]$(s)	Third time $[t_3]$(s)	AverageTime [M] (sec)
Slow				
Fast				

Note : For comparisons, have a friend create a table too or in the case of food comparisons, be sure to make a list of things you eat and their calories. On most items you can find the calories listed.

Calculations :

Be sure to Use Your Slide Rule!

You may need to convert things, so watch your units :

1 inch = 2.54 cm
1m = 100 cm
1 kg = 2.2 lbs.
1 Food Calorie = 4186 Joules

Number of kilojoules in Food when given Calories :

N = Calories*4.186

Average Calculation :

$$\textbf{Ave} = \frac{\textbf{Total for all Trials}}{\textbf{Number of Trials}}$$

Convert Weight from Pounds to Kilograms : __________

$$\textbf{Mass (kg)} = \frac{\textbf{Weight in Pounds}}{\textbf{2.2}}$$

$$\textbf{Weight in Newtons} = \textbf{Mass (kg)} * 9.8\,\frac{m}{s^2}$$

<u>Energy in a given Trial :</u>

<u>Gravitational Potential Energy in Climbing Stairs :</u>

PE = weight*height of stairs

PE = W*H

<u>Converting Joules to food Calories :</u>

$$\#\textbf{Cal} = \frac{\#\ \textbf{Joules}}{\textbf{4.186}}$$

<u>Comparison Formula :</u>

% or No. of Food Items energy expelled is equal to =

$$G = \frac{\text{PE (converted to Calories)}}{\text{Calories for item in question}}$$

(Note : To see how many times the stair climb needs to be done for a given food item in question, simply invert the formula – the numerator and denominator switch places)

<u>Power for Activity :</u>

Note : For proper units of Power, you need to convert the time values of seconds into hours. The unit of Power is Watts, which is Joules per second

1 hour = 60 minutes
1 minute = 60 seconds

Power = Work/Time = Change of Energy/Time = kilojoules/time(H)

$$\textbf{Power} = P = \frac{\text{Work}}{\text{time}} = \frac{\text{Wt (N)*Ht}}{\text{M}} * \frac{\text{60 min}}{\text{1 hour}} * \frac{\text{60 sec}}{\text{1 min}}$$

Conversion of Power to horsepower :

1 horsepower = 746 Watts

$$HP = P * \frac{1\ HP}{746\ Watts}$$

Conclusion :

Having fun and feeling empowered!

How do the values for Power compare? If you did the Activity with a friend, how do their values compare to yours? If you did it yourself only, how would the values change if you had a different weight – more weight or less weight – how would this affect the outcome? What about time – if you were to increase time or decrease time, how would the results change? If you did keep track of your food or have food for comparisons, how does your energy needs (in kJ) compare to say an apple, a sandwich, a glass of milk?

Breads	**Fruits & Vegetables**
Slice of regular bread 75 cal	Apple 95 cal
Bowl of Oatmeal 150 cal	Banana 111 cal
Bagel 245 cal	Grapes (1 cup) 104 cal
6 Saltine Crackers 78 cal	Dates (1) 20 cal
Meats	Orange 62 cal
Beef 1 steak 407 cal (164 g)	Orange Juice (8 oz) 111 cal
Chicken Breast 344 cal	Carrot (1) 25 cal
Ground beef hamburger 172	Asparagus 2 cal/spear
Pork Chop 257 cal (131 g)	Broccoli 207 cal per 1 bunch
Turkey leg 1136 cal (546 g)	Brussel Sprouts 8 cal per sprout
Ham slice 236 cal (145 g)	Corn 562 cal per 1 cup
Egg (1 egg) 78 cal	Green Beans 34 cal per 1 cup
Milk & Dairy	Peas 79 cal per 1 cup
Whole milk glass (8 oz) 149 cal	Cabbage 227 cal/head
Rice milk (glass) 133 cal	Tomato 20 cal per tomato
Plain Yogurn (standard cup) 138 cal	Potato 165 cal
Soy Milk (cup) 109 cal	French Fries 365 cal
Cottage Cheese (1 cup) 206 cal	Gnocchi 250 cal
Cheese (1 slice) 31 cal	Peanut Butter (2 tbsps) 188 cal

Activity #31
Various Rate of Heating Determinations
Grade Level : Middle School
Math Level : Calculating

Rate of Heating

All of us notice temperature changes and feel quite uncomfortable when too hot or too cold. The ream of heat extends beyond our personal thermostats and those in our homes and vehicles, however. Heat is its own topic in science – known as the realm of Thermodynamics and there are even specific laws in it which are cornerstones of all of science itself.

Why is the study of heat so vital? The obvious answer is nature and weather itself! Land and Water heat at different rates and the angle of the Earth's tilt with respect to the Sun plus its rotation creates a very complex system that results in different rates of heating along with the fact that the incoming radiation is partially reflected into space before even entering the atmosphere (about 30%). Different materials absorb at different rates and the energy can be redistributed, reabsorbed and retransmitted in the infrared (concerns about global warming due to materials even in the air such as CO_2, et al). The amount of water vapor, dust, along with the angle the sun strikes the surface, and the texture, and composition of the terrain plus the wind speed will affect the surface temperature. All of these affect the atmosphere's energy content, hence affecting pressure and pressure differences is the driver of the winds along with other factors too. This drives the weather!

Closer to home, heat and its transfer is critical in immobile structures (homes, buildings, et al) and even in mobile structures (cars, trucks, engines, et al). In buildings it is the retention of heat, particularly at cold times that is of importance. Yet when it is warm outside, the retention of cold air and the cooling of the air is critical. The measurement, dissipation, and control of the heat is very important. Even with a computer and its cooling fans and other appliances, such as refrigerators, and those that use heat like toasters, ovens, and coffee pots, heat once again becomes a factor of importance. The understanding, control, and use of heat is essential to many mechanical and biological items, aspects of life, and life itself.

The study of heat is a branch of physics called Thermodynamics . What of the definition of heat and the units of heat? In physics, heat is considered the energy transferred from one body to another due to thermal contact due to differences in temperature.

Naturally heat transfers from a body of warmer temperature to one of cooler temperature. When bodies are at the same temperature they are said to be in thermal equilibrium. When they are not the process of heat (energy) transfer can take place. The examination of heat looks not only at appliances and buildings, but also weather, the interior of the Earth, and other bodies in the universe such as the Sun and other stars. Heat is a measure of the quantity of energy of a system, hence there is no 'cold' per se, merely the absence of heat as compared to another body. Also the Heat or Energy can only transfer via one of 3 known means : Convection, Conduction, and Radiation.

Conduction is transfer of heat/energy from particle to particle in a material or between materials in contact with each other. Often conduction occurs in solids. Metals are good conductors while wood typically is not. Think of the question in a science class do you touch the metal spoon or wood spoon in the hot pot of boiling water – never the metal one! Conduction is vital in understanding heat transfer in various processes for both the natural world and manufacturing.

Convection is the transfer of heat/energy due to the movement of the particles of the substance itself carrying heated particles, such as currents in fluid materials (which includes liquids and gases). Convection is very common in the atmosphere and ocean but also found in molten layers of the Earth below the solid ground and even in stars like the Sun where these moving cells of heated plasma since electrified even generate magnetic fields too. The convection currents are very important to all things on all levels.

Radiation is energy/heat transferred by electromagnetic waves. This includes Visible Light, along with all the other forms – Gamma Rays, X-Rays, Ultraviolet, Infrared, Microwaves, and Radio Waves. Think of all the uses for these various waves we have in our world – TVs, radio, cellular phones, ovens, medical applications, light that can be turned into electricity with solar cells (there is an Activity in this book – Galaxy Girl's Inquisitive Pioneer Adventures) and so on. Plus these waves carry information from stars, planets, and nebulae to convey to us information about the universe at large.

The units of heat are : Joule, calorie, kilocalorie. In fact, the factor noted before is the conversion for calories and joules : 1 calorie is 4.186 joules. Note that these are not the food calories we consume, each of these is 1000 of the base unit calories, hence the food ones are often written with an upper case 'C' as Calories. We could then convert our food calories to kilojoules by multiplying by the conversion factor. In many European countries, cans and boxes of food do not note the Calories, but instead the kilojoules instead.

Heat is something which may be transferred from one body to another according to the 2nd law of thermodynamics – essentially it moves from a higher temperature to a region of lower temperature naturally (i.e. the noted natural process and via one of the aforementioned means or if it does go from a place of lower energy to one of higher energy, it requires the input of work).

In our following Activity we set out to explore the energy of heat and more specifically the rate of heating of different items. Most of us have encountered and/or read about this. It is classically illustrated with a pot of warmed water and in it are two spoons – one metal with its handle out of the water and the other wood with its handle out of the water. In a short time frame the metal spoon is far warmer than the wood one, hence we classify metal as a good conductor of heat, while wood is rather poor.

As with all Activities, have your parent's permission and help in doing these Activities. Always employ safety in doing the Activities.

Since we are looking at Rates, this means a number of the Activities will reach a conclusion of making a graph since we are looking for trends – regular changes in times as things are occurring due to the experiment we are conducting.

The key to each of these Activities as with most Rate type of Activities is that there is an Independent Variable that you control (the x-variable) and the Dependent Variable (the y-variable). Here the Change in Temperature in each case is the Dependent Variable.

Activity I : Rate of Heating based on Color of Object

Purpose : To compare the rates of heating of objects of different colors

Methods :

A) Painted sealed bottles or cans with water in them placed in sunlight (colors : silver & black)

B) Color tiles, felt cloth, plastic of same 4 colors (black, blue, green, white). Placed in pan of crushed ice (acts as snow). Light on it is constant for all. Measure depth with time for each color.

Notes for All Activities involving Light or Heat Sources : Safety first. Do not touch, stare, or put things on lit bulbs. Try to control the environment as much as possible by using a small pen light as your source when setting up and have the experiments in darkened areas as needed (the activity will tell you when this is critical, but it is fairly obvious).

Activity I : Rate of Heating based on Color of Object

Materials :

- **Activity IA :**
- Tin cans – radiation cans best (black, white, silver),
- Light bulb (about 100 W or greater) and a lamp where the shade can be removed,
- Water,
- Stopwatch,
- 3 Thermometers,
- Measuring Cup,
- Graph paper,
- Slide Rule
- **Activity IB :**
- Rectangular pan (glass preferable),
- Ice shavings or snow (if available),
- Ruler,
- Pieces of Felt Cloth approx, 2″ on a side, colors : black, white, green, blue, (note : all materials are of same type and size),
- Overhead Lamp or sunlight,
- Note : The overhead lamp needs to have a flexible head to point the light source at the pan of ice shavings and strike an area equally where the cloth patches will be placed,
- Stopwatch,
- Slide Rule

Note : Activity IA : if school lab radiation cans unavailable, then use home aluminum or tin cans which had food at one time (such as canned vegetable 16 oz cans), now are empty, cleaned out, and made smooth edged in the area where a can opener operated. Next paint the 2 cans, 1 silver & 1 black & 1 white with these, be sure to also create lids for them, such as using cardboard and poke a hole in the middle so that the thermometer can go through.

Note : Activity IB : Ice shavings can come from ice one makes at home and carefully smashing it up. A blender is a good idea instead. The use of sunlight is best in this Activity since it will provide consistent lighting on all the cloth pieces simultaneously, but if not possible, use one or more overhead lights that illuminate the ice surface with the cloth on it as equally as possible. Desk lamps with lamp shade cones to direct the light may be of use here (use appropriate equipment and do not place things on lights directly)

Note : When using lights – employ common sense safety. Do not look into the lights, do not touch the light bulbs, do not touch items to the light bulbs.

<u>Procedure :</u>

1) With any and all Activities read through, then set up the process for the Activity and run it without measurement and watch where you may have bumps in the road so as to anticipate these. Be patient – Repeat processes. Once all is well when measuring see to the values – do you notice patterns or trends – what do you predict will happen next as you change one thing or another from one trial in an experiment to the next. Always Explore & Enjoy. Be Brilliant!
2) **<u>Activity IA :</u>**
3) Each of the cans (either lab equipment or they are your own – see note above) must have the same amount of water. It is best if they are about half full.
4) They are to sit on small but equal stacks of books so that an unshielded light source sits in the middle of them where all the cans are equidistant from the light source. Initially the light is off.
5) Let them sit in the dark for an hour with the thermometers in them and then record the initial temperature readings for each of the cans in your table under the time '0' spot.
6) At this point activate the light. You have chosen your time frame for measurement (5 minutes works well). – Note it is best to shield the thermometer from direct light during the experiment so devise some shielding from cardboard or paper.
7) Check and record the temperature every 5 minutes for at least 1 hour.
8) Turn off the light. Put equipment away.
9) Graph the results as line graphs with Temperature (T) as the y-axis and Time (t) as the x-axis.
10) For each of the cans data, draw a best fit line (should be linear) and determine slope.
11) Compare and contrast results of the cans.
12) **Activity IB :**

13) Spread out the ice shavings in the pan to a uniform depth.
14) Turn the light on and in the circle of light on the ice shavings place the patches of cloth in a circular manner, but separated from each other so as not to affect each other. Be sure they are in the same radius region of the light so that they are receiving the same amount of light.
15) Every 5 minutes, measure the depth of the cloth patch from the top surface to the nearest 0.1 cm and record these measurements.
16) Record these measures for one hour.
17) When done, clean up and put away equipment.
18) Graph the data as separate lines on a line graph where Depth (D) is on the y-axis and time (t) is on the x-axis.
19) Draw the best fit line and calculate slope for comparison.

Data :

Activity I : Comparative Rates of Heating :

Activity IA : Heating Rate of Water in Can

Time Unit should be around 5 minutes, but consistent

Can	Temperature (°C)							
	Time 1	Time 2	Time 3	Time 4	Time 5	Time 6	Time 7	Time 8
Silver								
Black								
White								

Data :

Activity IB : Melting Rate of Ice under Cloth of a given Color

Time Unit should be from 5 to 10 minutes, but consistent

Depth (cm)

Can	Time 1	Time 2	Time 3	Time 4	Time 5	Time 6	Time 7	Time 8
White								
Green								
Blue								
Black								

Calculations :

Be sure to use your Slide Rule!

$$\textbf{Slope} = \frac{\Delta Y}{\Delta X}$$

Conclusion :

What do your results show you when it comes to color and heating rates for materials?

Other Ideas for Heating Rates Comparisons :

Instead use the same color can (pick black for example) and examine the rate of heating for A) different amounts of water B) different types of water (each starts at a different temperature – use the sink and one is hot, the other cold), C) different types of water (distilled, tap, tap with some salt (measured amount in it), tap water with food coloring in it. Each has all the other variables the same and use similar but modified data tables to correspond to the measured dependent variable. One could also consider for any of these experiments instead to use different powered light bulbs or different color light bulbs of the same wattage as the independent variable.

Activity #II : Rate of Heating with Distance Activity

Materials :

- Desk lamp with 100 W bulb and no shade,
- Measuring Tape,
- Tape,
- Measuring Cup (size can be ½ or ¾ cup),
- Water,
- 3 to 5 clear Glasses or clear plastic Cups,
- As many Thermometers as glasses (minimally 3),
- Slide Rule

Note : Safety first, do not look into the light and do not touch the light nor place anything on the light such as the cups or the thermometers!

Note : In all cases, if using plastic glasses, be sure to have sufficient distance between the bulb and the cup so as to not have any damage due to heat.

Note : If 3 Thermometers, they will be placed in glasses with water at distances of 15 cm, 30 cm, and 45 cm

Note : If 4 Thermometers, they will be placed in glasses with water at distances of 10 cm, 20 cm, 30 cm, and 40 cm.

Procedure :

1) The best set up is done in a room with no other lights affecting the activity other than the lamp with the bulb in it, so in the initial set up use a small pen light as a source.
2) See the Notes above with regards to safety and glass placement with thermometers.
3) In setting up the glasses let the thermometers sit in the water for some time (1-2 hours) in water with the lights off.
4) Once activated measure the temperatures at regular time intervals (for example 5 or 10 minutes is good). Record the data for each of the distances in question.
5) Be sure to measure the initial temperature of the thermometers once the light is activated.
6) After 1 hour, stop measurements. Clean up and put away equipment.
7) From the table of data, Graph all of the results. Each distance measured will have its own unique line of Temperature (y-axis) vs. Time (x-axis).
8) For each draw a best fit line, particularly through the areas of linear heating.
9) Calculate Slope of the lines and compare them.
10) First calculate the change of temperature (Final Temp at a given distance – Initial Temp at a given distance)

11) Next, Graph the Change of Temperature (y-axis) vs. Distance (x-axis). Draw a best fit curve through this curve (it should look like an inverse-square relation).
12) Determine the log(change of temperature) and the log(distance) for each of the values.
13) Graph the log(change of temperature) on the y-axis and the log(distance) on the x-axis.
14) Draw a best fit line and determine slope. (If it is an inverse-square relation, the slope should be -2).

Data :

Time Unit : _________ (minutes) (approx 5 minutes)

Record Temperature (°C)

Time Unit	Distance 1	Distance 2	Distance 3	Distance 4
0				
1				
2				
N				

Calculations :

Be sure to use your Slide Rule!

$$\textbf{Slope} = \frac{\Delta Y}{\Delta X}$$

Conclusion :

What relation did your data uncover for distance and heating from a source? Why do you think that this may be the case (see Inverse-Square Law of Light Activity for ideas)

Activity #III : Rate of Heating Comparison for Land & Water Activity

Materials :

- Two small Baking Pans of same size and composition (glass is best),
- Sand or potting Soil (enough to fill one of the pans),
- Water,
- 2 Plastic sandwich or snack bags,
- 2 Thermometers,
- Desk Lamp with no shade and 100W bulb,
- Stopwatch or clock to act as timer,
- Slide Rule

Procedure :

1) Like all of the experiments involving light on the object, find a place that has darkness for set up and let only the controlled light of the experiment be the only one affecting the soil/sand and water areas with their thermometers. For set up use a small pen light as a source.
2) In one pan fill to about 2/3 full with water while in the other loosely place the soil or sand (no need to pack it down) to the same depth.
3) Let the pans of sand/soil and water sit in a dark room for an hour or so before the actual measurements.
4) Place the thermometers in the plastic bags being used and put into the sand/soil. In the case of the water, you may have to tape the end down (duct tape is best – enough of it can protect the thermometer from the direct light when used here).
5) Be sure to leave the tops of the thermometers and tops of the bags exposed so as to have easy access to the thermometers to take out to read and be able to replace them.
6) Activate the light so that the pans are equally illuminated (hence the same distance. Be sure the thermometers are placed in some sort of biased way so as to be affected.
7) After 2 minutes start the initial reading of the thermometers. This is at time zero.
8) Start measuring time (every 5 minutes is good). At which times you record the temperature readings on the thermometers.
9) Record temperatures for at nearly an hour.
10) Turn off the light, dump out or put away materials.
11) For each set of data, plot Temperature (y-axis) vs. Time (x-axis).
12) For each separately draw a best fit line and compute slope.
13) Compare the rates of heating for each to decide which heats faster, land or sea.

Data :

Time Unit : _________ (mins) (choose a time interval like 5 minutes)

Record Temperature (°C)

Time Unit	Soil	Water
0		
1		
2		
N		

Calculations :

Be sure to use your Slide Rule!

$$\textbf{Slope} = \frac{\Delta Y}{\Delta X}$$

Conclusion :

Can you draw any basic conclusions about the heating rate of water versus soil/sand? What if the water were deeper?
As a possibility, try this experiment outdoors in sunlight.

Activity #IV : Rate of Heating with Color of Light Activity

Materials :

- 4 Light Bulbs (white, blue, red, green or yellow) (Note : all of the same wattage value),
- Lamp,
- Thermometer,
- Measuring Tape,
- Stopwatch,
- Slide Rule

Procedure :

1) In this activity, it is best to do each of the measurements on separate days, but under the same basic conditions with the exception of the color of the light bulb being used.
2) Create an order to the bulbs to be tested (for example : white, red, then blue).
3) Set up the lamp with the bulb in question to be used. Plug it in. Note that you should not stare into the bulb when active nor touch it when active.
4) With the bulb off, but some sort of back light on to allow view (a small pen light is sufficient) set up the thermometer at a standard distance that will be used in all trials. (for example : 20 cm)
5) Note, you may have to mark this distance somehow and be sure to set up the thermometer in the same manner each time for each of the successive trials.
6) Turn on the light and after 2 minutes write down the temperature on the thermometer. This will be the initial temperature.
7) Every 5 minutes for 1 hour take the reading of the temperature. If it levels off before then, then discontinue.
8) Do this same set of procedures for each of the bulbs in turn.
9) For each set of results, graph the results : Temperature (y-axis) vs. Time (x-axis).
10) Draw a best fit line for the linear and climbing portion of the graph. Determine the slope of each of these lines.
11) What conclusions can you reach concerning the rate of heating and the color of light?

Data :

Chosen Time Unit : ______ (mins) (choose 3 to 5 minutes)

Record Temperature (°C)

Time Unit	White	Blue	Red	Green
0				
1				
2				
N				

Calculations :

Be sure to use your Slide Rule!

$$\textbf{Slope} = \frac{\Delta Y}{\Delta X}$$

Conclusion :

What effects does color of light have on heating?

Conclusion to All Activities you were able to do in this Activity :

Wow! All things have a story to tell us – we have to be creative in thought, curious in nature, and passionate about our need to explore to learn and be Brilliant!

Stellar exploration wasn't it?! Wow! All sorts of questions to consider and look at and maybe you started to create others to look up or read about in your explorations here. The concept of heat has so many avenues to explore and importance it cannot be listed here.

Activity #32
Finding the Rate of Cooling
Grade Level : High School
Math Level : Challenging

Determining the Rate of Cooling Activity –

Ms Briana Andromeda Purcell here - Consider the question, 'how long will it take my cup of coffee to cool off?'. Quickly any of us would answer that it depends on many factors, or variables speaking scientifically. The amount of coffee, the initial temperature of the coffee, and the type of cup it is in and whether this cup is covered or not would be the first quick responses. With thought, we might add that the ambient temperature of the surroundings plays into it, as well as whether we have added anything to the coffee, such as cream.

The question of cooling is not just the thought of the coffee aficionado, but one of science since the amount of energy a material has and the transfer of energy (heat) concerns many of us whether it involves our homes, businesses, and other buildings, not to mention heat transfer in motors and other mechanical devices, furnaces, and understanding nature, such as questions of how long does it take a lake to freeze or thaw? The natural rate of heat flow is a science unto itself (thermodynamics) and involves topics in the areas of physics, chemistry, astronomy, geology, biology, as well as engineering. The science of thermodynamics is critical as it gives us the conservation of energy (energy cannot be created nor destroyed, only transferred from and/or transformed into other forms). It stipulates the maximum effective work and the limits to efficiency of a system, which coincidentally comes from the temperature difference of two items in a system (Maximum Work (W) $\sim \Delta T = T_H - T_C$)

The question of the Rate of Cooling was one considered by one of the most famous scientists known, Sir Isaac Newton who is more well known for his work in physics, astronomy, and mathematics with his laws of motion, law of gravity, describing the use of a prism to separate light, creating the first reflecting telescope and calculus, and here even his thoughts on 'cooling'. It resulted in what is referred to as "Newton's Law of Cooling". Newton made the observation in his statement that the rate of the heat flow from one body to another (always hot to cold naturally and requiring work to go the other way – this is one of the Laws of Thermodynamics) is proportional to the difference in temperature between the 'hot' and 'cold' objects and time. What this statement implies and seems to defy common sense is that very hot objects cool more rapidly than warmer objects in the same setting. What seems to be a simple expression in words, is actually complex mathematically as expressed below.

$$\frac{\Delta T}{t} = k\Delta T$$

At first glance this idea seems potentially linear, but it is not. It turns out to be a differential equation with a natural logarithmic 'decay' (meaning negative) rate. Without going into calculus details, we arrive at this equation :

$$T(t) = T_c + T_p e^{-kt}$$

t = time (measured in minutes)
T(t) = temperature of the hot object at time 't'
T_c = the initial ambient temperature of the environment (cold room)
T_h = the initial temperature of the hot object at time zero (hot water)
T_p = initial values ($T_h - T_c$)
k = constant for rate of cooling for this set of conditions (per minute)
It gives the value when multiplied by time creates the exponent that since it is negative indicates the rate of decrease of the hot temperature.

Our goal here is to monitor a hot cup of water and let it cool for some time. Our goal is to find the rate of cooling, or 'k'. Most of the other variables in that complex formula are easy, such as the initial temperatures of the hot water and the cold room it is in. From the data, we will graph it, first as it, and then graph the log of the temperature value (using the L scale) versus the time. Note, unlike the Kepler Activity, this is not a log-log graph, but is instead a log-linear graph. On this latter graph, we draw a best fit line and compute slope with our Slide Rule. This slope is still not the answer. To find 'k' we then take the slope value and divide by the log value of the natural log base value, e, which is approximately 2.71828 (on our rule it is at 2.72). With 'k' determined, we have solved a complex formula of the rate of cooling for the cup of hot water. With it, we can even predict how long it will take to reach a given temperature, such as nearly room temperature.

Our investigation of 'k' will depend on the amount of water, the type of cup it is in, the initial temperature of the room and the water itself, so that if we tried various other types of cups, say a paper cup (that can contain coffee) as compared to Styrofoam as compared to a ceramic mug, we would find different 'k' values. These are all possible ideas to explore in our Activity as a follow up thought experiment if we are curious. We can alter the amount of mass in one study, the initial temperature of the material in another study, the type of container in another study. Be sure to hold the other variables constant when varying the Independent Variable (and interestingly in each of these cases it is Time) in question here to note the Dependent Variable (namely the Temperature). Notice what we are doing here in this discussion – it is not merely a set of givens, but thoughts to ponder and how do they affect the outcome. That is the heart of science – curiosity and imagination in combination.

Purpose : To determine the rate of change of temperature for a cup of hot water in a cold room and find the mathematical equation approximating this relation.

Materials :

- Thermometer(s) (Note : these need to be safe for hot water temps),
- Water (can be hot water from tap),
- Cup (can be paper (but must be able to hold hot drinks), Styrofoam, ceramic is the best choice for safety reasons),
- Tea Kettle,
- Heating Source (stove or microwave, if carefully used),
- Room or Refrigerator,
- Clock or Timer,
- Graph Paper,
- Goggles,
- Oven mitts or hot pad if needed – note other than pouring the hot water, it does not need handling,
- Slide Rule (use scales C, D, L)

Note : In heating water exercise safety and have parental supervision and permission. In the case of microwaves, be very careful and do not overheat – only warm the liquid in question, it does not need to even be near boiling at all, only lukewarm – let your parents handle the cup – it is a good idea to use oven quality gloves and wear goggles.
Note : The cups must be designed to hold the hot liquids in question. Do not handle these with bare hands – the Activity is to measure temperature after they are poured and they are left to cool. When you go to empty them be sure to use appropriate safety, such as a cloth.
Note : The thermometer needs to be lab safe so that it operates over a wide range of temperatures, below freezing and above boiling points of water. temperatures, below freezing and above boiling of water.

Procedure :

1) With any and all Activities read through, then set up the process for the Activity and run it without measurement and watch where you may have bumps in the road so as to anticipate these. Be patient – Repeat processes. Once all is well when measuring see to the values – do you notice patterns or trends – what do you predict will happen next as you change one thing or another from one trial in an experiment to the next. Always Explore & Enjoy. Be Brilliant!
2) Though the data table notes mass, it is not essential to measure (unless this is a variable to examine in various trials to see how mass effects the outcome). In the case of mass, you can use the assumption that 1 g of water occupies 1 cc of volume and then use measuring cups to determine the volume which turns out to be the number of grams of the substance (if water is used). Otherwise, you may need a mass scale for more precise measures (again, if mass is a quantity under consideration).

3) The primary Activity is to measure a given quantity of hot liquid and measure its temperature at chosen intervals of time (1 minute increments is fine, but it can work with 2 minute intervals – note that it is best to measure temperature values for at least 20 minutes)
4) First decide the tools to use here. For example, if you want to simply have hot water from the tap without extra heating, then you can use any of the cup types noted and you do not need a source to heat it.
5) If you decide to heat the water, do so carefully so as not to burn yourself. If a child, have permission and observation by parents in doing this.
6) If you have decided to heat the water, there is no need to boil it. Simply warm it slightly. Also, it is best to use a ceramic mug if heating the water in the microwave. In microwave use, be careful and do not overheat. If the water is heated in a tea kettle, be sure to use a proper cup that takes hot liquids. Exercise caution when and where needed with a warm cup of liquid.
7) If you have two thermometers this is best, but if not let the thermometer set in the room or place you will have the hot liquid cooling. This can be simply sitting on the table in the room or for more extremes place it in the refrigerator. Note : Do not change positions during the experiment and be sure it is a place that remains constant throughout the measuring period which will extend for up to 30 minutes. In temperature measures read to the nearest 1/10th of a degree.
8) Take the temperature reading of the cooling place and record it (T_c).
9) Heat the water (not to boiling) and have it in its cup. Measure its temperature (T_h) and record this temperature.
10) Leave the thermometer in the cup of water and let it cool in the cooling place.
11) Every minute (t) record the reading on the thermometer (T) for up to 30 minutes.
12) When completed with data recording, first create the ΔT chart. We will assume that the environment's temperature has not changed, so from each of the readings subtract the temperature recorded at that time. This is ΔT.
13) Now Graph time (min) on the x-axis and change of temperature (ΔT) (°C or °F depending on thermometer) on the y-axis.
14) Create a Table of time (min) and log (Temp) by using the L Scale of the Slide Rule.
15) Graph time and log(Temp) on a new chart.
16) Draw a best fit line through the time vs. log (temp) data.
17) Select 2 points on the time vs log(temp) graph to determine slope with
18) And using the slide rule compute the slope from two points that this line passes through
19)
20) $$m = \frac{\text{change of Log (Temperature) at points}}{\text{change of time at points}}$$
21)
22) Determine the cooling constant 'k' from the relation of k and slope.

23) This is done by dividing your computed slope by log(e) which is taken as 2.72
24) Note that 'k' should turn out negative in value since the slope is negative in value. When placing it in the formula just place in the absolute value, since the formula given here has the negative value factored in already.
25)
26) Assemble the data to find the final formula.
27) If you want to estimate the amount of time for cooling, return to your graph of time vs. log(temp) and extend the line until it crosses the x-axis. The time at this point is when the temperature difference is zero, which is when the water is the temperature of the environment.
28) Though the Slide Rule is a recommended tool, all of the calculations can be done with a graphing scientific calculator or the use of a spreadsheet program. In these calculations you have to generate a table of data, graph it, and then find the slope and/or equation of the best fit line for the data. Other formula calculations can be done with these tools as well.

Data :

Tc = ___________ (cold spot temperature initially and consistently)
Th = ___________ (hot water temperature initially)
Mass of Water = ___________ (g) (Note: Not needed, but may be considered)

Time (minute)	Temperature (°)
1	
2	
30	

Calculations :

Be sure to use your Slide Rule :

Change of Temperature value to generate graphing point

$$\Delta T = T_h - T_c$$

Slope of Line of Graph of time vs. log(change of temperature)

$$\mathbf{m = \frac{change\ of\ Temperature\ at\ points}{change\ of\ time\ at\ point}}$$

Changing slope 'm' to 'k'

$$\mathbf{k = \frac{slope\ m}{\log_{10}(2.72)}}$$

Final Formula form for Cooling Curve :

$$\mathbf{T(t) = T_c + T_p e^{-kt}}$$

t = time (measured in minutes)
T(t) = temperature of the hot object at time 't'
T_c = the initial ambient temperature of the environment (cold room)
T_h = the initial temperature of the hot object at time zero (hot water)
T_p = initial values ($T_h - T_c$)
k = constant for rate of cooling for this set of conditions (per minute)
It gives the value when multiplied by time creates the exponent that since it is negative indicates the rate of decrease of the hot temperature.

<u>Summary :</u>

In initially looking at your data, you might think that the first graph, time vs. change of temperature seems somewhat linear, but it is not. To check to see how accurate your results are type your initial table of data into a spreadsheet program (like Excel) and have it draw a scatter plot, add a trend-line and have it be exponential. The exponent should be similar to your 'k' value if you have done a good job with the data. Note that Slide Rules only have 2-3 significant digits of precision, hence there can be a fair amount of error in this. But look at the level of precision with respect to the computer's generated outcome. Consider what is effecting the error in your Activity and redo it one or more times trying to eliminate these problems.

You can check your answers once 'k' is determined by plugging in values in a step by step manner. First multiply the time value in question (t) by the determined 'k' value. If you can do this mentally, then continue and if you want to check your answer use the C & D scales for the multiplication. Be sure to keep track of the decimal point and the fact that the exponent is negative. We will call this value 'M' and use it in a moment. Now take the approximate value of 'e' at 2.72 (from 2.71828...) and multiply this by the coefficient in the formula which is initial ΔT. This product is then looked up on the D scale and a value for its log base 10 is found on the L scale. We will call this value 'N'. Note that both M and N are negative, so their product will be positive. This value, N, is then multiplied by the earlier determined value, M and its result is found on either the C or D scale (keep track of the decimal point here and this is multiplied by ΔT to find a product P that is added to T_C which should reasonably match the value in your table (given the precision of the Slide Rule).

Stellar studies here Inquisitive Pioneers – who would of thought a cooling cup of hot water and such can be a scientific curiosity and have so much complexity?

Alternative Activity Ideas :

Time, resources, and proper supervision permitting other ideas of the rate of cooling as being affected by initial and environment room temperatures (for example would the results be affected by being in the refrigerator while cooling ?), mass, type of container, the coverage of the container.

Alternative Ideas with a more advanced Slide Rule :

For those with LL scales on their Slide Rule, first find 'e' on LL3 which is aligned with the left index of the D scale. Reading along D to find a given value of 'k * t' (which can be first determined using the C & D scales and keep in mind that this exponent is negative) and at the value on the D scale now read to the other LL scales to find the outcome of the natural log raised to a negative power – this is often found on LLO1 or LLO2 scales. With this value, look it back up on the C & D scales and multiply it by the change of temperature coefficient and add this result to the environment's temperature.
This final outcome should be reasonably close to your table values.

Activity #33
Comparing the Rate of Cooling

Grade Level : High School
Math Level : Challenging

Comparing the Rate of Cooling Activity –

This Activity is a good follow up to the Rate of Cooling Activity (Activity # 32). For ideas on Heat Energy, its transfer look to the Heat Energy Activities (such as Hot & Cold Mixtures # 34, Rate of Heating #31).

For the prelude to this idea see Rate of Cooling Activity (Activity # 32)

Our goal here is to monitor a hot cup of water (as we did in the original Rate of Cooling Activity) in a glass and let it cool for some time in an environment that we control. Our goal is to find the rate of cooling, or 'k' for each of these environments and determine by comparison which affects the Rate of Cooling the most.

The basic set up is quite simple – a glass (can be glass or a thick plastic) which is set in a larger bowl. In the glass will go the hot water which comes from the faucet. The bowl itself will contain the environment we wish to consider. In the first trial we will use just plain cold water from the faucet. In the second trial, we will use an ice and cold water mixture. In the third trial we will then use cold water, ice, and salt. In all cases we measure both the initial temperature of the environment and the hot water (from the faucet) that goes into the glass. We merely monitor time and take temperature readings of the hot water system as it cools. We will take the initial value for the environment that the hot water will sit in (the material in the surrounding bowl), but will not monitor it and assume that it remains constant (though it does not) – Note for those with 2 thermometers, you can monitor both systems if you like.

It is easy to see that having several identical bowls and glasses would serve best here – but realize that you would need all the more thermometers and have to be quite focused on taking measurements. It is therefore recommended to do each trial separately, time-consuming that it may be.

Most of the other variables in the complex formula presented above are easy, such as the initial temperatures of the hot water and the cold environment it is in. From the data, we will graph it, first as it, and then graph the log of the temperature value (using the L scale) versus the time. Note, unlike the Kepler Activity, this is not a log-log graph, but is instead a log-linear graph. On this latter graph, we draw a best fit line and compute slope with our Slide Rule. This slope is still not the answer. To find 'k' we then take the slope value and divide by the log value of the natural log base value, e, which is approximately 2.71828 (on our rule it is at 2.72). With 'k' determined, we have solved a complex formula of the rate of cooling for the cup of hot water. With it, we can even predict how long it will take to reach a given temperature, such as nearly room temperature.
Our investigation of 'k' will only depend on the environment as the key variable under consideration since the amount of water, the type of cup it is in, and the approximate initial temperature of the water in the glass will be held constant.

Purpose : To determine the rate of change of temperature for a cup of hot water in a given environment to find the mathematical equation approximating this relation.

Materials :

- Thermometer(s) (Note : these need to be safe for hot water temps – Can be done with 1, but 2 is useful),
- Water (both the cold and hot will be used from the tap),
- Measuring Cup (so that each trial has the same amount of water),
- Ice,
- Salt,
- Glass (can be glass or plastic),
- Timer,
- Graph Paper,
- Slide Rule (use scales C, D, L)

Note : the thermometer needs to be lab safe so that it operates over a wide range of temperatures, below freezing and above boiling of water.

Procedure :

1. With any and all Activities read through, then set up the process for the Activity and run it without measurement and watch where you may have bumps in the road so as to anticipate these. Be patient – Repeat processes. Once all is well when measuring see to the values – do you notice patterns or trends – what do you predict will happen next as you change one thing or another from one trial in an experiment to the next. Always Explore & Enjoy. Be Brilliant!
2. Set Up :
3.
4. First find the best tools to use here – the glass used has to sit in the bowl and the bowl needs a flat bottom in an area that the glass can sit.
5. Next, just using water from the faucet fill the bowl while it is in the sink about 2/3 full (the amount will depend on the displacement caused by the glass) and holding the glass firmly displace the water and be sure that it does not go over the edge.
6. Next fill the glass about ¾ full (or more if needed) so that it has sufficient mass to remain stable in the bowl when it has water by itself.
7. You need to measure or mark the amounts needed both for the bowl and the glass and this is used for each trial. You are ready once you know how much is needed.
8.
9. The Activity :
10.
11. The order of the Trials is best suited for this order of materials in the bowl : 1) Cold Water, 2) Cold Water & Ice, 3) Water, Ice & Salt. You can even have more studies where you have incremental amounts of salt in further studies if interested. – Note in the 3rd trial, the amount of salt used is decided upon by you, but depending on the size of the bowl being used, you might consider a few tablespoons.

12. In a given Trial, fill the bowl with the material to the mark (so that it will not overflow when the glass is in it as determined in the Set Up).
13. After the material in the bowl has set for about 2 minutes take its temperature and record this (T_c) .
14. If you have two thermometers this is best – you can then leave this first thermometer in the material. If you do not, now take it out of the material and set it aside.
15. Using the measuring cup, fill to your determined mark or level in the glass the needed amount of hot water from the faucet (be careful not to put your hands in it).
16. Place the thermometer in the hot water and once the thermometer reaches its temperature (in about a minute) record the initial (T_h) temperature in the data table. Leave the thermometer in this glass.
17. As you read and record this thermometer value, place the glass with the hot water into the bowl with its material.
18. Every minute (t) record the reading on the thermometer (T) in the hot water in the glass for up to 30 minutes.
19. Follow all of the steps #10-18 for all Trials. Record all the data.
20.
21. Calculations :
22.
23. Determine the mass of hot water being used from the conversion chart (unless you had a mass scale present and wanted to measure it). Be sure that it is the same in all Trials.
24. For each trial do all of the following calculations separately :
25. When completed with data recording, first create the ΔT chart. We will assume that the environment's temperature has not changed (that is the bowl with the material), so from each of the readings subtract the temperature recorded at that time. This is ΔT.
26. Now Graph time (min) on the x-axis and change of temperature (ΔT) (°C or °F depending on thermometer) on the y-axis.
27. Create a Table of time (min) and log (Temp) by using the L Scale of the Slide Rule.
28. Graph time [x-axis] and log(Temp) [y-axis] on a new chart.
29. Draw a best fit line through the time vs. log (temp) data and using the slide rule compute the slope from two points that this line passes through
30.
31. $$m = \frac{\text{change of Temperature at points}}{\text{change of time at point}}$$
32.
33. Determine the cooling constant 'k' from the relation of k and slope by dividing slope by log(e) or 2.72.
34. Assemble the data to find the final formula.
35. If you want to estimate the amount of time for cooling, return to your graph of time vs. log(temp) and extend the line until it crosses the x-axis. The time at this point is when the temperature difference is zero, which is when the water is the temperature of the environment.
36. Compare the 'k' values and the rate of change of temperature for the various environments. Speculate as to how or why this is the case.

37. Note :
38. Though the Slide Rule is a recommended tool, all of the calculations can be done with a graphing scientific calculator or the use of a spreadsheet program. In these calculations you have to generate a table of data, graph it, and then find the slope and/or equation of the best fit line for the data. Other formula calculations can be done with these tools as well.

Data :

Type of Environment : ________________
(cold water, ice, ice & salt)

Mass of hot water used : ___________ g

T_c = ___________ (cold environment temperature initially and assumed consistent – unless being monitored)

T_h = ___________ (hot water temperature initially)

Time (minute)	Temperature (°)
1	
2	
30	

Calculations :

Be sure to use your slide rule!

Change of Temperature value to generate graphing point

$$\Delta \mathbf{T} = \mathbf{T_h} - \mathbf{T_c}$$

Slope of Line of Graph of time vs. log(change of temperature)

$$\mathbf{m} = \frac{\textbf{change of Temperature at points}}{\textbf{change of time at point}}$$

Changing slope 'm' to 'k'

$$\mathbf{k} = \frac{\textbf{slope m}}{\log(2.72)}$$

Final Formula form for Cooling Curve :

$$T(t) = T_c + T_p e^{-kt}$$

t = time (measured in minutes)
T(t) = temperature of the hot object at time 't'
T_c = initial ambient temperature of the environment (cold room)
T_h = initial temperature of the hot object at time zero (hot water)
T_p = initial values ($T_h - T_c$)
k = constant for rate of cooling for this set of conditions (per minute)

It gives the value when multiplied by time creates the exponent that since it is negative indicates the rate of decrease of the hot temperature.

Summary :

In initially looking at your data, you might think that the first graph, time vs. change of temperature seems somewhat linear, but it is not. To check to see how accurate your results are type your initial table of data into a spreadsheet program (like Excel) and have it draw a scatter plot, add a trendline and have it be exponential. The exponent should be similar to your 'k' value if you have done a good job with the data.

How do your various 'k' values compare? Do you think that in each case the environment had an effect on the rate of cooling? Did each environment have the same specific heat capacity value? What other factors may have had an influence?

In summation, then, first it can be said that even the simplest of events, the cooling of a glass of hot water can be rather mathematically complex. Second, a simple stick with numbers, the slide rule is just the tool to find the answer to how this happens. This opens doors to how scientists and engineers of yesteryear did the work that they did.

Activity #34
Predicting Final Temperatures of hot & cold water mixtures

Grade Level : Middle School
Math Level : Calculating

Mixing Hot & Cold Water Mixture Activity
Comparison of Measured and Calculated Outcomes using the Slide Rule

What if one summer you encounter a swimming pool nearly filled, but the water was just too cold? Would adding hot water to it change its temperature? What if it was only 1 glass of hot water? How about 1 bucket? Clearly not!

Now consider the idea on a smaller scale. You have two buckets of water, one with cold water and the other with hot water. You dump them both into a still larger, yet initially empty pale. What is the final temperature of the mixture? Is it ever greater than the original temperature of the hot water? Is it ever cooler than the cold water? No. The answer will always be somewhere between the two initial temperatures. Exactly in the middle if equal amounts of the two are added seems like a logical answer.

That is the intent and goal of this activity – to mix hot and cold water and find their final temperature and determine through measurement and observation how mass affects this outcome by varying the amounts of each of the types of water.

Though apparently simple, this lab has a subtle complexity to it in that to solve it mathematically one must employ the concept of **conservation of energy.** We begin with the idea that the amount of heat gained by one system is the amount lost by another. That is to say that the total energy before mixing the two is exactly the same as afterward. Energy itself cannot be created nor destroyed.

Heat Gained = Heat Lost

$\Delta Q = 0$

$mc\Delta T_{hot} + mc\Delta T_{cold} = 0$

The conservation of energy is one of the most important pillars of science and the basis of many of the ideas and their analysis.

Purpose : To predict and measure the final temperature of a measured mixture of various predetermined amounts of water into a final mixture where the initial amounts have different amounts and temperatures.

Materials :

- Styrofoam cups (16 oz or larger is best),
- cold water (from refrigerator),
- warm water (use spigot, not boiling),
- 3 thermometers* (lab quality),
- Full sets of measuring cups**,
- Slide Rule

*Note : This can be done with 1 personal Thermometer *but requires* that the refrigerator where the cold water is sitting at has an thermometer in it so as to measure the cold water temperature while the 1 personal thermometer is initially in the hot water to which the cold water is added.
Be sure in this case to record both the initial temperatures of the cold and hot water, mix them and let the mixture sit, with gentle stirring, for about 1-2 minutes and then record the mixture's final temperature. Also Temperatures are assumed to be in °C, but if you need to convert : $^{o}C = \frac{(^{o}F - 32^{o})}{1.8}$

**Note : Notice we are NOT using a Mass Scale. Though the specific heat formula notes mass ($Q = m*c*\Delta T$), we will make a critical assumption in our calculations. We will first convert the ounces of our measuring cups to milliliters (if not marked that way). The numbers we will use come from cooking where 1 cup = 240 mL, so ¼ c = 60 mL, ½ c = 120 mL, and ¾ c = 180 mL (Note : the value is closer to 1 cup = 236.6 mL, so decide in using the Slide Rule in your calculations where you will place your mark).
Next we will assume that the density of the water being used is ideal at 1.0 g/cc. Since 1 cc = 1 mL, and using the formula for density, $\rho = \frac{m}{V}$, it can be seen that the number of milliliters is the same as the number of grams of water being used! Also, we are using the conventional specific heat capacity of water and have it at $C_W = 1\frac{cal}{g*^{o}C}$. Better still since both the object giving off the heat and taking it in are the same substance, this specific heat capacity value cancels out of the equation as it is.

For those who want greater accuracy and have a mass scale available, simply measure the mass of the empty measuring cup and subtract this from the mass of the measuring cup with water in it to determine the actual mass of the water used.

Procedure :

1) With any and all Activities read through, then set up the process for the Activity and run it without measurement and watch where you may have bumps in the road so as to anticipate these. Be patient – Repeat processes. Once all is well when measuring see to the values – do you notice patterns or trends – what do you predict will happen next as you change one thing or another from one trial in an experiment to the next. Always Explore & Enjoy. Be Brilliant!
2) Note that the Data Tables are incomplete and you must fill them in yourself after copying them and expanding them to meet your needs for the number of trials conducted.
3) In the refrigerator have at least one large pitcher (two is best) [or use a few water bottles] all filled with water and they have been in there overnight. Have a thermometer in the refrigerator with them (unless the refrigerator has one and simply use it).
4) The large Styrofoam cup is the mixture cup. Place a thermometer in this cup.
5) **First Series of Trials :**
6) Use the measuring cups,
7) Start with cold water from one of your sources and begin with the 1 cup measurement.
8) Place this 1 cup of cold water into the mixture cup.
9) Record the initial temperature of the Cold Water (T_C).
10) Next, in sequence of trials add hot water to the mixture :
11) Before you do, however, measure the temperature of each of the Hot Water amounts while in the measuring cup with a thermometer and record its initial temperature (T_H)
12) Use first ¼, then ½, ¾, then 1 cup of hot water in each trial. Record this under Volume as size (Volume) – Note these are separate trials!
13) Note : you can use what you have available, but generate a sequence to your trials.
14) Then mix them in the large Styrofoam cup. Do not stir too much or let stand too long. Measure the temperature with the thermometer in place after no more than 1 minute. Record this value (T_f).
15) Dump out the mixture and let the cup sit for up to 2 minutes after each trial.
16) **The Second Series of Trials :**
17) Now redo the entire process but instead start with a set amount of hot water (1 cup) and
18) Then use a sequential amount of Cold Water, such as ¼ cup, ½ cup, and ¾ cup.
19) In each trial record the Initial Temperatures of the Hot Water and Cold Water and then the Final Temperature of the Mixture.
20) Calculation Steps :
21) First for each trial determine the Mass of water used for the mixture (see Note above for conversion 1 c = 240 mL).
22) Use the Law of Conservation of Energy which when used where you calculate with a Slide Rule will result in the predicted final Temperature of each of the Mixtures created.
23) Compare your predicted results from the formula to the experimental or measured outcomes.

Data :

Trial	Volume Hot Water (size)	Volume of Cold Water (size)	T_H Initial Temp of Hot Water (°C)	T_C Initial Temp of Cold Water (°C)	T_f Final Temp of Mixture (°C)
1					
2					
3					
Etc					

Trial	Mass of Hot Water (g)	Mass of Cold Water (g)	Initial Temp of Hot Water (°C)	Initial Temp of Cold Water (°C)	Final Temp of Mixture (°C)	T_F Calculated Temp of Mixture (°C)

Calculations :

Use Conservation of Energy : Heat Lost = Heat Gained

$$\Delta Q = 0$$

$$mc\Delta T_{hot} + mc\Delta T_{cold} = 0$$

$$[\ \Delta T = T_f - T_i\ ]$$

$$m_H T_f + m_C T_f = m_H T_i + m_C T_i$$

Formula for Final Temperature Calculation :

$$T_F = \frac{m_H * T_i + m_c * T_i}{m_H + m_C}$$

Conclusion :

Amazing simplicity yet interesting outcomes! The best analysis begins with a comparison of predicted versus actual outcomes for the temperatures and an explanation as to the differences between them. What are some of the key assumptions and possible energy loss areas in this lab?

Activity #35
Determine the Specific Heat Capacity of Water
Grade Level : High School
Math Level : Calculating

Specific Heat Capacity of Water Activity

Consider for a moment the old adage : 'A watched pot never boils'. Where does it come from and why do we think that? The nature of water is the reason. It heats up very slowly. In fact, if we had two spoons in that same water as it heated, say a metal one and a wooden one, we know which to handle with care. The metal one! Why? In all of the cases, the slowly heating water, the almost as slowly heating wooden spoon and the more rapidly heating metal spoon is due to what is known as the specific heat capacity of the material. Water is so good it was even used in early tubing flowing for astronauts to act as a layer of insulation in their suits.

Some of the common factors we might consider beyond composition are the amount of matter, the initial temperature of the material along with the temperature of the surroundings (see Newton's Law of Cooling in the Rate of Cooling Activity #32), along with the color of the material (why we often wear dark colors in the winter as opposed to the summer with lighter colors). Even the state of matter will determine the rate of heating it. Take for example H_2O. It is ice in a solid state, water in a liquid state and steam in a gaseous state. In each of these states each has not only a different amount of energy per unit amount, but also a different specific heat capacity based on the arrangement of the molecules themselves. In fact, there is another consideration in going from one state to another and that is a whole other lab consideration not examined here. All of the aforementioned are not specific heat capacity itself. This property is directly connected to the nature of the material itself.

The way to determine the physical property of the amount of energy per unit mass and change of temperature of a substance is known as the specific heat capacity of a substance.

$$Q = mc\Delta T$$

The enegy, Q, is measured in either Joules or calories (depending on the chosen units). In fact, a calorie is defined by this latter formula as coming from water. The specific heat capacity of water is :

$$\mathbf{C_W = \frac{Q}{m_W * \Delta T}}$$

For water, for each gram of substance, the change of 1°C requires what is defined as 1 calorie (4.186 Joules). This value determination is the goal of the Activity.

The energy, Q, does not only go into the change of temperature, which is directly related to the translational motion of the particles of the substance. It can also go into the rotational motion of the particles, as well as the internal vibrations of the particles bonds as well.

To determine the specific heat capacity of a material the calculation is rather straightforward. Merely measure the amount of energy needed to change a given amount of mass by a specified change in temperature. The ratio (noted above) of the energy to the product of the mass and change in temperature is the specific heat capacity of the material. Metals tend to have values less than one. The lower the value, the easier it is to change a given mass by a given amount of temperature change. Say, for example, a known metal has a specific heat capacity of $0.5\frac{\text{Cal}}{\text{g*°C}}$. this means that it requires 0.5 calories for each gram of substance and its change of temperature of 1°C.

Compare this to water, which is $1\,\frac{\text{Cal}}{\text{g*°C}} = 4.186\,\frac{\text{J}}{\text{g*°C}}$, indicating it needs twice the amount of energy for this same outcome. This might be where the old adage noted , 'a watched pot never boils' may arise from.

The energy value, Q, in the equation tells how much is needed in one consideration or can be given off by a substance at a given temperature too. This is important to this activity since to find the specific heat capacity we employ the unknown metal in question and water that we know plus the critical basic idea of conservation of energy :

Heat Lost = Heat Gained

Sum of all Heat Q = 0

$Q_W + Q_M = 0$

The heat lost by a material, here we are using an electric kettle/boiler (best to use a travel heater that can run on AC) and we assume that all of its energy is given to the water. We use this idea to find how much the temperature of some given amount of water at a measured initial temperature changes during a measured time interval. Each of these becomes a point on our graph (energy used versus mass and temperature change). The slope of the best fit line in this case reveals the specific heat capacity of water.

Why is the study of heat so vital? The obvious answer is nature and weather itself! Land and Water heat at different rates and the angle of the Earth's tilt with respect to the Sun plus its rotation creates a very complex system that results in different rates of heating along with the fact that the incoming radiation is partially reflected into space before even entering the atmosphere (about 30%). Different materials absorb at different rates and the energy can be redistributed, reabsorbed and retransmitted in the infrared (concerns about global warming due to materials even in the air such as CO_2, et al). The amount of water vapor, dust, along with the angle the sun strikes the surface, and the texture, and composition of the terrain plus the wind speed will affect the surface temperature. All of these affect the atmosphere's energy content, hence affecting pressure and pressure differences is the driver of the winds along with other factors too. This drives the weather!

Closer to home, heat and its transfer is critical in immobile structures (homes, buildings, et al) and even in mobile structures (cars, trucks, engines, et al). In buildings it is the retention of heat, particularly at cold times that is of importance. Yet when it is warm outside, the retention of cold air and the cooling of the air is critical. The measurement, dissipation, and control of the heat is very important. Even with a computer and its cooling fans and other appliances, such as refrigerators, and those that use heat like toasters, ovens, and coffee pots, heat once again becomes a factor of importance.

This is where Specific Heat Capacity comes into consideration. Each material has its own value for the amount of energy needed to change a given mass unit by a given amount of temperature change. The higher the value (such as with water at $4.186 \frac{J}{g{*}°C}$ means it takes a great deal of energy to change the temperature of water. Compare this to most metals which it requires less energy to change its temperature. For example aluminum has a value of $0.897 \frac{J}{g{*}°C}$ while common wood has a value of $1.70\frac{J}{g{*}°C}$. both are less than water hence heat up faster, but notice how much less the metal is as compared to the wood. All the more reason why a metal spoon in the hot pot of water is dangerous to touch and very hot very quickly as compared to the wooden spoon in the same pot (do not try this).

This physical property of matter, the specific heat capacity, is but one of many ways to examine matter. Other ways can include density, rate of expansion when heated, electrical conductivity, and the like. These properties, such as the specific heat capacity, can be used to determine its value or usefulness in various applications, such as when employed as a conductor or an insulator.

The study of heat is a branch of physics called thermodynamics . What of the definition of heat and the units of heat? In physics, heat is considered the energy transferred from one body to another due to thermal contact due to differences in temperature. Naturally heat transfers from a body of warmer temperature to one of cooler temperature. When bodies are at the same temperature they are said to be in thermal equilibrium. When they are not the process of heat (energy) transfer can take place.

The examination of heat looks not only at appliances and buildings, but also weather, the interior of the Earth, and other bodies in the universe such as the Sun and other stars. Heat is a measure of the quantity of energy of a system, hence there is no 'cold' per se, merely the absence of heat as compared to another body. Also the Heat or Energy can only transfer via one of 3 known means : Convection, Conduction, and Radiation. The units of heat are : Joule, calorie, kilocalorie. In fact, the factor noted before is the conversion for calories and joules : 1 calorie is 4.186 joules. Note that these are not the food calories we consume, each of these is 1000 of the base unit calories, hence the food ones are often written with an upper case 'C' as Calories. We could then convert our food calories to kilojoules by multiplying by the conversion factor. In many European countries, cans and boxes of food do not note the Calories, but instead the kilojoules instead.

Heat is something which may be transferred from one body to another according to the 2nd law of thermodynamics (i.e. the noted natural process and via one of the aforementioned means or if it does go from a place of lower energy to one of higher energy, it requires the input of work).

Returning to the goal of the activity : Heat will be transferred to the water and we want to determine its ability to take in and manifest that energy due to its mass and the change of temperature it undergoes. This demonstration of energy is thermal and is illustrative of its nature. From it we can determine the property of specific heat capacity. One of many heat factors we need to know to examine any of the other dimensions of heat from the atomic to the macroscopic in all of the noted fields.

How cosmic is this adventure Inquisitive Pioneers! We are looking for a property of a material, water in this case, much like finding its density as we did in another Activity (#13) and here it is the Specific Heat Capacity. Awesome Sauce and Too Cool. Be Brilliant.

Purpose : To determine the specific heat capacity of water

Materials :

- Tap Water,
- Thermometer (lab quality),
- Stopwatch,
- Home model (typically travel) electric kettle or boiler,
- Measuring Cup with metric readings,
- Graph Paper,
- Ruler,
- Slide Rule

Notes :

1) In handling hot water be sure to have permission from adults or are supervised by adults if necessary plus exercise caution.
2) In this Activity we are assuming that tap water is free of dissolved materials and has a density of 1 g/cc. For better results, use distilled water.

Procedure :

1) With any and all Activities read through, then set up the process for the Activity and run it without measurement and watch where you may have bumps in the road so as to anticipate these. Be patient – Repeat processes. Once all is well when measuring see to the values – do you notice patterns or trends – what do you predict will happen next as you change one thing or another from one trial in an experiment to the next. Always Explore & Enjoy. Be Brilliant!
2) Record the Power Rating for the electric kettle/boiler in the data table (P) which is found on a label on the device or on the box it came in.
3) Fill to its normal level the electric kettle/boiler and have it run through its time to boiling the water and pour it out (this is to heat up the system).
4) In the time of the kettle/boiler heating, set aside a measured amount of tap water in a measuring cup.
5) Record the amount (V) [measured in mL which are cc] and record this as the amount of grams in mass (M) too since M=D*V, where D=1g/cc here.
6) Also for the water, measure the initial temp (T_i) [in °C] and record it on the data table. Note for all initial temps, record it once in the device, but it is not plugged in or activated. Let it sit for about 1-2 min in it then record the initial temperature.
7) Note that you must operate the system as noted in the directions and do so safely. Do not put things in there while it is active. After the temperature reading, the thermometer is removed and the system is then plugged in and activated. Do not do this yet, read through all of the directions first. Keep in mind you need to time the time for heating the water to boil. It may be a good idea to wear goggles (have adult permission and supervision).
8) Depending on the size capacity of the kettle boiler you will use 4 different amounts, starting with the smallest amount, for example 400 mL and going by 100mL per trial

up to 700 mL in the final trial. The amount depends on the size of the electric heater you are using. Choose the best range for this.

9) In each case record the mass and the initial temp of the water used.
10) Each of these amounts is a Trial on the Table below. Once a trial is done, pour out the water and let the electric heater sit for a few minutes. Then repeat the process by putting the next amount in and recording the values as noted in the following directions. Be careful with hot water.
11) For each trial let the kettle/boiler operate where it heats up to boiling. It will be assumed that the final temperature is 100°C.
12) For each trial, measure the amount of time (t) [measured in sec] with a stopwatch it takes for that given amount of water to heat to boiling (predict, how should time change as mass is increased for each trial?) and record it in the data table.
13) Compute the amount of Energy (Q) needed to heat the water to boiling due to the electric kettle/boiler and the amount of time needed. This calculation involves the system's power (P) and the amount of time (t) needed to heat the water.
14) Convert the Energy (Q) from Joules to calories!
15) Compute the quantity mass*change in temperature ($m*\Delta T$) and call this the mass factor (N).
16) Graph as coordinate points the mass factor and the Energy needed for that factor (N, Q) this is (x,y).
17) Draw a best fit line through these 4 points and compute the slope of the line (c). The slope is your determination of the specific heat capacity of water!
18) Determine the percent error from your experiment with the accepted value.
19) Though the Slide Rule is a recommended tool, all of the calculations can be done with a graphing scientific calculator or the use of a spreadsheet program. In these calculations you have to generate a table of data, graph it, and then find the slope and/or equation of the best fit line for the data. Other formula calculations can be done with these tools as well.

\

Data :

Power Rating of kettle/boiler : (P) = ____________ W

Final Temp. (T_f) = 100°C

Trial	V (cc)	m (g)	T_i (°C)	ΔT (°C)	t (sec)
1					
2					
3					
4					

Trial	Mass factor N (g*°C)	Energy Q (cal)
1		
2		
3		
4		

Calculations :

Be sure to use a Slide Rule!

General Equation of Specific Heat Capacity for a material :

$$Q = m*c*\Delta T$$

Change of Temperature : $\Delta T = T_f - T_i$

Mass Factor : $N = m*\Delta T$

Energy : $Q = P*t$

Solving for Specific Heat Capacity (c) :

$$c = \frac{Q}{m*\Delta T}$$

$$c = \frac{\Delta Q}{\Delta N}$$

Slope : $m = \frac{\Delta Y}{\Delta X}$

water has a specific heat capacity of : $\mathbf{1 \frac{Cal}{g*°C} = 4.186 \frac{J}{g*°C}}$

- use the above value for converting from Joules to calories

Percent Error : (use the value noted here)

$$\%E = \frac{[\text{Experimental Value-Accepted Value}]}{\text{Accepted Value}} * 100\%$$

<u>Conclusion :</u>

How galactic was this adventure! All things have a story to tell us – we have to be creative in thought, to be curious in nature, and passionate about our need to explore to learn and be Brilliant!

The matter is clear here, what was the value obtained and how does it compare to the known value? If different, can we find and control some of the issues at hand and can there be refined measurements?

Notes :
As can be seen there are a number of assumptions being made here, but it should provide a reasonable and safe outcome to examine this idea.

Activity # 36
Determining the Specific Heat Capacity of a Metal
Grade Level : High School
Math Level : Calculating

Specific Heat Capacity of a Material Activity

Ms Briana Andromeda Purcell here and opening another Activity on Heat and a particular property of materials namely the Specific Heat Capacity and in this case we are looking at this value when it comes to Metals.

If you are starting here there are a number of other areas to find further information on Heat topics such as the discussion of Heat and Specific Heat specifically is found in the Specific Heat Capacity of Water Activity as well as Rate of Heating and Rate of Cooling Activities as well.

Matter has a variety of properties which are commonly classified as either physical or chemical. The activities on the web site have looked a couple of the physical properties such as the activity involving density, while another looks at the relation of volume & pressure, and this one here looks at the physical property known as the specific heat capacity of a material.

Think for a moment as to what factors determine properties such as the heating (or cooling) of a material and the overall energy content of a material at a given temperature or even the energy needed to change the temperature of a material from one temperature to another. Our first thought typically notes that mass, the amount of material must be important. Certainly. Does a small pan with water heat to a given temperature in the same amount of time as a large pan with a lot of water? The small one becomes hot very fast. Next, the state of matter matters too. The molecule H_2O as a solid (ice) has a much lower temperature than room temperature water (liquid) and still farther from the energetic (gaseous) form steam. The Activities on Heating & Cooling examines some of these ideas. This activity, however, considers the other main property of matter, the type of material, which is just as important as the others in determining the energy involved in a substance and its circumstances – and that is the material itself.

Imagine there is a pan with water heating in it on a stove. Note : this is an imagined or thought experiment and I am not saying to do it. In the pan is placed two spoons with the handles sticking out for stirring – one of wood, the other metal. In a short time, will they be the same temperature? All of us know that metals heat up quickly as compared to things such as wood or even the rate of heating of the water. In a short time, the metal spoon will be the same temperature as the water in the pan. (All the more reason to be cautious around metals when exposed to heating elements).

The way to determine the physical property of the amount of energy per unit mass and change of temperature of a substance is known as the specific heat capacity of a substance.

$$Q = mc\Delta T$$

The enegy, Q, is measured in either Joules or calories (depending on the chosen units). Here ΔT is the change in temperature, m (as usual) is mass.

In fact, a calorie is defined by this latter formula as coming from water. The specific heat capacity of water is :

$$C_W = \frac{Q}{m_W * \Delta T}$$

For water, for each gram of substance, the change of 1°C requires what is defined as 1 calorie (4.186 Joules). The activity specified here is the specific heat of a metal sample, where we will use water (hence you can consider first doing the specific heat of water or use its assumed posted value) to determine the specific heat capacity of the metal.

The energy, Q, does not only go into the change of temperature, which is directly related to the translational motion of the particles of the substance. It can also go into the rotational motion of the particles, as well as the internal vibrations of the particles bonds as well.

To determine the specific heat capacity of a material the calculation is rather straightforward. Merely measure the amount of energy needed to change a given amount of mass by a specified change in temperature. The ratio (noted above) of the energy to the product of the mass and change in temperature is the specific heat capacity of the material. Metals tend to have values less than one. The lower the value, the easier it is to change a given mass by a given amount of temperature change. Say, for example, a known metal has a specific heat capacity of $0.5\frac{\text{Cal}}{\text{g*°C}}$. This means that it requires 0.5 calories for each gram of substance and its change of temperature of 1°C.

Compare this to water, which is $1\ \frac{\text{Cal}}{\text{g*°C}} = 4.186\ \frac{\text{J}}{\text{g*°C}}$, indicating it needs twice the amount of energy for this same outcome. This might be where the old adage, 'a watched pot never boils' may arise from.

The energy value, Q, in the equation tells how much is needed in one consideration or can be given off by a substance at a given temperature too. This is important to this activity since to find the specific heat capacity we employ the unknown metal in question and water that we know plus the critical basic idea of conservation of energy :

Heat Lost = Heat Gained

$\Sigma Q = 0$
$Q_W + Q_M = 0$

Here W is Water, M is Metal

The heat lost by the metal will be transferred to some water :

Heat gained by the water = Mass of Water *Specific Heat of the water* Change in Temperature

$$Q = m_W * c_W * \Delta T$$

$$\text{Specific Heat of Metal } (c_p) = \frac{\text{Heat gained by water (J)}}{\text{Mass of Metal(g)*Change of Temp of metal (°C)}}$$

$$\mathbf{c_p = - \frac{m_w * c_w * \Delta T}{m_m * \Delta T}}$$

The basic idea of this activity is simple : Heat water to boiling with the unknown metal in it of known mass. Take the metal from this energy source of known initial temperature (while wearing goggles and using tongs as well as being careful around the boiling water so as to not get steam on you – Note : you must have parental permission and supervision to do this activity and it is recommended that they handle the materials when needed) and place the heated piece of metal in an insulated cup we will consider as a calorimeter. In this cup is already a known amount of water at a known temperature. We then Measure the change in temperature of this calorimeter cup system and from our measurements then determine the specific heat capacity of the metal in question!

Note : Calorimetry is an experimental procedure to measure the amount of energy transfer from one substance to another via heat. And yes it can be used to find the calories in the foods we eat as well.

Note : It may first appear that the specific heat capacity of the metal in the formula above will be negative. No, it is not. Realize that the ΔT of the metal will be negative since T_{iM} is 100°C and the final temperature, T_f, will be the final temperature of the water in the cup (and realize that its initial temperature is not 100°C, but instead the initial room temperature water, T_{iW}).

Purpose : To determine the specific heat capacity of a metal.

Materials :

- Heat Source (electric or gas stove),
- Small Pan to heat water in,
- Thermally insulated Cup (can be created by two nested Styrofoam cups if a metal coffee tumbler is not available which is the best choice) – if possible have a plastic lid for the cup also,
- Long Metal Tongs,
- ¾ Measuring Cup,
- 2 Lab Quality Thermometers (that can be used in boiling water),
- Water from faucet,
- Mass Scale,
- Nuts and/or Bolts known to be made of a metal, for example Iron or Aluminum are good choices,
- Plastic or Wooden Stirring Stick,
- Goggles,
- Oven Gloves (if needed),
- Slide Rule

Procedure : Determining the Specific Heat Capacity of a Metal

1) With any and all Activities read through, then set up the process for the Activity and run it without measurement and watch where you may have bumps in the road so as to anticipate these. Be patient – Repeat processes. Once all is well when measuring see to the values – do you notice patterns or trends – what do you predict will happen next as you change one thing or another from one trial in an experiment to the next. Always Explore & Enjoy. Be Brilliant!
2) Note : For the activity – first read through the directions. Next be sure to have parental permission and supervision in doing these steps. Always exercise safety – such as wearing your goggles, keeping away from flame or heat source, do not stand over the steam from boiling water, do not touch objects in the boiling water or touch the boiling water.
3) Measure the mass of the metal piece with the mass scale (M). Record this value.
4) Now Half fill the small pan with water and place on the heat source.
5) Before turning on the heat, place the metal piece (nut or bolt) in question into the water.
6) Now Activate the heat source and heat the water to boiling.
7) Note : A lab quality thermometer can be used to verify temperature in this case (must be handled carefully – note that steam can burn – this should be done by an adult). Never touch the heated pan or beaker either. When in presence of heating materials, use the glove(s) and wear your goggles! –
8) Note : The above step is not a necessity – we can reasonably assume that the water will boil at exactly 100°C if we wish – of course, this will affect the precision of our measurements and the final accuracy of our results.
9) Take the mass of a ¾ measuring cup empty.

10) Using the ¾ cup fill it with water and determine the mass of the water by taking the total mass of the system and subtracting the mass of the cup itself. Record the mass of the water being used (M_W)
11) Note be sure to use enough water so that when the metal nut or bolt is placed in the Calorimeter Cup System the water completely covers it.
12) Place this water in the Calorimeter Cup System, cover and place a thermometer in it through the plastic lid. Also have the stirring stick in it.
13) After a few minutes, Record this initial temperature (T_{iW}).
14) Now that the metal piece has been in the boiling water and temperature has been determined to be stable for about 2-3 minutes, it is assumed that the metal is the same temperature as the water. Carefully, using the tongs and wearing goggles plus being sure to avoid heat and steam, extract the metal piece. (keep hands away from the steam, wearing oven gloves may be useful if necessary).
15) Place the metal quickly into the Calorimeter Cup System and cover.
16) Now turn off the boiling water heating source.
17) Every minute record the temperature of the metal in the water in the Calorimeter Cup System on a scrap table for monitoring purposes.
18) Record Temperatures until they begin to decline once again and record in the table the maximum temperature value reached. This may take a few minutes (up to 10-15 min).
19) Once done, the materials can be removed and the 'lab gear' cleaned and properly stored. Note that the water in the pan is still hot, so handle it carefully.
20) Calculations :
21) Using the conservation of energy law, determine the specific heat of the metal using a slide rule.
22) If you know the metal to be pure or suspect it is, examine the table below to see how close your value is to the accepted value for specific heat capacity of the metal.
23) If you do not know the type of metal you are using, perhaps the table can point you in a direction of its type. It would be best to perform the experiment at least twice to verify your results and then follow up with a density determination of your substance (you might employ the Archimedes method of density determination for this) to further characterize your sample piece and determine its makeup.
24) Though the Slide Rule is a recommended tool, all of these calculations can be done with a regular or scientific calculator. Some scientific ones even have built-in averaging formulae. For those who like spreadsheets, the data can be typed in and the formulae then also be typed in its own cell where the formula references each of the measured variables in their respective cells, for example B1..BN has the measurements and values used in the equation while BN+1 has the formula for all of these variables (why not A? Simple – use it to label you variables)

Data :

For each metal sample use the data system below :

Mass of metal piece (M_M): ____________ (g)

Mass of Water in Calorimeter (M_W) : ___________ (g)
Note : This is determined by the difference of the cup+water – cup masses

Initial Temperature of Water in Calorimeter (T_{iW}): __________ (°C)

Initial Temperature of Metal (Ti_M) : ___________ (°C)
Note : This may be assumed or carefully measured by the directions

Final Temperature of Calorimeter & Metal (T_f) : __________ (°C)

Heat Gained by the Water in the Calorimeter : _________ (J)

Calculations :

Be sure to use a Slide Rule!
(W = water, M = metal)

$$Q_W + Q_M = 0$$

$$Q = m * c * \Delta T$$

$$\Delta T = T_f - T_i$$

Goal : Determine the Specific Heat of the Metal :

$$C_p = - \frac{m_w * c_w * \Delta T}{m_m * \Delta T}$$

If the metal is known, determine the percent error :

$$\%E = \frac{[\text{Experimental Value-Accepted Value}]}{\text{Accepted Value}} * 100\%$$

Conclusion :

The first thing to notice about your specific heat capacities for metals is that the value is : 1) positive, 2) less than one. Next use the table (if you know you used a metal sample from those listed) below to determine how close your values are to the accepted values for specific heat capacity for those metals.

Another thing to consider in the conclusion is what does this value mean? Since water has a specific heat capacity of : $1 \frac{Cal}{g*°C} = 4.186 \frac{J}{g*°C}$, this is clearly a lot more than the metals. Do metals heat up more quickly or slowly than water? Consider large lakes or pools, do they heat up quickly or not? Therefore, what can be said of two values one for specific heat capacity, if one of them is much less than the other, does it heat up more quickly or slowly in the presence of a heat source as compared to the other larger value one? Extend this thought to park benches and ones made of metal or wood on a sunny day. Which of them heats up more quickly and is hotter to the touch? What then can you say about the specific heat capacity of wood?

Unless you have done the Heating/Cooling Activity, consider next what happens as summer approaches in an area with bodies of water, such as lakes and associated shoreline and land. Does the land heat up more quickly or slowly than the water? Think of the sand on the beach as compared to the water. (For more of a measurement of this look at the aforementioned activity and see what turns out).

Summary :

Table of Metal Information (for here and other activities)

Material	Density (g/cc)	Specific Heat Capacity ($\frac{J}{g*°C}$)
Iron	7.86	0.4494
Aluminum	2.7	0.9025
Copper	8.8	0.3845
Brass	8.5	0.380

Activity #37
Finding the Number of Calories of a Food Item
Grade Level : High School
Math Level : Calculating

Measuring the Calories Activity

This Activity is about Energy – for more information on the general topic of Energy, read the Mechanical Energy Activity as this has a comprehensive Prelude to the topic of Energy and some of its generalities, such as the fact that it cannot be created nor destroyed. In order to do work there must be a change in the energy of a given system ($W = \Delta E$).

Energy comes in many forms : Heat, Light, Electricity, et al. More specifically this Activity examines Chemical Potential Energy which is the energy stored in the chemical bonds of a substance due to the arrangement of the atoms and molecules and in the particular case of Food.

Chemical Potential Energy is released or absorbed as Heat during chemical processes or reactions. A simple example is burning wood or paper. The chemical bonds of the wood or paper once activated begin the process of oxidation where the carbon atoms in the wood or paper combine with the oxygen in the air, release heat and light, while leaving behind residual ash.

Like all things (living or mechanical), we need energy. We derive our energy from the chemical potential energy in the food we eat. Digestion along with the cells in our bodies break down the food and in a series of chemical processes gradually oxidize the food so that they release their energy for use by the cells in our bodies.

Oftentimes the Chemical Potential Energy of a substance is examined by using the formula : $Q = m*c*\Delta T$. Where 'Q' is the heat energy either released or absorbed, 'm' is the mass in question that is either absorbing or emitting the energy in question, 'c' is the specific heat capacity of that material and is a value unique to that given material (see Specific Heat of Water #35 and Specific Heat of Metal #36 Activities for more information), and 'ΔT' is the change of temperature in question. A device known as a Calorimeter (Is a device for determining the heat of reaction or other thermal properties) is used to measure the energy that a food substance contains. The branch of science (which is at the cross-section of Chemistry and Physics) is Calorimetry. Calorimetry – Is the technique or process of determining the heat of reaction or related properties such as the energy value of food.

In our Activity, the 'Q' will be the energy given off by the food burning. The mass 'm' will be water in a fondue pot that we have measured. We will also assume we know the specific heat capacity of the water being used ('c'). Using a lab grade thermometer, we will measure the change in temperature ('ΔT')the water undergoes during the emission of the heat energy of the burning food. From this we can determine the number of calories given off (roughly) by the food. We can compare our results to a table of known values.

Food Energy is measured in what we call Calories. Notice the use of a capital letter 'C' here. It is intentional. A Calorie (or Food Calorie, if you like) is the amount of energy needed to raise the temperature of 1 kg of water by 1°C. The more common calorie (with a lower case 'c') is the amount of energy needed to raise the temperature of 1 g of water by 1°C. Hence 1 Food Calorie is 1,000 calories. This, however, is not the only measures of food energy. Some European food labels do not list Calories and instead use the more common physics base units of energy, namely the kiloJoule (kJ), which is 1,000 J. For comparison, in our Activity, we can convert the Calories of our food samples into kilojoules as well.

This Activity will use a sort of 'calorimeter' to measure the energy in one or more samples of food we choose to examine. The 'calorimeter' will be a ceramic fondue pot that is normally heated by a tea light candle. Instead of the tea light candle, we will use the food as the heat source. Unlike the genuine lab article, our open fondue pot will therefore have a number of built-in errors in the measurement process. Not all of the energy will go from the food to the water in the pot to heat it – some of it is given off as light and heat that also radiates in all directions.

In this Activity we are going to examine small samples of foods that we know and may have had on occasion. Our burning process will oxidize the food much more rapidly than our bodies (and clearly in a different manner – though the release of chemical potential energy is the same result).

Note : There are a number of safety concerns in this Activity that must be used and observed. Always have parental permission and supervision in the Activity. In using fire, employ extreme care and safety. See the list of Notes that follow the Materials list for further safety considerations to be acted on.

Note : In Activity #30 Individual Power and Energy, we determined the amount of Joules, kilojoules, and Calories needed to climb a flight of stairs – It may be fun for you to compare your results here to that activity!

Purpose : To have an estimate of the number of Calories (and Joules) contained in common food items (though various nuts are recommended others such as a potato chip may be substituted) through the use of water by using a fondue pot as a 'calorimeter'.

Materials :

- Fondue Pot (or Pot for Chocolate or Cheese) – Note this type can be heated by a Tea light Candle but here you are not using the candle system – the food becomes the energy heat source,
- Aluminum Foil,
- Paper Clip,
- If available : Crucible (this will act as the heating vessel),
- Lab Quality Thermometer (non-mercury – temp range -20°C to 110°C),
- Various Nut Types to Consider : dry roasted Peanuts, dry roasted Almonds, dry roasted Cashews, and Walnuts. (Notes : Your sample to use needs to be small. Next – for anyone with food allergies do not use items that are associated with them – such as nuts – consider small pieces of potato chips),
- Barbecue Lighter,
- Trivet or Glass Pie Pan baking dish,
- Water,
- Measuring Cup with lip for pouring,
- Metal Tongs,
- Oven Mitts,
- Goggles,
- Mass Scale,
- Stirring Stick (normally for coffee cup),
- Clear Tape,
- Note : Extinguisher (if needed for safety),
- Slide Rule

Safety Notes : (Be sure to read before any Activity)

1. First and foremost – you must have parent permission and supervision for this Activity. Your parents should not only monitor but also do the process of setting to flame the sample food materials.
2. Be sure to be aware of food allergies and do not use materials that may affect these allergies (such as some types of nuts). Also in the case of all food sources, use a small sample (though measurable by your scale).

3. In any use of fire such as here, exercise great care and caution. Do not touch or handle hot objects. Do not get close to any flame. Watch your clothing and hair when it comes to flame – roll up sleeves, keep loose garments away from heat, tie back hair and keep it away from flame. Be sure the flame is out and objects have cooled for some time and still use oven mitts to take care of equipment that became hot during the Activity. Always have a safety plan in place to deal with any emergencies.
4. Do not heat objects in places prone to motion, falling, or catching fire. Be sure the crucible is stable. Have it set on the trivet or glass baking pie pan which is on a surface that is safe.
5. Though these items are regularly seen as food, in this case they are not seen as food in the lab. You never eat items used in any lab Activity.
6. Be sure to inspect your equipment and ascertain it is proper to use for this Activity. Your materials may be a different manufacturer than those depicted, but they must have the same level of safety to be considered for use in the Activity.
7. Be wary of any smoke produced in the course of this Activity and hence the possibility of your smoke detector going off.

Set Up Procedure :

1. With any and all Activities read through, then set up the process for the Activity and run it without measurement and watch where you may have bumps in the road so as to anticipate these. Be patient – Repeat processes. Once all is well when measuring see to the values – do you notice patterns or trends – what do you predict will happen next as you change one thing or another from one trial in an experiment to the next. Always Explore & Enjoy. Be Brilliant!
2.
3. Be sure to have read the Safety Notes.
4.
5. Selection of Food Samples :
6.
7. The first step is to determine which food samples you are going to use. This will also determine the number of trials you conduct. Note that it may be a good idea to do more than one trial per food sample (for example have 3 trials for almonds) so that an average value can be determined. Note that each piece should be whole and of sufficient size so that you can measure it on your mass scale individually.
8. For each sample arrange them so you know the order of use. Measure the mass of each sample (m_f) and record this in the Data Table.
9.
10. Set Up of Fondue Pot, Aluminum Foil, and Paper Clip :
11.
12. Your fondue pot (which is our homemade Calorimeter) more than likely is not the same as depicted. The key is that it is made to be used as a fondue pot, has a ceramic bowl, and is normally heated by a tea light candle.
13. Set the fondue pot and its base on a glass pie plate or other non-flammable surface that could deal with some amount of temperature change. Be sure it is in a stable area and keep other things away.
14. The Aluminum Foil is just a precaution and will serve as a barrier between the base and the paper clip (which is the next step) in which the food sample will be. In the

case of the Aluminum Foil fold a small strip a couple of times over and fit it into the base of the fondue pot and be sure that it is flat and stable (see photos).

15. The Paper Clip needs to be bent in such a way (see photos below) that one loop acts as a base that will sit within the fondue pot base beneath the fondue pot itself and the other loop is bent towards the fondue pot (not touching it) but bend the other branch of the loop back across the loop so as to make a sort of 'alpha' figure and be sure that this is level to the surface that the fondue pot and base are sitting on. This loop will serve as the place to put the food sample in your trials. (see photos of this arrangement).
16.
17. *Note : If doing more than one trial then all of the procedures for a given trial must be redone in each trial. This is true for all of the following Set Up Procedure Steps as well (redo each of the following for each trial) below – All of the prior as assumed ready to go :*
18.
19. <u>Water for the Calorimeter :</u>
20.
21. Make a choice in the case of the water used. Either you will assume its mass from its volume measurement, or you will measure its mass. If going by assumption, use the conversion chart below for fluid ounces into grams. Note it is best to use a ¾ cup size. If going by measurement, place the empty measuring cup on your mass scale and zero it out. Now add water by pouring it from another cup into it and measuring its mass. Record this value in the Data Table (m_w).
22. When you have the water amount measured, transfer this to the ceramic fondue pot. Note that the pot does not need to be in the base at this time.
23.
24. <u>Recording the Initial Temperature of the Water :</u>
25.
26. Place the Thermometer into the water. Typically it may not readily stabilize itself lying in the fondue pot bowl, so you may have to use some clear tape to hold it in place. In any case, be sure to be able to read the numbers. Let the thermometer sit for a few minutes and then record the initial temperature (T_i) of the water in the Data Table.

<u>Procedure :</u>

1) Before these set of Steps, be sure to have followed and completed all of the Set Up Procedure Steps and have read and acted upon the Safety Notes.
2) As noted you must have parental permission and supervision in this Activity. It is best to have parents light the food samples when it is time and be responsible to check on the burnt out food samples. Be sure to employ safety at all times and remain focused.
3) Be sure to have your sleeves rolled up, loose garments out of the way, hair tied back and away from any potential areas with flame and be sure to have your goggles on throughout the entire Activity.
4) At this point, we have all of the food sample masses measured and set aside and have the initial temperature of the water for the first sample to be tested.

5) Carefully place the food sample in the bent paper clip holder and place it atop the aluminum foil barrier in the fondue pot base. (see photo). The key is to have the food sample a little below the fondue pot so that when lit the heart of the flame reaches the fondue pot.
6) Carefully ignite the food. It is best to use a barbecue lighter since it is long and can be positioned just below the food sample.
7) Once the food is lit, carefully place the ceramic fondue pot bowl with the water and thermometer in it on the base. Be wary of the flame.
8) Watch carefully from the side wearing goggles and not getting too close. If the food sample goes out before it is fully consumed you have a couple of options – A) quickly re-ignite it (noting that you are contributing to the heat received by the water so this may affect your outcome), B) discard it and try another piece.
9) As the food sample burns watch the Temperature. From time to time, slowly stir the water with the stirring rod. Watch the Temperature for the highest Temperature it reaches and record this as the Final Temperature (T_f) on the Data Table.
10) To have a complete trial the food sample is fully burned out. Do not touch, taste or handle it in any way. Use the tongs to remove the food sample and the paper clip. Note – only remove it when no flame is visible. It is best to have parents handle this step. You can immerse the mass in a cup of water to ensure it is out. Discard the piece. – Once cooled set aside the fondue pot and water – if you could not get to the paper clip with it in the way, first let it cool, then move it. You can discard the water and reset the system as found in the Set Up. Do not handle with bare hands, first let it cool for some time and still afterwards use oven glove or mitts.
11) Follow all of the noted directions in the Set Up Procedure and the steps 1-10 in the Procedure for any and all separate food samples. Record the data. Note that it is best to use fresh water in each trial and be sure to let the thermometer reach an equilibrium temperature in the initial phase. Note too that all temperature measures should be to the nearest 1/10th of a degree.
12)
13) <u>Calculations :</u>
14)
15) The first calculation is to determine the Change in Temperature for each of the Trials (ΔT).
16) Determine the Amount of Heat Energy given off from the burning food sample (Q) by using the Heat Absorption/Emission Formula where you use the Mass of the Water (m_w), the Specific Heat of the Water (C_W), and the Change of Temperature (ΔT) for a given Trial for each calculation.
17) If you have done a series of trials and each of these are the same food substance (for example, all almonds) determine the Average Value for the Amount of Heat Energy given off by the food sample.
18) If you have determined the number of calories for the amount of heat energy given off, convert these into Joules for each of the trials separately.
19) For each of the Trials, determine the Energy Density by using the determined Amount of Heat Energy given off value for each trial divided by the mass of the food sample (m_f) in the given trial.

20) If you have the data for the number of expected Calories for your food sample, check your values by using the Percent Error Formula and find how much of an error your results have. Why do you think this occurred?

Photos :

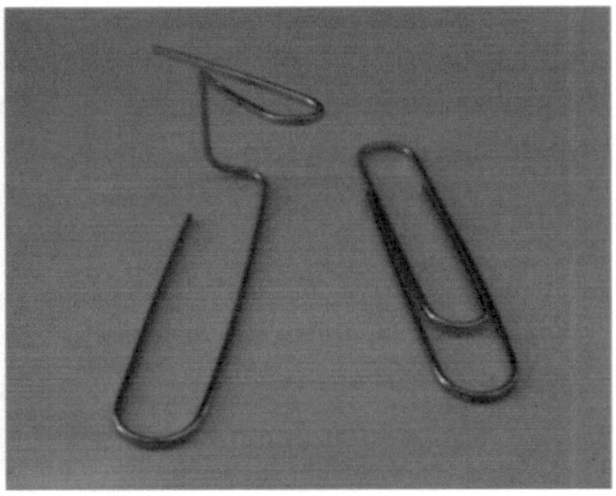

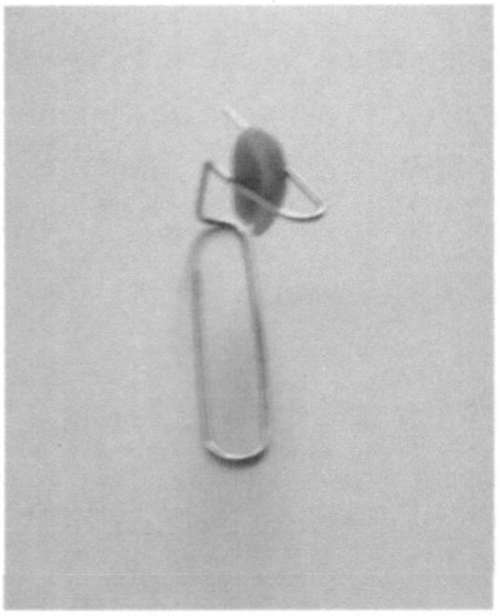

Data :

If a Packaged Food Item fill in the following data :

Item Name : ____________________________
Serving Size : ___________________________
Calories : _______________________________

Data Needed for Activity : (Note copy this below for each item)

Mass of Food Sample Used (m_f) : _________________ g
Mass of Water Used (m_w) : _____________________ g

Final Temperature of Water (T_f) : ____________ ^oC
Initial Temperature of Water (T_i) : ____________ ^oC
Change of Temperature of Water (ΔT) : __________ ^oC

Calculations :

Be sure to use your Slide Rule!

Heat Energy Given Off by Food :

$$Q = m*c*\Delta T$$

(Q = heat energy, m = mass, c = specific heat capacity of material, ΔT = change of temperature of material)

Energy Density :

$$N = \frac{Q}{m}$$

(Q = heat energy, m = mass of material giving off the heat energy)

Average formula (used in all Activities) :

$$x_{ave} = \frac{\text{Total of all trial values}}{\text{Number of Trials}}$$

$$x_{ave} = \frac{\Sigma x_i}{n}$$

Percent Error :

$$\%E = \frac{[\text{ Measured Value-Accepted Value }]}{\text{Accepted Value}} *100\%$$

Temperature Conversion (if needed) :

$$^{o}C = \frac{5}{9}*(\ ^{o}F - 32^{o}\ )$$

Constants that may be helpful :

1 mL = 1 cc
1 kg = 1,000 g
1 oz. = 28.3 g
1 fl.oz. = 29.6 mL
1 lb = 16 oz.
1 cal = 4.186 J
1 Food Calorie = 1,000 calories
1 Food Calorie = 4186 Joules

Specific Heat Capacity of Water

$$c_w = 1.0 \frac{cal}{g*°C} = 4.186 \frac{J}{g*°C}$$

Density of Water :

$$\rho_w = 1.0 \frac{g}{cc}$$

Food Facts for Comparison (if needed) :

Food	Portion Size (g)	Calories (Cal)	Fats (g)	Carbohydrates (g)	Proteins (g)
Peanuts, dry roast	28.3	166	14.1	6.1	6.7
Almond, dry roast	28.3	169	15.0	5.5	6.3
Cashews, dry roast	28.3	162	13.1	9.3	4.3
Walnuts	32.0	200	20.0	4.0	5.0

Conclusion :

How Stellar – All things have a story to tell us – we have to be creative in thought, curious in nature, and passionate about our need to explore to learn and be Brilliant!

Who would imagine – energy in food – quite literally! Wow!

Did your answers as to how many calories a small amount of food contains surprise you? How do your results compare to the materials (if packaged and providing information)? If you were able to test more than one sample of a given food (recommended for comparison) how do the results compare to each other? If you were able to test more than one food type, do all foods have the same amount of energy density (calories per gram)? Where in your experimental design are there 'losses' of energy that your measurements miss – for example, if you have information about the food source you are using, such as its calories and number of grams in a serving – how does this compare to your results – if yours are less, why do you think so?

Activity # 38
Estimating the Quantity of Electrostatic Charge

Grade Level : High School
Math Level : Challenging

Determining Charge with a Simple Electroscope Activity

Hey – Ms Briana Andromeda Purcell here and have you ever scuffed your feet, especially in winter, across carpeted floors and then got a zap from a doorknob and such? Yow – kinda fun and what an electrically charged time – pardon the pun.

The idea of a 'charged' object has been around for many centuries. The term 'electron' comes to modern times from ancient Greece where the ancient Greeks found that amber when rubbed with fur had the ability to attract bits and pieces of other fur, et al. The very term electricity comes from the Greek term for amber.

Though known on a very informal level the idea of electricity would take many centuries to be studied. The problem with further investigation on this are mostly due to the fact that the realm of the atom is very small and electrons are even smaller, since they are part of an atom. In fact, even today, we may know the mass of the electron (9.11×10^{-31} kg) and its charge (1.6×10^{-19} C) but there is no known dimensions of it, and some models even treat the electron as a wave instead of a particle. The properties of mass and charge took until the early 1900s with the Milliken Oil Drop Experiment to find these values. (See the Marbles Size Determination to relate to the size of the Atom Activity #20).

By the 1700s, notable persons, like Benjamin Franklin, were investigating the phenomena of electrified objects and their properties. Items, like the Leyden Jar were crafted and used to store electricity. Though called 'batteries' by Ben, today they are seen more as a Capacitor rather than a battery. The battery, however, also arose at these times with work by Volta. (Here in Ms Briana Andromeda Purcell's book Galaxy Girl's Inquisitive Pioneer Adventures there are Activities to make a battery (Activity #39) – too cool).

One of the chief realizations was that there are two types of charge, called positive and negative, by none other than Ben himself. The next main realization came from Charles Augustin de Coulomb (where we come by the term, Coulomb's Force) who came to quantitatively determine that the amount of force between any two charges varies directly with the amount of charge and the force varies as the inverse-square of the distance separating these forces. He used a torsion balance with charged spheres to measure this effect.

$$\mathbf{F = \frac{k*q_1*q_2}{d^2}}$$

F is the force, q_1 and q_2 are the charges in question and d is the distance between them.

Look at the structure of this formula. It is interestingly similar to the force due to gravitational attraction between any two masses as constructed by Newton over 100 years earlier than Coulomb's work. Each has a direct relation on a given property (the electrostatic force has charge, while the gravitational force has mass) and both are related to the distance by an inverse-square of the distance between the objects. (see Inverse-Square Law Activity #47)

When considering objects, we can readily divide them into two very general categories of materials : insulators and conductors (Note : There are items called semi-conductors, such as photovoltaic cells – which has its own Activity #49). In the case of insulators, though they are resistant to electrical charge flow, does not mean that charges cannot be deposited and built up on them. We all know this phenomena, such as running a comb through our hair on a dry day and having that 'static cling' effect, which can also be a part of clothes – especially coming from a dryer. We have played with this idea at some point in our childhood by rubbing a balloon in our hair and having it stick to a wall (what happens here is that the electrons now on the balloon as rubbed off our hair have a negative charge and induce a positive charge on the wall, since the somewhat mobile electrons on the surface are repelled and are driven away from that region that the balloon approaches – which now creates an attractive force between these opposing charges and leads to a temporary electrostatic bond), or better still scuffing our feet on carpeting and then touching an unsuspecting friend to cause a small shock.

In all cases of charge, it is the electrons that are transferred from one object to the next. The one that receives them builds up a negative charge while the one that loses them develops a positive charge. When combing our hair, electrons are transferred from us to the comb, hence our hair develops a temporary positive charge (all the more reason it is attracted to the comb as we pull it away).

The sum of the charges gained and the charges lost in these exchanges is zero. That is to say, **charge is conserved**. *This is one of the fundamental laws of nature of conservation.*

Some questions might arise in one's mind over these things, though. For a given insulator used as a charge receiver, how many electrons are on it? How large of a charge can be built up on it? (All objects have a maximum capacity – this is not explored here and when dealing with electricity, even in this static electricity lab, always exercise caution) How do the charges distribute over the surface? Does the shape or type of insulator affect the amount of charge that is on it? (Charges tend to have a certain amount of separation since they are the same charge, and the same charges repel each other – much like the idea that different charges attract each other) - Amongst many others.

The first question, though, that we are going to investigate is this : How much charge is on the object we have chosen to use in this case – the double pith ball electroscope? Despite its descriptive name, it essentially is two pieces of cork, Styrofoam, or some other insulating material suspended on two adjacent strings. On these we deposit a charge on from some other source and then take basic measurements of in order to find the quantity of charge on them. Knowing the amount of charge and the amount of charge on an electron, we can estimate the number of electrons on a given pith ball – something impossible to do for anyone before the early 1900s, despite the existence of simple desk-top objects like the electroscope since the time of Ben Franklin.

Our mathematics here is rather sophisticated and requires knowledge of vector analysis and some knowledge of trigonometry plus an understanding of Newton's Universal Law of Gravitation as well as Coulomb's Electrostatic Force Law. The formula is derived for you, but the preliminary steps are there so that one could do it for oneself as well. Also note, this Activity only uses charges built up on balloons, combs, and the like and has no connection to electrically conducting materials, electrical current, appliances and the like.

Purpose : To use a basic double pith-ball electroscope plus some measures of Distance or angle and mass in order to determine the amount of electrical charge and the number of electrons on each of the pith balls.

Materials :

- Double Pith-Ball Electroscope (see Note & Photo),
- Ruler,
- Protractor,
- Balloon (or Comb or some other insulator that can hold a static charge),
- Mass Scale (that can read to 0.01 g – but see note in procedure if unavailable),
- Slide Rule

Note : One could buy one of these or they are easy to make by using items such as Styrofoam peanuts as the pith balls. A better choice over regular string is to use dental floss. For the system that holds the pith balls, use a wire hanger.

Photo :

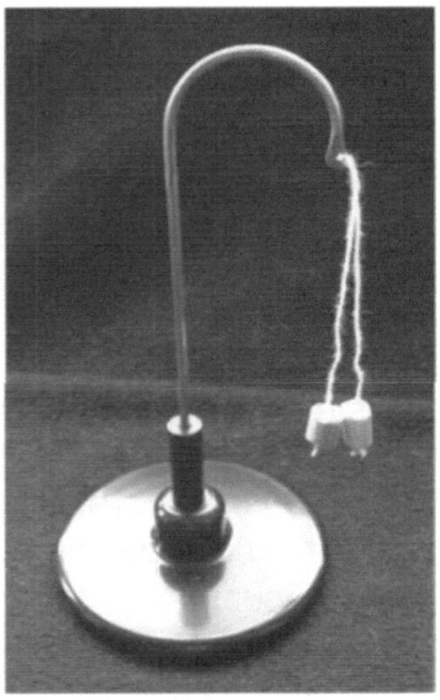

Procedure :

1. With any and all Activities read through, then set up the process for the Activity and run it without measurement and watch where you may have bumps in the road so as to anticipate these. Be patient – Repeat processes. Once all is well when measuring see to the values – do you notice patterns or trends – what do you predict will happen next as you change one thing or another from one trial in an experiment to the next. Always Explore & Enjoy. Be Brilliant!
2. First measure the length of the string attached to the pith balls. It is best to measure them to the same point on each one (the top, for example – though technically the middle is best since in Newtonian Mechanics, solid masses are treated as point masses acting through their center of mass) (L). Note, with two length values, take the average and use this as the length of the string.
3. The second measurement that needs to be taken is the total mass of the pith balls. The equation will use the average value for the mass and treat each as if they have the same mass.
4. Note : You will need a scale that can read to the nearest 0.01 g to have reasonably accurate results, but there is a way to determine the average mass, if you have your own pith balls.
5. If, for example, you are using Styrofoam peanuts, choose a number of them and cut them to the same shape with a pair of scissors. Take a large enough group of them to register a reading on the scale you are using. To determine the average value, take this total mass and divide by the number of peanuts on the scale.
6. With the average mass determined (m_{av}), now it is time to charge the electroscope.
7. Before doing so, you need to measure the equilibrium angle of the system. Use a ruler pointed away from the base of the system and choose a number (such as 4 cm or 2 inches) from which you will hold the protractor in an inverted fashion so that the vertex aligns with the strings of the pith balls and you can measure its equilibrium angle measure (Θe). Note the distance is done so that there is less chance of discharging or effecting the charged objects.
8. To charge the pith balls :
9. You can use a comb or a balloon – in either case pass it through your hair. Note : the drier the air and time of the year, the better the charge build up.
10. Deposit the static electric charge on both the pith balls. Try not to touch them as this will discharge them. You may have to do this a few times. You have succeeded when there is enough charge on them so that they repel each other and now have a new angle of separation to be measured.
11. Again hold the inverted protractor at the decided upon distance and measure the angle of apparent separation (Θs).
12. Determine the actual angle of separation of the pith balls (Θ) and record this result.
13. At this point you can discharge the system and begin again for other measurements. Note each set of measurements are a separate calculation.
14. Calculation steps :
15. You can choose to use the standard vector representation of the system and the basic formulae (and up to the intermediate steps) to derive your own equation for the charge on a pith ball – or simply use the one provided.

16. Note that this is a complex equation and will require several steps in using the slide rule, but can be done continuously. Like any algebraic equation, start from the inside and work it out.
17. Here it is best to start with the sine of the angle (first realize you have to determine this angle by dividing your angle of separation by 2. The sine of the angle is found on the C or D scale. The square of it is then found on the A or B scale.
18. Take this A/B scale value back to the D scale and multiply it by 4, then by L^2 (visually or with a cursor find what L^2 is if it cannot be done mentally) and finally by the tangent of half the angle of separation (this value, like sine is found on the C/D scale).
19. When done with all of that, divide by 9.
20. Through all of this be sure to keep track of the decimal values and the proper units used. L is measured in meters, for example. The constant 'k' has a power of 10^9. The sine and tangent values are decimal values.
21. Once a quotient is now determined, take the square root of this value by first finding it on the A/B scale and looking down onto the C/D scale. In this case, it is important to know where the decimal is since this will determine which side (left – odd (number of zeroes) or right – even (number of zeroes)) of the A/B scale is read. Hint : Charges are very small things and will typically have a negative exponent.
22. You can now determine the number of electrons making up this charge by dividing your result by the charge on a single electron.
23. Note : Though the Slide Rule is a recommended tool, all of these calculations can be done with a regular or scientific calculator. Some scientific ones even have built-in averaging formulae. For those who like spreadsheets, the data can be typed in and the formulae then also be typed in its own cell where the formula references each of the measured variables in their respective cells, for example B1..BN has the measurements and values used in the equation while BN+1 has the formula for all of these variables (why not the A cells ? Simple – use it to label you variables)

Data :

Total Mass (m) of both pith balls : _______ (g)
Average Mass (m_{av}) of a pith ball : _______ (g)

Angle of Equilibrium (Θe) : ________ (°)
Angle of Apparent Separation (Θs) : _________ (°)

Angle (Θ) of charged & separated pith balls : _____ (°)
Length of string from pith ball to vertex : ________ (m)

Depiction of Vectors involved :

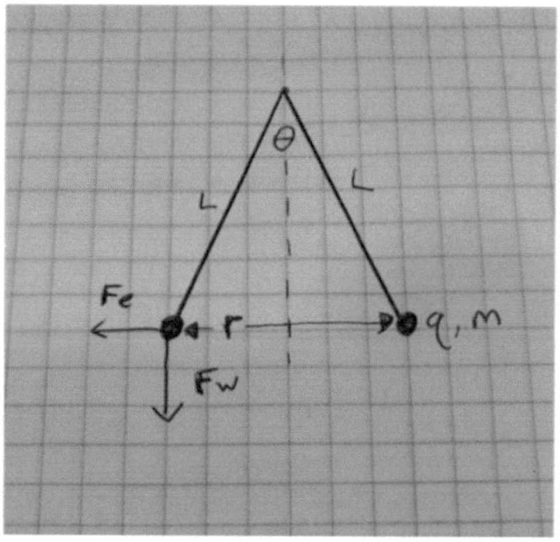

Calculations :

Be sure to use your Slide Rule!

Basic Formulae :

$\Sigma F_x = 0$
$\Sigma F_y = 0$

$F_w = m*g$

$F_E = \frac{k*Q_1*Q_2}{d^2}$

Intermediate Step :

$T*\sin(\frac{\theta}{2}) = F_E$

$T*\cos(\frac{\theta}{2}) = F_w$

$r = 2*L*\sin(\frac{\theta}{2})$

Angle of Separation :

$\Theta = \Theta s - \Theta e$

Derived Formulae :

$$Q = \left(\frac{4 * L^2 * m * g * \tan \frac{\theta}{2} * \sin^2(\frac{\theta}{2})}{k} \right)^{1/2}$$

Note : we are assuming that each of the pith balls has the same mass and the same charge

Average Mass :

$$m_{av} = \frac{\sum m}{n}$$

m : mass of a given piece, n : number of pieces

Constants & Conversion Factors :

$g = 9.8\ m/s^2$
$k = 9 \times 10^9\ (N*m^2)/(C^2)$
$1\ kg = 1000\ g$
charge on an electron : $1\ e- = 1.6 \times 10^{-19}\ C$

Conclusion :

Awesome Sauce! Searching for the total charge we developed and then how many electrons can be on this tiny item – wow !

What do your results show? Does it seem like a large number of electrons? What should happen to the pith balls as more charge is added? Does the size of the pith balls affect the amount of charge that can be held on them?

Activity # 39
Making and Measuring a Battery
Grade Level : High School
Math Level : Calculating

Factors Affecting a Simple Battery and the Slide Rule Activity

History of the Battery :

Hey Inquisitive Pioneers – Ms Briana Andromeda Purcell here and here in this Activity like some others (like Mechanical Energy which has the story of Energy p.) here is the tale of batteries and some ideas on electricity as a prelude.

The following is only about the history of the people and the events leading to the development of the batter. At no point are you to engage in these sort of activities or explorations. It is merely presented for historical reference and the operation of the basic battery. There are many ideas here involving chemistry and biology, some of which may seem unusual, even unethical today. This is only about information – and not to be done by anyone.

Following the history – We have two different battery explorations to engage in. One is inorganic whilst the other is organic. In our first Activity we explore a simplified battery composed of aluminum foil, pennies and salt water. In the second we consider one or more fruits and/or vegetables to help create electrical potential for a battery with some simple ingenuity – too cool.

Think about it - Today the world would find it very difficult, to say the least, and perhaps impossible to operate without batteries. We have them in our cell-phones, hand-held non-cord power tools both hand-held and much larger, in our cars and even cars powered by them exclusively, used in numerous hand-held games and toys, in submarines, on the space shuttle and other space based systems, and in medical implants for our heart and ears. This is only a partial list as the number of devices can go on and on. Not only that, in the world today there is a greater need for them since there are efforts afoot to have vehicles use them as a power source itself.

Interestingly the term "battery" first comes to us from none other than Benjamin Franklin. He actually used the term to describe a series of interconnected Leyden Jars (which actually operate not as a battery, but as a charge storing device called today a Capacitor) since it resembled a battery of cannons. (In fact it was Franklin who gave us the terms negative and positive in referring to the charges of an atom). Today in the strictest of definitions the term, battery, refers to a collection of two or more cells, but through popular usage it also refers to a single electric cell.

There was a lot of electrical experiments in the 1700s, many of them done by Franklin amongst others. Here, the scientists of the day began to lay the foundations for later work in electricity of the mid to late 1800s which led to such innovations as the electric motor and electric dynamo, AC and DC systems, numerous electrical devices including the incandescent light bulb perfected by Thomas Edison and all of these ideas were fueled by Michael Faraday's Law of Induction (the connection of changing electric and magnetic fields) and the crafting of the

Laws of Electromagnetism by James Clerk Maxwell. Before all of these though there were names such as Luigi Galvani, an Italian anatomist and physiologist (where we today use his name as galvanic response) where in 1780 he notices that a dissected frog's leg would twitch in the presence of a spark from a Leyden Jar. Later in 1786 he noticed the same behavior during electrical or lightning storms. Further experimentation by him he tried two different metals with the leg to create an electrical circuit.

Alessandro Volta read of these accounts and decides that the frog's moist tissues can be replaced with cardboard soaked in salt water. Instead of a twitching leg to register response, he incorporates non-living things to respond to the electrical current by using a Galvanic Cell.

In the case of a given cell, the terminal voltage of a cell that is not discharging is called its electromotive force (emf) and has the same unit as electrical potential, named (voltage) and measured in volts, in honor of Volta.

In 1800, Volta makes the first battery by piling many voltaic cells in series. The larger the pile, the greater the emf. His pile would produce about 50 V for a 32-cell pile. There are still those that refer to the battery (in Europe) as **piles**.

Like many discoveries, the case and effect of the system was not fully understood for some time. Even Volta disagreed with many on the matter, for example, by wrongfully thinking that this was some sort of free inexhaustible energy. Today we know this is not the case and all things in the universe fall under the conservation of energy laws found to be the fabric of the cosmos.

All circuits require Voltage which is a measure of the Electrical potential energy per Coulomb (J/C) which is the definition of a Volt. This is to say that one volt is the electrical potential difference across which one coulomb of charge will gain or lose one joule of energy. This acts as the 'push' for the charges in a circuit. If the voltage is too low for a given circuit to operate, it will not. Common Voltage sources are DC (direct current) related such as dry cells and wet cells (the basis of this simplified activity presented here) and even extends to AC (alternating current) such as an electric generator (aka dynamo) and electric motors.

$$\Delta \mathbf{V} = \frac{\Delta}{\quad}$$

It was **Michael Faraday** who found in 1834 that batteries were chemical reactions. According to Faraday, cations (positively charged ions) are attracted to the cathode and anions (negatively charged ions) are attracted to the anode.

These early batteries were great for experiments but not commercial development. Their voltages fluctuated and they could not produce a sustainable ample current for use. By 1836, what is called the Daniell cell was developed and even employed in industry for stationary items, such as telegraph networks. All of these initial batteries (often in glass) are what are called wet cells. In the late 1800s, a paste is made that operates in place of the wet environment and hence is born the dry cell battery. This makes portability much better and eliminates the fragility of the early wet cell system.

The electrical battery is a combination of materials which are electrochemical cells. These are used to convert stored chemical potential energy into electrical energy (which is often described as kinetic). Batteries consist of a number of **voltaic cells**, where each is composed of two half cells connected in series by a conductive electrolyte containing **anions** and **cations**. One half-cell includes electrolyte and the electrode to which anions (negatively charged ions) migrate (called the **anode** or negative electrode). The other half-cell includes electrolyte and the electrode to which cations (positively charged ions) migrate (called the **cathode** or positive electrode).

From chemistry, the battery is a **redox** (from oxidation-reduction) reaction. That is to say there is a reduction (addition of electrons, the reduction comes from a decrease in the oxidation number) which occurs to cations at the cathode, while oxidation (removal of electrons) occurs to anions at the anode. In the battery, the electrodes do not touch each other but are electrically connected by the electrolyte.

What this is saying and makes this occur is two metals that basically react with each other in the presence of the electrolyte. The two types of material, metals, in the voltaic pile are called electrodes. The reason for the potential difference is actually simple then, one of the metals reacts more vigorously than the other hence one metal becomes positively charged (the positive electrode) and the other becomes more negatively charged (the negative electrode). The potential difference is the voltage. This creates the pressure, force or push to allow current to flow when there is a complete circuit.

Each half cell has an **electromotive force** (emf) that drives electric current from the interior to the exterior of the cell. The net emf of the cell is the difference between the emfs of its half-cells (net emf = $E_2 - E_1$).

The electrical driving force or ΔV_{bat} across the terminals of a cell is known as the **Terminal Voltage (Difference)** and is measured in Volts. When it is not charging nor discharging, this value equals the emf of the cell and is called the open-circuit voltage. (The reason that the two values do not equal each other is that there is internal resistance in the battery, so that once connected the terminal voltage is a little less than the emf) [Note, for many labs in a regular physics class, internal resistance is neglected].

Despite differences in chemistry, there are categories of potential that have similarity for batteries. In one category, alkaline and carbon-zinc cells have approximately 1.5 V. In another set NiCd and NiMH (though different chemistries too) have 1.2 V. Finally, lithium cells can have emfs of 3 V or more.

There are two main classes of batteries : **Primary Batteries**, aka Disposable Batteries and **Secondary Batteries**, aka Rechargeable Batteries.
Disposable batteries are designed to be used once then discarded. Typically used in devices with low current needs with infrequent use. These cannot be recharged. Common forms of it are zinc-carbon and alkaline batteries.
Rechargeable batteries need to be charged before use and can be recharged by applying current to them to reverse the chemical reactions that occur from use. The oldest type of these is the lead-acid battery (used in cars often today). These types have large capacity (10Ah) and have peak currents of 450 A. Replacing many car ones in modern times is the sealed valve regulated lead acid (VRLA) battery. Some of these use a gel, or semi-solid electrolyte versus the older liquid acid ones or another type called the Absorbed Glass Mat form that has a fiberglass matting for absorbing the electrolyte.
There are a handful of Battery Cell Types which extends from the types of electrochemical cells that exist. The two most common will be discussed here : the **Wet Cell** and the **Dry Cell**.

The **Wet Cell Battery** (aka flooded cell, vented cell) has a liquid electrolyte. The other names comes from the fact that the electrolytic solution covers all the internal parts and these need to be vented since gases are produced during operation. These are the oldest major form of battery known. (Some of the names are : Daniell cell, Leclanche cell, Grove cell, Bunsen cell to name a few). Interestingly they may fall into either category of primary or secondary cells.
The **Dry Cell Battery** has a rather immobilized electrolyte that is in the form of a paste. There is some moisture but this is to allow current to flow. One of the distinct advantages of these is that they can operate in any position, unlike the wet cell (if tipped, it will spill the contents). Since the electrolyte is immobile this can lead to small batteries where there is little concern from leakage. The most common form of these type is the zinc-carbon battery (aka Leclanche cell) with a nominal voltage of 1.5 V which is the same as a alkaline battery. This is because each uses the same zinc-manganese dioxide combination. The standard makeup of the dry cell is a zinc anode (negative pole), usually in the form of a cylindrical pot, with a carbon cathode (positive pole) in the form of a central rod. The electrolyte is ammonium chloride in the form of a paste next to the zinc anode. The remainder of the space between the electrolyte and the carbon cathode is taken up by a second paste of the same material
with manganese dioxide, where this latter material acts as a depolarizer. Modern 'high power' batteries replace the ammonium chloride with zinc chloride.

Description of the Activity :

This Activity is the construction of a simple battery and an examination of what will increase its voltage and overall power. We will use simple materials : pennies, nickels, aluminum foil, salt, water, paper towel, and the like for our construction. With a multimeter we will measure voltage and amperage of the battery we construct. Graph the results to see if there are changes with the changes to your experiment.

Activity : The Voltaic Penny-Foil-Pile Battery

Purpose : To construct a simple battery and measure its power with changes in the amounts of materials comprising the battery system.

Note for Safety : Always have adult supervision and permission for any activity. Do not involve outlets in any manner.

Materials :

- Multi-meter,
- 20 regular Pennies
- Sheet of Aluminum Foil (8" x 11"),
- 2 sheets Paper Towel,
- Dish Pan,
- Water,
- Salt,
- Scissors,
- Graph Paper,
- Slide Rule

Alternative or Extended Activity Materials :

- 20 Nickels,
- 20 Dimes,
- 20 older Pennies (pre 1982),
- ¼ c Vinegar,
- ¼ c Lemon Juice,
- ¼ c Soda Water,
- Thin Cardboard (from back of pad of paper),

Note : You need to have adult supervision and permission for this Activity. At no time are you to use any electrical device, regular batteries or any electrical outlet. The only tool that involves electricity is the multimeter, which is a galvanometer and responds to voltages and currents.

Procedure :

1) With any and all Activities read through, then set up the process for the Activity and run it without measurement and watch where you may have bumps in the road so as to anticipate these. Be patient – Repeat processes. Once all is well when measuring see to the values – do you notice patterns or trends – what do you predict will happen next as you change one thing or another from one trial in an experiment to the next. Always Explore & Enjoy. Be Brilliant!
2) Rule No. 1 : See Note and be safe. This Activity is about the properties of materials, like aluminum foil, pennies, and salt water and does not involve electrical devices, regular batteries, and the like.
3) Prepare the electrolyte solution (water and salt) in a dishpan.
4) Side Note on Solution : Be willing to experiment with the concentration and perhaps instead try vinegar in a salt solution.
5) From the aluminum sheet, cut a 1" by 1" square and set aside.
6) Use the scissors to cut both the remaining aluminum foil and 1 sheet of paper towel into small squares the size of a penny. If possible cut as a circle instead to match the penny.
7) Side Note : You want as close a match as possible so that the layers are distinct and do not cross over as this may cause a 'short' in your system.
8) Soak, then lightly squeeze out the small 'squares' of paper towel.
9) Note : Take these pieces and place them on a easily tractable surface, like another sheet of aluminum foil for example for easy access.
10) Start the first pile with the aluminum foil on a flat table followed by the square (circle) of wet paper towel and then a penny.
11) This is your first pile.
12) Measure its voltage and current. With no readings, redo the experiment, or put on a second pile and try again. Measurement is done by using the multi-meter where one of the probes touches the aluminum foil and the other touches the penny on top of the stack of piles. Note that you have to rearrange the multi-meter to measure either in voltage or current.
13) Note : With multi-meters, the values will fluctuate a lot. Try to have one with a 'hold' button. Once the value seems to vary slightly around a given value, then hit the hold and accept that as the value.
14) Continue to create layers using the foil squares (circles) followed by the wet paper towel squares (circle) and the penny for each cell.
15) Measure after each cell is created both the voltage and the amperage and record the information in the table.
16) Once done, graph Voltage vs. Pile Number, draw a best fit line, determine slope.
17) Note : Determine the slope of the straightest part of the line, as it may vary to a curve possibly. For the more mathematically inclined : take the log of both the voltage and number of piles and graph these on a regular line graph. The slope of this line might reveal a power relation of voltage and number of piles.

18)

19) Calculate the Power of each Pile. Note : Each of the reading is in milli- so this means $\mathbf{10^{-3}}$. When calculating Power in watts this needs to be considered.

20) Graph Power vs. Pile Number. Like the Voltage vs. Pile Number Graph, determine the slope of the straightest part of the line.

21) Try other variations listed below for comparison.

<u>Data :</u>

Note : A Voltaic Pile is : Penny – saturated towel – Foil

Number of Piles	Voltage (mV)	Current (mA)

<u>Calculations :</u>

Be sure to use your Slide Rule!

P = V*I

(P = Power, V = Voltage, I = Current)

Slope :

$$\mathbf{m = \frac{\Delta Y\text{-dependent variable}}{\Delta X\text{-independent variable}}}$$

<u>Conclusion :</u>

Wow – Awesome Sauce – what adventures in Battery Building we have done here and still there is another one to do to!

Wow! – All things have a story to tell us – we have to be creative in thought, curious in nature, and passionate about our need to explore, to learn, and be Brilliant!

Here the conclusion first comes from the calculation of Power for each of the Voltaic Piles created. How does this change? What did your graph illustrate to you? What of the change in Voltage with changes in the Pile size?

Side Note : One might notice that there seems to be no 'true' circuit when measuring current, which is often the process – that is to say, we should have some resistance or a load in place for our battery. We can insert a very low resistance (10 ohm) if we like between the multimeter and the battery, but from my observations of this test with and without one I find that it merely results in a smaller current value but does not affect its overall character. This is because there is resistance in the test wires already from the multimeter and they are quite low.

Extended Activities :

Each Activity is a Question to Explore :

1) From the Materials List, Try a different electrolyte concentration. We used 1 tbsp of salt to 6 cups of water. Now try 2 tbsp, then 3 tbsp to see how this affects the results.
2) From the Alternative Materials List, use a different electrolytic blotter. Instead of paper towel, use thin cardboard.
3) From the Alternative Materials List, use different types of piles : Penny & Nickel, Penny & Dime, Nickel & Dime, to test the outcome of the voltage and amperage of the battery produced.
4) From the Alternative Materials List, test new vs. old pennies (the older ones have a higher percentage of copper).
5) From the Alternative Materials List, use one or more of the other electrolytes, such as lemon juice, soda water, or vinegar and test how different electrolytes affect the outcome. Be sure to include distilled water as a control.

Other Batter Activities :

Factors Affecting a 'Lemon' Battery and the Slide Rule Activity

For History of the Battery see the Factors Affecting a Simple Battery Prelude.

This Activity is the construction of a simple battery and an examination of what will increase its voltage and overall power. We will use simple materials : copper wire (here it may be a good idea to rely on pennies, preferably older than 1982), steel wire (here really using a large paper clip), and the body of the battery will be one or more of the following (depending on your resources) : lemon, lime, banana, potato, or cucumber for our construction of a battery.
With a multimeter we will measure voltage and amperage of the battery we construct. Graph the results to see if there are changes with the changes to your experiment.

Activity : The Voltaic Fruit/Vegetable Battery

Purpose : To construct a simple battery and measure its power with changes in the amounts of materials comprising the battery system.

Note for Safety : Always have adult supervision and permission for any activity. Do not involve outlets in any manner.

Materials :

- Multi-meter,
- Alligator clips on wires set,
- Pre-1982 Pennies or Copper Wire (18 gauge is good – needs to be stripped) – note : it is far easier and safer to use pennies,
- Large Metal Paper Clips,
- Wire Clippers (can have the wire stripper too) – Note : only needed if not using pennies,
- Fruit and/or Vegetables : Use older ones since these will not be consumed (never a good idea to consume materials used in a lab) : Choices can be items such as : Potato, Banana, Cucumber, Lime, Lemon, Orange (Note : Can have more than one of each of the one(s) chosen since we will test the idea of connecting batteries together in series too),
- Small Kitchen Knife,
- Graph Paper,
- Slide Rule

Note : You need to have adult supervision and permission for this Activity. At no time are you to use any electrical device, regular batteries or any electrical outlet. The only tool that involves electricity is the multimeter, which is a galvanometer and responds to voltages and currents.

Procedure :

1. With any and all Activities read through, then set up the process for the Activity and run it without measurement and watch where you may have bumps in the road so as to anticipate these. Be patient – Repeat processes. Once all is well when measuring see to the values – do you notice patterns or trends – what do you predict will happen next as you change one thing or another from one trial in an experiment to the next. Always Explore & Enjoy. Be Brilliant!
2. Rule No. 1 : See Note and be safe. This Activity is about the properties of the fruits and/or vegetables we have chosen to make our battery and does not involve electrical devices, regular batteries, and the like.

3. Pre- Activity Preparation :
4. Note : The following Only needs to be done if using copper wire that needs to be stripped. It is suggested that pennies are the best alternative to this.

5. Note : It is best to have adults (parents) to do the preparation. The key is to be safe.
6. If possible in buying copper wire that is bare. If not possible, then you have to strip the pieces you are going to use. Each piece will be 2-3 inches (5-8 cm). It is probably easiest to strip a small stretch of wire from the main coil and then cut off a bare piece at the appropriate length.
7. Note : The number of pieces will depend on how many batteries you create. Each piece of fruit or vegetable used is a cell or a battery and each will need one piece of copper wire. A good Activity might be up to 4 at most – resources permitting (but can be fewer or more if interested).

8. The Set up of the Batter in the Activity :

9. As it is not known what item(s) you are using for your Activity (lemon, potato and the like) I will write this from the point of view of a lemon being the main item. You simply insert what you are using in place of the term Lemon.
10. Also I will assume that you are using pennies and large paper clips.
11. The Lemon, one penny, and one large paper clip are all that is needed.
12. Note that each piece of fruit solutions is a Trial.
13. If possible It is best to have at least 2 or 3 pieces of fruit (the Lemon) to make batteries with. There is another low cost solution, though. A given lemon can be cut in half, hence acting as two batteries!
14. The Battery :
15. Before cutting a Lemon so that it can be two batteries, roll it firmly on the table (not so hard as to rupture the skin of the Lemon).
16. Have your parents cut the Lemon so that it can sit flatly on the table (this can be done by cutting it in half and then cutting off each of the ends).
17. With the first piece insert into the actual exposed Lemon the penny and one of the paper clips. Have some distance between them (about 1-2 cm is sufficient).
18. In inserting the penny and paper clip leave enough out so that the multimeter can be used to touch them in order to take measurements.
19. Amazingly this is the first battery! :)
20. With sufficient materials set up two or three more if you are able to.

21. Measurements in the Activity

22. Be sure to know how to use the multimeter and its proper arrangement of testing probes and which setting (probably mV and mA). Run through its operation to be certain of how to do this.
23. We are going to measure each of the batteries separately (give them names or numbers to distinguish them in the table of data).
24. Measurement is done by using the multi-meter where one of the probes touches the penny and the other probe touches the paper clip in the same Lemon.
25. Record these values voltages and currents in the table for each of the batteries.
26. Now With the individual measurements done, we are now going to look at connecting our batteries in series and see how this affects the outcome.

27. Series connection :

28. Presently all the batteries are in a row where one of the poles is on the left (for example the penny) while the other is on the right (the paper clip following our example). – Be patient, the pennies are probably the hardest to connect.
29. Using the alligator clip wires connect the paper clip of the first battery to the penny of the next second battery.
30. The next alligator clip now goes on this second battery on the paper clip.
31. These two batteries are now in series and we can measure the voltage and current of two batteries and place these values in our data table.
32. To measure these batteries in series one probe of the multimeter touches the penny of the first battery and the other probe touches the paper clip of the second battery.
33. Now, leaving the first two batteries connected now put a third battery in series by attaching the paper clip of the second battery to the penny of the third battery.
34. Measure the data for this set of batteries. Continue this for all of your batteries so that all end up in series. Each new one results in two new measurements of the total voltage and total current of these batteries. Recognize that you always place the probes on the open battery posts (that is there is no alligator clips on them).
35. Note : Once done the first battery has a battery with no alligator clip and the last battery has paper clip with no alligator clip.

36. Calculations :

37. Once done with all of the measurements first Calculate Power for each of the batteries individually. Note : Each of the reading is probably in milli- so this means $\mathbf{10^{-3}}$. When calculating Power in watts this needs to be considered.
38. For the series connection : For each series (2 batteries, 3 batteries, et al) calculate the Power of this set of batteries.
39. In the series connection : Graph Voltage vs. Number of Batteries in series. Draw a best fit line and determine the slope of it.
40. In the series connection : Graph Current vs. Number of Batteries in series. Draw a best fit line and determine the slope of it.
41. Try other pieces of fruit and/or vegetables for comparison.

Data :

Battery Name or Number	Voltage (mV)	Current (mA)

Number of Batteries in Series	Voltage (mV)	Current (mA)

Calculations :
Be sure to use your Slide Rule!

P = V*I

(P = Power, V = Voltage, I = Current)

Slope :

$$\mathbf{m = \frac{\Delta Y\text{-dependent variable}}{\Delta X\text{-independent variable}}}$$

Conclusion :

Cosmic! – Foods with battery potential – who would have thought!?

Here the conclusion first comes from the calculation of Power for each of the Batteries created. How does this change? What did your graph illustrate to you? Did you try other pieces of fruit and/or vegetables and how do each of these compare? If you were able to create a series of lemon batteries what does your graph show you in terms of more and more batteries in series with each other?

Side Note : One might notice that there seems to be no 'true' circuit when measuring current, which is often the process – that is to say, we should have some resistance or a load in place for our battery. We can insert a very low resistance (10 ohm) if we like between the multimeter and the battery, but from my observations of this test with and without one I find that it merely results in a smaller current value but does not affect its overall character. This is because there is resistance in the test wires already from the multimeter and they are quite low.

Activity #40
Electromagnet Graphing Explorations
Grade Level : High School
Math Level : Calculating

Electromagnet

Ms Briana Andromeda Purcell here and think back to younger days – and picking up a magnet which was one of the earliest items of fascination for us. Hold this item where invisible forces emanate from it – and can be strong enough to pass through us, pick up paper clips through paper or cardboard effortlessly when strong enough. It was the mystery of a compass and how the Earth itself moved the needle that attracted young Albert Einstein to develop a deeper curiosity for nature.

Further examination of magnets we find that they attract only certain types of matter (iron containing) and yet not others (paper, plastic, glass). Further investigation we uncover that there are two poles, North and South and the simple rules for them : Opposites attract and Likes repel. Further reading and learning and the magnet is not only an item on our refrigerators, they are found in many items : seals for doors, speakers, microphones, electric motors, electric generators, and so on.

Further we did not invent them, they are a natural part of nature. The Earth has a magnetic field generated in its current-carrying outer liquid metallic core, which extends far into space beyond the surface of the Earth and acts as a protective shield from harmful ionized particles from the Sun. Even the Sun and a number of the planets have magnetic fields, some like Jupiter's are much stronger than the Earth's field. Many of our inventions stem from this – we find in nature things that we have to create from other materials to bring it into a realm of scale and use for ourselves.

There are 3 basic categories of magnets : permanent, temporary, and electromagnet. All of them exhibit magnetic behavior due to moving electric currents (on a small or a large scale). The permanent ones have outer electrons in the metals that can move from one atom to the next in small loops called domains. This generates an electric current due to the electric field and perpendicular to it is a magnetic field. In the case of the electromagnet, the magnetic field is found and described by **Ampere's Law** which paraphrased states that the magnetic field strength around an electric current is directly proportional to the electric current in that wire or system.

First note that the electric current (field) and the magnetic field are at right angles to each other. A current carrying wire has a magnetic field around it. Use your right hand and wrap it around the wire and allow your thumb to point in the direction of the current. Your fingers wrapped around the wire are in the direction of the magnetic field. (This is classically called the Right Hand Rule)

Why then are the electric lines in our homes not acting as magnets? Simple there is a line going into the device and one coming out in the same wire. The magnetic field points in opposite directions and nearly cancels out.

The best way to make the electromagnet is not to use a straight wire, but instead to loop the wire. A coiled wire with current flowing will have a magnetic field where all the magnetic fields of the portions of the wire add together and generate a magnetic field. The more loops, the greater the strength of the magnetic field. Note that the electromagnet is only active when the electric current is flowing. This type of device is often called a solenoid.

The **use of the electromagnet** is extensive and found in large electromagnets to move metallic masses, cathode ray tubes or more commonly called conventional TV tubes where the electron beams are moved by electromagnets, galvanometers which are devices used to detect current / voltage by using a coil of wire in a magnetic field and the current in question passes through the coil to generate an opposing magnetic field to cause the deflection of a needle in the meter to render a reading (hence the basis of a voltmeter, ammeter, multi-meter), magnetically levitated trains due to electromagnets, electric transformers (where a current carrying-wire generates a magnetic field and this in turn generates an electric current of a different amount in another wire)[note this occurs in speakers too], and even in electric motors where current-carrying wires have magnetic fields which interact with the magnets in the motor to cause the rotor to spin (through the basic principles of magnetism – opposites attract and likes repel). Notice it is not only the idea of electricity and magnetism interacting but the creative and inventive mind that thinks of new ideas and applications.

As can be noted by this last idea, electricity and magnetism are inter-related phenomena. One causes the other and vice versa. **This is the basis of the 2nd Industrial Revolution and the development of electricity, all of its motors, appliances, tools, and applications in the modern world. Wow! These ideas and things clearly affected all that we not only did but certainly do even today. Think of electric motors in battery powered vehicles for example.**

Our activity involves the creation of an electromagnet from a current-carrying wire supplied by a power source (battery) and investigating the factors that may affect the overall strength of the electromagnet.

This sort of activity should inspire questions, though. If a current-carrying wire can generate a magnetic field, can a magnetic field induce a current in a wire? The answer is yes! This is the basis of **Faraday's Law**. Michael Faraday noticed that a wire when moving in a magnetic field will have a current generated or induced in it. It can be paraphrased thus :

In a place of a changing magnetic field there will be generated or induced an electric field. The strength of this electric field will be directly proportional to the rate of change of the varying magnetic field.

What this means is that the strength of the voltage will vary with the rate of the change of the magnetic field that the material experiencing the magnetic field will have created in it. A further analysis of this law of induction finds that the number of coils of wire will directly affect the induced electric field also known as emf (electromotive force) outcome as well. The larger the number of coils, the stronger the stronger the electric field.

This idea is the basis of the **electric motor** and **electric generator**. Each involves rotating wires and magnetic field. The motor converts electrical energy to mechanical energy while the generator converts mechanical energy into electrical energy. Each relies on the interaction of changing magnetic fields as caused by moving wires where one has current put into it (the motor) while the other has the current induced in it due to rotation (the generator).

Note : See Safety Notes below. This Activity creates a basic Electromagnet, which can cause batteries to heat up. This must be monitored carefully. It is only done with adult permission and supervision.

Note : It is imperative to be able to break or stop any circuit you create for this activity – i.e. a switch or the means to disconnect a wire – due to the fact that this type of circuit can heat batteries and cause damage to them. Leave no circuit unattended or active for too long a time – only short time uses.

Overall Purpose : To use the ampere's law to create a magnet from an electric current and test its strength by changing the variables that affect the formation of the magnet.

Purpose : To use an electromagnet and measure the strength of the magnetic field through changes in the number of windings of the wire making up the magnet.

Purpose : To use an electromagnet and measure the strength of the magnetic field through changes in the number of batteries being used in the magnet.

Materials :

- Iron Nails [as many as 3 different lengths are good here],
- Two D batteries & one 9V battery,
- Insulated Copper Wire (22 gauge type and 1-1.5 m length needed),
- Knife-switch Box (for circuit activation),
- Alligator Clips Wires (2),
- Wire Stripper,
- Battery Connection Boxes (single, double),
- Box of standard Paper Clips,
- Graph Paper,
- Ruler,
- slide rule

Procedure :

Activity I : The Number of Wire Windings vs Electromagnet Strength

1) With any and all Activities read through, then set up the process for the Activity and run it without measurement and watch where you may have bumps in the road so as to anticipate these. Be patient – Repeat processes. Once all is well when measuring see to the values – do you notice patterns or trends – what do you predict will happen next as you change one thing or another from one trial in an experiment to the next. Always Explore & Enjoy. Be Brilliant!
2) Choose an iron nail will be used for this activity.
3) For this activity it is best to use 1 D size battery. Put the battery in the connection battery box and connect it to the knife switch with alligator clip wires on both ends. One will go to the knife-switch and then another will lead away. This one will attach to one end of the wire used to wrap the iron nail while the other end of that wire attaches to the alligator clip coming from the battery connection box.
4) Take the length of wire and wrap tightly the metal rod (nail) some 8 times and leave the other ends out. Record the number on the chart. Record the number of times wrapped (W).
5) Strip about 1" of the ends of the wire so that they can be connected to the alligator clips.
6) Connect the ends of the wire to the clips and then close the switch to activate the circuit. Have the dish of paper clips (all separated) and bring the electromagnet near them.
7) For each trial decide on a distance to be (0.5 cm) and go over the surface of the materials to pick up. Decide on the number of passes (1 or 2).
8) Pull the collection on the magnet off to the side over a cup or dish and deactivate the switch. Brush off the materials into the empty dish and then count the number of items picked up. Record this number (N).
9) Redo the wrapping process in step #3 now increasing the number of times (go from 8 to 12 to 15 to 20 to 25 to 30, perhaps up to 50 if possible). A good suggestion is a logical sequential increment such as 3 or 4 per trial.
10) For each trial record the results – (both the number of windings, W, and the value of paper clips picked up, N).
11) Graph the results where the Strength of the Electromagnet (as determined by the number of paper clips) is the y-axis while the Number of Wire Windings on the Nail is the x-axis.
12) From the graph, draw a best fit line and calculate the slope using the slide rule.

Activity II : The Amount of Voltage vs. Electromagnet Strength

1) In this activity you have chosen a base number of windings for the nail. Look at your data in Activity 1 to make this choice. Typically 15 to 20 is a good value.
2) Be sure to have the battery connection box attached to the knife switch and open. See Activity 1 for directions to the initial set up. You are creating a series circuit basically in each of these cases.
3) Set up the wire-wound rod and attach it to the power source. Activate the switch.

4) Pass the electromagnet over the materials to be picked up (paper clips or ball bearings). Be sure with each situation of data collection you maintain the same number of passes (predetermined as in activity 1) and the distance for each of the passes (predetermined as in activity 1).
5) Collect, count, and record this number (N).
6) Use the next size battery connection box (2 batteries). You will have to reattach the alligator clip from the single box to the 2 battery box.
7) Redo the work and test the strength of the electromagnet once again.
8) When completed with the D batteries, now use the 9 volt battery. For this one, the alligator clips can come from the wires themselves and attach to the 9 volt battery directly.
9) Record all the data,
10) Graph the data in the same manner as activity 1 (here the x-axis is the increase in voltages) and determine the slope of the best fit line with a slide rule.
11) Note :
12) Though the Slide Rule is a recommended tool, all of the calculations can be done with a graphing scientific calculator or the use of a spreadsheet program. In these calculations you have to generate a table of data, graph it, and then find the slope and/or equation of the best fit line for the data. Other formula calculations can be done with these tools as well.

Safety :

1. Adult Supervision and Safety are always important in activities.
2. Keep voltages low and Never use high voltage here as it can be dangerous and electrocute you.
3. It is important to activate the switch after making the loops and do not activate it without loops since the current in the wire will heat the battery which could lead to an explosion.
4. Monitor the temperature of the batteries. Do not leave them connected and do not have them on too long. They will heat up.
5. Activate the circuit to create the electromagnet for only short periods of time and monitor the battery so that it does not become too hot.
6. Never insert the wires into any other electrical outlet or devices.

Data :

Activity I : The effect of the number of wire windings on the strength of the electromagnet.

Voltage Used : ____________
Number of Times Passed Over Paper Clips : _______
Distance of Each Pass Over Paper Clips : ________

Number of Wire Windings (W)	Number of Paper Clips Picked Up (N)

Activity II : The effect of the number of batteries on the strength of the electromagnet.

Number of Wire Windings Used : ____________
Number of Times Passed Over Paper Clips : _______
Distance of Each Pass Over Paper Clips : ________

Voltage Used (V)	Number of Paper Clips Picked Up (N)

Calculations :

Be sure to use a Slide Rule!

For all Activities, Graph the Data recorded where the y-axis is always the Number of Paper Clips Picked Up (as it is the Dependent Variable), and the Variable controlled (i.e. the Independent Variable) is the other one on the x-axis – which will either the number of windings or the voltage used. Draw a best fit line and determine the slope of the line in each case by using a Slide Rule.

Slope = Rate of Change of Magnetic Field Strength (note field strength is indicated by the amount of mass it affects here in this activity)

$$\textbf{Slope} = \frac{\textbf{Number of Paper Clips Picked Up}}{\textbf{the Controlled Variable}}$$

Conclusion :

Wow – How Cosmic was this investigation! Using common items, wires – batteries – nails to create magnets to pick up paper clips and make a magnet! Too cool!

All things certainly have a story to tell us – we have to be creative in thought, curious in nature and passionate about our need to explore, to learn, and be Brilliant!

Examine the data and determine which, if any, of the variables has an impact on the strength of the magnetic field. Can you write a statement that generalizes the relations you may have uncovered. As the number of windings increased, what happened to the number of paper clips that could be picked up? How does the number of paper clips picked up relate to the inferred strength of the magnetic field generated?

Activity #41
Rate of Spin of a simple Electric Motor measure
Grade Level : High School
Math Level : Calculating

Electric Motors Activity

Ms Briana Andromeda Purcell here once again and we are carrying ideas from one Activity to another. From the Electromagnet Activity (# 40) we use the idea of electricity in a wire to create a magnetic field once again but for specific reasons and ends here in this Activity. We have the awesome journey to use that wire with a magnetic field in the presence of a permanent magnet and then create essentially an Electric Motor!

The Electric Motor is basically the reverse of the Electromagnetic Generator, Alternator, or Dynamo. In the case of the generator either a wire coil is spun between the poles of a magnet or vice versa so that an electric current is induced in the wire coil that is taken advantage of. The power source is typically steam, water, or wind which converts mechanical energy into electrical energy. Read more of the ideas of electromagnetic induction which is a part of the history of the electric motor as well in the Electromagnet Activity #40.

The **Electric Motor** is essentially the opposite of the generator. A power source, usually electricity causes the wire coil (set between two permanent magnets) has its current create a temporary magnetic field. The pieces of the motor are organized so that the current flow that creates this magnetic field is opposite the magnetic field of the permanent magnets. Of course, like poles in a magnet repel, so this causes a force to act on the wire coil that is set along an axis of rotation. This force acting along a distance from a central axis is a torque which sets the coil in rotational motion. Once half-way around the current is made to reverse direction by a split ring and this recreates the process again and maintains the rotation back around where the process starts again. – In essence we use the properties of magnetic fields to set matter, the wires in our case the small scale version of the coils of wire in a regular electric motor, into motion – in this case spinning so we can do things with it!

Hence the electric motor turns electrical energy into mechanical energy! Just by its simple description it is easy to see that energy must come from a source and is used to do something else. Since nothing is 100% efficient, there are losses (hence the motor gets hot). This helps to understand conservation of energy (see Mechanical Energy Activity #27).

Electrical Motors have a wide array of uses and can range in size from the size of smaller than a fingernail and used in a watch to larger than a car and are used to propel ships, power industrial fans, and have ratings in the millions of watts. Most are handheld in size and power such things as fans, pumps, household appliances (washer, dryer, refrigerator), handheld power tools (drills, table saws, et al). They are one of the chief result of the 2nd Industrial Revolution which was centered on electricity and its development.

The ideas of electricity and magnetism date to speculations in the 1700s but were worked on in the 1800s and came into mathematical relationships and descriptions with Michael Faraday as early as 1821. Faraday did a public demonstration of a free-hanging wire in a pool of mercury through which a current was passed. In the pool of mercury was a permanent magnet. With the current flowing, the wire rotated about the magnet. This meant that the current-carrying wire had a magnetic field encircling it. He developed Faraday's Law. Basically it states that the induced voltage in a coil of wire is proportional to the product of the number of loops of wire and the rate at which the magnetic field passing through the loops is changing. More generally it states that an electric field is generated in any region of space in which there is a changing magnetic field with time.

The best way to think of it is this : A changing electric field induces a magnetic field and a changing magnetic field induces an electric field. This description connects to light even, since it is one of the many electromagnetic waves (radio, ultraviolet, x-rays, infrared to name a few) and the wave that it is.

By 1827 with experimentation Anyos Jednik developed a device he called a "lightning-magnetic self-rotor" which has the basic parts of the electric motor today, the stator, the rotor, and a commutator. However, it was not until 1832 when William Sturgeon invented an electric motor capable of turning machinery. Emily and Thomas Davenport patent a motor design in 1837 in America and it was made for commercial use. The early engines had limited success due to the lack of power availability. They were DC systems and at this time there were no significant battery systems in place.

In 1855 the first electric motor car by Anyos Jednik came into being. Not much else was done, however, for some time and the modern DC electric motor took full form in 1873 by Zenobe Gramme and it was used in industry. More developments followed, such as 1888 when Nikola Tesla created the first practical AC electric motor.

With further testing, experimenting, and new ideas and understanding of electrical and magnetic forces (Maxwell's equations in the 1880s) plus the practical application of ideas to design led to stronger, more efficient engines. With this, industry could literally go into high gear since now it had a workhorse that had high reliability, power, and was efficient. It is the electric motor that changed the workplace to reduce the number of needed people, animals, and to replace older hydraulic pressure systems to run and operate all levels of machinery. The home life changed drastically in terms of ease and convenience. Even today the electric motors used in homes and industry accounts for more than half of all the electricity being used. Even today in the 21st century the use of the electric motor is once again finding its way into discussions involving motorized vehicles for transportation.

Our goal is to build and test a simple DC electric motor only using battery wire, paper clips, a magnet, and batteries.

Notes on Safety : As with any electrical device or situation, exercise caution. Be sure to have adult permission and supervision. Be careful of sharp objects (such as with wires). Do not place wires in other electrical devices, outlets, and the like. Be sure to wear goggles.

Purpose : To measure the spin rate variance of a constructed electric motor due to the amount of applied voltage to the motor.

Alternate Purpose :

Purpose : To measure the spin rate variance of a constructed electric motor due to the number of windings of coil wire in the motor.

Materials :

- Magnet Wire (24 gauge or higher) (total 6 m in length),
- Ceramic Magnets (3 of them each approx. 3 cm^2 by 0.5 cm thick),
- Block of Styrofoam (approx. 5 cm^2 by 2 cm thick),
- Battery Holder Packs (1 AA, 2 AA, 4 AA),
- Batteries : four AA and one 9-Volt,
- Wires with Alligator Clips,
- 2 Large Paper Clips,
- Sandpaper (safest method) OR Craft Exacto Blade,
- Laser Photo Tachometer (to determine spin rate),
- Goggles,
- Slide Rule

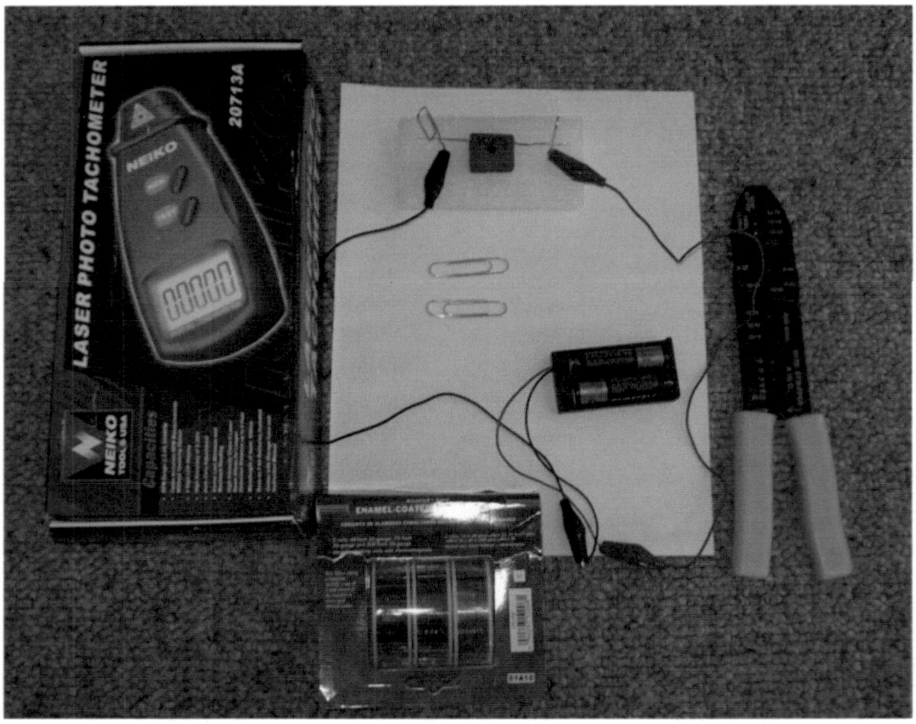

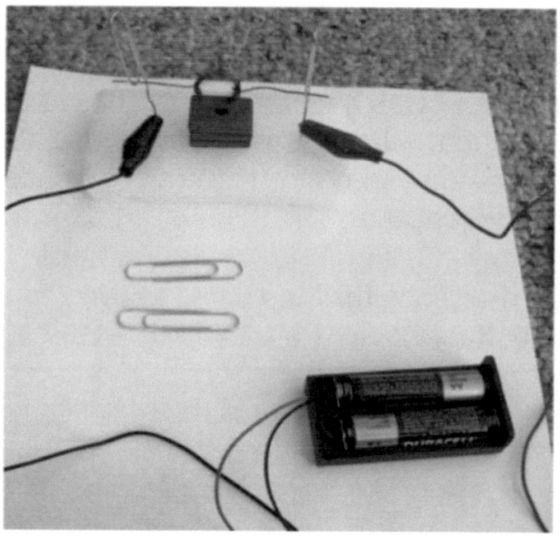

Set Up of Electric Motor : (Have parents do this)

1) Cut a long enough piece of magnet wire so that it can wrap around a broom handle or C battery about 10-20 times and have about 3 inches either side of this at most. (Length is around 36-48 inches total).
2) When the coil is created wrap the wire back around the coil so as to secure it and the the remainder of the wire project away from the edge of coil in opposite directions.
3) Be sure to refer to the pictures to help in the descriptions!
4) With the coil facing you flat face (like a coin) and the wires projecting at the sides place the projecting wire piece from each side in turn and use either sand paper or a sharp blade and scrape off the enamel coating on One Side of the Wire.
5) Do this to the other projecting wire and scrape The Same Side so that the top is scraped for both while the bottom is coated.
6) Into the foam block stick two large paper clips after straightening the outer coil of the outer loop. This creates two loops standing up.
7) Between the paper clips place the ceramic magnets.
8) Through the loops of the paper clips place the coiled loop of wire so that the projections pass through the paper clips and rests there.
9) For this and the procedure it is good to wear goggles.

Procedure :

1) With any and all Activities read through, then set up the process for the Activity and run it without measurement and watch where you may have bumps in the road so as to anticipate these. Be patient – Repeat processes. Once all is well when measuring see to the values – do you notice patterns or trends – what do you predict will happen next as you change one thing or another from one trial in an experiment to the next. Always Explore & Enjoy. Be Brilliant!
2) Follow all of the assembly directions for the engine above.
3) Place the batteries in the battery holder (start with the least and work your way up in each of the trials.
4) Over the coil of wire on the motor use the tachometer reflective coil tape (be sure that the underside is coated with dark permanent marker)
5) Attach alligator clips to the base of the paper clips and the other ends to the battery pack.
6) With some agitation the motor should operate. Recheck all the work of the engine and test it before the reflective tape if needed.
7) Use the tachometer to obtain readings of rotation rate for a given voltage. Record these results.
8) With each trial try another voltage.
9) Graph the data and calculate slope of the best fit line of Spin Rate vs. Voltage for the Motor.
10) Note :
11) Though the Slide Rule is a recommended tool, all of the calculations can be done with a graphing scientific calculator or the use of a spreadsheet program. In these calculations you have to generate a table of data, graph it, and then find the slope and/or equation of the best fit line for the data. Other formula calculations can be done with these tools as well.
12) Alternative Trial
13) With more battery wire create another loop and run the tests again but here see how the effect of the number of windings affects the outcome.
14) Note here the voltage is kept constant, since the variable under consideration is the number of windings.
15) Perform the same basic calculations only here it is spin rate vs number of windings.

Data :

Activity I : Spin Rate with Varying Voltage

Number of Windings : ______________ (20 recommended)

Trial	Voltage	Spin Rate
1	1.5 V	
2	3.0 V	
3	6.0 V	
4	9.0 V	

Alternate Activity :
Activity II : Spin Rate with Varying Number of Windings

Chosen Voltage : ______________ (3.0 V recommended)

Trial	# Windings	Spin Rate
1	10	
2	20	
3	40	
4	60	

Calculations :

Be sure to use your Slide Rule!

Graph the Measured Data with the Independent Variable (the number of windings or voltage) on the x-axis and the Dependent Variable (the motor's spin rate) on the y-axis.
Draw a Best Fit Line.

$$\text{Slope} = \frac{\Delta Y}{\Delta X}$$

$$\textbf{Slope of Line} = \frac{\Delta \textbf{Spin Rate}}{\Delta \textbf{Independent Variable}}$$

Conclusion :

Amazing and Cool! To make a simplified Electric Motor and measure it too!
What did your results show? Do you think there are upper limits to this?
Why or why not?

The power and use of the electric motor cannot be underestimated and can be a great source of ideas, inspiration, and other lab activities. There are different materials and slightly different means of arrangement to achieve the same ends and even kits to build them.

Activity #42
Simple Magnetic Field Strength determination

Grade Level : High School
Math Level : Challenging

The Basic Magnetic Field Strength Investigation Activity

For a good introduction to Magnetism reread a number of Activities prior to this one since Magnetism is related to Electricity and a wealth of information is contained there. So read the introductions to Electric Charge Activity (Activity #38) and the Electromagnet (Activity # 40) and Electric Motor Activities (Activity #41) for more information, in general, on the connection of Electricity to Magnetism.

It essentially turns out that a moving current has an associated magnetic field. (Much like a changing magnetic field generates a current). Though in this Activity we are instead using a permanent magnet in this Activity, the moving current is the aligned outer electrons in the material which result in the magnetic field for the magnet.

Like other forces (gravitational and electric), the magnetic force on a large enough scale should exhibit an inverse-square law characteristic. That is to say, as distance from the magnet, the force should decrease as the inverse-square of the distance. For more on the idea of inverse-square laws, see the Inverse Square-Law of Light Activity (Activity #47)(where light illumination from a point source acts in the same manner).

In this Activity we explore Magnetic Field strength in a rather simple manner. We use a Magnet separated from a piece of metal it can affect by some poster board paper barrier. The piece of metal is a part of a Tension Scale so that we can pull the scale and measure the amount of force that the magnet is exerting to hold the metal in place. We use different thicknesses of poster board paper, measure, and graph our results.

In the Activity, be sure to have parental permission and supervision. Be sure to wear goggles. Act carefully in a slow and deliberate manner for safety-sake as well as taking measurements.

Purpose : To measure the pull (force) of a magnet on a magnetic material at varying distances so as to uncover the relation of distance and field intensity through graphing and analysis.

Materials :

- 1-3 small ceramic or neodymium (strong) Magnets,
- 100g Tension Scale (depends on magnet strength),
- Poster Board,
- Scissors,
- Caliper or Ruler,
- Masking Tape,
- Graph Paper,
- Goggles,
- Slide Rule

Photos :

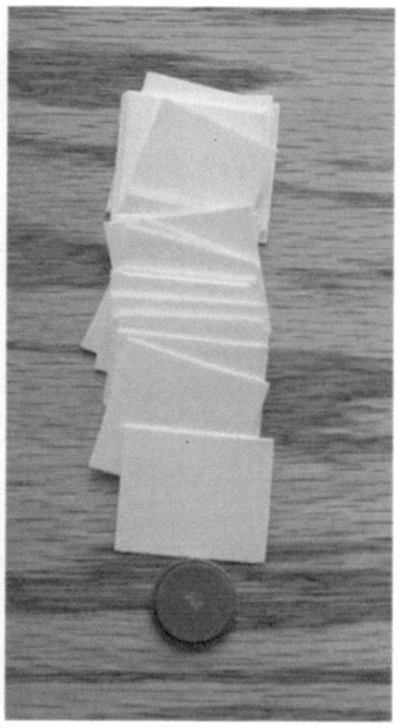

Procedure :

1. With any and all Activities read through, then set up the process for the Activity and run it without measurement and watch where you may have bumps in the road so as to anticipate these. Be patient – Repeat processes. Once all is well when measuring see to the values – do you notice patterns or trends – what do you predict will happen next as you change one thing or another from one trial in an experiment to the next. Always Explore & Enjoy. Be Brilliant!
2. In the Set Up do the following :
3. Be sure to have a strong magnet that can effect items, like paper clips. through a number of sheets of paper. This is because we are using poster board (or something thicker than regular paper) in this activity.
4. Next test the Tension Scale and whether or not the hook or clip at the end is attracted to a magnet. If not, attach securely a paper clip.
5. Before the next step, and any step involving the magnet being attached to the metal hook/clip of the scale (even through the poster board, which is the investigation in this activity) you should wear goggles in case the hook/clip should spring back too quickly and/or possibly detach. Knowing this can happen, be aware and work safely. Also examine the tension scale – it is recommended that the hook/clip is securely attached. Have a parent supervise the activity and inspect the tension scale.
6. Test the Tension Scale through several layers of paper (or even a piece of cardboard) with regards to the magnet attracting it and holding it.
7. In testing the Tension Scale, now slowly pull the scale so that it extends while attached to the magnet. At some point it should release. If you can fully extend the scale, then the thickness is not enough for the barrier between the magnet and the scale hook/clip or the scale needs to be of a higher order (1000 g instead of 100 g)
8. The final step in the set up is to cut small squares out of the poster board that are a bit larger than the magnet being used (see photo). The number will depend on the strength of the magnet – a good estimate is 25.
9. With the tests done, onto the Activity :
10. Zero out the Tension Scale.
11. It is recommended during and between given trials when it can be checked it is a good idea to check this and reset it as needed.
12. In each trial you will do the following :
13. Have chosen a given number of poster board barriers (referred to as barrier pieces in the data table) that will be between the magnet and the tension scale hook/clip.
14. It is best to start with the maximum and then diminish this number (as will be seen in the continuation of measurements) (for example 25). Record this number (N).
15. Stack the number of chosen pieces atop the magnet that is on the table top and tape them down. The tape will look like a bridge from the side and can easily be removed (as it will be in subsequent trials). Note that the top piece will probably be stuck to come along for each of the trials, but that is okay.
16. It is best to have this next to the table's edge, since you now need to use a ruler or a caliper to measure the thickness of the barrier and record this value (T) in centimeters.
17. Now bring the Tension Scale hook/clip down to the top of the stack and let it magnetically attach. Once secure slowly draw it away, all the while watching the reading on the Scale.

18. At some point the hook/clip will detach. Be sure to keep an eye on where the scale reading had gone to (M). Also be sure to watch your eyes by wearing goggles! Keep enough distance for safety.
19. Do the prior two steps 3 times, each time remembering and recording the tension scale reading. These values will be averages for a given trial ($\mathbf{M_{ave}}$).
20. In the next and subsequent trials, remove the tape and remove some of the poster board barrier pieces to decrease thickness, and re-tape the new stack.
21. It is logical to have a pattern to the number of pieces, for example if you started with 25, with each trial remove 5 of the pieces, so that the next trial is 20, then 15, and down to 5 or zero if you like.
22. Recognize that with zero pieces, you will more than likely need a scale that is of a higher caliber, since the force is quite large at this point. If you do not have one, simply note that it exceeds your measurement capabilities and goes off the scale.
23. Record the new values of pieces (N) and thickness (T) along with checking the calibration of the Tension Scale.
24. Repeat the process of measuring the amount of tension force (M) and averaging the reading for a given trial (M_{ave}).
25. Once all trials are done, put materials away.
26. Calculations
27. Though we have not technically measured force (unless your scale is marked as such), we can either use the readings as they are or first convert them to genuine force measures (F) from the M_{ave} values. (Use Force formula and conversions as needed).
28. Graph Tension Force (either M_{ave} or the newly computed F values – hereafter simply called F) on the y-axis vs. the thickness (T) (which can be in centimeters or meters) on the x-axis.
29. In the case of this graph it should not be straight and in fact, if ideal will have a slope of an inverse-square (see Inverse Square Law Activity #47 for further illustration) or proportional to — .
30. One of the first ways to see how close this relation is to an inverse-square is to graph calibrated Tension Force values (F) by dividing all of the table values by the largest value in the table (this means the largest one will have a value of 1, all others will be less).
31. Using the data for Thickness from the same data points (the one with the largest Tension Force value, use its distance value as the calibration value and divide all Thickness data points by this value, (hence the smallest value should be 1 and all the rest should be numbers greater than 1).
32. Now take the inverse-square of the calibrated Thickness ($\frac{1}{T^2}$).
33. Note that the above directions indicate that you must reformulate your table where the y-value column is the calibrated Tension Force value which is determined by taking a given Tension Force value and dividing by the largest value in the table. Do this for each of the values.
34. In the case of the Thickness values, take the inverse-square of each of the calibrated Thickness values.
35. Note that all of these actions can be done on the slide rule (as should all of the calculations). Here one can use the C1 and then the A scales after finding the T value

first on the D scale. Note that the C1 is very useful for inverses, and as for the divisions, simply employ the C & D scales.

36. On this graph we can draw a best fit line and determine slope, which ideally should be 1.
37. We now want to find out if it is indeed an inverse square, so now take our original data table and convert all values in it (F and T) into log values of each using the slide rule's L scale.
38. You will now have a table of Log(F) and Log(T) values.
39. Which you can now graph as well on a new graph of Log(F) on the y-axis and Log(T) on the x-axis.
40. Draw a best fit line through this graph and determine slope – it is ideally linear and should have a slope of -2.
41. You can do this same set of steps for other magnets (larger or more) and see if there is a difference.

Data :

Trial #	No. of barrier pieces : N	Thickness of barrier : T (cm)	Scale Reading M_1 (g)	Scale Reading M_2 (g)	Scale Reading M_3 (g)	Average Scale Reading : M_{ave} (g)	Force measure of scale : F (N)
1							
2							

Calculations :

Be sure to use your Slide Rule!

Average :

$$\mathbf{X_{ave} = \frac{\Sigma x}{n}}$$

(x's are the values measured and 'n' is the number of times)

Slope :

$$\mathbf{m = \frac{\Delta y}{\Delta x}}$$

$\Delta x = x_2 - x_1$

Force from Mass Measurements on Tension Scale :

$F = m*g$

(Note : 'm' here will be our M_{ave})

Conversions & Constants :

1 m = 100 cm.
1 kg = 1000 g.
$1 \text{ N} = 1 \frac{kg*m}{s^2}$
Acceleration due to Gravity : $g = 9.8 \text{ m/s}^2$

Conclusion :

Super – What adventures in Magnets we have done here with such simplicity!

What do your results show – that is, as distance increases, what happens to the magnetic force intensity ? Does it decrease in a linear manner, an inverse manner, an inverse-square manner? Will it matter if you change the type of barrier between the magnet and the scale? How did the force change when there were more or less magnets, different magnets? If you did the Inverse-square law activity, How does the graph of Force and Distance compare to the Light Intensity and Distance graph in the Inverse-square law of Light Activity?

Activity #43
Exploring Waves with a Spring
Grade Level : High School
Math Level : Calculating

Speed of a Wave on a Spring

Hello Inquisitive Pioneers! Ms Briana Andromeda Purcell here and we have a great Prelude here like the Prelude to Mechanical Energy Activity (#27), this Activity has a lengthy discussion on a topic, namely Waves. This is because like Energy, Waves are a critical cornerstone of all sciences particularly Physics with other disciplines such as Astronomy. You can start here for some information about waves in general or go right to the Activity afterwards.

Primary Reading about Wave Types and Wave Components :

A **Wave** is any disturbance that repeats regularly in space and time which is transmitted from one place to the next with no actual transport of matter. Waves transport energy and are used to transport information. For example, drop a pebble into a pond to create a set of waves. They radiate out from the impact spot of the stone on the surface along the surface of the water. The gravitational potential energy of the stone you dropped turned into kinetic energy which in turn became the sound, heat, and wave energy of the pond. The waves move out and can do work on something they encounter, such as erode the shoreline boundary. In the area of information, light waves are used to tell us about the stars in the universe (their temperature and composition, for example) and radio waves are used to transmit information. – The value of waves cannot be overestimated – they transfer energy and/or information.

There are 2 basic types of waves : **Mechanical and Electromagnetic** (which will be discussed) and each of these have component parts that have familiar terms.

Waves occur in nature with a great deal of regularity and we make use of them and generate them too for use. Waves on the ocean as noted at the shoreline are common to us all. Any sound is a wave. This includes those people can hear and those they cannot (infrasonic and ultrasonic). Sound waves, a mechanical wave, can travel through the Earth, such as when a earthquake occurs and is recorded even on the other side of the world. Light, an electromagnet wave, is a wave as are all of the other forms of radiation, such as radio waves, infrared radiation, ultraviolet radiation, x-rays, and gamma rays.

The classical description of a wave in the form of a picture shows a vibration pattern that oscillates with regularity with time on an x-y axis where x is time and y is displacement. The form is a sinusoidal wave typically (though there are other types).

One of the first parts of a wave to take notice of is what is called the **Wavelength (λ)**. The wavelength is the distance from one point on a given wave to its corresponding point on the next wave in succession. This can be from crest to crest (the high point of the wave) or trough to trough (the low point of the wave) and any other chosen point. The unit for wavelength is a distance unit and is typically meters (m).

The height of the wave has a name as well. **Amplitude (A)** is the distance from the midpoint of the wave to its crest (or trough). It is measured in distance units, like meters (m). The amplitude is related to the energy of a wave. For a Mechanical Wave, the energy of the wave is proportional to the square of the Amplitude. For example, this would be related to the Intensity of Sound, which is a measure of its energy and it is a mechanical wave.

Since a wave is something that occurs with regularity, there must be a measure of this behavior. Frequency is this measure. **Frequency (f)** is the number of events (cycles, vibrations, oscillations, or any repeating event) per unit time. This measure is how many vibrations occur in a second. A frequency of 20 is twenty times per second while 100 is one hundred times per second. This measure has the unit Hertz which is $\frac{1}{\text{seconds}}$ (Hz or s^{-1}).

If we know how many waves per second are occurring, then the inverse of this tells us the amount of time for one wave. **Period (T)** is the amount of time to complete one cycle. It is measured in time units, typically seconds (s).

The relationship of period and frequency can be expressed mathematically therefore. They are each the inverse of the other. Their product for a given wave is equal to one.

$$\textbf{frequency} = \frac{\textbf{1}}{\textbf{Period}} \qquad\qquad \mathbf{f = \frac{1}{T}}$$

The other key variable is the **speed of the wave**. It has the same definition as speed for any physical object, though here it is a wave. Wave Speed is the distance covered by the wave per unit time.

$$\mathbf{v = \frac{d}{t}}$$

The distance one wave travels (wavelength λ) is in the period of the wave (T), so we can write the regular equation as :

$$\mathbf{v} = \frac{\lambda}{\mathbf{T}}$$

But the Period is the inverse of frequency (f)
so we can rewrite the equation as :

$$\mathbf{v} = \lambda^{*}\mathbf{f}$$

These last two equations are known as the **Wave Equation**. Knowing the wavelength and the period and/or the frequency of a given wave, we can determine its speed.

Regardless of type of wave, all of these equations can be used for describing them.

The first major category of waves are **Mechanical Waves**. Mechanical Waves are waves that require a medium. A medium is a physical material that the wave or disturbance travels through. Mechanical Waves primarily include Sound Waves.

The sound waves we are familiar with are the ones in air that we can hear. These range in frequency from 20 Hz to 20,000 Hz. Despite this wide range of hearing capacity, there are still greater ranges. The **infrasonic** range is for those frequencies less than 20 Hz (infrasonic means below the range of hearing). Some animals, such as elephants have been found to transmit in this range. The frequencies above 20,000 Hz are called **ultrasonic** (meaning above the hearing sound range). These frequencies are used by bats, dolphins and others that use **sonar** (sound navigation ranging). We use them presently in medicine to create internal pictures of our bodies and other objects.

An important idea to note here is the necessity of a **medium**, such as air for sound. Imagine you place an electric battery-powered alarm clock in a bell jar and it goes off. Through the glass it is still audible. Now use a vacuum pump and draw out as much of the air as is possible (one cannot get all of it). What is noticed is that the sound seems to fade to nothing at all. If the alarm clock were the classic one where a striker is moving to and fro and hitting two bells, we see this still happening through the glass of the jar, but now there is no sound. If we could touch the clock we would feel the vibration, but now it has nothing to travel through, hence no sound!

This is similar in reasoning to why a person who has taken in some helium sounds so different (much higher pitch) than before when speaking in air. Helium is a gas, much like air is, but it has a much lower molecular weight than air. (Helium is the second lightest element, while air is composed of diatomic forms primarily of nitrogen and oxygen). The lower molecular weight of the gas allows our vocal chords to vibrate faster, hence at a higher frequency or pitch. So we have a 'squeaky' sounding voice with some helium passing over our vibrating vocal chords. **(Note : Do not do this as there may be more than helium in the helium-filled balloon, and too much of non-oxygenated gas can be harmful).**

Sound waves are not just sound in air, though, since sound can travel through liquids and solids just like gases. The most well known examples are directly from nature : Earthquakes and the Sonar and songs used by dolphins and whales. In the case of earthquakes the colliding plates of the Earth which can give way from time to time due to the rocks breaking sends a wave of energy (sound) through the solid rock itself. Seismographs around the world record these events and with geometry the location and intensity of the earthquake can be determined.

What is more important about Earthquake waves is that the entire Earth interior is not solid and part of it is liquid! We cannot venture there and for many years the interior of the Earth was a total mystery Earthquake waves are not of just one variety, but 3. The first type we may be familiar with are **Surface Waves**, as these are on the surface and effect the local surroundings of where the earthquake occurs. The next are the **Primary Waves** (aka **P Waves**). These are **Longitudinal Waves** (aka Compression Waves) – which are waves that vibrate along the direction of travel. These have regions of 'compression' and 'rarefaction' (higher and lower pressure and density regions). These can travel through both solids and liquids, hence can go through the Earth. The other notable wave is the **Shear Wave** (or **S Wave**) which is a Transverse Wave form. **Transverse Waves** are ones that vibrate at right angles to the direction of travel for the wave. These latter waves cannot travel through a liquid, hence there is a 'shadow zone' on the opposite side of the Earth from an earthquake where these waves do not show up as they outline the outer core region which is liquid.

In a solid the compression waves speed is determined by the medium's compressibility and density, while the speed of the shear waves in solids is determined by the material's stiffness, compressibility, and density.

An interesting fact is that waves in solids are faster than in liquids and these are faster than waves in gases (on the average). The speed of sound in air is about 331.5 m/s, while in helium it is 972 m/s. The speed of sound in water is about 1490 m/s yet in sea water (since there is salt and other minerals dissolved in it) it is 1530 m/s. The speed of sound in solid copper is 3560 m/s, while in solid aluminum it is 5100 m/s and in solid iron it is 5130 m/s. (Note that this is primarily true for solids that are crystalline in nature). It is not the density of the material that accounts for the differences, but is, instead, due to the stiffness and compressibility of the materials involved. The easier a material is to compress, the slower the speed. Solids are very hard to compress, hence a faster speed as compared to gases which are readily compressed so have a slower speed.

Noting the wave equation, the wavelength and the frequency are related directly to wave speed. The first thought is that since wavelength is part of the equation, then speed is determined by the wavelength of the wave. This is NOT the case! Imagine that a person is sitting 10 m away and is going to play the high pitch piccolo while 10 m in an opposite direction is a person going to play a low pitch tuba. With a signal by a light being turned on, they each simultaneously play a note. Which arrives at your first? Neither. They both arrive at the same time since they are not just equidistant, but they travel at the same speed. Interestingly, this fact is true for both Mechanical and Electromagnetic Waves. Though, each wave type does not travel at the same speed they have the same properties within their own realm. All sound waves in air at the same temperature and other atmospheric conditions, will travel at the same rate of speed (about 331.5 m/s). The speed will vary with temperature and yet be constant for a set of frequencies. This same idea as an experiment would show the same situation for light of different colors, for example blue and red. Of course the light would travel at the much faster speed of the speed of light (3×10^8 m/s).

This brings us briefly to the other type of wave, namely : **Electromagnetic Waves**. These are waves that do not require a medium and are transverse wave forms composed of vibrating electric and magnetic fields that are at right angles to each other and to the direction of travel. Light is the most well known electromagnetic wave and there is a whole family of them : radio waves, microwaves, infrared waves, ultraviolet rays, x-rays, and gamma rays. Though they each have their own unique wavelengths (and hence frequencies) they all travel at the same speed in a vacuum, namely what is called the speed of light. This is a cornerstone of Einstein's famous laws of Relativity (Special Relativity in particular). Regardless of the relative motion of the observer, all parties will agree that the speed of light is constant! Also it stipulates that the maximum speed in the universe is this value and that matter cannot go this speed, only energy of these waves (and gravitational waves if found to exist) can.

The consistency of the speed both for sound and of light is important to all of science, particularly in the form of the **Doppler Effect**. This comes from Austrian scientist Christian Doppler (1803 – 1853) who noted the apparent change in frequency of a passing train from the perspective of an observer at a station. As a train approaches and it sounds its whistle, the whistle's frequency is higher than if it were a stationary sounding whistle. Once passed, and the moving train again sounds its whistle, the frequency is lower than its normal frequency. Note that in all cases, stationary, approaching, or receding, the wave speed of the waves (for the whistle) are constant. Why then does the frequency change? As the train approaches, the sound wave crests from the whistle are closer together when they are encountered by the stationary observer at the station. Hence they are encountered more often than the stationary whistle, so a higher frequency or pitch in this case. When the train recedes (moves away) the wave crests are more spread out, so a lower rate of encountering the wave fronts, so a lower frequency hence lower pitch. Note that this is not the intensity of the wave. A far away object will sound less intense than one close by since the wave spreads out by the inverse-square law for distance (see Inverse-Square Law of Light Activity #47).

The Doppler Effect also applies to car horns and sirens as well. Today the Doppler Effect is used by bouncing radar beams off of moving objects, such as ice particles or water droplets in the air, baseballs, and cars to determine their speeds. In the former cases noted, this is for weather to find whether a system has precipitation, its size, speed, and other important factors about it. In sports, the speed of a ball or race car can be readily found. For law enforcement, the radar gun is used for determination of speed for cars on the roads.

Even more importantly for Light is the Doppler Effect. Any moving light source will give away its relative motion as compared to an observer by a frequency shift in its characteristic wavelengths. This is where we find the terms 'blue-shifted' and 'red-shifted'. If the naturally occurring waves are shifted towards the blue end of the spectrum, the object is said to be 'blue-shifted' and hence moving towards the observer. If the waves are found to be moved towards the red end of the spectrum it is then called 'red-shifted' and moving away from the observer. Note that it does not mean towards the red or blue end of the spectrum itself. Take for example radio waves. They too can be Doppler Shifted, as can all other forms of electromagnetic energy. To move away means they are shifted to a lower frequency than their natural frequency at a stationary mode and to move towards means they are shifted to a higher frequency from their stationary frequency. Not only is this useful for moving objects, but also rotating one too. The side moving towards the observer will be blue-shifted while the side rotating away is red-shifted. This information is also used to determine the relative speeds of objects in the Universe and was used by Hubble to find that the Universe itself is expanding when he was looking in the 1920s through the Hale telescope at distant galaxies.

The Activity Description

In the Activity there are two methods of approach for measuring the speed of waves in a spring (best to use a Slinky or some other item similar to it). One method involves tension in the spring (basically stretching it and holding it in place as best one can) while the other is to vibrate the spring so as to attain natural resonant frequencies.

Many objects have frequencies associated with them, hence this is why we make instruments out of different materials and they have certain shapes. All of this involves the wave behavior of what is referred to as Resonance. First, there is the concept of natural frequency. **Natural Frequency** is the frequency at which an elastic object, once energized, will vibrate. Look at guitar strings on a guitar sometime. They have different thicknesses. When plucked each vibrates at a different frequency than others of different thicknesses. Each has its own characteristic frequency. The frequency will depend with mass, length, and tension on the string. This is for all stringed items, guitars, violins, pianos, and harps for example. Not only do strings do this, but many other objects have natural frequencies associated with them. Items such as drums, bells, tuning forks, and the like all vibrate at particular frequencies depending on their material, shape, and other factors. This will be employed in our lab in the use of the spring.

Resonant Frequency is the minimum energy required to continue the vibration at that frequency. Resonance is a phenomenon that occurs when the frequency of forced vibrations on an object matches the object's natural frequency. Acoustic Resonance is the tendency of an acoustic system to absorb more energy when it is forced or driven at a frequency that matches one of its own natural frequencies of vibration (i.e. its resonance frequency).

By vibrating the spring in a back-and-forth manner with increasing amounts of effort (energy), the majority of the time the spring seems to have no 'rhythm' to it, but at certain levels of energy, the spring will vibrate back-and-forth at a resonant frequency equal to one of its natural frequencies.

The goals of this Activity are to measure and determine the wave speed in a spring, both longitudinal and transverse waves. Next these will be compared to a formula for wave speed in a spring when stretched to a given length and under a measured amount of tension.

Overall Purpose : To create and measure from a mechanical wave the distance and time of pulses on a spring to determine wave speed and to determine the role in tension on the spring as it relates to wave speed.

Activity I :
Purpose : To measure a compression wave speed on a stretched spring of measured tension and compare the measured wave speed to the predicted wave speed due to tension.

Activity II :
Purpose : To measure a transverse wave speed on a spring for comparison to other wave speeds determined on the spring.

Activity III :
Purpose : To use the property of standing waves for a spring to determine the speed of the waves traveling the spring.

Materials :

- Large Spring (Slinky or similar type works best),
- Timer,
- Long Measuring Tape,
- Tension Scale,
- Mass Scale (may be needed, but can use the Tension Scale),
- Tiled floor (best), (wood will work too),
- Graph Paper,
- Goggles,
- Slide Rule

Procedure :

For All Activities :

1) It is best to have 2 people for this activity since one will hold the spring, but it can be done with one as long as one end of the spring is secured in some manner.
2) Have adult permission and supervision. Wear safety goggles. Always also check the system for the spring so that it is held and remains secure throughout the activity.
3) For all activities, the spring is stretched out along the tiled or smooth floor. In the first activity, the length will vary with each trial. For the other two activities, the length is held constant (in fact it is best to have it the same for each of these exercises – perhaps 1.5 or 2.0 m)
4) Along the spring have the measuring tape open so as to be able to determine the length of the stretched spring.

Activity I : Longitudinal (Compression) Wave Speed Determination

1) Follow the directions for all activities above.
2) Use the scale and measure the mass of the spring (M).
3) In this activity, the spring will go from a shorter length to its maximum length (L) (the range could be 1.0 m, 1.5 m, 2.0 m or some other combination depending on the spring and space to do this activity).
4) With the spring set up use the tension scale attached to the end to measure the amount of tension required to hold the spring at this length. Record this value (F_T).
5) Note : Do not leave the scale attached to the system. You need to find a way to attach it – Books holding a plastic bar between them across the spring may work. In any case, be sure it is secure and be safe – wear goggles and do not stand close. Also do not use too much tension – if the spring were to come off its system, it should not be under so much tension so as to be a danger.
6) With the spring set up, send a compression pulse along the spring. It should bounce back from the 'fixed' end (someone else holding it or it is held in place by some means).
7) Do a practice run to note how long this takes. Use a timer.
8) Now conduct the actual trials and time how long it takes a wave pulse to travel the length and back on the spring. (t)
9) Do this 3 times for each length and average the values.
10) Redo this for 2 other lengths, each with 3 trials per length and average them. Be sure to record the new distance (L) and the new tension force (F_T).
11) Calculate Average Speed of the Wave from the distance and times.
12) Calculate the predicted speed from the Wave Speed due to Tension equation.
13) Compare results.

Activity II : Transverse Wave Speed Determination

1) Have the spring held at a constant distance for all trials in this part of the activity. (For example 1.5 m or 2.0 m). (L)
2) Send a Transverse Wave along the spring (that is, snap it) and test time the wave and its echo return wave.
3) Now conduct this trial 3 times – record the 3 times for the wave to travel the total distance (2*L) and average them. (t)
4) Use the Average Speed formula and calculate the speed of the wave.
5) If you want, try other lengths (for example the 3 distances used in Activity 1 here) for comparison.
6) Like Activity 1, be sure not to have too much tension, be sure to wear safety goggles, and be sure to secure the spring and monitor it.

Activity III : Standing Transverse Wave Speed Determination

1) For this activity, the spring must be held firmly in place like the other cases. Here, One person is merely the anchor and not moving, while the other is the agitator and energizes the spring with a back and forth motion to make the spring oscillate.
2) When the spring has a half wave resonance (you see one crest only) which then bounces back and becomes a trough, begin counting 10 oscillations which are timed. (t)
3) Note that one oscillation is a complete cycle, in the first frequency it will go from peak to trough to peak again – this is one. In essence it has a holding pattern that reverses itself and then returns to this pattern.
4) Do this for the next two fundamental frequencies for the spring at this length. The second pattern should require enough energy to have a complete wave. The third pattern will have 1.5 waves at one time.
5) You will have to work at it to find the right amount of energy for each case. Be sure to wear safety goggles.
6) From the data, determine the Period and the Frequency of the Wave.
7) Use the Wavelength and either the Period or Frequency in the Wave Equation and determine the Speed of the Wave.
8) Time permitting, try another length of the spring and redo the exercise.
9) From all 3 exercises compare the wave speeds and note similarities and differences.

Data :

Activity I : Longitudinal (Compression) Wave Speed Determination

Mass of Spring (M) : ____________ (kg)

For each Distance, measure the spring Tension and take the average speed of the wave pulse for 3 trials. Complete this 3 times (that is 3 different distances of stretched spring)

Each trial has its own length, tension force, and speed. Each will have its own speed. Compare this average speed to the computed speed using the wave speed due to tension force equation.

Distance of Spring [L] : ___________ (m)

Tension of Spring [F_T] : ____________ (N)

Trial	[D] Distance for Wave (2*L) travel	[t] Time for Wave Pulse Travel (s)
1A		
1B		
1C		

Activity II : Transverse Wave Speed Determination

For the given distance (L), maintain this for each trial. Do at least 3 trials and average them and then determine wave speed.
Note : Here you are using a Transverse Wave (pluck the spring perpendicular to its length to send a pulse along it).

Distance of Spring [L] : ___________ (m)

Trial	[D] Distance for Wave (2*L) travel	[t] Time for Wave Pulse Travel (s)
1		
2		
3		

Activity III : Standing Transverse Wave Speed Determination

Do this entire session twice, but maintain the stretched length of the system. Then average the corresponding results from each data set.

Stretched Length of System [L] : ____________ (m)

Trial	Wave Description	[λ] Wavelength (m)	[t] Time for 10 Oscillations (s)
1	2L		
2	L		
3	$\frac{2}{3}$L		

Trial	[T] Period (s)	[f] Frequency (s^{-1} , Hz)
1		
2		
3		

Calculations :

Be sure to use your Slide Rule!

$$v = \frac{\text{distance traveled}}{\text{time for traveled distance}}$$

Average Speed Formulae (know which variables to use)

$$v = \frac{d}{t}$$

Wave Equation :

$$v = \lambda * f$$

Wave Speed due to Tension Equations :

$$\mu = \frac{m}{L}$$

$$v = \left(\frac{F_T}{\mu} \right)^{1/2}$$

μ is mass per unit length (stretched distance), L is stretched distance, F_T is the tension force in the spring for that distance stretched

Frequency-Period Relation :

$$f = \frac{1}{T}$$

Frequency & Period :

$$f = \frac{\text{Number of Oscillations (n)}}{\text{Time for oscillations (t)}}$$

$$T = \frac{\text{Time for Oscillations (t)}}{\text{Number of Oscillations (n)}}$$

Conclusion :

Awesome Sauce and Cosmic – What a grand opening adventure into the world of Waves, their parts, some fundamental measurements, and an examination of their existence in Nature and our uses of them.

All Things certainly have a story to tell us – we have to be creative in thought, curious in nature, and passionate about our need to explore, to learn, and Be Brilliant!

In Activity I, examine how the speed changes (or not) as the length and hence tension on the spring changes. Also, how do these average speed calculations compare to the wave speed predicted by the tension force equation for speed?

From Activity II, how do the wave speeds compare in the trials to each other? How does this transverse wave speed compare to the compression wave speed in Activity I? How does the transverse wave speed compare to the predicted wave speed from the tension force equation for speed?

In Activity III, how do the wave speeds compare to each other for each of the standing wave patterns? How do these transverse wave speeds compare to the compression wave speeds of Activity I and the transverse waves of Activity II?

Activity # 44
Several Pendulum Characteristics Explorations
Grade Level : Middle School
Math Level : Calculating

The Period of the Pendulum and the Slide Rule Activity

Hey Inquisitive Pioneers! Galaxy Girl, Ms Briana Andromeda Purcell here and we are in the part of my book looking at Waves, Light, and such yet we will look at a physical science and Physics idea of Pendulums – an idea explored and explained by Galileo! Still the question as to why is this Activity here? The back-and-forth motion can be seen to be wave-like in its behavior for one reason – the pendulum if it had a pen on it or sand coming from it and a scrolling piece of paper was moving by underneath at right angles to the pendulum motion would trace out a really nice sinusoidal wave – or common wave pattern.

A simple Pendulum is a given mass at the end of a 'rod' (chain, string, solid rod) that is allowed to swing back and forth freely with respect to a fixed point at the end opposite the mass. Gravity is the main driving force for this system.

A simple natural question arises from this : What factor or factors affects the period of oscillation (the amount of time to swing back and forth)? Is it the mass, the type of mass, the type or length of the rod or something altogether different than these ?

This question came to Galileo and was recounted in his biography from accounts of him watching the swaying of bronze chandelier in the cathedral of Pisa and timing it with his own pulse (note that pocket watches did not yet exist). In his book, Two New Sciences, Galileo concludes from his experiments on pendulums that pendulums of the same length independent of mass are isochronous (taking the same amount of time). His mathematical analysis found that the square of the period varied with the length of the pendulum. He also thought that it was independent of amplitude (the size of the swing) which is not true and later investigations show that Galileo's conclusions are true for angles 15° and smaller.

These conclusions follow from Galileo's understanding of how the acceleration due to gravity does not depend on mass (all masses change speed at the same rate in a state of free fall). [This is examined further in the Acceleration due to Gravity Activity #23].

Pendulums have uses too. By the 1700s, parts for clocks can be made small enough to develop clocks that are mechanically powered by masses that fall and are kept in time by a swaying pendulum, or as they are more commonly called today 'grandfather clocks'.

A pendulum seen edge-on is akin to an orbiting satellite about its parent planet or a planet about its star. This type of back-and-forth motion has a name (when the restoring force causing the motion varies with the distance from equilibrium) and it is called Simple Harmonic Motion. A pendulum swaying back and forth is similar to a mass attached to a spring moving back and forth. This is another more math-related reason for the connection to waves.

Pendulums are a very good examination of conservation of energy. When at their highest points, the pendulum has zero velocity, hence zero kinetic energy. But it still has energy in the form of gravitational potential energy which is at its maximum since this increases with height. As it falls, the potential energy decreases, but the energy does not disappear. This turns into kinetic energy which manifests itself as a changing and increasing velocity. At its lowest point of the swing, which is the equilibrium position, it achieves maximum velocity hence maximum kinetic energy while having no relative potential energy. This also explains why the pendulum climbs back to its original height (as noted by Galileo too in this and studies of inclines). This is because energy cannot be created nor destroyed so it turns back into potential energy. With each swing, there is some energy loss due to friction (an idea described by Galileo) so in time the displacement decreases until it stops oscillating and the energy has all turned to heat and sound.

The importance of this constant motion which relates to Newton's Laws (see Activity #25), the Conservation of Energy and Conservation of Momentum gave Foucault, a French scientist in the 1800s an idea to make a large and massive pendulum system that would not move from its orbital plane in its motion but since the Earth spins an interesting phenomenon can be witnessed – Over the course of a day the swinging pendulum can be seen to knock down small blocks arranged in a circle and this is because the pendulum maintains its back and forth motion in the same plane it began in whilst the Earth is spinning beneath it! This illustrates that indeed the Earth rotates! How cosmic & awesome! These are named Foucault's Pendulum and there are many Science Museums that have one in them!

This activity explores the factors that could affect a pendulum in order to reach the same conclusions that Galileo did in 1638.

Activity I : Mass & Pendulum Period

Purpose : To investigate and mathematically uncover the relation of the period of a pendulum to its mass.

Activity II : Mass Type & Pendulum Period

Purpose : To investigate and mathematically uncover the relation of the period of a pendulum to its type of mass (solid or liquid).

Activity III : Rod Type & Pendulum Period

Purpose : To investigate and mathematically uncover the relation of the period of a pendulum to its type of pendulum rod (rigid or flexible).

Activity IV : Pendulum Rod Length & Pendulum Period

Purpose : To investigate and mathematically uncover the relation of the period of a pendulum to the length of a pendulum.

Materials :

- String,
- Nuts and bolt (eye form is best – large for both),
- Meter Stick or Measuring Tape,
- Mass Scale,
- 16 oz bottle,
- Water,
- Scissors,
- Wooden Dowel Rod (solid rod in Activity III),
- Eye Hook,
- wooden rod (1″ x 1″ x 2 ft),
- Chairs (2),
- Protractor,
- Timer,
- Graph Paper,
- Slide Rule

Preparation before Activity Procedures :

1. With any and all Activities read through, then set up the process for the Activity and run it without measurement and watch where you may have bumps in the road so as to anticipate these. Be patient – Repeat processes. Once all is well when measuring see to the values – do you notice patterns or trends – what do you predict will happen next as you change one thing or another from one trial in an experiment to the next. Always Explore & Enjoy. Be Brilliant!
2. Always have parental permission and help in Activities. Always employ safety in dealing with all materials.
3. Note that each of the data tables provided is minimal and needs more entries for each of the trials – a good number of trials is at least 3, but 5 is recommended.
4. Use eye bolts and nuts that can attach easily that are as large as possible. These can act as incremental masses by name instead of amount if no scale is present. The best choice is with a scale.
5. With a scale present, mass out the eye bolt individually followed by additional masses (attached nuts) one at a time. Keep these in order of use in the activity.
6. Have 2 conventional chairs with wood/plastic/metal bodies and flat top backs to act as platform to operate from. Place rectangular board across it and screw into it the hook which can be an open eye hook hanging down for pendulum rod attachment. Be sure to have board between chairs secured (tape, other ?).
7. If doing the solid rod pendulum, a closed eye hook on one end is recommended as this can readily attach to the system set up for the pendulums as noted in the prior steps.
8. For all pendulum situations, regardless of type, length, or mass all will be drawn back no more than 15°. To do this use the Protractor held beneath the board with the eyehook and the rod system (either string or rod) so that when drawn back the pendulum rod does not move more than a displacement of 15°.
9. Each of the Activities recommends 10 oscillations, since this makes the math much easier, but you do not have to use this number – but whatever number of oscillations are chosen, be consistent with each of the trials in a given Activity and it is best for

comparisons across Activities to have the same number of oscillations. Note that this can be an Activity unto itself – one where the number of oscillations is small such as 4 or 5 and one where it is larger such as 12 or 13 for comparison results – plus it gives you greater practice in the use of the slide rule.

10. For the solid rod to be used attach a small eyehook in the end carefully.

Procedure :

Activity I : Mass & Pendulum Period

1) For this activity, string will be the choice of pendulum rod type. Length of system can be a choice and should be between 20-40 cm. Each end of the string has loops tied in it for attachment to the hooks at either end.
2) The masses to be used are solid and will be the eyebolt and nuts. Read the preparation steps above. For setting up the pendulum system and having the masses ready to go.
3) Attach the string to the eyehook and attach the eyehook. Measure from the top attachment to the bottom of the eyehook. Record this in the data table.
4) Record the 1st mass as the eyehook mass in trial #1.
5) For each trial draw the pendulum back the same angle regardless of mass used.
6) When released start the timer.
7) Let the pendulum oscillate back and forth. One complete cycle is one oscillation or period.
8) Count 10 of these and then stop the timer.
9) Record the time of the timer in the data table.
10) Add more mass and repeat the cycle process of step 5.

Activity II : Mass Type & Pendulum Period

1) The type of rod to be used is string and the length will be constant for all measures in this activity (choose a given length of string between 20-40 cm).
2) The mass to be used is water and will be held by a pop bottle attached to the string. Attach the string so that the bottle is open or can be opened.
3) Begin with a measured amount of water mass either via scale measurement of water in measuring cups on a scale or simply use measuring cups and calculate the mass.
4) Use incremental amounts of water starting with ¼ cup and add ¼ cup increments to the system.
5) Draw back the pendulum the same amount of distance each time and let it swing 10 oscillations timing them with the timer. Record the time needed for each trial.
6) For each trial increase the amount of mass in water.

Activity III : Rod Type & Pendulum Period

1) In this Activity you must decide how to attach the given mass (which is constant for this activity) to the solid rod – you can consider duct tape – a possibility is to tape one of the bolts with the head at the bottom of the solid rod and the bolt points straight down from the rod so that nuts can be attached. Decide on the amount to be used.
2) In this activity, the mass is held constant as is the length of the pendulum rod. The string with the mass must be the same length as the mass on the solid rod.
3) There are only trials of the string pendulum and the solid rod pendulum systems. They can be done more than once to verify the overall results.
4) Always draw the pendulum back the same initial distance and it is less than 15°.
5) As with the other activities, let the pendulum swing or oscillate 10 complete times and record the amount of time for this.

Activity IV : Pendulum Rod Length & Pendulum Period

1) Choose a solid mass that acts as a constant for all trials in this case where the question of pendulum rod length is being investigated. The nuts and bolts are a good choice in this Activity.
2) Measure out approximately 50-60 cm of string and wind it around the main support rod for the pendulum system so that the initial distance is shortest.
3) Begin with the shortest distance of string for your trials, between 10-15 cm with the mass attached.
4) As with all trials draw back a short yet constant angular distance of less than 15° for each trial.
5) Let each pendulum length oscillate 10 complete times and record the time for this event.
6) Unwind more and more of the pendulum string so that the length of the pendulum increases in some sort of incremental fashion for each trial. Then follow steps 4 & 5 until all trials are done.

Data :

Activity I : Effect of Mass on Pendulum Period (solid bob)

Type of Pendulum Rod : ___________

Length of Pendulum Rod : __________

Trial	Mass (unit or g)	Time (t) for 10 oscillations (s)	Period (T) for 1 oscillation (s)
1			
2			

Activity II : Effect of Type of Mass on Pendulum Period (liquid bob)

Length of Pendulum Rod : __________

Trial	Mass (unit or g)	Time (t) for 10 oscillations (s)	Period (T) for 1 oscillation (s)
1			
2			

Activity III : Effect of Pendulum Rod Type on Pendulum Period (rigid & flexible)

Mass of Pendulum Bob Used : __________

Type of Pendulum Rod Used : ___________

Length of Pendulum Rod Used : _________

Trial	Length [L](cm)	Time for 10 oscillations (s)	Period for 1 oscillation (s)
1			
2			

Activity IV : Effect of Pendulum Length on Pendulum Period (string)

Type of Pendulum Bob : ____________

Mass of Pendulum Bob : __________

Trial	Length [L](cm)	Time for 10 oscillations (s)	Period for 1 oscillation (s)
1			
2			

Calculations :

Be sure to use a Slide Rule !

Watch your calculations when conversions, such as centimeters to meters are needed.

Procedure :

1) It is best to use the C & D scales of the Slide Rule for the calculations. If squaring is needed be sure to employ the A or B scales.
2) For each activity, determine the period of the pendulum by using the formula for period (T).
3) For Activities I & II, graph Period vs. Mass and draw a best fit line. Determine its slope. (It should be zero – what does this indicate?)
4) For Activity 4 graph Period vs. Length.
5) Recreate the Table for Activity IV and take the log value for the Period and the Length.
6) Graph the Log (Period) vs. Log (Length) and draw a best fit line and find the slope of this line to determine the exponential relation between Period and Length. (It should be: $L \sim T^2$). (The slope ideally is 2). Note the log value is found on the L scale of the slide rule.
7) One way to verify the relation is to graph Period^2 vs. Length which should turn out to be a straight line.
8) Use the ideal formula for period as determined by length and calculate what your period values should have been as compared to what you measured them as. Note that the ideal formula may vary – for most pendulums the simple formula suffices, but if the rod has a large mass as compared to the bob, then the complex formula for period should be used.

Calculations :

Be sure to use your Slide Rule!

Information :

1 mL water = 1 cc water = 1 g water

Formulae :

$$\mathbf{T = \frac{t}{10}}$$

(Note : Assumes using 10 oscillations – instead of 10 put your number if it is different)

$g = 9.8 \text{ m/s}^2$

p Formula for simple pendulum (mass concentrated in bob) :

$$T = 2*\pi*(L/g)^{1/2}$$

$$L = \frac{g*T^2}{4*\pi^2}$$

Formula for Complex pendulum (mass of rod is considerable as compared to bob) :

$$T = 2*\pi*((2*L)/(3*g))^{1/2}$$

$$L = \frac{2*g*T^2}{3*4*\pi^2}$$

Conclusion :

Cosmic – What adventures in Pendulums we have done here!
Wow! All things certainly have a story to tell us – we have to be creative in Thought, curious in nature, and passionate about our need to explore, to Learn and be Brilliant!

Examine your results to find which if any of the variables have affected the period of a pendulum. From the reading, results, and calculations it should be found that the period is related to the length of the pendulum.

It might be interesting to test the idea of whether the complex period formula is needed or not – have a rigid rod pendulum system activity where the mass of the bob is small as compared to the rod and one where the mass of the bob is considerable as compared to the rod and see if these results are similar or different.

Side Note : When two variables are graphed opposite each other and there is either a random pattern or a zero slope, then this indicates no relation to the variables. If instead the two graphed variables show a regular change (linear or not) there is an indication of a relation between them.

Summary :

For the teacher, parent or interested student :
All of the activities explore the relation of period to length and mass, but only for small angles of angular displacement. Clearly this is a hypothesis to explore as well. Choose a length and mass for a given set of trials, then change the angle at which the pendulum starts. For example, start at 10°, then 20°, then 40°, then 60°. Graph these results to see if there are any variations.

An important idea arises in this Activity. Note that several of the activities in it show no relation of the variables. One might think why go down this road at all and only pursue the other possibilities? Simple – Just because the idea from the hypothesis as extended from known ideas in science seems to indicate one answer (the length of the pendulum) does not mean that the question concerning the other variables (such as mass) have been answered. A failed experiment is just as important as a successful one. The process is about being true to the numbers and realizing what they are telling us.

Also, try other challenging ideas : make a pendulum with a certain Period, for example 1.0 seconds. Compare your outcome to the predicted value found in the Pendulum Period equation.

Interesting Use of the Pendulum :
As you found in your investigation, the period of the pendulum depends on length of the pendulum. With this idea in mind, one can construct a 1 s pendulum as noted above, but can test the formula from the activity for various lengths of pendulums and the period that is predicted for it. Construct and test these pendulums. Be sure to run several trials and take the average of the values. Also note, the shorter the pendulum, the greater the potential for error due to the fact that the period becomes shorter, hence your reflexes at measuring its period will vary considerably.

A Personal Stopwatch :
An interesting use of these pendulums is to use them when there is no stopwatch available. Though one might not have the exact time for the pendulum known (in our case, if tested we do), the pendulum can be used in place of a timer in any or all of the activities in the book. Note that the precision of the tool decreases considerably. In the case of using it, just count the whole number of swings that the pendulum goes through for a given timing. Since for each trial it has the same value, all of our measurements for that activity using our makeshift stopwatch are valid!

Activity #41
Determining the Wavelength of Laser Light
Grade Level : High School
Math Level : Challenging

Determining the Wavelength of Light in a Laser

The two main categories in Physics are these essentially in the general realm of ideas : Matte and Waves. Here in this Activity and many of the surrounding ones we will explore Waves in general, Light the most often (which btw can be thought of as a Particle too). Waves are certainly essential to all things – Light is the energy that begins the food pyramid, we use electromagnetic and sound waves of all sorts to do things – especially transit information. Waves as an energy source are being explored and beginning to be tapped into as well.

Waves, in general, are phenomena in nature that transport energy from one place to another. Sound waves are compression or longitudinal mechanical waves that are vibrations of the medium they travel through (solids, liquids, or gases). Electromagnetic Waves are not mechanical and hence do not need a medium to travel through. All electromagnetic forms, like light, travel at the speed of light which is approximately 3×10^8 m/s. [More description depth of Waves is found in the Exploring Waves with a Spring Activity #43] . There is another Activity even determining the Speed of Light (Activity # 46) – too Cool!

All waves also exhibit other unique behaviors in the presence of matter, such as reflection, refraction, and diffraction. In this Activity, the behavior of a wave passing through a narrow opening, commonly called diffraction. Defined **Diffraction** is the bending of a wave around a barrier, like the edges of a barrier. This is a behavior that helped Christian Huygens win the day over Newton's idea of how to characterize light. Newton had proposed a model in which light was composed of individual particles. If it were as it passed through a narrow opening, the most that would happen is that the majority would continue in a straight line, while some would create a fainter and fainter spray pattern from the center.

What was found, though, when light was passed through a single narrow opening, and in a later experiment using a pair of slits, was that there was a central bright region, called **the primary maxima**, *but on either side of it* were **bright spots separated by dark gaps**. The first pair either side of the primary maxima are called the 1st maxima, the next are the 2nd maxima and so on in terms of names for them. Huygens' Wave model of light could explain this phenomenon through **constructive and destructive interference of the waves as it passes through the narrow openings**.

The best way to describe why this occurs is to use **interference of waves**. **Waves can interfere in a constructive or destructive manner**. In Constructive Interference, this is the addition of two or more waves since their amplitudes are in the same direction. These result in the bright spots. In the case of Destructive Interference, two or more waves have amplitudes in different directions hence subtract or diminish the outcome. This corresponds to the dark region between the bright spots.

Instead of using a single or double slit arrangement, this activity uses what is called a diffraction grating which can have several hundred lines per millimeter! You have to know the number of lines per millimeter to successfully do this activity since its number will affect the pattern produced. A good thing to do is to have two different gratings and try the activity twice to see the outcome.

When using a diffraction grating, the large number of slits results in the same sort of diffraction pattern (dependent on the number of slits) that occurs with one or two slits. The geometry of this situation allows one to find the wavelength of the light itself! This is because the distance traveled by one beam through one slit corresponds to a distance equal to a multiple of the wavelength of the light from an adjacent slit. They are in what is called phase with each other. These waves will add up on the wall constructively where they hit and produce the bright spot.

This Activity takes advantage of the diffraction behavior of waves and their resulting interference in order to determine from geometry and careful measurement the wavelengths of light, such as that from a laser.

A quick question must be asked and addressed which is : Is laser light any different than ordinary light? Laser light is actually a terms that is an acronym. It stands for **Light Amplification by Stimulated Emission of Radiation**. It is a means for emitting electromagnetic radiation (like any ordinary light does) with a little more focus as compared to ordinary light. In essence, it is concentrated light. First, laser light has waves that are in step with each other, this is called coherence. Light from a bulb is not this way and given off in a random manner, hence might be called incoherent light. This coherent light is composed of waves of identical frequency, phase, wavelength, and polarization. Though this may seem different, it is not. We are using the best source of light so that we have consistency in the frequency and wavelength plus it is a naturally narrow beam. Caution must be practiced however so do not look into it or even let it bounce off and strike you in the eyes from highly reflective surfaces.

Purpose : To use the property of diffraction of light and geometric optics analysis to determine the wavelength of a common laser.

Safety Notes : Have Parent permission and supervision for this activity. Do not shine laser light into anyone's eyes. Always exercise caution when using a laser.

Materials :

- Laser (HeNe, Green, or other),
- Meter Stick,
- Diffraction Gratings (of known slit width, marked lines/mm),
- Measuring Tape,
- Table,
- Tape,
- 2 Pieces of cardboard or 2 toilet paper towel tube,
- Roll of Paper or taped together sheets to act as screen which will have marks made on it (dull finish to paper is best),
- Slide Rule

Laser Information & Other Pre-Preparations :

1. With any and all Activities read through, then set up the process for the Activity and run it without measurement and watch where you may have bumps in the road so as to anticipate these. Be patient – Repeat processes. Once all is well when measuring see to the values – do you notice patterns or trends – what do you predict will happen next as you change one thing or another from one trial in an experiment to the next. Always Explore & Enjoy. Be Brilliant!
2. Can use regular small lasers. The key is that it must be placed in a v-shaped box parallel to the table (see step 3).
3. Wavelengths of various common lasers that may be used : Red HeNe Laser : 633 nm, Green : 543.5 nm, Orange : 612 nm, and yellow : 594 nm ($1 \text{ nm} = 1 \times 10^{-9} \text{ m}$)
4. For the v-shaped holder : Either use a thin piece of cardboard folded in an accordion fashion so that it creates a v-pocket for the laser OR split a cylindrical toilet paper towel tube vertically and use each half circle as sides so that looking down on it there is a channel that is v-shaped.
5. Create a 2nd v-shaped form which the diffraction grating can set in. Note that when activated the laser has to hit and go through the grating, so work with the cardboard pieces or tubes so that this happens. A simple solution is to place the laser v-shaped platform on a thin book so as to elevate it slightly.
6. Diffraction gratings can be found on line easily at low cost. Average ones are $2-$3 each and have 13,500 lines per inch (1 in. = 2.54 cm) – Note : Depending on cost, using 2 different gratings in 2 separate trials (i.e. different lines per mm) is a good idea for comparative purposes. Record this value (q) on your data table.
7. Place the Table adjacent to a wall long-wise. On the wall tape the paper roll to act as a screen across greater than the width of the table (over 1 m and up to 2 m in length). (Best to use dull finish paper).

8. Note : Test the system so that a pattern is seen and works well. There is no need for further adjustments once the pieces are all in place and operational.
9. **SAFETY at all times – Must have parent permission and supervision for doing this activity. Lasers are not to be looked at or directed into anyone's eyes. Also be cautionary with laser light bouncing off metallic or shiny surfaces.**

Procedure :

1) Read through and conduct the pre-preparations so that the laser goes through the diffraction grating and has various maxima on the screen. Be sure to have measured the distance from the diffraction grating to the central maxima **(L).**
2) Note : Do not measure with the laser on. The distance **(L)** is measured with the laser off.
3) It is best to mark where the central maxima is with a pencil on the paper screen and the central portion of the 2 left and 2 right maxima (you can do more if interested). Be careful in marking and do not look into the laser. Use of a dull finish paper is best so as to avoid reflection. Face only the paper when marking it and do not look back to the laser.
4) Turn off the laser and take down the paper screen for measurements.
5) Note : You are not taking measurements with the laser on. You only mark the paper where the maxima fall. Measurements are done after the laser is deactivated and the paper is taken down to be measured.
6) Once all markings are done fill in the data table from measurements taken. Though it is noted to the meter, realize that you are measuring with a tool that can measure to the 0.1 of a mm.
7) The Data your are recording here is the distance from the central point to each of the subsequent maxima **(x).**
8) Follow the directions in the calculation section to determine the wavelength of light for the laser and compute percent error.
9) Calculations :
10) Though the Slide Rule is a recommended tool, all of these calculations can be done with a regular or scientific calculator. Some scientific ones even have built-in averaging formulae. For those who like spreadsheets, the data can be typed in and the formulae then also be typed in its own cell where the formula references each of the measured variables in their respective cells, for example B1..BN has the measurements and values used in the equation while BN+1 has the formula for all of these variables (why not A? Simple – use it to label you variables)
11) Determine the distance between the lines on the diffraction grating (d). (Be sure to convert to m).
12) Determine the distance from the diffraction grating to each of the maxima **(L_o**) (which is the hypotenuse) using the Pythagorean Theorem (the other sides are L and a given x)
13) You can either determine the sine of the angle in question from your data are merely move ahead and use the wavelength formula (since the sine is now expressed as a ratio of the measured and calculated sides of the triangle – see diagram).

14) With a calculated wavelength **(λ)** compare it to the known values and compute percent error.
15) An alternative is to determine all of the wavelength values for the maxima used and then average the values as well.
16) If other diffraction gratings : Other Experiments : If you are using more than one diffraction grating , redo all the steps and have a new screen.

Photo for Set Up :

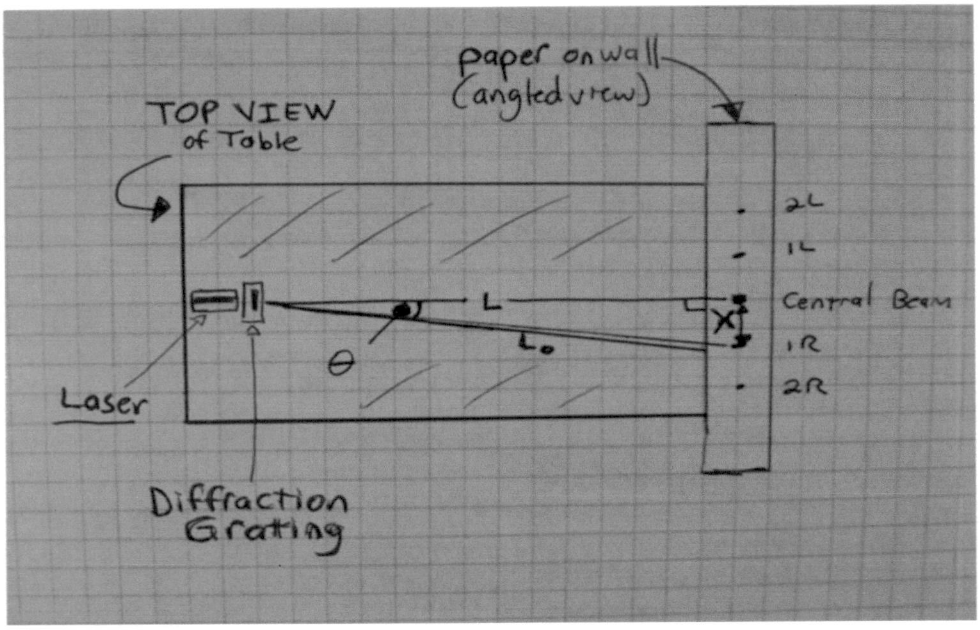

Data :

Number of lines per meter on the diffraction grating : _____ [q]
(Note – read the grating, conversion needed)

Spacing of Lines in Grating [d] : __________

Distance from Grating to Screen [L] : ___________ m

N	x (m)
1 (L)	
1 (R)	
2 (L)	
2 (R)	

L : Left of center, R: right of center
Note that 'n' is 1 or 2 in the given case.

Calculations :

Be sure to use your Slide Rule!

1 m = 1000 mm
d : The spacing between lines on the diffraction grating (m)
L : The distance from the diffraction grating to the screen (m)
L_o : The hypotenuse distance from the diffraction grating to a given maxima on the screen (m)
x : The measure of the distance from the central initial maxima to the maxima in question (m)
n : Is the order of the maxima (n=1,2,et al as needed)
Θ : The angle as measured from the central maxima at which the next maxima occurs as determined by similar triangles from analysis of the constructive and destructive interference of light waves passing through the diffraction grating. (° or rad)
λ : The Wavelength of light in question (m)

Needed Formulae to Find the Wavelength :

$$d = \frac{1}{q}$$

$$L_o^2 = L^2 + x^2$$

$$\text{Sin}\ (\Theta) = \frac{x}{L_o}$$

$$n*\lambda = d*\sin(\Theta)$$

Wavelength Formula (derived) :

$$\lambda = \frac{d}{n} * \sin\ (\Theta) = \frac{d*x}{n*L_o}$$

Percent Error Formula :

$$\%E = \frac{[\text{Experimental Value-Accepted Value}]}{\text{Accepted Value}} * 100\%$$

Average :

t : number of trials

$$\lambda_{ave} = \frac{\Sigma\lambda}{t}$$

Possible and Good Approximations Formulae in this Activity :

1) If Θ is used in radians, then $\Theta_R \sim \sin\Theta^o$
   To use – find Θ from

   tanΘ = -

   $$\Theta_R = \frac{2*\pi}{360°}$$

2) sinΘ ~ tanΘ for small angles
   This means for small angles, the ST scale on the slide rule can be used. What is being done here is we are using L instead of L_o and saying :

   $$\frac{x}{L_o} = \frac{x}{L}$$

<u>Conclusion :</u>

Imagine that – a pointer laser, a measuring tape, and a diffraction grating plus our skills and we are able to find the wavelength of light itself! Awesome Sauce.

The best way to conduct the activity is to do it more than once and be careful and certain of your numbers. Know that there is a certain level of precision in your measurements, hence this affects the accuracy in your results. The best measure is to determine percent error in your wavelength determinations.

Activity # 46
Using a Microwave to Determine the Speed of Light Activity
Grade Level : High School
Math Level : Calculating

Ms Briana Andromeda Purcell here once again and this Activity is Awesome! Imagine using a microwave oven, a paper plate, a ruler, and a slide of cheese and we are going to measure through calculation the Speed of Light! Awesome Sauce and Brilliant!

The key to this Activity is we are combining facts about nature (things called waves and standing waves), objects (microwaves), looking at a basic equation for speed (in this case for waves and their speed), and the fact that in nature all Electromagnetic Waves travel at the same speed in a vacuum in order to connect our idea back to the Speed of Light. This is incredible genius and creativity combined – too fun!

For those who want a refresher on Waves and ideas about Waves in my book here : Galaxy Girl's Inquisitive Pioneer Adventures there is considerable prelude information about waves in the Spring Wave Speed Activity in general, so it may be best to start there for an overall understanding of waves in general and electromagnetic waves in particular. Some of the ideas directly related to the topic at hand are noted below in the writing.

Electromagnetic Waves are actually quite familiar to all of us. We see them in the form of Visible Light, which represents only a small portion of the electromagnetic spectrum (Note first determined to be white light when combined as noted by Newton – for more on this and an Activity on light see Spectrum Analysis). There are many others, however, we hear of often and are used very frequently by humans in their devices and daily lives. We use X-rays to examine the interiors of items, even ourselves when it comes to our bones. We hear of the harm and damage of Ultraviolet Rays and the need for sun block. Infrared rays are used for looking for people, animals, and the like at night. All of radio, TV, satellite phones and the like use a portion of the electromagnetic spectrum to send and receive messages and information which is called collectively Radio Waves. Also in there are Microwaves, not just the device but the waves the device uses to heat water molecules (to get them to vibrate) to heat food and cups of coffee and such.

Other than the fact that all of the waves mentioned are composed of the same parts – a magnetic wave and an electric field wave – there is one other unique feature all Electromagnetic Waves share in common which was found in Maxwell's Equations describing these waves and found to be a cornerstone of the Universe by Einstein and that is :

In a vacuum all electromagnetic waves travel at what is classically called 'the Speed of Light'. Though this term is used, it is only because of human study of light first and the others later on. This value is listed in the Calculations section below and will be used in the Activity that follows and is normally seen as 3.0×10^8 m/s or 186,000 mps (that is miles per second not hour as we are normally expecting). If one could send a beam in a circular path around the world, it would circle the globe 7.5 times in one second!

The Microwave uses a device called a magnetron to generate microwaves from electrical energy usage. These waves are contained in the oven chamber even with the perforated screen at front because these holes are smaller than the wavelengths of the microwaves, so they cannot escape.

We might think that the waves in the microwave just randomly bounce around but they do not. The waves set up what is referred to as a Standing Wave Pattern. This is much like taking a slinky spring or a jump rope and making it move with just enough energy so that the item moves in a rhythmic and stable fashion, such as when two friends are spinning the jump rope just right to make what appears to be a half-wave from the side. It appears that the wave is rather stationary and not moving, hence its name. In the case of the jumping rope example, The two people are at what are called nodes and the high point in the middle is the wave crest. What if the two people with the jump rope (or the person moving the slinky spring back and forth) do so even faster. They can reach another stable speed where a node appears in the middle between them and now there is a complete wave cycle present as viewed from the side. Try this with a string, a jump rope if you can attach it at the other end or have someone work with you or try the slinky to see the increasing number of wave crests/troughs with nodes for an idea of this concept.

All of this discussion of standing waves applies to microwaves as well. The waves in the microwave set up a standing pattern. If one had a very small instrument that could measure at different spots across the microwave one would find that microwave energy would rise to a peak, then decrease and peak yet again as one moves across the microwave chamber. The peaks are where the wave crests of the microwaves are while the areas of low to no energy are the nodes. This is why microwaves have a rotating table in them! This way something that is being heated will pass through these regions of peaks and valleys repeatedly. Without rotation there would only be hot spots on the food while the other areas remain rather cold.
We wish to take advantage of these peaks and valleys, so we must disable the rotation of the platter by taking it out as well as the post (or at least covering the post with an inverted yet rimmed paper or Styrofoam plate so that it acts as a non-moving surface to place our second plate with the cheese on) that the platter sits upon. Now when we go to heat something, which we will do in the Activity, there will be hot and cold spots. We will use something like chocolate or marshmallows (see list below in Activity) and only heat it a very short time to cause just the initial melting of the surface. What we are after is the distance between these hot spots! That distance corresponds to ½ of the wavelength of the microwave wave. We merely have to double this distance and we are half-way to determining the speed of light for these waves.

We have to use what is classically called the Wave Equation.

$$c = \lambda * f$$

It states that the speed of a wave is equal to the product of its wavelength and the wave's frequency. Normally the variable v is used instead of c since this equation is for all wave types (electromagnetic and mechanical alike) but when 'c' is used it is referring to light and all of its associated electromagnetic wave forms since they all travel at a constant speed all the time as Einstein predicted and found.

Also don't let these letters let you think that speed equations change drastically for physical objects like cars and people as compared to waves like sound and light. They actually convey the same ideas only using different letters or variables – here wavelength (λ) is the length or distance of the wave much like there is distance covered in a speed formula ($v = d/t$ - here d is distance). Notice too that in the speed equation we divide by time (t) so we need time in the wave equation that is from the frequency (f) – this is defined as the number of waves per second – the unit here is essentially 1 over time 1/t. So from the wave equation we have distance divided by time just as it is for speed! Just different letters for different items measured but the same idea! Wow!

In our case of our Activity, We have the wavelength, but what of the frequency? This can be found on the manufacturer's sticker for the microwave (be sure to have parental support and help in this) found on the backside or underside of the device. Most microwaves have a frequency of 2450 MHz (2450×10^6 Hz) or one could simply go to the internet and type in the model name and number and with some research uncover the number needed.

As noted this is called the Speed of Light, as this applies to any and all Electromagnetic Waves in a vacuum. One could readily note that there is air in the microwave, and this will technically affect its speed, but not much at all – in fact if properly done your answer should be within a small percentage of error and somewhat reliable to one perhaps even two significant figures. By the way be sure to follow all the directions below, have parental permission and help, plus use food items that one will consume – don't be wasteful. It is important to note on the side, however, that this Activity is one of the rare ones – in virtually no case does one ever consume anything in a lab exercise, but here the Activity is designed to be safe and allow for that outcome!

Enjoy and Explore :)

Be Brilliant!

Purpose : To determine the speed of light from distances between melting spots on food in a microwave due to heating by a microwave standing wave pattern and multiplied by the frequency of those waves

Materials :

- Microwave,
- Microwave-safe baking or casserole dish,
- Food Item to Use : Choose one from : Chocolate bar(s), Chocolate morsels, bag of mini-marshmallows, cheese slices - (note for choice – must be sufficient to span if not cover the dish being used in a uniform layer – part of this depends on the food's use afterwards – good idea, don't waste good food),
- Ruler,
- Toothpicks,
- Goggles,
- Slide Rule

Important Notes :

Note 1 : Must have parental help and permission in this Activity. There are a number of steps which should involve parental help, such as disabling the rotating dish and finding the frequency of the microwave on the back or bottom of the machine (Note – you might find it on the internet with some research on your model), as well as properly handling heating of food items. The key is that you have a microwave where the center post of the rotating mechanism easily comes out – if not, then you cannot – most lift right out. Be sure to read all directions before doing the Activity. Note if it does not come out – this is okay – simply have a rimmed Styrofoam or paper plate that can be inverted on the rotating post so that it does not interfere with the plate which now remains stationary for our use!

Note 2 : Do not overheat the Food Item being used. The range of time does depend on microwave power and the item chosen, but typically ranges from a handful of seconds most of the time and depends on what you use (slices of cheese need little to no time, other thicker things like a layer of chocolate especially if initially cold will be longer) up to not longer than about 40 seconds. The key is watching the item while heating and once melting is detected, then stop the process. – Also do not touch it – it may be too hot and you are going to measure it, you don't want to ruin it.

Note 3 : Goggles are noted and should be worn as a good practice for science lab work, but also in the remote chance of the food item being too hot (which if done correctly it should not be) and it bubbles.

Note 4 : Food Choice is by preference and should be done in the knowledge that this food will be consumed or used in some sort of recipe as it is.

Note 5 : The microwave-safe dish is best to be a dish that has a wall so that a rigid ruler can be put across it and one can read looking down at the food beneath the ruler above it spanning the dish rim.

Note 6 : There are others who have written about this lab and have used whipped egg whites or very-buttered bread instead of the list noted above.

Set-Up :

- Be sure to follow all of the Notes above in your procedure
- Before using the microwave it is best to record the frequency of it found on the sticker label (typically on the back of it), f. Record this in your data table. If unable to find it, perhaps try the internet or simply use the suggested value in the table in Calculations as this is the most common value
- In preparing the microwave for use in the lab, you are taking out the turntable and its post (typically the post is plastic and lifts right out) – This is so a dish can sit in there and not rotate when the microwave is on
- Be sure to use a microwave-safe dish and it is best if it covers as much of the microwave floor as possible but also have a span that can be spanned by a ruler that can rest on its rim
- In terms of the ruler, it is best to use a solid one (that is not flexible)
- In choice of food item to use, use one you plan to make use of

Procedure :

1. With any and all Activities read through, then set up the process for the Activity and run it without measurement and watch where you may have bumps in the road so as to anticipate these. Be patient – Repeat processes. Once all is well when measuring see to the values – do you notice patterns or trends – what do you predict will happen next as you change one thing or another from one trial in an experiment to the next. Always Explore & Enjoy. Be Brilliant!
2. Be sure to follow all of the Notes listed prior to the Procedure, follow safety guidelines, such as wearing goggles, and also have completed the Set-Up before continuing.
3. Whatever your choice of food item, it will be referred to simply as chocolate from this point on for reference purposes in these directions.
4. Using a microwave-safe baking dish place chocolate in the bottom (note it is best to use solid-type with no other items in it and that it is reasonably thick). It will cover as much of the area as possible, particularly spanning from one side to the other.
5. Place the dish in the microwave in the center, close the door and turn it on. The amount of time will depend on the food type, its thickness, and power setting chosen, but in virtually all cases will be less than 40s and normally only run in the neighborhood of 20s.
6. The key to the amount of time heating is not only that it is short in time, but will depend on you keeping a careful eye on the food item heating. At the first signs of melting in a couple of spots, it is time to stop it. Do not let it overheat, bubble or even liquefy to any appreciable amount.
7. Be sure not to touch the item or move it too much to disturb it since you must measure the distance between the spots.
8. The number of measurements will depend on the number of spots. Each set of two spots in sequence are to have their distance between them measured. This is referred to as a Trial in the Data section and denoted by the letter S.
9. So this means Trial 1 is for the distance between spots 1 & 2, Trial 2 is for the distance between spots 2 & 3, and so on.
10. The best way to measure is to place the ruler across the rim of the dish and read by looking directly down where the centimeter side is above the spots in question.

11. Realize you are not starting at '0' with each measurement. For example let's say spot 1 is at 4.3 cm while spot 2 is seen at 10.8 cm – how far are they apart. You need to find the difference of these numbers and this is S (for this example it is 6.5 cm).
12. To help in determining where the 'spot' is use the toothpicks and poke them into the center of the melted spot on the chocolate, if soft, and if not, merely touch down to the center-point of the spot in question when reading its value on the ruler. Note do not force it into the chocolate as this may affect the readings of not only this spot but others. – In essence the toothpicks are guide posts. – Use only if needed.
13. Once all the measurements are done, look at these values – they should be similar to each other if done correctly. Determine the average of these values (X_{ave}) this is the value of half of the wavelength of the microwave in question.
14. Now double the average value and this is the wavelength of the microwave wave for our equation ($\lambda = 2*X_{ave}$) [Be sure to convert this to meters]
15. With the wavelength determined, we can determine the speed of light from the wave equation since we have either read the frequency from the manufacturer on the microwave or are using the standard value provided in the table below.
16. How did your results turn out? Use the percent error formula to determine how close you came expressed as a percentage. Be sure to only use two significant figures as this is the extent of your measurements as well.
17. Be sure to properly clean up and reset the microwave. It is more fun to use food that is being used in a fun recipe. Enjoy! :)

Data :

Frequency of Microwave [f] : ___________ (Hz)

Trial	S (cm to nearest 0.1) Distance between melting spots
1	
2	
Average	

Calculations :

Be sure to Use Your Slide Rule! :)

Average Value :

$$\mathbf{X_{ave} = \frac{\sum S}{n}}$$

(S is the distance between points, n is the number of trials)

Wavelength :

$$\lambda = 2 * X_{ave}$$

Wave Speed : (Speed is Wavelength times frequency)

$$v = \lambda * f$$
$$c = \lambda * f$$

Percent Error :

$$\%E = \frac{[\text{ Actual-Experimental }]}{\text{Actual}} * 100\%$$

Constants & Conversions :

$1M = 10^6$
$1G = 10^9$
$1m = 10^2$ cm
$c = 3.0 \times 10^8$ m/s (value to use here)
[$c = 2.99792458 \times 10^8$ m/s]
f = 2450 MHz (most microwaves are this)
[use this value if unable to find or read yours]

Conclusion :

Wow! All things have a story to tell us – we have to be creative in thought, curious in nature, and passionate about our need to explore to learn and be Brilliant!

How Cosmic – Use cheese, a ruler, a microwave to find the Speed of Light!

How did your results turn out (percent-error wise)? How similar were the melting points in the measured distances between them?

Activity # 47
Graphically investigating the Inverse-Square Law of Light

Grade Level : High School
Math Level : Challenging

Inverse-square Law and Light Activity –

The heart of all of science is to not just notice phenomena, but the connections that a given variable has with other related variables. For example – say you exert a given force on an object and it accelerates – you have measured both he force and acceleration. Now you double the force, what should the acceleration do? Think about Newton's Laws – these two variables are directly related to one another. So doubling the force, should effectively double the acceleration! In science there are direct, indirect (that is to say one over the variable so that it is in the denominator) and there is another common one, the Inverse-Square Law or relation. One of the most universal relations is known as the **Inverse-Square Law**.

In Physics, as well as all of the sciences, any physical law that states that some given physical quantity or intensity is inversely proportional to the square of the distance from the source of that physical quantity is said to be an **inverse-square law**.

The Inverse-Square Law basically applies when a given force, energy, or other conserved quantity is radiated outward in a radial manner from a source so that the surface area of this radiated area is spherical ($4*\pi*r^2$) hence proportional to the square of the radius, the force, energy or other quantity must spread out over this area. so this quantity must too diminish in intensity in an inverse proportional manner with the distance squared. It turns out that there are a number of phenomena that follow this general relation.

This activity explores the illumination of light, which all of us notice naturally. When we are close to a light it is clearly much brighter than when we are at a distance from it. **(Note do not stare into active light sources as this can damage ones eyes)**.

One of the first noted and mathematically explained **inverse-square relations** was the invisible force of gravity itself noted by Sir Isaac Newton in 1687. In his work, the Principia, he not only outlines his concepts of motion but discusses mathematically the force of gravity. He notes that the force of attraction of gravity falls off with the inverse-square of the distance between the centers of masses, such as the force between the planets and the Sun.

For example, this means that if there are 2 equally massed planets, yet one is twice as far from the Sun as the other, it would have only ¼ the gravitational force acting on it as compared to the other ($1 / 2^2 = ¼$). For the same mass at 3x the distance the force of gravity acting on it as compared to the closest one is only 1/9 as strong.

If on the other hand, let's consider 2 equal masses and move one mass towards the other, the amount of force rapidly increases. At half the original distance ($1/(1/2)^2$), the force of gravitational attraction between them is 4 times what the original force was! Recall that a net force necessitates acceleration (Newton's 2nd Law) – all the more reason planets closer to the Sun move faster than those farther away!

Newton's Universal Law of Gravitation :

$$F = G*\frac{m_1*m_2}{d^2}$$

Considering a Force with all other factors other than distance taken out :

$$\mathbf{F \sim \frac{1}{d^2}}$$

What is important about this relation in terms of mathematical analysis is that the product of the inverse-square of the distance and a given quantity (here force) will be a constant. We will make use of this in the Activity in terms of Light Intensity and the inverse-square of the Distances we use.

Later in history from Newton's time, August Coulomb found in 1785 that the attractive and repulsive force between unlike and like charges behaved much like gravity with regards to distance. That is to say, the amount of electrostatic force varies as the inverse-square of the distance between the charged objects.

Notice how fundamental forces like Gravity and the Electrical Force (2 of the 4 primary forces of the Universe) act in this manner. Imagine the surprise, excitement and wonder of the scientists who uncovered such things in those times. Nature was measurable, it was knowable, it made sense and seemed to have a simplicity and harmony to it.

It turns out that the **Intensity of Light** (which includes all forms of Electromagnetic Radiation) *has this same relation between intensity and distance from the radiating source! With distance, the intensity of light falls off as the square of the distance. Wow!*

Our goal is to examine the Inverse-Square Law and its application to the Intensity of Light. In the Activity, we use a Solar Cell (photovoltaic cell) connected to a Multi-meter that is at a measured distance from a luminous source (an incandescent lamp with a light that is on). We measure the Current (I) for the Solar Cell at the given distance (d) when illuminated by a light source. (Note : We are not using Voltage, since it will remain fairly constant with distances that are similar to each other as in our experiment (test this for yourself)).

It turns out that there is a relation between the square of the current and the power of the light. How this relation comes about is this way : Power for an electrical system can be determined as voltage time current ($P = V*I$) and from Ohm's Law ($V = I*R$) we can substitute for V the terms I*R and obtain ($P = I^2*R$). The resistance (R) here is treated as internal and constant to the solar cell (photovoltaic) and factored out, so $P \sim I^2$.

In the case of light sources, Illumination (B) is the number of lumens per meter squared ($B = \frac{L}{4*\pi*D^2}$)(the units of Illumination are lux). The Luminance (L) aka Luminous Flux is proportional to the Power of the Light Source ($L \sim P$) hence $L \sim I^2$. Though we measure light sources in terms of Power (Watts) they can be considered in terms of their luminance. In fact for a given wattage of a incandescent bulb, they average about 15 lumens per Watt while fluorescent bulbs (depending on power rating) range between 50-100 lumens per Watt.

Our Activity will place a simple incandescent light source at a distance from a photovoltaic cell connected to a multi-meter set up to measure the current (in milliamps) at varying distances from the light source. Both the Distance (D) and the current (I) are recorded. This data is graphed, mathematically recalculated, graphed again in two ways to determine the relation between light intensity (taken as current) and distance from the source to find the inverse-square law relation!

It is useful to examine this phenomena since it involves the idea of inverse -square relations, graphical and mathematical analysis, plus in the case of light, it is practical since when coupled with other interactions of light with matter (particularly reflection, absorption & re-emission, as well as refraction) it helps in understanding light and can be employed for light intensity needs for indoor as well as outdoor lighting considerations.

Ms Briana Andromeda Purcell here and commenting on the prior analysis and presentation of ideas – keep in mind we are merely seeking to validate and turn into a mathematical understanding something all of us know intuitively and from practical experience. Are lights close at hand brighter than ones farther away? Certainly! We all know this. This is where the Inquisitive Pioneer in you comes in and then asks – is there a relation of distance to brightness and can I find it? That is our quest. Be Brilliant and here as well Radiant!

Purpose : To investigate through graphical and numerical analysis from measurements using a photovoltaic cell and a multi-meter the relation of light intensity and distance from the source.

Materials :

- Short Lamp with bulb (60 watt) (take off shade),
- Measuring tape (metric measures are best),
- Multi-meter ,
- Photovoltaic solar cell,
- Paper Towel tube or Toilet Roll tube cut to 5 cm cylinder,
- Sheet of dark Construction Paper,
- Graph paper,
- Long table (kitchen will do),
- Inverted Crate or Stack of books (to set multi-meter and solar cell on),
- Darkened Room,
- Small LED light to allow for writing, reading of information,
- Clear Tape,
- Slide Rule

Procedure :

1) With any and all Activities read through, then set up the process for the Activity and run it without measurement and watch where you may have bumps in the road so as to anticipate these. Be patient – Repeat processes. Once all is well when measuring see to the values – do you notice patterns or trends – what do you predict will happen next as you change one thing or another from one trial in an experiment to the next. Always Explore & Enjoy. Be Brilliant!
2) Note : have the room as dark as possible, but have a small LED light to be able to do work as needed.
3) First cut the paper tube (paper towel tube or toilet roll tube) to a length of not more than 5 cm.
4) Attach to one end the dark construction paper and cut out the hole.
5) This tube with a shield will be placed in front of the solar cell so that only a constant area with some protection will act to only allow in the light along that path and keep out all other stray light sources.
6) With lights in the room on, set the lamp at the far end of a table.
7) Unfurl the measuring tape from the base of the lamp away from it across the table.
8) It is best to use an inverted crate (or stack of books if the crate is not available) to place the multi-meter and solar cell on.
9) The Solar Cell must be set up so as to be at the same height of the lamp and facing it (the tape can help hold it up) (books, magazines and the like can be used for either the lamp or the crate to establish a level environment so that the solar cell and light bulb are on the same level).
10) Set the solar cell so that it is the prescribed distance as noted in the table for data.
11) Note that the tube assembly created in steps 2,3,4 is placed in front of the solar cell.
12) Attach the multi-meter so that it will read current (I) measured in milliamps.

13) Turn on the bulb and turn out the other lights in the room.
14) Note : Do NOT stare directly into the light when taking measurements.
15) For each reading, always let the system sit and stabilize for a few seconds. The values may oscillate around a couple of numbers – take the one that seems to be the most frequent.
16) Once you are done with all of the measurements, turn on the room lights and turn out the lamp. Disconnect the multi-meter and solar cell.
17) Follow the remainder of the directions in the Calculations portion.

Data :

Measuring the Inverse-Square Law Indirectly

(through Multi-meter Current Readings & Power Calculations)

Distance (cm)	Current [I] (mA)	Current^2 [I^2]	$\frac{1}{D^2}$
10			
20			
30			
40			
50			
60			
80			
100			

Calculations :

Be sure to use a Slide Rule for all of your calculations!

1) Use the Slide Rule and find the squared value of the current (I^2) and place this in the appropriate column. (This uses readings from the D Scale and looking at the A Scale).
2) Note that Electrical Power is proportional to the square of the current in the circuit. ($P \sim I^2$).
3) First use the D Scale looking up the distances used in the Activity and find the corresponding value on the C1 Scale (its inverse). Be sure to watch the exponents here! Fill in this portion of the Data Table.
4) Create a Table of log(distance) and the log (current^2) values as read from the Slide Rule. (Read the data from the D Scale and find the log values on the L Scale).
5) Graph on the Y-axis Current^2 (which is related to Power which is related to Light Intensity) vs. Distance on the X-axis. This curve should approximate an inverse-square relation.
6) To test the idea of the relation in the Activity :

7) Calculate the Product of Current^2 and $\frac{1}{\text{Distance}^2}$ values. If done correctly, they should be approximately the same. Add these values up and divide by the number of data points to determine an Average value.

8) Graph I^2 (Y-axis) vs. $\frac{1}{D^2}$ (X-axis).

9) Draw a best fit line and determine slope for this graph. It should be approximately the same as the average of the product!

10) Graph on the Y-axis the log(current^2) and on the X-axis log(distance).

11) Through the log-log graph data, draw a best fit line and calculate the slope of this line (which should be near -2).

12) Your relation will reveal the connection between the variables. If it is at -2, this means that the light intensity (which is measured by current^2) is inversely related (hence the negative sign) to the square of the distance

(light intensity $\sim \frac{1}{\text{distance}^2}$)

13) Though the Slide Rule is a recommended tool, all of the calculations can be done with a graphing scientific calculator or the use of a spreadsheet program. In these calculations you have to generate a table of data, graph it, and then find the slope and/or equation of the best fit line for the data. Other formula calculations can be done with these tools as well.

Formulae :

$P = V*I = I^2/R$

(Note 'R' here is assumed to be constant and factored out – we are noting that $P \sim I^2$ and that P ~ Light Intensity)

$$\textbf{Product} = I^2 * \frac{1}{D^2}$$

$$\textbf{Average} = \frac{\Sigma(\text{products})}{\text{number of products}}$$

$$\text{slope } m = \frac{\Delta Y}{\Delta X}$$

$$m = \frac{\Delta(I^2)}{\Delta(\frac{1}{D^2})}$$

$$m = \frac{\Delta \log(I^2)}{\Delta \log(D)}$$

Conclusion :

How Cosmic was this investigation! Finding that a simple observation that the light seems dimmer with distance actually has a mathematical relation we can find and using common tools!

Wow! All things certainly have a story to tell us – we have to be creative in thought, curious in nature, and passionate about our need to explore to learn and be Brilliant!

Examining the data and the graphs one should be able to find the inverse-square law relation of light intensity and distance. Be sure to take into account all sources of error and redo this as necessary.

Activity #48
Estimating the Size of the Moon
Grade Level : Middle School
Math Level : Calculating

Size of the Moon Activity –

On a given full Moon night have you thought about how far away the Moon is, how big the Moon is? Well Inquisitive Pioneers, Ms Briana Andromeda Purcell here and we will use some basic math to use the distance to the Moon and a simple observational trick to then find what the diameter of the Moon is! Imagine that! The interesting part (beyond the math) is the fact we are merely observing the Moon and taking a few measurements as well as using extremely common and readily available items – A meter or yard stick, a 3 x 5 card, a hole punch, scissors, and tape.

In our Activity, we will use a Proportion where we take the unknown diameter of the moon to its distance from the Earth as a ratio and set this equal to the known diameter of a punched out hole in an index card over a measured distance along a meter stick that the hole is from us when observing the Moon so that it just fits the hole exactly.

We will only use 2 significant figures, though a Slide Rule can go to 3 for this exercise. Those who want to take it further, use the more precise data.

Materials :

- Hole Punch,
- Clear Tape,
- Meter Stick,
- Index Card,
- Scissors,
- Full Moon,
- Distance to Moon : **2.4×10^5 mi (3.8×10^5 km),**
- Paper, writing tool (to record measurements),
- Slide Rule

More Precise Data :

Earth-Moon Distance :
2.39×10^5 mi (3.84×10^5 km)

Data for Comparison (the Diameter of the Moon) :

Note : Be sure to go past this section until after you have taken your measurements and calculated the estimated diameter of the Moon from your data!

2 significant figure value : 2.2×10^3 mi (3.5×10^3 km)

3 significant figure value : 2.16×10^3 mi (3.47×10^3 km)

Procedure :

1) With any and all Activities read through, then set up the process for the Activity and run it without measurement and watch where you may have bumps in the road so as to anticipate these. Be patient – Repeat processes. Once all is well when measuring see to the values – do you notice patterns or trends – what do you predict will happen next as you change one thing or another from one trial in an experiment to the next. Always Explore & Enjoy. Be Brilliant!
2) Use a ruler and measure the width of the meter stick (yard stick) you are going to use and write it down on scrap paper for use.
3) Knowing the width of the meter stick, measure this amount twice from the edge of a 3x5 card so as to mark the 3" side with two marks.
4) Draw these as two parallel straight lines across the card. (see 1st photo)
5) All of the remainder of the card beyond the second mark line is to be cut away and discarded.
6) In the middle one of the drawn strips, place a piece of clear tape that is about 1 cm square in size.
7) In the center of the tape, punch out a hole.
8) With the ruler measure the diameter of the hole (H) and record it in the data table.
9) Note : In your measurements you only have to be consistent. That is both the hole diameter and its distance must be in the same units. You can use the metric system or the English system.
10) Cut along the line between the two strips so that only enough of the middle remains on the top of the meter stick and the remainder acts as flaps that wrap around the meter stick.
11) Place the meter stick flat on a table. On top of it place the card so that the hole is centered on the stick.
12) Fold along the line so that the hole is standing up perpendicular to the meter stick.
13) Wrap the other strip in contact with the meter stick around the meter stick and tape it so that it becomes a firm wrap that can be slid along the stick. Cut away excess so that it is a firm sleeve for the meter stick. (see 2nd photo)
14) If correctly done you have made a stand up hole to view through from the end of the meter stick which can be slid along the stick as needed. (see 3rd photo)

15) When using it be sure to keep the hole portion perpendicular to the surface of the meter stick.
16) On a clear night with the full Moon out go outside and point the stick at the Moon so that you have to look along the stick and through the hole in the card at the Moon.
17) Be sure to use one eye looking along the stick and through the hole at the Moon.
18) Move the hole so that it just matches the size of the full Moon. Be careful not to move this once in place since its distance (D) is needed in the data table. Record this value.
19) Note : Be sure to measure it correctly. The point that the hole is over is the distance from the end that is zero (if looking from that end).
20) With data in place calculate the diameter of the Moon using the formula provided and using your Slide Rule.
21) Note the ease of using the Slide Rule since the formula is a proportion!
22) Note : In using the Moon Diameter Formula, pick which units you want your answer to be in by choosing the distance to the Moon measure as either kilometers or miles. Whichever is chosen is how the answer will come out.
23) Check to see how accurate your results are by using the percent error formula (again using the Slide Rule).
24) Note : Though the Slide Rule is a recommended tool, all of these calculations can be done with a regular or scientific calculator. Some scientific ones even have built-in averaging formulae. For those who like spreadsheets, the data can be typed in and the formulae then also be typed in its own cell where the formula references each of the measured variables in their respective cells.

Photos

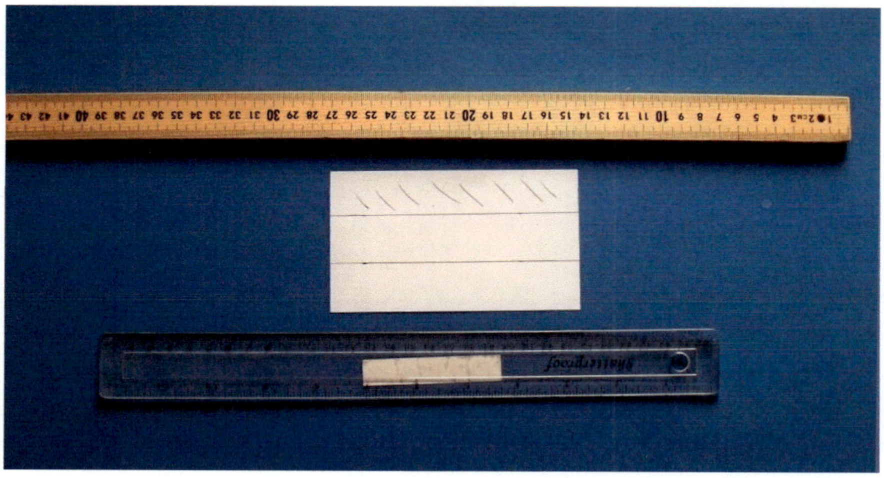

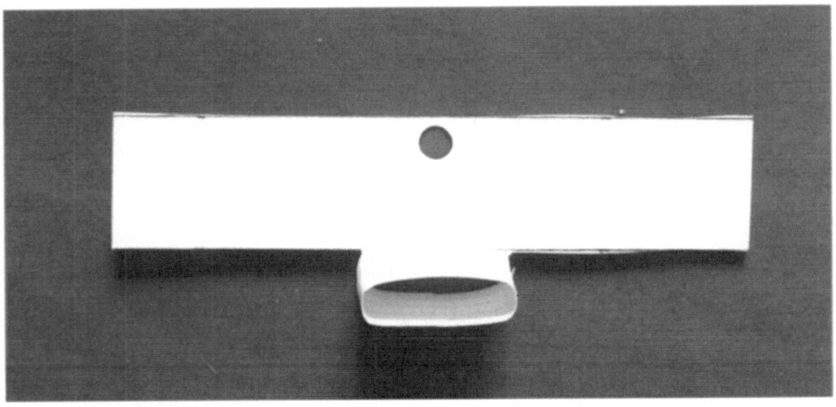

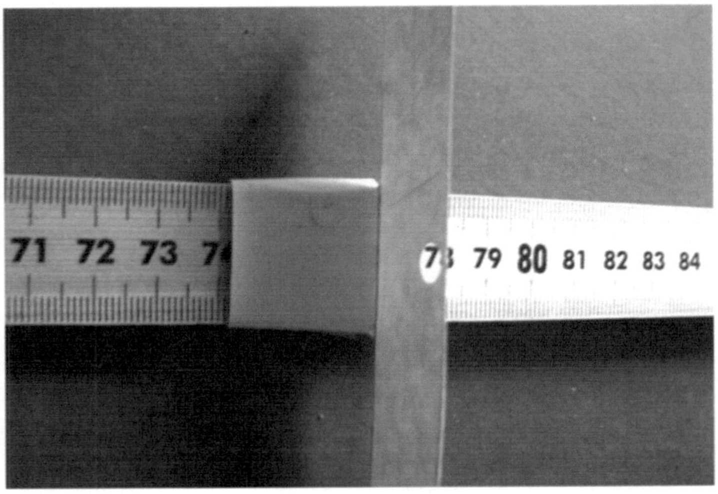
71 72 73
79 80 81 82 83 84

Formulae :

Be sure to use your Slide Rule!

Moon Diameter Proportion Formula :

$$\frac{H}{D} = \frac{\textbf{Unknown Moon Diameter}}{\textbf{Known Moon Distance}}$$

(H : hole diameter, D : distance of hole to eye on meter stick)

Percent Error Formula :

$$\%E = 100\% * \frac{[\text{Accepted Moon Diameter - Calculated Moon Diameter}]}{\text{Actual Moon Diameter}}$$

Conclusion :

Cosmic Cool here Inquisitive Pioneers! Such basic ideas from math and simple tools and we can determine the size of the Moon! Awesome Sauce!

Here is an alternative or an add-on to the Activity: Some of typical measures for this Activity are these : 0.5 to 0.8 cm for the hole diameter and the typical distance range for how far the hole needs to be is > 55 cm and < 95 cm. This is because the distance for this ratio of Earth-Moon distance to the Moon diameter is about 111 times.

This can be seen as asking the question, 'how far would a dime seen face on need to be in order to cover the full Moon?' The dime is nearly 2 cm (around 1.8 cm) so it would have to be nearly 200 cm (2 m) distant.

Explore & Enjoy!
Be Brilliant!

Activity #49
Solar Cell Examinations
Grade Level : High School
Math Level : Calculating

Solar Cell Measures Activity :

Hey Inquisitive Pioneers – Ms Briana Andromeda Purcell here with yet another incredible Activity. This one, like the Mechanical Energy (#27) one and the Wave one (# 43) has a bit of a story prelude which relates to the main tool we are using and measuring its ability, namely the Photovoltaic Cell. Since these technology tools use Sunlight there will be some light look at the energy of the Sun too! So read on this following Prelude on Solar Energy and Photovoltaic Cells for those interested in the subject. Otherwise you can move ahead to the Activity that follows concerning an investigation of these photovoltaic cells.

The Sun provides an enormous amount of energy for the Earth. The energy is discussed with more details and numbers in the Solar Constant Activity #53, but it can be said that the Sun's energy reaching the Earth's upper atmosphere is on the order of 174 petawatts of power (peta meaning 10^{15}). This energy drives the wind, the waves, provides heat, light, and energy for life forms of our planet and we are utilizing tools to directly turn light energy into electrical energy.

Solar Energy in many forms, especially in the present world of technology, is used for many reasons. The most obvious is light, which is utilized for everyday visual uses. The most obvious employment of the Sun outside of regular light and heat is the growing of crops by humans which takes advantage of photosynthesis by plants which converts carbon dioxide (CO_2) and water (H_2O) into molecules to provide energy, namely sugar. Also sunlight is the base of the food pyramid both on land and in the ocean – we are not the only ones who use it – in fact we learned it from nature (which is really what science is all about). The heat of the Sun will be discussed more thoroughly in the Heating Activities #31 (if employing the Sun), both in human application and its appearance in nature, such as weather. Also the prelude on Energy in Mechanical Energy (#27) has some information about the Sun's energy too.

When looking at our technological use of Solar Energy Humans have basically utilized light in two ways categorically : Passive and Active Solar Energy Technologies. These classifications result from the way the sun's energy is captured, converted, and distributed. The Passive Solar Energy Technology will be explored in more detail in the Rate of Cooling Activity #32 indirectly, but it can be said here that it is based upon the thermal properties of materials positioned in favorable ways to take advantage of the Sun' energy, heat, and light. This is to say we let nature take its course – use certain materials to let them heat up from sunlight and then tap into that heat reservoir.

Active Solar Energy ideas are ones that take the sunlight and convert it into other forms of energy (the most common being electricity) for use. In this technology there are two chief types : 1) concentrating solar power - focused sunlight and the use of Stirling engines or for boiling water to use with steam turbines 2) utilizing the photoelectric effect - photovoltaics cell use (which is the focus of this Activity).

In concentrating solar power, here optics such as lenses or parabolic mirrors are used to focus the sun's light energy onto a region or point to act as a heat source to actively power a power station, such as a Stirling engine. These heating systems often heat a fluid of some form to cause it to do work, which can then activate an electric generator. Others can be used to boil water to turn turbines as well.

The other type of Active Solar Technology are Solar Cells – aka Photovoltaic cells (the latter term can apply to any and all light whereas the former applies to sunlight) which convert sunlight directly into electricity through the photoelectric effect.

The term 'photovoltaic' comes from 'photo' Greek for 'light' and 'voltaic' derived from the Italian physicist Volta and the term 'volt' came from which is the unit of electro-motive force, so 'voltaic' is intended to mean 'electric'. This idea was first written about but not constructed in 1839 by the French physicist A. E. Becquerel. The first created one came in 1883 by Charles Fritts, who used coated the semiconductor selenium with a thin layer of gold at the junctions. This first one was on 1% efficient. Not long after the first solar cell based on Heinrich Hertz 1887 photoelectric effect came from the Russian physicist Aleksandr Stoletov. Even Albert Einstein played a part in the history of the solar cell with his 1905 paper explaining the photoelectric effect for which he won a Nobel Peace Prize in 1921. In time more research and modifications were made to the solar cell. A much more efficient form using a diffused silicon p-n junction was developed by Chapin, Fuller, and Pearson in 1954. Some solar cells today have reached values of 24%, and even 42% absorption efficiency (the latter one with concentrated light), well above the average in the industry which is at 12-18%. (Note that this stands in contrast to sunlight-to-electricity efficient, which is lower). Today there are even Megawatt solar power generating plants being built.

This type of electrical energy production is not only found in small devices, such as arrays for charging batteries and uses in calculators, but is always in use in space (satellite and space station power sources) and it is finding a growing market worldwide here on the ground. The efficiency of them is increasing, their fragility is decreasing, and their costs are coming down. There are even bendable solar cell arrays that are used as shingles on roofs these days. As of

2010 power generated by them is going on in over 100 countries. It is estimated that 4800 GW could be harnessed this way yet only some 21 GW are being used at present. Since 2002, photovoltaic production has risen by 20% per year making it the fastest-growing energy technology. Historically the first commercial use of it was in 1966 on Ogami Island in Japan to change the Ogami Lighthouse from gas torch to fully-sufficent electrical power. The three leading countries in use of this technology are Japan, Germany, and the United States. Presently most stations range from 10-60 MW capacities and in the near future 150 MW or more are projected. All of these worldwide efforts are driven by the need for renewable energy sources and their utilization.

(Note Giga is 10^9 (billion) and Mega is 10^6 (million)).

Solar Cells are made of semiconducting materials (these are elements that are in the metalloid region of the Periodic Table between metals and nonmetals). Over the cell is a clear covering of glass or plastic to allow light in, but to protect the semiconducting wafers. These cells are encapsulated and care connected in series and/or parallel connections depending on the need for greater increases in voltage and/or current and for the overall power of the system. The arrangement is called an array.

The solar cell generates a DC (direct current). The power output is measured in watts or kilowatts (depending on the number of cells in the array). To determine the number of cells needed and the arrangement of cells needed, the energy needs calculated in watt-hours, kilowatt-hours, or even a energy per day as kilowatt-hours per day are often used. A quick mental rule is this : Average Power is equal to 20% of Peak Power.

The power from a solar cell array can be fed directly into the electric power grid through inverters. If these arrays are stand-alone the energy is often stored in rechargeable batteries if it is not immediately being used (such as powering a device). Smaller panels can be used as chargers and direct power sources for cellular phone chargers, solar-powered calculators, solar lights for things like bikes and solar-charged camping lanterns.

How do they work? When photons encounter the surface, some pass right through (since their energy has no effect on the materials), some reflect off the surface of the cell, while some photons have the right amount of energy that corresponds to the 'silicon band gap'.

These photons that hit the solar panel and are absorbed by the semiconducting materials energize the electrons in the valence band. These electrons are regularly tightly bound in covalent bonds between neighboring atoms. The energy they are given excites them into the conduction band where it can move more freely in the semiconductor. Where the electron was, now has what is called a 'hole'. This creates a potential for electrons in surrounding molecules to move into it. Hence electrons move one way while the holes move in the more-or-less opposite direction. Next these negatively-charged electrons are knocked loose from their atoms are allowed them to flow though the semiconducting material and produce electricity. The cells composition only allows for the electrons to move in one direction. Hence, when cells are connected to each other will have an overall voltage and current flow and produce a measurable and useable amount of DC electricity.

The most commonly known solar cell configuration is called the p-n junction and is made from silicon laced with other semiconducting materials. Basically it is a materials where a layer of n-type silicon is placed in direct contact with a layer of p-type silicon. That is just a mental model, while typically the layer of silicon has one side diffused with a n-type dopant into a p-type wafer (or vice versa) so that the effect is the same.

In the p-n junction type there is a diffusion of electrons from the region of high electron concentration (the n-type side of the junction) into the region of low electron concentration (the p-type side of the junction). When the electrons diffuse across the p-n junction, they recombine with holes on the p-type side. This does not continue indefinitely, since a charge build up results in an electric field, which in turn, creates a diode that promotes charge flow (aka drift current) which opposed and eventually balances out the diffusion of electrons and holes. The area that no longer contains any mobile charge carriers is called the space charge region.

The solar cell is connected to an external load at its positive and negative points. The voltage measured is equal to the difference in the electrons in the p-type portion and the holes in the n-type portion. With voltage, when connected to any load (which has resistance) then a current can occur (Ohm's Law).

The amount of Power from a solar cell easily can be seen from the amount of light falling on it, which is one of our Activities. Also, the number of solar cells used and how they are arranged will affect the amount of power available from a solar array. The two common types of arrangements are called : Series & Parallel arrangements.

Like in the Ohm's Law where Resistors are connected one after another and hence are a Series, the same is true for solar cells. Here the positive lead connects to the next solar cell's negative lead and so on. What is left is the first one has a negative lead not connected, and the final one has a positive lead not connected. These can be connected to a device or a multimeter for determining its readings. One should find that the voltages should sum up and hence increase, while the current values should remain constant.

In the Parallel arrangement, the positive leads are all connected as are the negatives to each other in two lines across all of the solar cells. From the lead or final solar cell to additional lines are connected for connection to a device or a multimeter. Here all the voltages are the same, so it will not increase, while the currents will all sum up and have a larger value.

One could also be more creative, possibly wanting both greater voltage and current and create, for example, two separate lines where each line has two solar cells in series with each other while the two lines are in parallel to each other. This system will add up in a way so that it has both increases in voltage and in current values.

Our investigations will take the solar cell and concentrate on constant light sources at different distances, angles, and even the amount and/or type of light reaching the solar cell. The last activities examines the connections of more than one solar cell and looks at series and parallel arrangements.

One idea not explored here is distance of the light source from the solar cell, since this addressed in the Inverse-square Law of Light Activity. Also the solar cell can be used in the Solar Constant Activity #53.

The Solar Cell Investigations :

I. **Solar Cell Power due to Exposed Area of Cell**
II. **Solar Cell Power due to Angle of Light Exposure**
III. **Solar Cell Power due to Light Intensity**
IV. **Solar Cell Power due to Wavelength of Light**
V. **Solar Cell Power when connected in Series**
VI. **Solar Cell Power when connected in Parallel**

Activity I : Solar Cell Power due to Exposed Area of Cell
Purpose : To measure and determine the effect on voltage and current measurements (to calculate power) for a constant light source at a given distance while changing the amount of solar cell surface area exposed to the light.

Activity II : Solar Cell Power due to Angle of Light Exposure
Purpose : To measure and determine the effect on voltage and current measurements (to calculate power) for a constant light source at a given distance while changing the angle that the light strikes the solar cell.

Activity III : Solar Cell Power due to Light Intensity
Purpose : To measure and determine the effect on voltage and current measurement (to calculate power) for a constant distance light source of varying intensity light emitted and received by the solar cell.

Activity IV : Solar Cell Power due to Wavelength of Light
Purpose : To measure and determine the effect on voltage and current measurements (to calculate power) for a light source emission filtered by color filters to allow certain wavelength access to the solar cell.

Activity V : Solar Cell Power when connected in Series
Purpose : To measure and determine the effect on voltage and current measurements (to calculate power) when a solar cell array has its cells connected in a series circuit manner when receiving a varying light source intensity.

Activity VI : Solar Cell Power when connected in Parallel
Purpose : To measure and determine the effect on voltage and current measurements (to calculate power) when a solar cell array has its cells connected in a parallel circuit manner when receiving a varying light source intensity.

Materials :

- Three 1.5 V Solar Cells (only one if not doing series or parallel),
- 6 Alligator Clip wires or wires to connect the solar cells (2 for one),
- Multimeter,
- Bulbs : 25 W, 40W, 60W, 75W, 100W,
- 2 Small Lamps (regular post type and flexible head),
- Protractor,
- Meter Stick or Measuring Tape,
- Color Filters (R, G, B),
- Piece of Dark construction paper or cardboard,
- Scissors,
- Tape,
- Graph Paper,
- Small penlight to read and write by,
- Item(s) to create angle for solar cell to rest on and be at – can be a stack of books/magazines, use of poster board or cardboard folded, or some other imaginative item to get the job done,
- Slide Rule

Procedure :

1) With any and all Activities read through, then set up the process for the Activity and run it without measurement and watch where you may have bumps in the road so as to anticipate these. Be patient – Repeat processes. Once all is well when measuring see to the values – do you notice patterns or trends – what do you predict will happen next as you change one thing or another from one trial in an experiment to the next. Always Explore & Enjoy. Be Brilliant!
2) All Activities use a Light Source (Lamp with at least one bulb used), the multimeter, and at least one solar cell, and all have their calculations done with the slide rule.
3) Recommended : Use rectangular, 1.5 V solar cell
4) Note that since the effect of the light on the solar cell is the primary measure of interest, try to do most of the measurements with as little external light as possible. (have the small light available to write things down).
5) Exercise safety in using the lamp : Do not touch the bulb, Leave it unplugged until the bulbs are properly in place and then activate it, do not place items near or on the bulb (such as the solar cell or multimeter).
6) It is best between trials to turn off the light, turn off the multimeter, and then set it up to operate for that given trial.
7) Recognize that not only do you turn the dial (typically) to change from reading voltage to reading current on the multimeter, you must also change the configuration of the test wires. Be sure of the instructions on how to properly operate the multimeter for voltage and current readings.
8) Since the solar cell is part of every experiment, always have the positive and negative leads coming from one (or more) the solar cell(s). If properly done, there are only two wires, one positive and one negative to touch or connect to with the multimeter leads.

9) Note : Lamps are used so as to control the amount of light, its distance, angle, et al. But once there is an understanding of the results in using a solar cell, one could use their intended source, namely the Sun on sunny days and try these experiments using it when and where possible. Be sure to not look directly into the Sun. Use shadows of pencils act as guides for pointing items directly at the Sun. Also employ safety when using bulbs as well – do not touch active bulbs nor even when turned off – do not look directly into the light. Employ common sense safety procedures.

10)

11) Activity I : Solar Cell Power due to Exposed Area of Cell

12) Measure the Length and Width of the Solar Cell. Calculate its Area.

13) Use the dark construction paper. Measure and draw a rectangle 3 times the width by one length of the solar cell dimensions.

14) Cut the first piece to be the length and width of the solar cell.

15) For the second, cut a rectangle the width and for the length measure 75% of the length and cut it there.

16) Save the 25% piece too since this is the final piece.

17) For the third, cut a rectangle the width and for the length measure 50% of the length and cut it there.

18) Each of the rectangles are used to cover the exposed area of the solar cell in turn as noted in the data table.

19) For this Activity, it is best to use the flexible light since it can be directed onto the solar cell lying flat on the table or floor. If not, just be sure not to move the lamp once in place so that the amount of light available is the same for each trial.

20) In setting up the solar cell, be sure to connect the wires so that you can connect a meter to take readings.

21) Choose a Wattage bulb (60 W recommended) and a distance for the light (about 25 cm recommended). Plug in and activate it.

22) Choose an order (like the one noted in the data table) for percent exposed.

23) Start with the Voltage readings [V]. (note readings might not be in volts (V), depending on distance and bulb intensity)

24) After each reading, record it, turn off the light and place the appropriate cover on the solar cell for the next reading.

25) Reactivate the light and continue the readings.

26) Next configure the multimeter to read Current [I]. Check to see it has the proper setting (note that readings may not be in amps (A) so be sure to correct for this).

27) Note : 1000 mX = 1 X (X can be volts or amps)

28) Go through the process and record all of the current values.

29) Calculate the Power using your slide rule.

30) Create 2 graphs of this : First a Bar graph of Power versus Percent Exposed. Second a Line graph of the same variables and compute the slope of the best fit line for this case.

31) A good follow up to this activity is to try other wattage bulbs to see if the results vary or not.

32) **Activity II : Solar Cell Power due to Angle of Light Exposure**
33) In this Activity, use the lamp (60 W bulb recommended), the solar cell, the multimeter, a stack of books and the protractor.
34) The best set up is using a standard post lamp set at a constant distance (25 cm to 50 cm).
35) The solar cell is atop a duel set of books and goes from lying horizontally to standing vertically, propped up by paper/magazines/cardboard.
36) The key is to have the center of the bulb and the solar cell when standing up directly in line (this is referred to as vertical or 90°)
37) The angle that the solar cell is at is measured (see data table below)
38) The angle is with respect to the ground and is the compliment of the angle with respect to the solar cell. Determine this.
39) At each angle, measure the Voltage and the Current using the multimeter.
40) When done, calculate the Power of the solar cell for each of the angles.
41) Create both a Bar Graph and a Line Graph of Power versus the Angle and compute the slope of the best fit line for the line graph.
42) Like the other Activities, try other wattage bulbs for comparison.
43)
44) **Activity III :**
45) For this Activity, use the standard post-style lamp, have all the bulbs available, and use the meter stick to determine distance between the lamp and the solar cell.
46) As in Activity 2, you could have the solar cell stand vertically and facing the bulb (though no matter its orientation to the light if held constant it will be okay).
47) Choose a distance between the lamp and the solar cell that is held constant (recommended 25 cm to 50 cm).
48) Set up the lamp with the lowest wattage bulb first and work your way through to the highest wattage bulb for each of the trials.
49) With a given bulb in and on measure both the voltage and current produced in the solar cell and record your results in the table.
50) When done with the data table, calculate the Power for each of the bulbs.
51) Graph in a Bar Graph and a Line Graph the results of the measurements with Power as the y-axis and Bulb Wattage as the x-axis.
52) For the line graph, draw a best fit line and compute slope.
53)
54) **Activity IV : Solar Cell Power due to Wavelength of Light**
55) In this activity use one bulb type (60 W recommended) and placed in the lamp at a constant distance (25 cm to 50 cm recommended).
56) Begin with the measurements of the solar cell for voltage and current with no filter in place.
57) Between each trial turn off the light and set up for the next set of measurements.
58) For each of the trials, place one of the color filters (red, green, blue) on the photovoltaic cell and then activate the light and take measurements of voltage and current with that filter in place.
59) For each of the trials, white light and the filter, calculate the Power being produced by that light as measured.
60) Compare results to each other (one way is to let the white light be the 100% base and divide this value into each of the others to find its percentage).
61)

62) Activity V : Solar Cell Power when connected in Series

63) In this Activity, 3 Solar Cells are connected to each other in a Series Circuit fashion (that is there is one path for the current to flow). To do this : use the alligator clip wires and connect the positive terminal of one solar cell to the negative of the next in line, then its positive to the negative of the next.

64) In the final step have one alligator clip attached to the first solar cell's negative terminal while another alligator clip is attached to the last solar cell's positive terminal. These are the terminals to measure from with the multimeter.

65) Choose a bulb wattage to begin with at a constant distance (25 cm to 50 cm recommended).

66) Choose a way to configure the solar cells so that they receive approximately the same amount of light (examples : 1) they can lie flat in a circle around the light if the wires are long enough, 2) the best choice is to stand them up as we did in Activity 3 since we are going to not only compute this Power but also compare this outcome to Activity 7 when connected in Parallel).

67) Once set up, activate the light and take measurements of voltage and current using the multimeter. For the measurements record the overall circuit and each of the solar cells in turn as well.

68) Note : Be sure to connect the multimeter correctly when measuring voltage and current. For example – the voltage can be done by touching across each set of terminals in the array, while the current is found by connecting the ammeter setting of the multimeter in series with the circuit at one spot (all of the currents will be the same – try a couple different spots to be sure).

69) Also for comparison purposes, after completing the first data set, you redo this exercise with a different size bulb.

70) Before calculation, what does the data show you? (If properly done, when in Series, the voltage should be the sum of all the voltages).

71) From the data set, calculate the Power of the solar cells connected in series. This will be compared to the Parallel circuit case next.

72)

73) Activity VI : Solar Cell Power when connected in Parallel

74) In this Activity, 3 Solar Cells are connected to each other in a Parallel Circuit fashion (that is there is a unique path for each of the solar cells current to flow). To do this : use the alligator clip wires and connect the positive terminal of one solar cell to the positive of the next in line, then connect an alligator clip wire between the negative terminal of the first solar cell to the second. Connect the third to the second in exactly the same fashion. Hence all of the positives have a line and all of the negatives have a line, much like a ladder.

75) In the final step have one alligator clip attached to the first solar cell's negative terminal while another alligator clip is attached to the last solar cell's positive terminal. These are the terminals to measure from with the multimeter.

76) Choose a bulb wattage to begin with at a constant distance (25 cm to 50 cm recommended).

77) Choose a way to configure the solar cells so that they receive approximately the same amount of light (examples : 1) they can lie flat in a circle around the light if the wires are long enough, 2) the best choice is to stand them up as we did in Activity 3 since we are going to not only compute this Power but also compare this outcome to Activity 6 when connected in Series).

78) Once set up, activate the light and take measurements of voltage and current using the multimeter. For the measurements record the overall circuit and each of the solar cells in turn as well.

79) Note : Be sure to connect the multimeter correctly when measuring voltage and current. For example – the voltage can be done by touching across each set of terminals of any member in the array, while the current is found by connecting the ammeter setting of the multimeter in series with the circuit at one spot – that is be a part of the path for a given solar cell (all of the voltages will be the same – and the currents should be the same too since they are the same solar cell, but this is not always the case - try a couple different spots to be sure).

80) Also for comparison purposes, after completing the first data set, you redo this exercise with a different size bulb.

81) Before calculation, what does the data show you? (If properly done, when in Parallel, the circuits overall current should be the sum of all the currents of the separate solar cells).

82) From the data set, calculate the Power of the solar cells connected in series. This will be compared to the Series circuit case in the last Activity.

83) Notes for all : When constructing Bar Graphs, note that the area of the bar when the axes are Voltage and Current will become Power!

Data :

Activity I : Solar Cell Power due to Exposed Area of Cell

Bulb Wattage Used : ____________ W
Recommended : 60 W

Area of Photovoltaic Cell : ________ cm^2

Distance of Light : ___________ cm
Recommended : 25 cm to 50 cm

% Exposed	Area Exposed (cm^2)	Voltage (V)	Current (A)
100			
75			
50			
25			
0			

Activity II : Solar Cell Power due to Angle of Light Exposure

Bulb Wattage Used : ___________ W
Recommended : 60 W

Distance of solar cell nearest base to lamp : __________ cm
Recommended : 25 cm to 50 cm

Angle (°)	Angle to Light (°)	Voltage (V)	Current (A)
0 (horiz.)			
15			
30			
45			
60			
90 (vertical)			

Activity III : Solar Cell Power due to Light Intensity

Distance of Light Source : ____________ cm
Recommended : 25 cm to 50 cm

Wattage Used (W)	Voltage (V)	Current (A)
25		
40		
60		
75		
100		

Activity IV : Solar Cell Power due to Wavelength of Light

Bulb Wattage Used : ____________ W
Recommended : 60 W

Distance of Light Source : __________ cm
Recommended : 25 cm to 50 cm

Wavelength color	Voltage (V)	Current (A)
White		
Red		
Green		
Blue		

Activity V : Solar Cell Power when connected in Series

Distance of Light Source : __________ cm
Recommended : 25 cm to 50 cm

Wattage Used (W)	Voltage (V)	Current (A)
25		
40		
60		
75		
100		

Bulb Wattage Used : ___________ W

Solar Cell	Voltage (V)	Current (A)
1		
2		
3		

Activity VI : Solar Cell Power when connected in Parallel

Distance of Light Source : ____________ cm
Recommended : 25 cm to 50 cm

Note : This table is for the overall circuit values and not for the individual solar cells in the array. The table below that is for each of the solar cells.

Wattage Used (W)	Voltage (V)	Current (A)
25		
40		
60		
75		
100		

Bulb Wattage Used : ____________ W

Solar Cell	Voltage (V)	Current (A)
1		
2		
3		

<u>Calculations :</u>

Se sure to use your Slide Rule !

Area of Solar Cell : **A = L * W**
(rectangular area : Area = Length x Width)

Slope : $\mathbf{m} = \frac{\Delta y}{\Delta x}$

90° = Angle(1)+ Angle(2)

Electrical Power : (P is in Watts)

$$P = V*I$$

Conclusion :

Wow – what cosmic adventures in using technology and turning light energy into electricity! So cool! Notice when it comes to technology and using it in these labs you may be using your imagination and wondering where can this be used and begin to dream up places and ideas. Or perhaps you are more interested in the physics and chemistry of the solar cells themselves and ask are there ways to change and/or improve them. Today there are different ones other than the classic ones I presented here in this lab – some are trying tiles to act as shingles on the roof – some are rigid while others are thin and flexible, still others are making transparent glass have imbedded materials to act as solar cells so that tall buildings which can have many square feet of glass become giant sunlight collecting surfaces – science lets the imagination soar!

Explore & Enjoy! Be Brillant!

The conclusion depends on which of the Activity(ies) were undertaken. In all cases, there is a calculation of Power and it is examined as compared to some other changing variable (light intensity, distance, et al). What do the results show and how do they compare to science-based expectations?

Summary and Alternate Ideas :

Besides using a constant light source as we did here, try and use the Sun. Be sure to understand its changing position throughout the day and other factors that may affect outcome as well. This makes for a good comparison to the light bulbs used as well.

Activity #50
Graphically determining Kepler's 3rd Law
Grade Level : High School
Math Level : Challenging

Kepler's Law Calculations Activity:

Ms Briana Andromeda Purcell here and this Activity begins with an essay I created discussing the importance of logarithms and their use by Kepler in uncovering natural science laws and rules we still employ today. With his 3rd law there is a way to uncover what he found to be true of the period of an orbiting body and its distance from the main body that it orbits using a slide rule and employing the L scale.

Kepler's contribution to Logarithms and the first application of them in Astronomy

Many a student and scholar has heard the name Johannes Kepler who is famously associated with laws bearing his name, Kepler's 3 Laws of Planetary Motion, but there is much more to this gifted mathematician, scientist, and astronomer than is summarized in a few pages of a descriptive astronomy course text. One of the key note features of his life is the contribution to logarithms as a math form to employ and its use in deriving his still used laws. The following explores Kepler and connects him to science and math as well as explores a way to examine his laws and any others for that matter in the classroom through logarithms.

Johannes Kepler (1571 – 1630) is called by many today as the last astrologer and the first astronomer. Along with that he has a number of remarkable firsts one might not know : he is first to investigate and describe the formation of pictures with a pinhole projection system, pave the way for image formation description with ideas of real, virtual, upright and inverted images from lenses and dealing with magnification, recognize and explain that depth perception comes from the use of both eyes, describe the process of our vision via refraction, formulate designs for eyeglasses for both nearsightedness and farsightedness, and explain how a telescope works. It was his works in Astronomia nova (the New Astronomy), Harmonics Mundi (The Harmony of the World), and Epitome Astronomiae which contains his laws that Newton read and acted as the spark to explore the more refined laws of motion to explain both celestial and terrestrial mechanics and leading to the description of the universal law of gravity which can be used to derive all of Kepler's Laws. In addition to these things and others is his derivation of logarithms purely from mathematics and their use in analyzing Tycho's data that was used to derive these laws.

A summary of Kepler's Laws :

1) The orbital paths of the Planets are Elliptical with the Sun at one focus.
2) A radius vector from the Sun to a given Planet will sweep out an equal area in an equal amount of time . (AKA the Equal Area Law)
3) The square of the Planet's Orbital Period is proportional to the cube of its semimajor axis distance. (This distance can be considered as the mean distance).

The importance of these laws, when taken in conjunction with Galileo's observations through a telescope in 1609, finally topples the long-held geocentric model of the universe in favor of the heliocentric model. Kepler's 1st Law eliminates the notion of circular orbits and smaller circles on these circles to compensate for errors in measurement. Kepler's 2nd Law for example removes the idea that planets move at uniform speed. For the Equal Areas in Equal Times to work for an Ellipse, the planet moves fastest nearest the Sun and slowest when farthest from the Sun. Newton's Law of Gravity gives the cause to this effect. Beyond this when combining Kepler's 3rd Law and Newton's Universal Law of Gravitation and examining binary star systems one can determine relative masses of stars. When applying this idea to our Solar System we are able to determine the mass of planets and the Sun too. Wow Science from those times paving the way to space, planets, and stars which began with curious minds.

Kepler's first two laws were published together in Astronomia nova (A New Astronomy) in 1609 and had taken him years to arrive at by examining many mathematical models and restarting the process over and over. His dedication to calculation matching reality cannot be overlooked, though. His math models at one point only differed by 8 arcminutes (the full Moon is approx. 30 arcminutes) from Tycho's observations. He would not accept this. This process of focus on actual measurements and objective diligence is what we call the 'scientific method' today. These years of calculation was called by him his 'war with Mars' since Mars was his primary set of data under consideration. Had logarithms been available, these years of processing could have been done in hours by an individual.

Still motivated, Kepler felt that there must be a connection between all of these numbers. This third law was even more tiresome and was not published until 1619 in Harmonices Mundi. An interesting question arises, how did the notion of the square of one variable value for planetary data being equal to the cube of yet another variable seems surprising to say the least. To help frame this time, realize that John Napier of Scotland had published in 1614 his work on logarithms called Mirifici Logarighmorum Canonis Descripto. Napier is noted to have written in the instroductory comments to his own tables " a logarithmic table is a small table by the use of which we can obtain a knowledge of all geometrical dimensions and motions in space (from http://www.mathpages.com/rr/s8-01/8-01.htm p. 3) It is reported that Kepler had seen this work by 1616.(http://www.mathpages.com/rr/s8-01/8-01.htm on p. 2) Kepler's own interest in logarithms is even revealed in his 1621 publication on them. Further examination this history reveals the following, though: In 1620 in Kepler's Ephemerides he not only republishes his 3rd Law but also dedicates it to Napier (who had died in 1617). Further Kepler was working on the Rudolphine Tables and published with them in 1628 were not only the astronomical tables to predict star and planet positions were tables of 8 figure log tables as well as a response to the mathematican Maslin who admonishes Kepler for his use of this new idea in math since no one was clear on its validity. Kepler uses Euclid's Elements Book 5 to demonstrate clearly the mathematical validity of logarithms. Further the tables generated in the book upheld Kepler's ideas since they accurately predicted observational outcomes for many decades hence adding to the strength of the heliocentric theory. (A quick aside is the fact that in the time frame of 1620s to 1630 with 1620 and Gunter's Scale, and around 1622 with William Oughtred's creation of his Circles of Proportion otherwise known as the Slide Rule based on logarithms and their math laws).

It is here then that much credit should be bestowed upon Kepler for his dedication not only to science but to mathematics and the opening of the door to the world of logarithms which can be considered the prime power in calculation that comes to all parts of the now-emerging industrializing, voyaging, and investigating world. Logarithms as noted by Florian Cajori states : "The miraculous powers of modern computation are largely due to the invention of logarithms" (p. 1) which begins his work on the history of the prime tool to come from them, the Slide Rule in the book: A History of the Logarithmic Slide Rule and Allied Instruments (first published in 1910). All fields of science, mathematics, engineering have benefitted from them.

As an educator I am interested in sharing all of these ideas, but too much information, too many lectures can lead to a lack of interest and even understanding. Also it appears that Kepler's 3rd Law seems to come from nowhere and has no rationale whatsoever. Instead why not provide (or have access to) a table of basic planetary data such as their distances from the Sun and periods of revolution about the Sun. That is the basis of the following Demonstration of Kepler's 3rd Law Activity. This same idea appears in my book of activities to be done by using a slide rule – that is this book here : **Galaxy Girl's Inquisitive Pioneer Adventures** by Briana Andromeda Purcell and also found with many other ideas at my website : http://galaxygirl-brianaandromedapurcell.com/index.html.

Students can then be lead with as many directions as deemed necessary to investigate these numbers. First choose a planet to act as one to compare to. The Earth is a natural choice. Then put all of the data in terms of the Earth. The use of either log tables or the slide rule for this activity can be very useful. In the case of logarithms, since the value is a ratio, one only has to subtract the table values of numerator and denominator to find a solution to look up and put down the antilog of in the table. The level of the student will determine where to place the work here.

Next, take this relative table of values and look up their log values (or again use a slide rule). Since the Earth was chosen as the base to measure from, the planets Mercury and Venus will have negative values. The Earth will become the values of zero and zero since it is the log (1) for each of the values and will be the point (0,0) on the upcoming graph of this data.

Now graph the Log of the Orbital Period (Earth Years) versus the Log of the Semimajor Axis (in Astronomical Units). This can be done by hand or done in Excel for teachers wanting to employ computer technology. In fact the calculations can be done with Excel as well. I suggest the old-fashioned way. Some data points are Venus (-0.41, -0.211), Earth (0,0), Mars (0.183, 0.274) for example. Drawing a best fit line and computing the slope of this linear line by choosing two points reveals a value of 1.5. (Be careful with Excel as to which column is the Y and which is the X axis and the computation of the slope from this- my results yield 1.497 so at 2 or even 3 significant figures it is 1.5).

Since this the ratio of log of period (P) to log of semimajor axis (A) this means that the numerical ratio is 1.5 to 1 means 3 to 2 in terms of whole numbers. Use cross products and then raise each side to the power of 10 to eliminate the log notation, we find the cube of the semimajor axis is equal to the square of the orbital period ($A^3 = P^2$) just as Kepler had found!

The great thing with astronomy and physics are that there are a number of logarithmic relations that can be explored in such a manner, such as the relation of distance to apparent magnitude and absolute magnitude ($m\text{-}M = 5*\log_{10}(D/10 \text{ parsecs})$), star mass and luminosity ($L \propto M^4$), luminosity and radius ($L \propto r^2$), luminosity and surface temperature ($L \propto T^4$), mass and star lifetime (Star Lifetime $\propto M^{-3}$). In fact any and all data tables can be graphed in the log fashion employed here to uncover the relations between variables and derive the needed equations. This can help to develop math skills and connect them to science and technology. In conclusion, Kepler clearly has a place at the table of not just scientists, but also mathematicians and was critical in the realm of logarithms to their calculation, mathematical demonstration of, and practical connection of them to natural phenomena.

- This next activity is spaced-based. In Astronomy there are some useful general relations for bodies (stars and/or planets) that orbit each other called Kepler's Laws.
- You may recall the reference to him in the above article and his contributions to logarithms.
- I go through how to use the slide rule and reading the scales to find the various values, but you can go right ahead and use the table afterwards and find the log value for each period and distance value, plot these as points, find the best fit line and determine the slope which in turns yields the power relation of these variables.
-
- From the Earth we know it takes 1 year for the Earth to orbit the Sun and watching carefully, we find that it takes **Venus 0.615** of our years and **Mars 1.881** of our years to orbit the same star.
- Kepler's Third Law basically states that the square of a Period (P) (the amount of time to orbit another body) is directly proportional to the cube of its semi-major axis distance (A) (here we will use average distance which is a good approximation for orbits that are not too eccentric).
- So our formula looks like this :
- **A^3 is proportional to P^2**
- ***Our question is thus : In terms of Earth's distance from the Sun, how far away are Mars and Venus respectively from the Sun?***
-
- We can make the approximately equal sign into an equal sign by using the Earth as the comparison and setting the problem up this way :
- A^3 (Mars) / P^2 (Mars) = 1 = A^3 (Earth)/ P^2 (Earth)
- So **$A^3 = P^2$**
- (When dealing with units relative to another body, such as the Earth)
- Solve for each of the Planet's Relative Distances in turn.
- We will start with **Mars**.
- We were given the Period of the Planet at 1.881 years. We can only find 1.88 on our D scale which is where we begin.
- Look from 1.88 on D scale to the A scale to find the Period squared. The value falls just short of 3.55 on A scale, so we can say 3.54.
- Now we need to find the cube root of this number.
- The formula states that the square of the period (which we have) equals the cube of the planet's distance.
- We now need to use our K scale as this is the cubing and cube-rooting scale.
- Since the squared value equals the cubed value, we have that, we need to find the cube root of 3.54 from the K scale.
- But first a problem?
- Which one of the 3.54 values, there are 3 of them as you read across the K scale since it goes from 1 to 1 a total of 3 times!?

- Much like for the A scale where the numbers go from 1 to 10 and then 10 to 100, the same principle holds for the K scale, only here it is 3 times. So it goes from 1 to 10, then from 10 to 100, then from 100 to 1000.
- So we find 3.54 between the first and second 1. Be sure to estimate the placement answer to 4 place values (first three are significant and the fourth is estimated).
- Read from it down to the D scale to take the cube root of this value. It aligns with 1.525
- This is our answer in terms of Earth's distance from the Sun, which means mars is on average 1.525 x farther away than we are.
- (The given actual answer is 1.524, which places our value in very good company.)
-

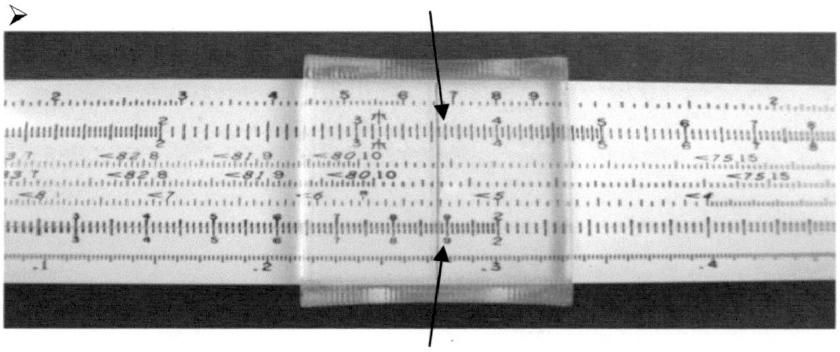

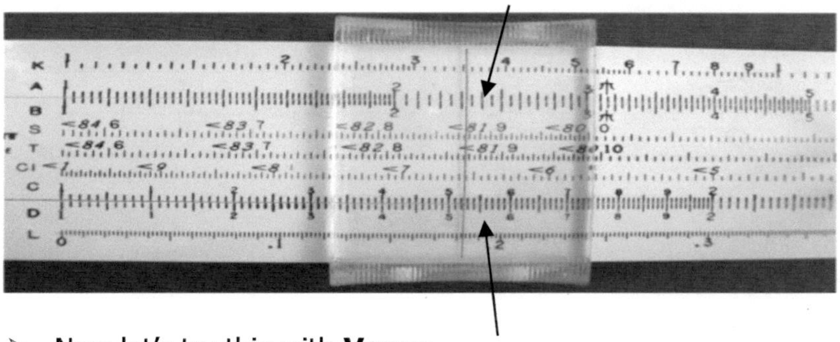

- Now let's try this with **Venus**.
- Start on D and disregard the decimal for now. Find 615 and read the square of it on A and come up with 378.
- But now the decimal. Realize that 6 x 6 is 36 and if we think of the number in scientific notation, 0.615 is $\mathbf{6.15 \times 10^{-1}}$. We have to square this number. For the exponents, we add the exponents and come up with -2, so our answer is $\mathbf{37.8 \times 10^{-2}}$ or 0.378
- But how to read this on the K scale. Recall I mentioned 1-10, etc, what of decimals?!
- Think of what it is : 378 thousandths. We can use this fact to look at the scale, so we look at the K scale between the last two ones, or in the last third of the scale. This is the place between 100 and 1000.
- Find 385 there and read down to D scale once again (see photo) and find the value. 722
- We read this as Venus having a distance that is 0.722 of Earth's distance to the Sun
- (the actual standard answer given to this level of accuracy is 0.723).

-
- **LOG-LOG PLOT ACTIVITY**
-
- Given the process of how to do this, now turn to the available table below of the planets known to the ancients and take the log values of both the period and distance for each planet.
- Treat each as a pair of data points (log of distance, log of period) and graph these on a x-y axis.
- Draw a best fit line and then determine the slope of this line.
- You should find that in whole number ratio it should come to the ratio of 2 to 3 since these are the powers of those values.
-
- ***ASIDE***
- *A further exercise, if you wish, is that the Earth's average distance is 150 million kilometers or 93 million miles.*
- *Given our results, we can easily compute the numeric distance values for these planets.*

Orbital Data

	Semimajor Axis (10^6 km)	Semimajor Axis (Earth = 1AU)	Orbital Period (d)	Orbital Period (Earth years=1)
Mercury	57.9	0.387	88.0	0.241
Venus	108.2	0.723	225	0.615
Earth	149.6	1.000	365	1.00
Mars	227.9	1.523	687	1.88
Jupiter	778.6	5.205	4332	11.9
Saturn	1434	9.586	10761	29.5

Conclusion :

Super Cosmic! What does your graph show – Isn't it incredible that all things have a story to tell us – we have to be creative in thought, curious in nature, and passionate about our need to explore, to learn, and be Brilliant!

Explore & Enjoy

Activity #51
Spectrum Analysis
Grade Level : High School
Math Level : Calculating

Hey there Inquisitive Pioneers! Ever see a Rainbow – or when soap bubbles have a rainbow pattern in them or when glass and plastic pieces designed to break up light into its spectrum produces a rainbow spectrum cast near or even on us? Here we will break up light as did Sir Isaac Newton and begin to examine it itself – each of these colors as best we can with what resources we have available and reach some general conclusions and ideas about light, its energy and the science behind it all – which is the story of nature that we explore. Explore & Enjoy! Be Brilliant!

Many of us have read or heard about Sir Isaac Newton's work with the spectrum of light. He used a glass prism to separate light into its usual rainbow pattern we are all familiar with. What made his observation notable, though, was that he put other prisms in the path of each of the colors to find that you could not separate the light further. He rightfully concluded that white light from the Sun was indeed a combination of these colors, we often use the acronym false name of : ROY G BIV (Red, Orange, Yellow, Green, Blue, Indigo, Violet).

This spectrum of color is not the only components of the electromagnetic spectrum given off by the Sun, however. Many of us have heard about and encountered in one way or another other electromagnetic waves with names such as Ultraviolet and Infrared. To be 'ultraviolet' literally means 'above violet' while in the case of 'infrared' means to be 'below red'. From these descriptions it is easy to see that they are directly above and below the color spectrum that we can visually perceive.

This Activity will explore the Infrared and whether or not we can detect it through temperature measurements, though we cannot see it.

The Infrared portion of the spectrum today has many uses. We use infrared cameras in military, police, and security areas in order to locate people and animals in the dark due to their heat signature patterns. These cameras are used in heave smoke and fire situations to locate people and animals as well. Infrared imaging is used to examine heat loss from structures such as buildings and machinery. This is to study for defects, inefficiencies, and stresses. Infrared imaging data analysis is used in meteorology for looking at ocean currents and convection currents in the atmosphere. This is for weather both in the short term and long term. When used in imaging the Earth, it can be used to uncover ancient roadways and civilizations. While other analysis of the Earth data can reveal minerals as well as vegetation pattern. It is used in medicine to examine the body in a non-intrusive way. In Astronomy infrared imaging is used to analyze various objects and phenomena such as being able to penetrate clouds of dust to examine cradles of star formation.

The beginning of the journey in the Infrared can be traced back to Sir Frederick William Herschel (1738 – 1822), who was an astronomer who worked with his sister Caroline making catalogs of celestial phenomena such as nebulae and double stars and who also discovered the planet Uranus (1781). In 1800 he decided to follow up on Newton's spectrum work and was interested in how much heat energy did each of the colors of the spectrum contain. Much like our Activity, he used a glass prism to separate the spectrum into its respective colors and used thermometers with blackened bulb ends to see the effect of each portion of the spectrum on the change in temperature in a given time. To have a control in his experiment, he placed one outside the range of the visible spectrum. He found that each of the colors did indeed change the temperature more so than the control as would be expected. However there was an interesting pattern that emerged. From violet to red the temperature kept increasing and more surprising was when he placed the control thermometer below red it was the warmest of all and had the greatest change in temperature. In essence, he had uncovered the infrared portion of the spectrum. He first called these invisible rays 'calorific rays' from the Latin word for heat. Today, of course, we call them Infrared Rays. His further experiments should that these rays had the same characteristics of other waves – they could be absorbed and transmitted, they could be refracted, and they could be reflected.

Our goal is to do a similar experiment to Herschel's and examine the infrared portion of the spectrum as well. Enjoy :)

Purpose : To use basic materials to analyze the Sun's spectrum of light while examining the possibility of other electromagnetic waves passing through a prism from the Sun that are not visible (in this case infrared) and to measure the rate of change of temperature each portion of the analyzed spectrum causes.

Materials :

- Glass Prism (only glass, plastic is not effective, about 4" or more),
- Cardboard Box (best size is box size for photocopier paper),
- Alcohol-type Thermometers (up to 5 is best for the Activity),
- Black Paint or Permanent Black Marker,
- White Paper,
- Stopwatch or timer (measurements every 2 minutes),
- Graph Paper,
- Scissors and/or cutting tools for the cardboard box,
- Tape,
- Stack of thin books or magazines (for propping the box with regards to the Sun),
- Pencils, Rulers, other (for propping up thermometers in box to read),
- Table (card table for outdoors),
- Slide Rule

Note : This Activity is conducted outdoors and requires as clear of a sunny day as possible for best results.

Note : Do not look at the Sun. Be sure to have parental permission and supervision in the set up and in the Activity. For example, cutting the box for fitting the Prism will most likely require parental help and efforts.

Note : Though the number of Thermometers seems quite large and may be a problem in obtaining, one could still use as few as three if needed. See Set Up for more information.

Set Up :

1. The basic conclusion to the Set Up is as follows :
2. In using the completed spectrum device : It is best to have a table outside with the box on it.
3. The following gives general directions that you need to create a strategy for in accomplishing. It is best to refer to the photo below the Set Up for reference.
4. In the upper edge of the box on one of the short sides is a place cut out to fit the Prism in a snug fashion (can use some tape to help hold it in place).
5. It is best to have the prism towards the corner of this short side since you need to fit the thermometers inside the box and have the bulbs of the thermometers in the spectral light.
6. It is best to test the prism to see how it needs to be seated so as to allow light to pass through. Note that as time changes, you will have to change the angle of the prism since the Earth is rotating and the Sun's position is therefore changing.
7. Inside the box on the table or box bottom (depending on whether the box is inverted or not) are the thermometers all parallel to each other and each set in the spectrum portion under investigation. If it is not inverted, then it must have a lid since you want only a portion of sunlight coming from the prism to strike each of the thermometers.
8. On one of the long sides of the box is a hole cut in it to view the thermometers so that they can be read (may need a flashlight). Recognize that it is best to have the thermometers slightly angled so as to be readable (not too sharp of an angle). (See photos of drawings).
9. Note the size of holes depends on your arrangement and size of the prism. If the cardboard of the box is not too supportive of the prism, be sure to employ a small amount of tape (that is not across the place where the light is coming through). Other suggestions might be rubber or plastic stoppers or other pieces of cut cardboard or folded index cards cut in small enough pieces.
10. The thermometers have their own preparation. They need to have their bulbs blackened in appearance to facilitate absorption of light. This is done by either the black paint (can use black nail polish) or permanent black marker you have.
11. Note : The hole to see the thermometers is best to make it like a window that can open and close between readings (just cut 3 sides and fold). Recognize that you will always have to conduct the experiment at a similar time of day when first testing the box if your windows are too small in width.
12. In the completed set up, the bottom of the box should be lined with white paper.
13. An important decision comes from what part of the spectrum is to be analyzed, which depends on the quantity of equipment available – namely the number of thermometers. Clearly it is best to have one that will be in the Infrared region and in any good experiment it is best to have a Control to measure against for comparison which is the thermometer at the far end of the box out of the spectrum, but it is also good to have at least a third thermometer for at least one of the color portions of the spectrum (say red or blue, for example). – Realize though one can conduct the experiment a number of times in a short time frame and do different portions of the spectrum with your given supplies.

14. It is best to experiment and test the box and complete set up several trial runs without data to have a good idea as to a plan to conduct the experiment and record data. For example, there will be a considerable effort to find just the right position for the prism so as to catch the sunlight at its given angle (due to time of day) in order to project a spectrum on the floor of the box. – Also determine what is best to use to prop up the thermometers to read them easily, as well as slightly propping up the box when it is facing the sun in order to facilitate the sunlight passing through the prism. Essentially it is practice, patience, and focus. Enjoy :)

Photo of Set Up :

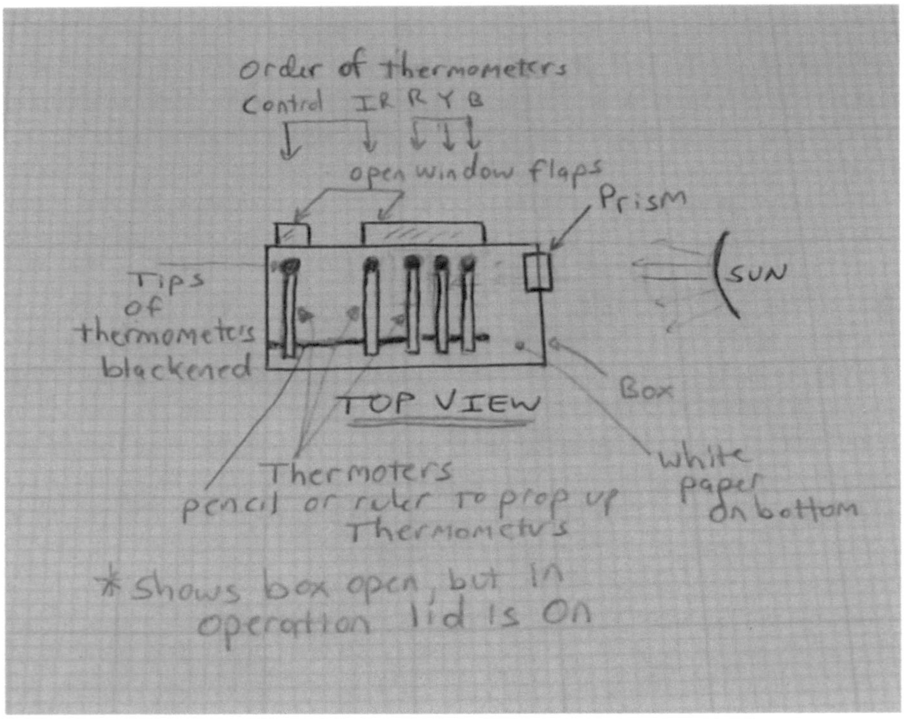

Procedure :

1. With any and all Activities read through, then set up the process for the Activity and run it without measurement and watch where you may have bumps in the road so as to anticipate these. Be patient – Repeat processes. Once all is well when measuring see to the values – do you notice patterns or trends – what do you predict will happen next as you change one thing or another from one trial in an experiment to the next. Always Explore & Enjoy. Be Brilliant!
2. At this point you have read all of the materials, even the Procedure at least one time through. You have constructed your box to use a prism to separate the sun's light into its spectrum and have tested it from the Set Up section.
3. Always be sure to have parental support, permission, supervision, and help whenever necessary. Do not look into the Sun. Be sure to be careful when in the Sun, such as wearing sunblock.

4. To initiate the Activity, have all the materials set up and ready to go – sunny day, table, spectrum analyzer box all set up.
5. At this point though you have put the thermometers in the box, it is entirely sealed (that is to say, block any incoming light) to allow the thermometers to reach an equilibrium temperature with the outdoors. This may take 10 or more minutes.
6. Record the initial temperatures of each of the thermometers (they may be slightly different) on separate scratch paper, calculate an average and place this answer in the Initial Temperature reading space on the Data Table.
7. Set the box so that the incoming light produces the spectrum you want and need to place your thermometers where they belong. Record each of the thermometer's temperatures separately for time 0 in the data table.
8. Start your timer and every two minutes record the temperature reading from each of the thermometers in the data table. Do this for at least 6 minutes, 10 is preferable.
9. With all of your readings accomplished, you can put away your spectrum analyzer device.
10. Calculations :
11. For each of the Thermometers separately determine the Average Temperature for a given portion of the spectrum, or the Control.
12. Graph Time (x-axis) vs. Temperature (y-axis) points for each of the thermometers separately.
13. For each thermometer data set of points, draw a best fit line through those points (recognize it will intercept the y-axis at a point and not 0). There are as many lines as there are thermometers.
14. For each line, determine the slope of that line.
15. For comparison, look at the various colors studied along with the infrared (IR) and the Control slopes and note any conclusions you might reach. (Which is rising at the fastest rate?).
16. As with most experiments, it is oftentimes best to repeat this study for comparison to other days.

Data :

Initial Temperature of Thermometers (Average) in shade for 10 min. ________

Part of Spectrum		Temperature at	This time :		
	0 minutes	2 minutes	4 minutes	6 minutes	Average Temp.
Blue					
Yellow					
Red					
Infrared					
Out of Spectrum - Control					

Calculations :

Be sure to use Your Slide Rule! :)

Slope :

$$\mathbf{m = \frac{\Delta Y}{\Delta X}}$$

Average :

$$\mathbf{Xave = \frac{\sum x}{n}}$$

Conclusion :

Cosmic Light Activity here!
Wow! All things certainly have a story to tell us – we have to be creative in thought, curious in nature, and passionate about our need to explore, to learn and be Brilliant!

What do your results show you about different parts of the spectrum that we can see in terms of the rate of change of temperature per unit time? What would it mean if a given part of the spectrum warms something up faster?
What did your experiment show for areas outside the visible spectrum? Can you conclude that there are regions beyond the visible – if so, how?

Activity #52

Grade Level : High School
Math Level : Challenging

The Efficiency of a Complex Electrical & Mechanical System

Hey Inquisitive Pioneers! Ms Briana Andromeda Purcell here and this Activity is a combination of many of the Activities found in my book : Galaxy Girl's Inquisitive Pioneer Adventures by me. It looks to ideas in Acceleration on an Incline (# 22), Mechanical Energy (#27) and others including the possibility of using electricity and light too plus heating activities. We are making a system of Cause-and-Effect situations so that we have input energy into our system and determine how much is 'lost' at each step so as to estimate how efficient our system is. Awesome Sauce!

Efficiency is simply the Ratio of Useful Work Output to Work Input. It is a natural consequence stemming from the Laws of Thermodynamics (see associated Rate of Heating and Cooling Activities #s 31-37). Often in our Activities we have assumed a perfect transference of energy, such as gravitational Potential Energy to Kinetic Energy (such as in the Mechanical Energy Activity), but there is some loss of energy to heat and sound (such as the sound of a rolling marble and the small increase of temperature of the surface it rolls across). Reading the story of Energy in the Mechanical Energy Activity (#27) is a good guide for the right realm of science reasoning and thought to see this Activity from.

This Activity is much like a Rube Goldberg device, though far less sophisticated, where there are a series of steps. We use a light source (perhaps a desk lamp or a flashlight – torch to United Kingdom users, or the Sun itself) to power some solar cells which in turn is connected to a small electric motor which is attached to a pulley with a string attached to a bag of masses (marbles perhaps) to lift them a given distance.

Each step is its own system we have looked at in some manner in other Activities, such as the light striking the solar cells or the lifted marbles being then propelled down a ramp, but here we have combined these.

In each case there will be Work Input by some energy source and Work Output. The ratio of Output to Input ideally would be 100%, but in reality, will be less as our results will show. Better still, in combination the final efficiency of our system will be the combined product of each of the steps multiplied together. Test this idea for yourself.

Efficiency is a very critical component of our everyday lives both in real measured amounts and in the processes we engage in – such as when we go out in the car we do several errands and connect the steps in a loop so as not to do multiple trips or re-crossing our path. When it comes to mechanical and electrical processes we find ways to increase efficiency – such as lubricated joints, use of ball bearings in motors, and the like to reduce friction, hence reduce heat therefore wear and tear on systems and convey the energy from one system to another more effectively with less losses. The better we do this, such as turning the stored energy of coal into steam then into electricity that powers our homes and factories, then the less waste there is.

Purpose : To measure the Work Input & Work Output of various subsystems of a complex system (electrical and/or mechanical) to determine Efficiency at each step of the system with its energy transformations.

Materials :

- Bag of Marbles,
- Ruler,
- Meter Stick,
- Batteries (if needed for the flashlight – if using a flashlight),
- Known Lumens value Flashlight (the higher the better, 1200+ best) OR,
- Light Bulb (75 W or 100 W) and Desk Lamp (best choice - see Note),
- Solar Cell(s) (the more used the better the system – 3 is good),
- Small Model Electric Motor (1.5 V – 3.0 V, can operate around 80 mA),
- Alligator Clip Wires,
- Pulley that can attach to motor,
- Lab Quality Thermometer (Note : instead can use a multimeter with thermocouple),
- Mass Scale,
- Multimeter (possibly with various applications *),
- Tachometer,
- Oven Pad,
- Plastic Crate,
- Styrofoam Sheets (to cover in 2 layers the plastic crate), OR
- Cooler – (Styrofoam is good, small Plastic one with drain plug best),
- Light Socket with Wires and Plug,
- Timer,
- String,
- Tape and Duct Tape,
- Scissors,
- Goggles,
- Stack of Books,
- Rubber Bands,
- Slide Rule

Note : You must have parent permission and supervision in this series of Activities. Be sure to always act with caution and care. Throughout the process wear goggles, know how the equipment works, test materials, be patient, and be willing to redo the tests as needed.

Note : This set of Activities are a series of steps that begins with a light source (one of the following : the Sun, a lamp and light bulb system, or a very strong flashlight) that activates a set of solar cells which in turn are connected to an electric motor which turns a pulley with an attached string that lifts a small mass (small bag of marbles). In each step the energy going in (the work in as it is known) is measured / calculated and the energy out (called the useful work output) is determined. The ratio of the latter quantity to the former one is the efficiency of that step of the situation.

Note : This Activity has a number of steps and it may be best to test each individually to see it in operation and then join the subsystems to have a continuous stream of actions. Also this means you may have to run a number of trials to have it operate correctly.

Note : If using the lamp and light bulb system (hence the air calorimeter) be sure that your design does not overheat any surface. Exercise caution and let the system cool after operating for a given time. Do not place anything in contact with the bulb, especially when in operation.

Note : In the case of the desk lamp, it may be best to have a base socket for the bulb (see photo) and a plug that can attach (see photo). This like the other components of the activity require parent consent and supervision. This is best as compared to the desk lamp since it sits low to the ground, hence you may not need stacks of books and the like when using the air calorimeter.

Note : The air calorimeter has two choices : Either construct a make-shift system or use a sufficiently large cooler (the crate can be plastic or Styrofoam). The make-shift one uses a plastic crate (see photo) that has at least one layer of Styrofoam (can have two, one inside and one outside). In any case the key is this : when the light is positioned in the air calorimeter the active bulb must have clearance so as not to overheat any surface. Clearly do not touch the active bulb and let it cool sufficiently before moving. The goal is to have a system that can be 'closed' off and a temperature-measuring system is inserted and read for a short period of time as the air temperature changes (such as with a thermometer or using a thermocouple attached to a multimeter).

Note : * The recommended Multimeter has the conventional applications such as AC & DC measures for voltage, current, as well as for Resistance, but can also have a Photometer for light intensity (and is actually needed if using a strong flashlight as noted above). The model : Mastech MS 8229 is very well suited for this.

Note : If using the lamp and bulb system, you do not need a photometer in your multimeter. The light output will be determined indirectly with an air-calorimeter noted in the directions below.

Photos of Set Up and Other Components :
Note that the last photo of the Air Calorimeter is a work in progress

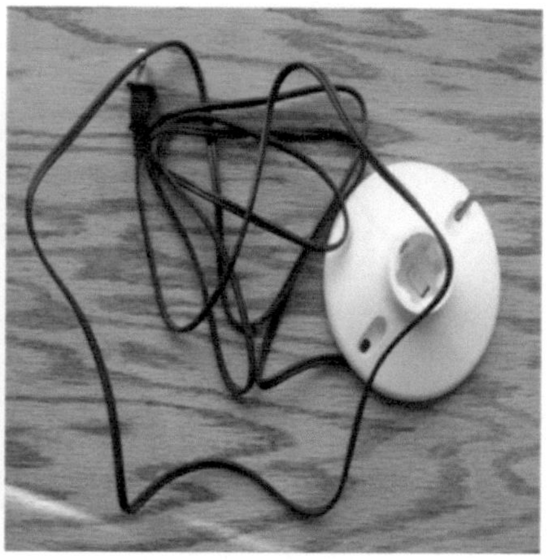

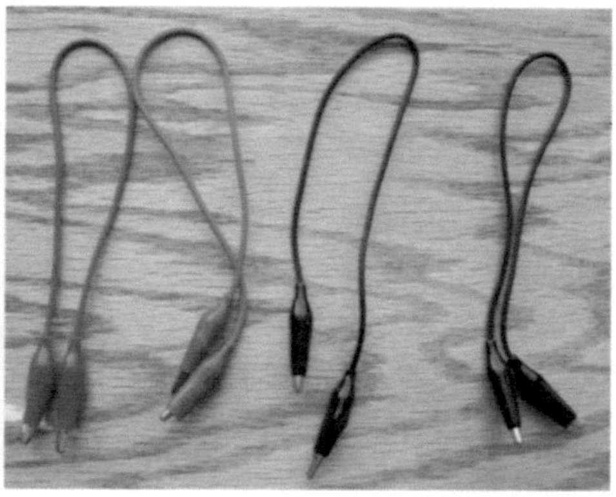

Procedure :

- With any and all Activities read through, then set up the process for the Activity and run it without measurement and watch where you may have bumps in the road so as to anticipate these. Be patient – Repeat processes. Once all is well when measuring see to the values – do you notice patterns or trends – what do you predict will happen next as you change one thing or another from one trial in an experiment to the next. Always Explore & Enjoy. Be Brilliant!
-
- **SET UP :**
-
- Note : With all of the Activities noted here – Always have parental permission and supervision. It is best to wear goggles. Always employ safe procedures.
- Note : All the Set Up and Measurement Procedures only set up the materials and take measurements, and in each section do the calculation for power in each section then in the Data Analysis section following all of these is where the efficiency for the data is addressed.
- This means that you may have to read the directions in a manner that satisfies what you are doing and where you are at. For example, if you had done Activity 1B (as an example illustration), then there is no need to read 1A nor 1C.
- Next you can then do computations for 1B in the Data Analysis OR continue in sequence for each of the Activities (2, 3, etc) until set up and measurements are completed.
- Note : You may have to do repeated trials and be sure to test each step of the path before having it operate. It is best to look at each component as a step. –
- In the case of multiple trials, it is often best to do a given trial as many as 5 times. Then eliminate both the highest and lowest values and then take the average of the other 3 where this average acts as the intended data measurement value.
-
- NOTE : Your light source determines which Activity 1 (A,B,C) you do :
- Choosing the subset of Activity 1 is determined by how you are able to measure the light source being used :
-

- If you have a Photometer, then you can do any of them,
-
- If you do not have a Photometer, then you must do Activity 1B,
-
- If you use a very strong flashlight, then start with 1A,

- If you use a light bulb and lamp, then start with 1B (recommended),
-
- **<u>Set Up of the Air Calorimeter (used only with Activity 1B):</u>**
-
- Note : The initial Air Calorimeter is shown in the set of photos and is the last one. Note that this is not complete, yet.
- What is needed is an open grate crate with thin Styrofoam sheets (can use foam-core poster board instead).
- The Styrofoam is cut so as to line both the inside and the outside of the crate along the plastic.
- It can be held in place by duct tape.
- Note to leave enough to make a 'lid' as well.
- Special Note : If this takes too much time, money, resources, one could instead simply use a Styrofoam Cooler in place of all the aforementioned items!
- An even better choice is a small plastic cooler with a drain plug – the need for a hole is because a thermometer will be inserted to measure the change in air temperature with a lit light bulb inside.
- Returning to the Air Calorimeter Set Up :
- With either the Styrofoam-lined crate or the Styrofoam Cooler, now invert it and pick a corner through which either a Thermometer will be inserted OR where the Thermocouple will be inserted. It is best to make a hole with a thin pencil or better still a knitting needle if available.
- Re-invert the box and use a piece of Styrofoam to construct a partial barrier that can be held in place with tape so that when the Thermometer or Thermocouple is in the Air Calorimeter it is not going to be in direct line of light from the light bulb which will sit opposite the temperature probe slot at the 'top' of the air calorimeter.
- It can be seen that there is a reference to the 'top' in a quotation-mark manner. This is because ultimately the box will be inverted and placed atop the 'lid' which lies flat on the ground and on top of the 'lid' in the center of it is the light bulb in a small light fixture (see photos) and connected with an appropriate electrical wire system so that it can be plugged in.
- The first main measurements to make here are the Length (L), Width (W), and Height (H) of the interior of the Air Calorimeter space. Record these measurements. Measure these to the nearest $1/10^{th}$ of a centimeter.
- This completes the Air Calorimeter Set Up and initial measurements –
- The Air Calorimeter needs the Volume (V) calculated from it and when coupled with the Density of Air (ρ_A) will yield the Mass of air in the Air Calorimeter (m_{air}) (do this calculation).
- Use this Air Calorimeter calculation with Activity 1B next
-
- If you use the Sun, then start with 1C,
-

- **The goal of Activity 1 is to measure the power of the light source, which will be the Input Work for Activity 2**
-
- **Activity 1A : A Strong Flashlight as the Light Source**
-
- Note : In order to use a flashlight, it must be a very luminous type, such as 1200 lumens or more. These can be expensive, so it is recommended to do Activity 1B or 1C.
- Note : If using a flashlight, it is best to only use an LED type and not other luminous source lamps which can be very hot. The flashlights described here are strictly battery operated and LED type and this is the only type recommended.
- Note : Since most, if not all very luminous lamps are LED types, it would be very difficult at best to use a simple air calorimeter, so the best measurement system would be to use a photometer to determine lux output of the light and then to translate this into Power.
- Note : The fastest way to test whether or not you can use this path is to quickly test your solar cells – connect them to the motor and see if this type of light can power the system. If it cannot, then you need to try either another method or find still a brighter light (if accessible – note that the more luminous types cost more)
- The first thing to measure is the Power Input into the light source.
- Use the Multimeter to measure the batteries both for Voltage (V_{1A}) and Current (I_{1A}) when not connected to the lamp. Be sure to know how to operate the multimeter to take these measurements. It is best to have the batteries connected in the same manner as they are in the flashlight (in series essentially) and measure them operating together. Record these results. The results from these measurements will be calculated and are considered Work Input #1.
- Now to measure the flashlight output and translate the light output into Work Output #1 (which in turn acts as Work Input #2).
- The most effective measurement of measuring the light is to use a photometer.
- Have the flashlight positioned so that it has an effective cone of light and the multimeter photometer sensor will be centered facing it at 1.0 m distance.
- The key to the measurement is to have as little outside light as possible, so you must find a means to read the meter (most digital have illuminated screens) such as drawn shades, etc.
- The reason for 1.0 m distance is the fact that the lux (lx) meter will be measuring lumens/meter-squared (lm/m^2)
- Record this lux measurement (L_{1A})
- Move on to Activity 2
-
- **Activity 1B : A regular Electric Lamp as the Light Source**
-
- With the Set Up completed first for the Air Calorimeter (see above) now we can measure indirectly the Power Input and Power Output of the Light Bulb as a Light source in our Activity
- Note as with any bulb, do not touch it when active nor handle it for some time after it is has been active since it is hot. Do not place objects on or too near the active light bulb. Be sure to have parental permission and supervision.
- First measurements are to determine the power going into the lamp.

- We can simply take the measure of the light bulb, say 75 W (a good recommendation), as the measure of the input power to the system.
- This value will be considered the Work Input #1 (W_I) and can be written on the table. If you are not measuring the light bulb's input power directly (requiring a specialized multimeter) now move on to the light output measure as work output #2.
- Light output as Work Output #2 Measurement :
- The output power can be measured either directly, using a Photometer as we did in the flashlight case (only here we use a light bulb), so if this is your method, read the directions in section 1A and replace flash light with the light bulb and do the same thing.
- The other measurement method is indirectly by measuring the heat the light generates and use this to subtract from the input power to find how much useful power is turned into light.
- Side Note : It may be a good experiment on its own to try both methods to see how similar or different these values are and consider why they are different!
- Indirectly measuring Power Output of the Light Bulb using the Air Calorimeter :
- Note that the Air Calorimeter must be completed (see directions above) for this next section.
- Here the bulb is in the socket and the Air Calorimeter is placed over it.
- Initially do not activate it.
- Use either a lab quality Thermometer inserted through the hole in the top (note that it needs to have a barrier to block the direct light upon the thermometer, such as a piece of white poster board or Styrofoam) or you can use a thermocouple attached to a multimeter for temperature readings.
- Take an initial temperature reading with the light off (T_i).
- Now start the timer, and take temperature readings every 2 minutes for at least 10-16 minutes total time (t).
- At this point you have enough measurements to be able to compute the amount of energy given off as heat by the light and then with the time (and proper conversion factors) convert this into watts.
- This is done by creating a graph of Temperature (y-axis) versus Time (x-axis), draw a best fit line and determine slope.
- With the slope you can now determine the heat energy of the system by using the specific heat capacity of air and the volume of air with this slope value – determine this power value.
- Note : This is NOT the power output of the light that we are going to use. We need to take this value and subtract from the Power Input value for the light and the difference is the power output of the light! This difference is the power value for the light we are going to use.
-
- **Activity 1C : The Sun as the Light Source**
-
- We have two choices here :
- Either make an assumption as to the power output of the Sun OR
- We can determine the value of the power output of the Sun by doing the Solar Constant Activity noted in the book.

- Note that the Efficiency in Activity 1 cannot be computed since the input is considered to be the same as the output (though not technically true, one could do the computations of how much of the Sun's energy is given off as light as compared to the amount of energy generated by fusion in the core, but that is another matter)
-
- Regardless of which of Activity 1 (A,B,C) you have done, now move on to the rest of the Activities in sequence :
-
- **Activity 2 : The Light Source acting on the Solar Cell**
-
- Regardless of which light source you have used, you now have a Power or Work Output #2 for a given light source to examine in Activity 2.
- This value will now act as Power or Work Input #2 for Activity 2.
- That logic will continue for each of the Activities, where the Work Output from the prior step becomes the Work Input for the next step.
- When it comes to Efficiency for a given step, it becomes the ratio of the Work Output for that step over the Work Input to that step.
-
- Now we are going to examine Power Output#2 for Activity 2, which comes from the use of one or more Solar Cells.
- Note : It is best to use 3 solar cells rated at 1.5 V and connected in series with each other for later power needs of the electric motor.
- With the Solar Cells connected properly and receiving the light from the light source, now use your multimeter to do the following :
- First, with proper settings on the DC Voltage, and connections of the cells to the meter (it is connected in parallel with the cells) measure the voltage of your solar cells with the light acting on them.
- Record this voltage value (V_{SC})
- Second, now reconnect properly the multimeter so as to measure DC Current (it will be in series with the cells) and be sure to have it on the proper setting so that the circuit does not blow – it is best to begin at the highest setting and then turn down to incrementally lower settings and noting the values along the way.
- Record the current value (I_{SC})
- You can readily calculate Power Output#2 here and do so, but can easily continue to the next Activity (3) where this acts as Power Input#3
-
- **Activity 3 : The Solar Cell acting on the Electric Motor**
-
- From Activity 2 we now have the Power Input#3 (which is Power Output#2) for this Activity.
- Now connect the Solar Cells to your Electric Motor.
- Test it to make sure it operates correctly
- Note, it is best to have a stand for the Electric Motor. A good idea is a stack of heavy books where a ruler is sandwiched somewhere near the top of the stack and jutting out a couple of inches. The Electric Motor is then held in place by a rubber band or two on the motor. Be sure that as little vibration is going on as possible when it is in operation. Also do not run the motor too long in any of the Activities so as to prevent overheating.

- Once the motor is operational connect small cut-out thin poster board circular disk (about ¾" diameter) to the axle of the motor so that it rotates.
- Measure and record the diameter of this disk in centimeters (d)
- Use the black reflective tape that comes with your Tachometer and attach a small piece to one edge of the disk –
- And to promote balance, attach a small piece of masking tape of approximately the same size and mass to the other edge of the disk so that it rotates in a balanced fashion
- Before using the disk with the tape on it, you need to measure its mass (m_d) and record this value in the table.
- Let the motor operate and use the Tachometer to measure the angular speed of the system. (ω)
- Note : We are going to assume that the motor has 'no load' on it, though there is one, hence this will affect the outcome slightly. In our next set of measures we are placing a significant load on the motor, though.
- Now Use the mass, diameter, and angular speed to determine the rotational kinetic energy of the system (KE_R)
- In the calculation section we are going to use this calculation for rotational kinetic energy as the power of the system to determine efficiency.
- Note : We are then going to make some critical assumptions :
- We will assume that the rotational kinetic energy started at zero
- We will assume that in each moment of time the rotational kinetic energy is constant
- With the value of rotational kinetic energy determined, we can then say that the system has this output of energy per unit time (1 second) and hence this same number of watts as the Power Output#3
-
- **<u>Activity 4 : The Electric Motor acting on the Load</u>**
-
- The Power Input#4 is the same as the Power Output#3 from above
- Now to the Electric Motor attach a pulley system (solid rotating disk – which may be purchased, but if unavailable it is best to use poster board cut into 2 disks approx. 1-2" diameter which are glued onto a central foam-core poster board disk of a smaller diameter (½" to 1.5" depending on the outer disks) leaving at least ½" to the outer rims of the make-shift pulley) – On the pulley you make as well you might want to cover the foam-core area with a thin piece of masking tape.
- Next have a piece of string less than 1.0 m in length but it should be longer than 50 cm.
- The length will be affected by how far in the air the motor is based upon the use of the stack of books as noted in the prior Activity.
- Test to be sure the pulley system is operational
- With an operational system we need to do a few measurements :
- Measure the diameter of the inner foam-core disk (ds) and record this.
- Measure the mass of the entire pulley disk system (mp) and record it.
- Now attach the string to the pulley and place a small mass on it (ideally use a plastic sandwich bag with a marble or two)
- Again test the system
- With further tests continue to add masses (marbles) until the system can still lift it but it requires some effort.
- Note : You may want to have back-up pulleys and be sure to keep the tension on your pulley system (a piece of masking tape may help) if needed

- When the necessary mass is in place, measure the entire mass (M) on the scale and record this value for the load of the system.
- Before activating the motor, be sure to measure the distance it will travel from the ground to some given height (H) and record this value.
- With all in place, now Use a stopwatch and measure the amount of time (t) for a given trial for lifting the mass.
- Perform this measurement 3 times and average the time (t_{ave}) which is used.
- From the measurements, first :
- Calculate the new rotational kinetic energy of the motor using the pulley system by turning the linear kinetic energy of the system into rotational kinetic energy.
- The linear kinetic energy is determined from the distance traveled by the bag of marbles divided by the amount of time it takes for them to travel.
- The gravitational potential energy of the system is determined by the mass of the bag of marbles times the acceleration due to gravity times the distance (height) traveled.
- Here the change of gravitational potential energy (PE_g) over the amount of time (t_{ave}) as a power along with an additional factor of the rotational kinetic energy treated as a power of the new disk system will be used in combination to determine the Power Output#4 in watts for our system. (see calculation section for this efficiency calculation)
-
- **Analyzing the Data in All Activities :**
-
- Note : Each of the Sections will first be analyzed for their Power Input and Power Outputs as needed – this will be followed by the Efficiency Calculations for each of the Activities respectively since these are independent and require the Output Work from the prior Activity (which now acts as the Input to the Activity in question) and the Output in a given step is the measurement from that Activity :
-
- **Efficiency Calculations :**
-
- Initial Efficiency of Activity 1 :
- Note that this cannot be done if using the Sun
- For all other calculations one merely takes the ratio of the Work Output, which is the energy of the light source over the Work Input to the light source (such as the battery power or wattage of the bulb)
-
- Efficiency of Activity 1-2 :
-
- This is the ratio of the Work Output as determined from the Power of the Solar Cells as determined from the calculation of Power from the measurements of Voltage and Current to the Work Input which is the Work Output from the last Activity which is the Power of the light source
-
- Note : Independent of which light source used, you have determined its power which acted as initial work out #1 and now it will be used as Work In #2.
- In this calculation, use the readings from the Solar Cell (Voltage and Current) and determine Power from this. This Power will be used as Work Out #2
- Now use Work Out #2 & Work In #2 to determine Efficiency for this step of the Activity

-
- Efficiency of Activity 2-3 :
-
- This is the ratio of Work Output, which is the Rotational Kinetic Energy of the Motor being treated as Work to the Work Input of the Solar Cells to the Motor
-
- Efficiency of Activity 3-4 :
-
- This is the ratio of the combination (addition) of the gravitational potential energy of the bag of marbles lifted plus the rotational kinetic energy of the pulley disks of the motor to the input energy of the motor which is the initial rotational kinetic energy of the motor as determined in Activity 3
-

Data :

Air Calorimeter for Activity 1B :

Dimension	Measure (cm)
L : Length	
W : Width	
H : Heigth	
V : Volume	cm^3

Mass of Air in Air Calorimeter (m_{air}) : _________ g
Determined from air density and calorimeter volume

Activity 1A :

Flashlight :
Voltage (V_{1A}) : ___________ V
Current (I_{1A}) : ____________ A
Power (W_I) : ____________ Watts
Power from calculation and is Work Input#1

Distance of photometer : 1.0 m
Photometer Measure (L_{1A}) : _________ lux
Calculated Work Output #1 : __________ watts
Note that this is Work Input #2
This measurement is also for the Lamp if there is no Air Calorimeter

Activity 1B :

Noted Lamp Wattage (W_I) : ___________ watts
This can be Work Input #1

Calculated Work Input #1 (W_I) : __________ watts

Use of Air Calorimeter :

Time (min)	Temperature (°C)
0	T_i
2	
4	
10 – 16	T_f

Change of Temperature (ΔT) : __________ °C
Quantity of Heat added to Air Calorimeter : ___________ J

Calculated Slope of Line (ΔTemp / Δtime) : __________

Calculated Power Loss due to Heat in Air Calorimeter : ________ watts

Net Work Output of Light (W_{O1}) : ____________ watts
Calculated from $\mathbf{P_{Net} = P_{max\ input} - P_{loss}}$
Note that this is Work Input #2

Efficiency for Activity 1, 2, 3, 4 :

Note : To change this to all other Activities, simply change the 1 to 2 and so on as needed for each of the Activities (2, 3, 4)
This means you need to do this calculation for each Activity where the proceeding Activity is the Input Work and the Work measured in that Activity is the Output Work

$$\textbf{Efficiency} = \frac{\#\ ,\qquad ,}{\#\ ,\qquad ,}$$

Efficiency = _________

Activity 1C :

Use Solar Constant or Do Solar Constant Activity
This is Work Output #1 and is Work Input #2

Activity 2 :

From Activity 1, we now have Work Input #2

Solar Cells :

Total Voltage (V_{SC}) : _____________ V
Total Current (I_{SC}) : _____________ A

Total Work Output#2 (W_{O2}) : _________ watts
Acts as Work Input#3 (W_{I3})

Activity 3 :

Disk diameter (d) : ________ cm
Disk radius (r) : ________ cm
Disk mass (m_d) : ________ g

Angular Speed of Motor (ω) : __________ rev/s

Rotational Kinetic Energy of Motor (KE_R) : _________ J
Note : This is the assumed Power - Work Output #3
Note : This is the Work Input #4

Activity 4 :

Inner Disk :
Disk diameter (d) : ________ cm
Disk radius (r) : ________ cm
Disk total mass (m_T) : ________ g

Circumference of Disk (C) : _______ m

Length or Distance Traveled (H) : ________ m

Mass of Marbles used (M) : _________ kg

Potential Energy of System (PEg) : _________ J

Time to travel distance H (t) : ________ s

Calculated Linear Speed (v) : $\left(\frac{H}{t}\right)$: _________ m/s

Calculated Rotational Speed (ω) : _________ rev/s

Rotational Kinetic Energy (KE_R) : _________ J

Total Energy of System (E) : __________ J

Total Power of System (Work Output #4) : ($\frac{E}{t}$) : _______ watts

Calculations :

Be sure to use your Slide Rule!

Efficiency :

$$\mathbf{Eff = \frac{useful\ Energy\ Out}{Energy\ In}}$$

$$\mathbf{Eff = \frac{W_{out}}{W_{in}}} \qquad \mathbf{(W = Work)}$$

Power :

$$\mathbf{P = \frac{W}{t} = \frac{\Delta E}{t}}$$ (General Formula : W=Work, ΔE = Change of Energy, t=time)

$P = V*I$ (Electrical : V=voltage, I=Current)

$P = N*S$ (N = No. of Lux or L_{1A} and S is number of square meters)

$P_{Net} = P_{max\ input} - P_{loss}$ (Determines Net Power in Light Output)

Energy :

$W = F*d$ (General Formula for Work : W=Work, F=Force, d=distance)

$W = \Delta E$ (General Formula : W=Work, ΔE = Change of Energy)

$PE_g = m*g*h$ (Gravitational Potential Energy : m=mass, g=acceleration due to gravity, h=height)

$PE_{el} = ½*k*x^2$ (Elastic Potential Energy : k=spring constant, x=displacement of spring)

$KE = ½*m*v^2$ (Kinetic Energy : m=mass, v=velocity)

$KE_r = ½*I*\omega^2$ (Rotational Kinetic Energy = Moment of Inertia * Angular Speed)

$Q = m*c*\Delta T$ (Heat Needed-Emitted = Mass * Specific Heat Capacity of a Material * Change of Temperature)

Other Relations Needed :

$\Delta N = N_f - N_i$ (The change of a given variable is the difference of a final value and an initial value)

$$\mathbf{m = \frac{\Delta Y}{\Delta X}}$$ (Slope for a given line on a graph)

Area of a Square : A = L*W
Volume of a Rectangular Solid : V = L*W*H

$C = 2*\pi*r$ (Circumference from the radius of a circle)

$$r = \frac{d}{2}$$ (radius is half the diameter of a circle)

$F = m*a$ (Net Force = mass * net acceleration)

$F = k*x$ (Hooke's Law : Force = Spring Constant*Distance Stretched)

Speed :

$$V = \frac{\Delta d}{\Delta t}$$ (Linear Speed = Distance Traveled / Elapsed Time)

$$\omega = \frac{v}{r}$$ (Angular Speed = Linear Speed / Radial Distance)

Moment of Inertia Formulae :

$$I_{disc} = \tfrac{1}{2} * m * r^2$$ (m = mass, r = radius)

$$I_{rod\text{-}mid} = \frac{1}{12} * m * L^2$$ (L = length)

$$I_{rod\text{-}end} = \frac{1}{3} * m * L^2$$

Constants :

1 m = 100 cm
1 kg = 1000 g
1 lux = 1 lm/m^2
1 kg = 2.2 lbs
1 in = 2.54 cm
1 ft = 12 in
1 lb = 16 oz.
1 oz. = 28.3 g
1 cal = 4.184 J
1 Watt = 1 J/s
1 kW = 3,414 BTU/hr

Acceleration due to Gravity : $g = 9.8\ m/s^2 = 980\ m/s^2 = 32.4\ ft/s^2$

Specific Heat Capacity of Water

$$c = 1.0 \frac{cal}{g*^{o}C} = 4.186 \frac{J}{g*^{o}C}$$

Specific Heat Capacity of Air (at room temp of 25° C)

$$c = 0.242 \frac{cal}{g*^{o}K} = 1.012 \frac{J}{g*^{o}K}$$

Actual Solar Constant : 1.366 kW/m^2
Corrected Amount received at Earth for Solar Constant :
342 W/m^2

1 kW = 10^3 watts
1 kW = 3,414 BTU/hr
1 BTU/hr = 2.93 x 10^{-4} kW

Constant Used to Convert Light Into Power
1 lux = 1.46 x 10^{-7} Watts/sq.cm. (at 555 nm (green light))

1 lux = 1 x 10^{-4} lumen/sq.cm.
1 lux = 0.0929 lum/sq.ft. (also ft-candles)

Conclusion :

Wow – What a Cosmic Adventure this was – In many ways you have traveled through most of the book in this single Activity since it utilizes the ideas and Activities found in many of the sections of the book! Awesome Sauce!

Cosmic! All things certainly have a story to tell us – we have to be creative in thought, curious in nature, and passionate about our need to explore, to learn, and be Brilliant!

Explore & Enjoy!

One of the first set of questions to consider are these : What are the types of energy involved in your system? What type of energy transformations take place in this system? When it comes to efficiency, what becomes of the energy that seems to be 'lost' when going from one step to the next? How do you account for it (assuming conservation of energy applies)?

What do your results show for Efficiency for component parts? How do you think this relates to the whole system? Why do you think some subsystems were more efficient than others? Would do you think can be done (and perhaps tested) to change or effect efficiency of the whole system or some subcomponent of it? If the system were less complicated, would efficiency be a larger value or not (bearing in mind, would it depend on the components involved)?

Alternative Tests to consider for comparison (with parental permission and supervision) :

1) Try a different wattage bulb
2) Try a different type of bulb (fluorescent)
3) Try a different distance for a given bulb intensity
4) Try a different size solar cell, or more solar cells
5) Try a different engine size
6) Try a different number of marbles as the load to be lifted
7) Try a different initial energy input of your design and testing (with parental permission and supervision)

Activity # 53
Calculating the Solar Constant

Grade Level : High School
Math Level : Challenging

The Solar Constant, Solar Energy and the Slide Rule Activity –

Hey Inquisitive Pioneers! Ms Briana Andromeda Purcell here and this latest topic is the ever-so important and closest star to our home the Earth, namely the Sun! We will engage in a process to use simple tools, basic measurements and some math to find an estimate of just how much energy does the Sun give off! Wow! Imagine that – we will use a clock, some insulated perhaps Styrofoam cups, a measuring cup, thermometer, water and of course the Sun in our search for the power of the Sun! Explore & Enjoy and as always – Be Brilliant!

The Sun is the home star of the Earth and the whole of the Solar System. It is 1.4×10^6 km in diameter (approximately 109 x Earth's diameter), hence its volume could hold over 1 million Earths! The mass of the Sun is approximately 2×10^{30} kg (330,000 x that of the Earth).

Despite this the Sun is fairly average amongst the family of stars in the cosmos being classified as a G2V type. (Class ranges : O,B,A,F,G,K,M with each having 10 subcategories and the V is a Main Sequence star meaning it is in the prime of its life and generates energy from fusion of hydrogen into helium).

The Sun is composed primarily of Hydrogen (73.5 %) and Helium (25%) with the remainder being many common elements such as Oxygen, Carbon, Iron, Sulfur, and Nitrogen.

The Sun is not only radiant in visible light spanning the spectrum (ROYGBIV) as uncovered and described by Newton as being the components of white light) but also emits radiations at many other wavelengths, such as Radio Waves, X-Rays, Infrared Radiation, and Ultraviolet Rays.

In this Activity we examine the Sun's energy output directly.
Where does the Sun get all of this energy? Like other main sequence stars, the Sun has nuclear fusion taking place in its core. In this process in the 17 million degree core the Sun converts 630 billion kg of Hydrogen into 625.7 billion kg of Helium (a difference of some 4.3 billion kg of mass loss) per second. This results in a power of 3.9×10^{26} W given off by the Sun! Wow! How Cosmic! In time through the processes of conduction, convection, and radiation, the energy reaches the surface in some 20,000 years and even more and is then emitted into space as electromagnetic radiation. The photons (all of which that move at the speed of light 3×10^8 m/s) that reach the Earth (some 1.5×10^8 km) take over 8 minutes of time.

The importance of the Sun's energy cannot be overstated. It is the primary reason that the Earth is at the temperature it is at. The Sun drives our atmosphere and weather systems. Hence it is the energy source for all wind generators as well as ultimately those that use water to drive turbines since it is the primary energy source of the water cycle on Earth. All plants use it through chemical reactions involving chlorophyll to convert water and carbon dioxide into sugar molecules which enables plants to grow and act as the pivotal base of the food chain for all organisms on Earth. This means that all Sciences (Physics, Chemistry, Biology, Astronomy, et al) have direct and minimally indirect relation to the Sun and its effects on the Earth. Today there is a new look at the Sun and its power and its importance to our needs.

Like all measures in Science, the energy given off by the Sun is just one of those quantities. From measurements a term has been developed called the Solar Constant. It is a measure of the flux (the amount of incoming solar electromagnetic radiation per unit area that is incident on a plane perpendicular to the radiation at a distance of 1 AU (astronomical unit)). This measure is all forms of energy coming from the Sun and is not just visible light.

The current measured value of the Solar Constant is 1.366 kW/m^2 on the average (note it does vary with our place in the solar system since at times of the year we are closer while other times farther away from the Sun).

This value does and has changed over time. The reasons for this are not fully known. The science of the sun is multidimensional (studying mathematical models of the interior of the Sun, helioseismology, solar sunspot activity, the mystery of the missing neutrinos, and explanations of the superheated corona, and many other areas) and far from complete. Even in 2010 it was found that the Sun's surface can have cascading fluctuations in the magnetic field causing massive solar storms – none of which were known of or had been predicted before. This also includes coronal mass ejections which even caused an electrical power outage in Canada in 1989 since it affected the Earth's magnetic field which in turn affected transformers and power lines (recall from the Electromagnet and Electric Motor Activities that changing magnetic fields generates electric currents, which happened here).

When it comes to our measure of the Solar Constant, we are only looking at visible light and none of the other electromagnetic forms of radiation. Also we have to do this from the surface of the Earth instead of in space. This means in the Activity we are leaving out many of the other energy forms, but take note that the majority of the energy is in the visible light form, so our measured approximation is a good one. Our Activity also has to consider the fact that not all of the light reaches the Earth's surface (where our measurements are made) as well as the transparency of the atmosphere (is it hazy or not due to water molecules and/or dust) and where the measurements are taking place (the angle of latitude where one is at will affect the amount of sunlight that reaches the surface since it must pass through different amounts of atmosphere during the measurements). All of this is simplified by a 'constant' factored into the Activity to compensate for these things. More precision can be found by internet research for formulae factors to incorporate as well as more precise equipment and measurements.

Purpose : To determine the daily amount of solar energy that reaches the Earth's surface through the properties of the specific heat of water, the water's temperature change, measurement, and calculation.

Purpose : To determine an estimated value of the Solar Constant.

Materials :

- 2 Large Styrofoam cups,
- Water,
- Measuring Cup,
- Black Paint,
- Thermometer (lab quality),
- Clear Plastic Wrap,
- Flat Surface (large plate or pan or TV tray),
- Ruler,
- Paper,
- Tape,
- Clock or Stopwatch,
- Alternative Method Tools : Multimeter & Solar Cell,
- Slide Rule

Procedure :

Set Up & Needs :

1) The Radiometer (our solar energy collection device) is the two nested Styrofoam cups with the inside one having its entire interior painted black (be sure it is dry for the experiment).
2) The main need is a sunny day and the experiment is best ran in the middle of the day (between about 11 AM up to 2 PM).
3) You determine the angle to place your Radiometer at by placing the nested cups on a flat surface and elevating it so that at the time of the activity with the Sun shining on it there will be no shadow cast by the cup on the water in the cup. That is, it is directly facing the Sun.
4) For those who like numbers in this case, the angle at which the board will be to the level ground will be the same as the latitude of the location of the activity.
5) In the process of this set up determine how best to create an angled surface. A suggestion is a wood board or plastic TV tray (try not to use metal surfaces) as supported by books. Be sure to test this outside. You want the cup to hold in place, tape helps this.
6) On your angled surface place one of the Styrofoam cups and see how much water can be put in without spilling. The goal is to fill it in a way and at that angle so there is no shadow of the cup on the water's surface and the water does not spill out. (about 2/3 is good typically)

7) Measure this useful amount of water with a measuring cup. This will be the amount used in the Activity (V).
8) Note that the Thermometer will be placed in here too, so take that into consideration.
9) Always employ safety in all aspects of your Activity – do not look directly into the Sun.

Activity Procedure :

1) With any and all Activities read through, then set up the process for the Activity and run it without measurement and watch where you may have bumps in the road so as to anticipate these. Be patient – Repeat processes. Once all is well when measuring see to the values – do you notice patterns or trends – what do you predict will happen next as you change one thing or another from one trial in an experiment to the next. Always Explore & Enjoy. Be Brilliant!
2) About 30 minutes before the Activity fill the nested cups with a measured amount of water (V) as determined from the set up above.
3) Note it is best to use room temp water neither too cold nor hot).
4) Place them on the angled surface outside facing in the general direction of where the Sun will be in 1/2 hour from now. Do not take the temperature yet. The goal is to have the water be at the same temperature as the surroundings.
5) About 5 minutes to measuring time cover the cup with the plastic wrap. At this time insert the thermometer too but still no readings.
6) It is best to insert the thermometer and place it in a manner for easy reading (to be seen from the side so as not to block the Sun striking the water's surface).
7) When the Radiometer has been in place for 5 minutes take your first temperature reading (T).
8) Record temperature readings every 2 minutes for 30 minutes (or at least until there is a temperature change of about 5°C), Δt is all of the seconds elapsed.
9) Note that you may have to slightly shift the platform with the Radiometer on it to keep it in the sunlight since the Earth is spinning and the Sun therefore 'moves' across the sky.
10) When done with the measurements, carefully remove the plastic and thermometer.
11) Use the bright-colored paint (white, yellow, or silver) to mark a line on the interior of the cup at the water line all the way around).
12) Now you can dump out the water.
13) Use scissors to cut the interior cup down to the line you have drawn.
14) Invert this angled cut surface onto some paper (it can be regular or graph paper) and trace carefully the edge of the cup as close to the inner lip as possible without distorting the cup.
15) With the traced ellipse on the paper use a ruler to measure in centimeters both the long and short axes of the ellipse (A, B).
16) Calculations to Perform :
17) Though the Slide Rule is a recommended tool, all of these calculations can be done with a regular or scientific calculator. Some scientific ones even have built-in averaging formulae. For those who like spreadsheets, the data can be typed in and the formulae then also be typed in its own cell where the formula references each of the measured variables in their respective cells.

18) Use the Ellipse formula to determine the Area of the Water's Surface. Convert this answer to m^2.
19) Calculate the amount of mass in grams, then convert to kilograms, of water used. Note the volume should have been read in mL and 1mL of water equals 1 g of mass of water.
20) After determining ΔT, find the Experimental Solar Constant (Q) by plugging in all of your measured variables.
21) With the determined Solar Constant, find the percent error from the Known value of it. Compare this value to the corrected value in the table of information below.
22) Using the measured Solar Constant from the Activity, determine the amount of Power [N] (J/s or watts) reaching the Earth's surface itself.
23) Now determine the total Power [P] of the Sun's energy that reaches the Earth's orbit.
24) Since we have determined the total power (P) that reaches the sphere of Earth's radius we have indirectly found the total Power radiated by the Sun. How does this value compare to the known expected value for the incoming solar power ?
25) Alternative Comparison Method :
26) Use a Solar Cell (Photovoltaic Cell) connected to a Multimeter :
27) When angled at the Sun (be sure to not be behind glass indoors), measure the Voltage (V) with the Solar Cell (most are in the neighborhood of 1.5 V),
28) Now reconnect the multimeter to measure milliamps (mA). Depending on where you live, the size of the solar cell, the overall transparency of the atmosphere this value can possibly range to over 250 mA (be sure to note the range capability of your multimeter and have it set correctly for a reading).
29) From the Voltage and Amperage readings, calculate Power : $P = V*I$
30) Measure the surface area of the solar cell actually collecting sunlight (look carefully at the solar cell to find the area actually receiving and processing the light). Convert this measurement into square-meters.
31) Like the prior calculation for the Solar Constant divide the Power determined from the electrical readings by the area of the solar cell and as before factor in your correction factor (k = 0.5 in the denominator, which in turn is really multiplying it by 2 in the numerator) for you Experimental Solar Constant Value (Q).
32) This value is good for comparative purposes and an alternative method to the aforementioned activity!

Data :

Volume of Water : ____________ mL

Mass of Water : ____________ kg

Ellipse : A : ____________ cm
B : ____________ cm

Time (in Minutes)	Temperature (^oC)
0	
2	
30 (?)	

Calculations :

Be sure to use your Slide Rule!

Constants to be used :

1 kg = 10^3 g
1 km = 10^3 m
water specific heat capacity (c) of : 1 $\frac{Cal}{g*^oC}$ = 4.186 $\frac{J}{g*^oC}$

for calculations use 4.2 $\frac{J}{g*^oC}$

Whole Surface Area of Earth : 5.1 x 10^8 km^2
Half of Earth Surface Area (SA) : 2.6 x 10^8 km^2
Earth-Sun average distance (R): 1.5 x 10^8 km
Correction Factor for Solar Constant Formula (K) : 0.5
(note this can range from 0.05 to 0.95)

Formulae :

Mass of Water : (1 mL = 1 cc, 1g H_2O is 1 mL)

$$M = D*V$$

Area of Ellipse :

$$W = \frac{\pi*A*B}{4}$$

A & B are the measures of the axes

Solar Constant

$$Q = \frac{m*c*\Delta T}{K*W*\Delta t}$$

Power Received by the Earth on its Sun-facing Surface :

$$N = Q*(SA)$$

Total Power radiated by the Sun :

$$P = 4*\pi*Q*R^2$$

Percent Error : (use the values noted here)

$$\%E = \frac{[\text{Experimental Value-Accepted Value}]}{\text{Accepted Value}}*100\%$$

Constants to compare to :

Actual Solar Constant : 1.366 kW/m^2
Corrected Amount received at Earth for Solar Constant :
342 W/m^2
Total amount of Solar Power reaching Earth's orbit :
1.74 x 10^{17} W
Energy per unit time & area received at Earth :
1.96 cal/(min*cm^2)

Conclusion :

How Solar and Cosmic was this Adventure! Here we have explored in the realms of physics and astronomy to touch our star, the Sun and learn about it! How radiant!

Wow! – All things certainly have a story to tell us – we have to be creative in thought, curious in nature, and passionate about our need to explore, to learn, and Be Brilliant!

How well did your values match the expectations of the Solar Constant and if not, what things affected your measures?
In trying the Alternative Method, how do the values compare?

Project
Personal Slide Rule Template

MAKE A PAPER 6" SLIDE RULE

Ms Briana Andromeda Purcell here and here is one of the most interesting things you can do with my book : **Galaxy Girl's Inquisitive Pioneer Adventures**. You can actually make a functioning calculator out of paper! Your own Slide Rule! The scales on it will be able to help you in essentially all of the Activities in the book too – Wow – Awesome Sauce!

On the following two pages after the directions are two different templates than enable you to make a 6" slide rule. Both are needed so read the directions carefully but you have to take time to watch your folds since there are no fold lines and such – sorry.

Your Slide Rule will have these scales for use :
C, D, πC, πD, C1, D1, CF, DF, C1F, A, B, S, T, ST, K, L

Things Needed : Both of the Templates, Scissors, A good grade or thickness of paper – find at Office Supply stores or Art stores, Ruler

Steps to making a slide rule :

1. Make a Copy of both Pages – they are needed to make the Slide Rule.
2. For the first one where all the scales are present (page A) cut out the box – this will be the Slide itself – leave the sides on your box since you have to move it through your Stator pieces which are made from the other set of Scales on the other page (page B). It is best to begin with page B though since you may want to leave a little extra paper at the top and/or bottom of this slider piece so that it fits well in the page B stator page piece(s).
3. On Page B look to the third set of scales down the page and cut off the rest of the paper along the bottom edge of the bottom set of scales. Interestingly this I scale is the Top Stator.
4. Next on Page B on the top set of scales but off the excess paper along the line above the top scale and only here straight across as you did for the bottom set of scales only there along the bottom. This top scale on Page B is our bottom Stator.
5. Now place the Page A piece you have cut out facing out on the backside of the page B piece you have cut out so that they face away from each other.
6. Fold the top and bottom pieces of the page B piece so that it folds over to the side where page A is and line up the scales there and fold it well.
7. Your scales should have the πx scale from page B above the πx scale on the page A insert or slide piece now and the corresponding x scales should align below too.

8. Now we need a cursor, The simplest way is to use a ruler if you wish to lay across it, but be careful so that it aligns well.
9. It may take some adjusting, practice, and maybe have to make another one so be patient. – there is an Art to Math and Science!
10. Explore and Enjoy

These scales came from the web site : The International Slide Rule Museum (sliderulemuseum.com) found on the Slide Rule Reference Scales tab and set with graphics by Andrew Nikitin

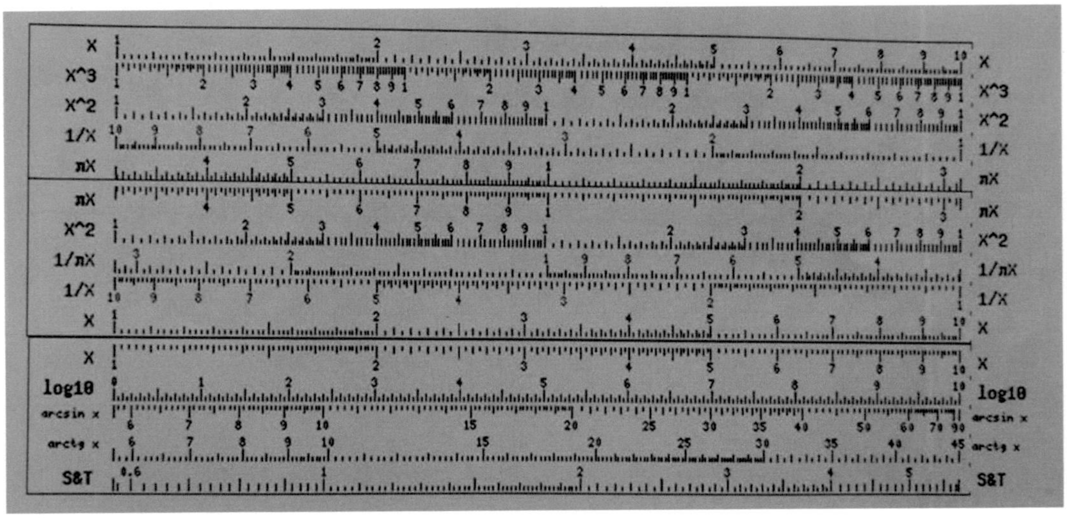

Page A

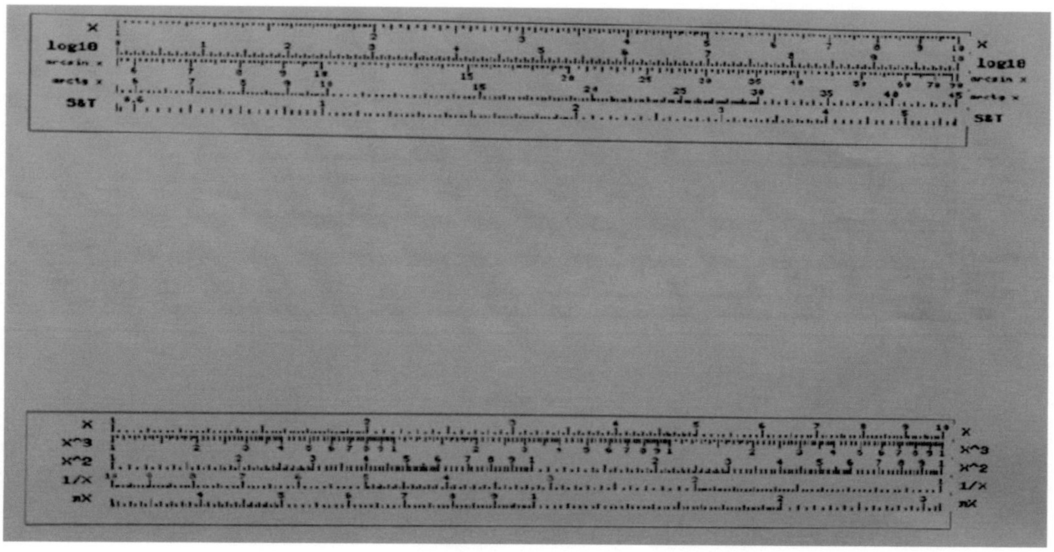

Page B

Long Scale Version of Linear Slide Rule

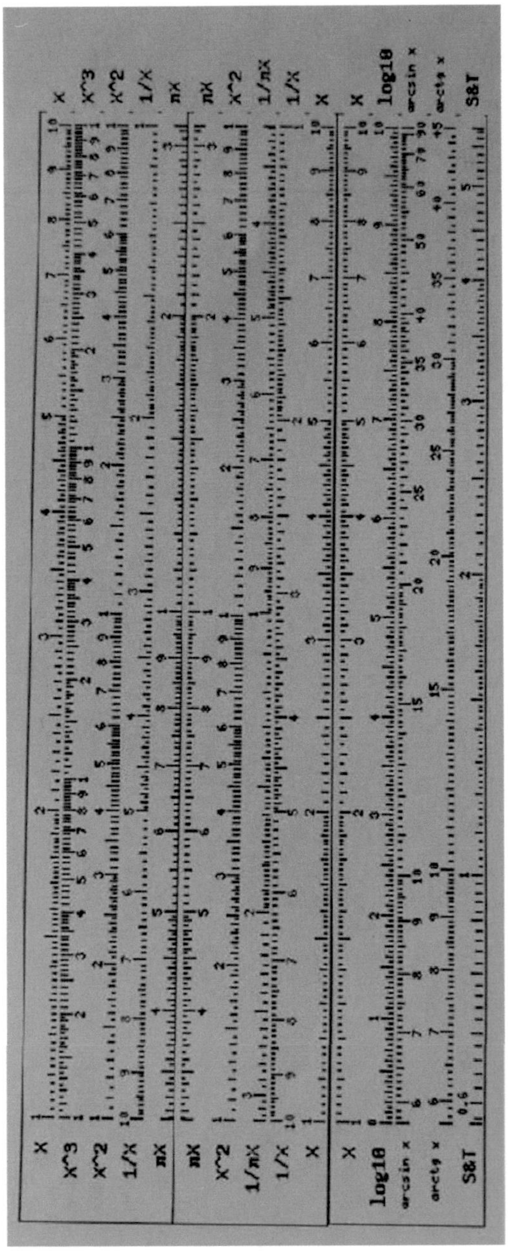

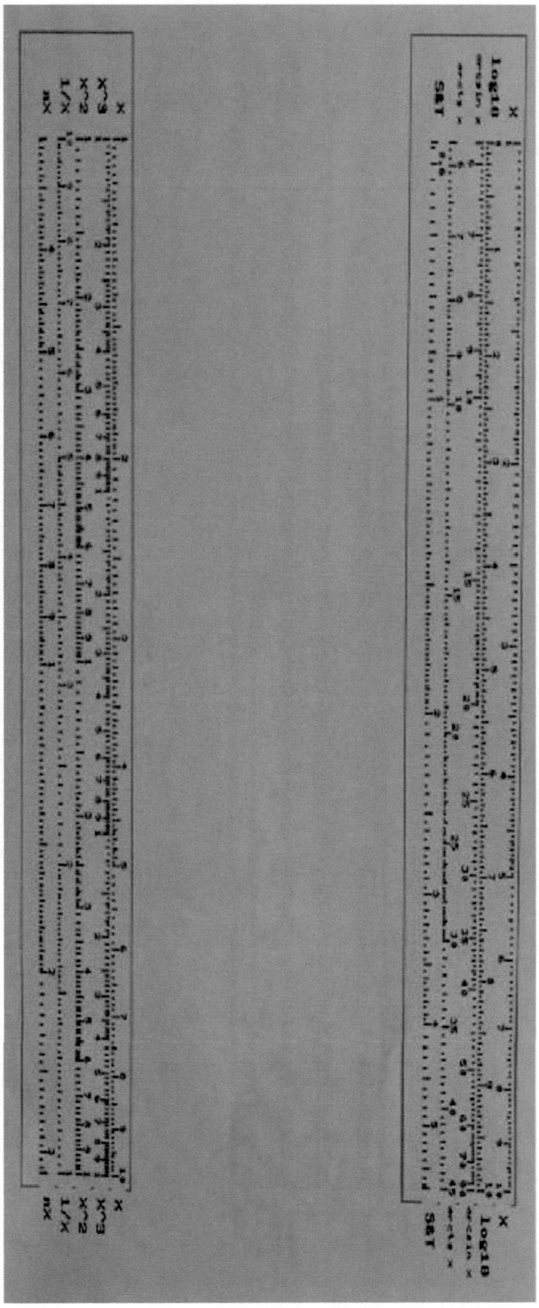

Fowler Paper Template Image

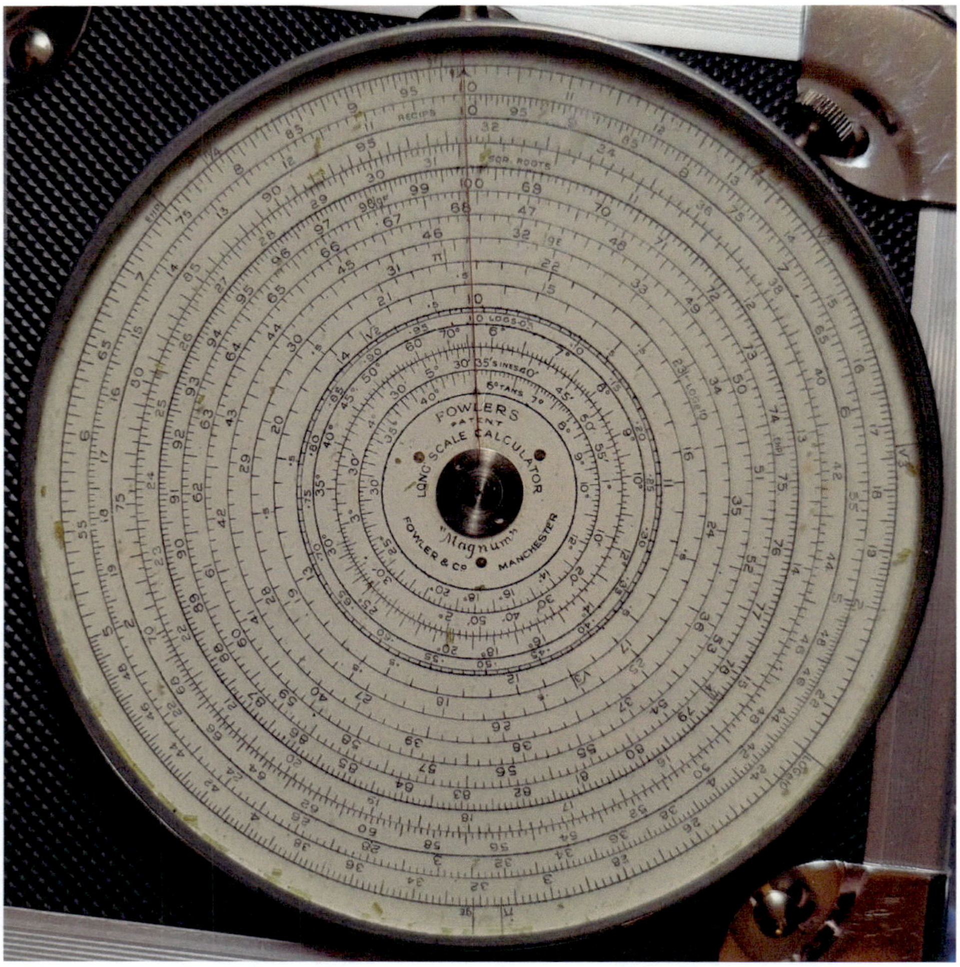

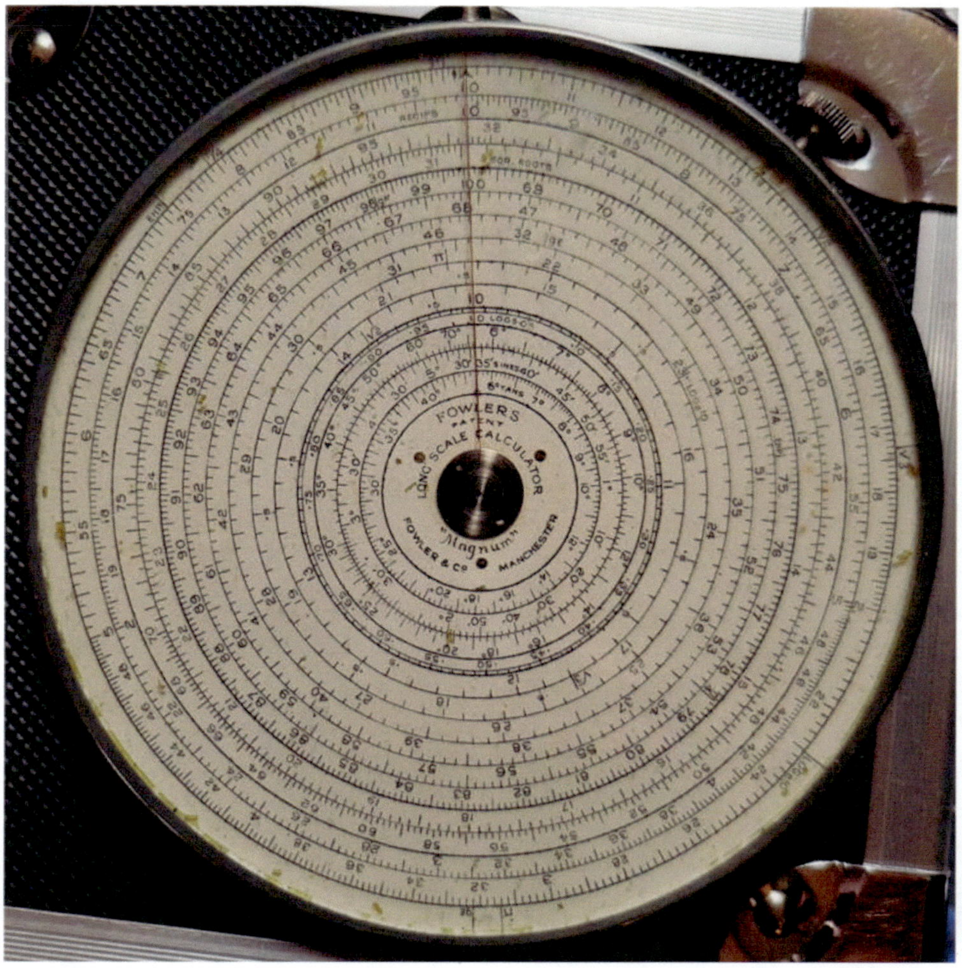
FOWLERS
PATENT
LONG SCALE CALCULATOR
"Magnum"
FOWLER & Co
MANCHESTER

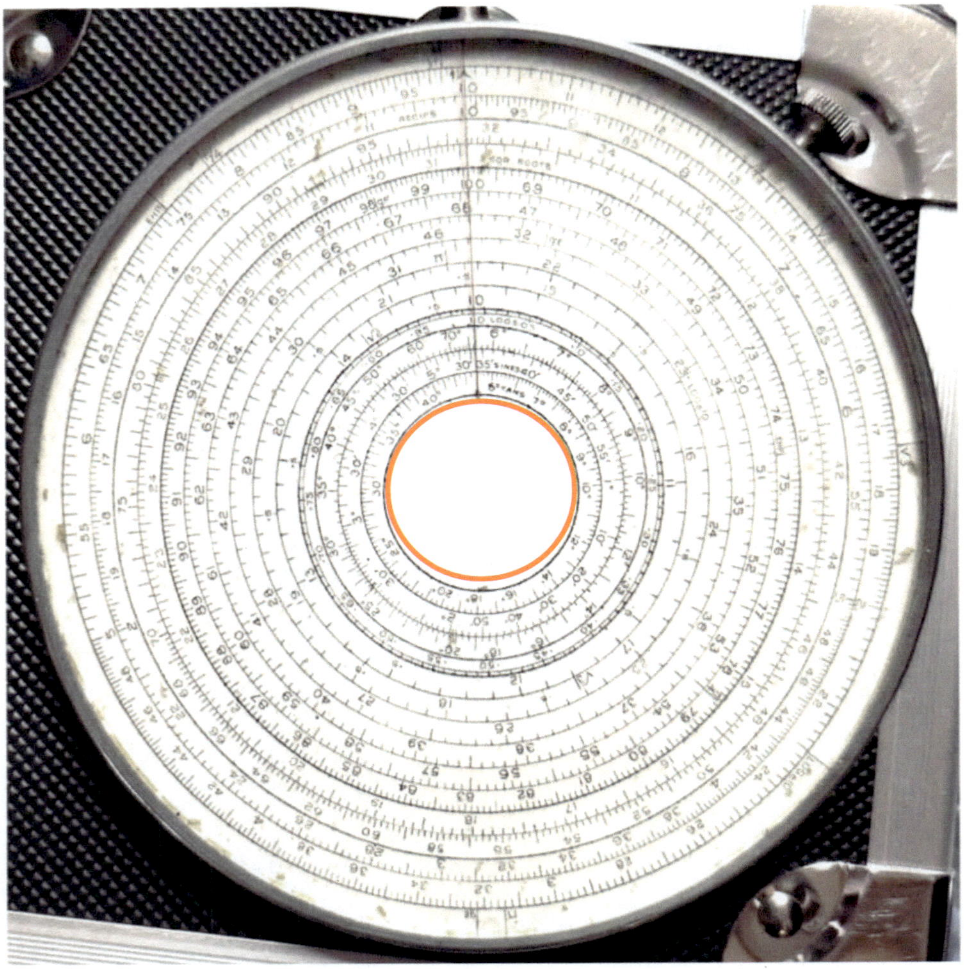

Fowler Circular Calculator Slide Rule Directions to Assembly

Notes from Briana Andromeda on All these Slide Rule Projects
So you have a choice – a 6" model, the 10" model (each more or less but close) and here is the third possibility – the Circular Calculator Slide rule modeled after a Circular Slide Rule over 100 years old – Imagine that!
In the first two slide rules you clearly need plastic to make a Cursor but here it is a bit more complicated – there are two Cursors needed to operate the Circular Slide Rule
I have used markers, a fastener that I poked through the template I printed and glued to a backing board – and used a circular compass to make what is needed. See photos for this. How to use this and the others – keep in mind there are chapters on How to Use a Slide Rule plus the article that follows this last project tells the tale of this Slide Rule and its uses – Notice it has 8 scales – must not be too powerful – but in my article I illustrate that it can be used to have 25 scales – can you believe it?! Wow – Brilliant

Also following that article I will give you a golden ticket opportunity
It is a coupon since you have been gracious enough to have purchased my book

First you will receive a 9 scale Pickett 120 either Yellow or White as selected by me and is some 60 years old and fully functional and is the one I photographed in this book for illustrations

Second I will send you the Paper Templates to all of the Slide Rules so you do not have to bend the book

Third I will send the needed plastic for all the cursors

You have to supply markers, the compass, the scissors, and tape and glue

There is a cost despite the coupon

For the shipping, handling, and the supplies I ask for a check or money order for $15

This is actually less than my costs so there is a slight net loss on this but that is okay

You wish to undertake an adventure

Make a Slide Rule of Paper that is as strong as a basic Scientific Calculator is today

You hold infinity and infinite creativity and thought in your hands

A very worthwhile journey and adventure

A Brilliant Slide Rule
The Fowler Magnum
Large Pocket Circular Slide Rule Calculator
Briana Andromeda Purcell

The overall Sphere of the Slide Rule World has Rectilinear Slide Rules, Cylindrical Slide Rules and Circular Slide Rules. Interestingly William Oughtred crafted the first slide rule in the 1600s and it was Circular and distances between the numbers on the logarithmically spaced numbers on the scale were spanned by calipers.
Even some 300 years later though the majority of created and crafted Slide Rules were Rectilinear or Linear there were a number of interesting Circular, sometimes referred to as Pocket Slide Rules and called Calculators by the Fowler Company of England. One of the most intriguing, unique, powerful and beautiful pieces is the Fowler Magnum Long Scale Circular Slide Rule which were crafted in the age range of 1927 to 1939, so if you have one of these (and I have several since I love and cherish and use them so much) this is the era from which you are working and partaking in with your adventures and

interests as I do. This is the time of a new interest in Egyptian look, Mechanical looking items with lines and geometric form – a sort of new classic era as well and also known as Art Deco – and the Fowler Magnum is part of these eras and looks – very office appearing, this girl sees it on a classic wood desk which has a classic Sherlock magnifier at hand, wood and metal office desk tools, an ink well and fountain pen – one smells the wood, feels the leather, touches the metals and wood and unique stone desk pieces whilst the warmth of the sunlight gently enters and makes the room glow – perhaps a candle or two, maybe an oil lamp still there but also the introduction and connection to a wire on the wall of an electric desk lamp and in the far of the room a radio – voices from other parts of the world, planes and cars now outside the window, large steam engines and large liners travel the landscape and oceans in search of new horizons and the Fowler Magnum in your hands and a warm cup of tea on the desk and an idea or two cooking in your heart and mind for a new adventure.

The chief question is this – what are your intentions for your Slide Rule in the long run? A good general one has 9 Scales and more advanced ones can have up to about 30 scales (see my other Articles about the B1, the 2/83n, and the N3 for more information) but then too there is the aesthetic, the attraction, the twinkle in your eye about a piece. It may be its styling, its unique nature and/or Scale arrangement, the number and uniqueness of its Scales, the clever types of Scales, its history and story for that piece, its construction, its age, and other eclectic sort of electric and magnetic if not magical qualities it possesses. This is why I choose what I own and use – particularly the only Circular Slide Rule (other than my 2 Scale Slide Rule Breitling Cosmonaute 24 hour face, Navitimer ie Mechanical with no batteries watch I wear daily) is the Fowler Mangnum – the most awesome piece and I love her. She is the Galaxy Girl model I treasure and it has all of the aforementioned unique aspects – very old and antique (made in the 1930s), High end Construction – Metal and Glass – a gorgeous piece lovely to hold, admire, and feel as if you are in another era with it, and has two 20 inch slide rules that sits in the span of your hand, is very easy to use and has unique Scale interactions with each other that creates still other unnamed Scales that are present and is as powerful as a basic Scientific Calculator today and requires no external power other than your thoughts and imagination to operate it. I call it Galaxy Girl since I am Briana Andromeda by name and in this very same era the Andromeda Galaxy was scientifically examined in the 1920s and it was found using the Doppler Effect to be moving away from us as all the galaxies out there hence the size and scale of the Universe jumped exponentially as well as Cepheid Variables were used (and stars were analyzed using Logarithms btw) to further determine the actual distance of some 2.2 million light years away – I do call the B1 bamboo Pickett Slide Rule I call the Andromeda as well.

What makes this piece not just Mathematical is that one might go as far as saying it is Magical and involves the keen interests, efforts, thoughts, and imagination of the user such a fine tool. First let's look at its design. Very classic made of an all metal body or having the earliest form of plastic, bakelite for a back and the front is glass with a transparent cursor moved by one of the two knobs on the slide rule – whereas the other moves the disk hence it has two cursors since one is permanent on the glass front whilst the other is mobile – so acting in the manner that Oughtred had done as a spanner making calculations easy and convenient. This beautiful model looks good on a desk and belongs in a nice case, the original case is leather but there are hard cases for it which are often felt lined and looking very classy as does the piece herself. Because it is metal and glass it has good mass and due to its size – some 4.5 inches diameter (11.3 cm dia) she is a very good size especially as compared to the history of the Fowler pieces through the years and is 1.5 inches wider than their average models which were more truly pocket models whilst this one fits in the span of one's hand. But this size makes for easy to read, and much longer scales.

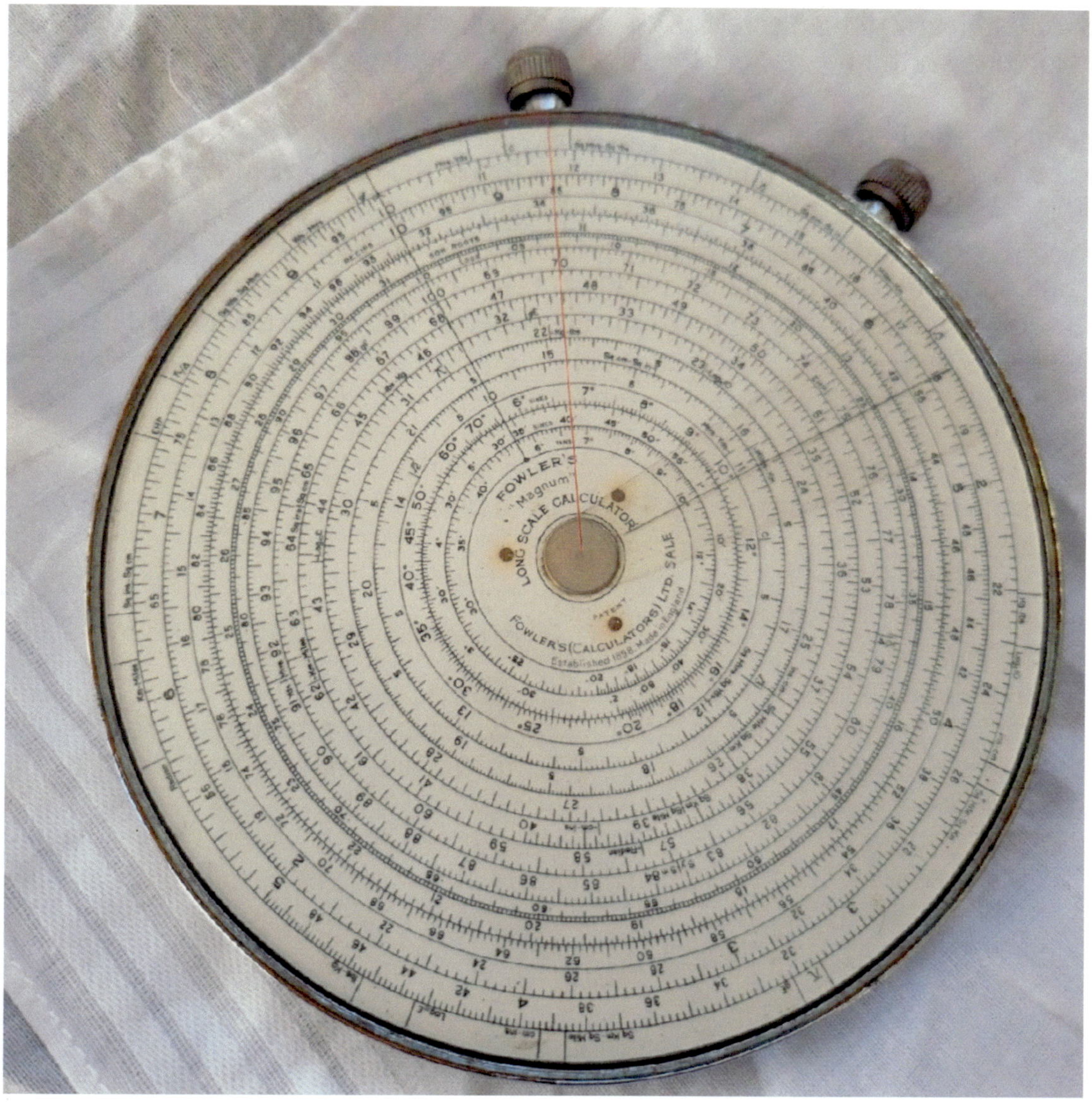

As I explore not only the Mathematical but also the Magical I will note some of the unique and even I dare say Brilliant aspects of this piece (which I feel would have merited awards if such things existed for this piece) Note it has not one but two more than 20" slide rules to use hence 3 if not 4 significant figures available in calculations both on the same rule – imagine that in the palm of your hand – compare that to the 2/83n some 50 years later a rectilinear piece with two W scales one on each Stator and the others are mobile on the Slide so as to have a 20" slide rule – yet here they are not one but two – wow – awesome sauce. Better still when used together these two scales, the R scale and the Long C as I call them one creates the K scale and hence has a Cube, Cube Root system that is fast, efficient, and easy to use – plus note I mention the R scale which when read against the short C scale on the other ring, you have the ability to have Squares and Square Roots, a very needed and common element in a well made slide rule since many relations in Science and Engineering require that sort of mathematical work of either Squaring or Square Rooting and here it is fast and easy. (

Read on and learn all the secrets of the wonderful world of the Fowler Magnum – It is an adventure worth undertaking I assure you – I am amazed daily by this classic delight).
Notice in passing in the prior paragraph the idea that Scales operate together and when doing so create and invoke still other scales – this is the first and most important realm of the Magic, the Mystery, and the Power of the Mathematics of the Fowler Magnum – there are 8 scales contained in 14 rings on this gorgeous treasure yet I make the claim here and now – this results in 20+ (and I mean 20 plus – can be as many as 24 perhaps more scales) scales – no this is no mistype – there are indeed 20+ scales and I will explore them all here. Talk about potent portability for mathematical magic – this is Brilliant I dare say yet again.

Let us continue our introduction – Reading from the Outer Ring to the Innermost ones these are the Scales present on the Fowler Magnum :

> Short C, C1 , R1 & R2 (square roots), L Log Scale , Long C (having 6 Rings), Sine S scale, T / S (tangents and sines) scale for small angles , and finally the innermost T scale for Tangents
>
> What the Fowler Magnum can do Scale-wise :
>
> Note the primary reason is that the two cursor system makes any and all Scales a set of mobile compliments for calculations ! Super Cosmic Cool and Brilliant

Scale Name	Radius (cm)	Radius (in)	Diameter (cm)	Circumference (cm)	Scale Length (cm)
Short C	5.5	2.125	11.0	34.5	
C1	5.0		10.0	31.4	
R2	4.5		9.0	28.2	
R1	4.5		9.0	28.2	56.4
L	4.2		8.4	26.4	
Long C 6	4.0		8.0	25.1	
Long C 5	3.8		7.6	23.8	
Long C 4	3.5		7.0	22.0	
Long C 3	3.1		6.2	19.5	
Long C 2	2.9		5.8	18.2	
Long C 1	2.5		5.0	15.7	124.3
Sin (6°-75°)	2.0		4.0	12.6	
Sin / Tan (38′ to 5°30′)	2.0		4.0	12.6	
Tan (6° – 45°)	1.7		3.4	10.7	

Notes : 6″ Slide Rule (ie Pocket) is 12.2 cm
so the Trig Scales for the most part are Pocket Slide Rule Size
This is more than effective since these are only Scales to Read and not typically to directly employ in the calculation as one would do with paired scales (which this one has due to the Cursor Dancing Pair) and the accuracy is sufficient at 2 significant digits for the given angle
Besides one could always carry a small pocket book of Sines, Tangents and such if really needed for more digits since their values can readily be dialed into the Long C if desired for greater accuracy
10″ Scale (Regular Rectilinear Slide Rule) : 25.4 cm
Notice the Short C is 1 and 1/3 of a 10 inch Slide Rule
The 20″ Slide Rule is 50.8 cm
The R1 and R2 Square Root Scales which can be used
as a Slide Rule is 56.4 cm – so this acts as more than a
20″ Slide Rule (the first one)
Look then at the Long C Scale – 124.3 cm, 2.5 times the length of a
20″ Slide Rule – can you believe it!? How Cosmic and Awesome is this –
in the palm of your hand is not one but two 20+ inch Slide Rules,
3 Active Slide Rules and the smallest is longer than a 10″ one too

Here is both the List and Explanation with Examples of the Possible Lists of Scales to be found with Creative Imaginative and Direct Thought as well on

the Fowler Magnum Circular Slide Rule Calculator – and Note the amazing fact
There are 20 Scales – Brilliant!!

Short C, Short D, Short CF, Short DF, C1, D1, C1F, D1F, L, the L Scale and the Short C can be used for any and all powers and roots and changing bases (see my excellent examples of how to to this) plus therefore the L Scale acts as a Generic LL +/ - Scale as I call it, Long C, Long D, Long CF, Long DF, Sine, Cosine, Sine / Tangent for small Angles, Tangent, and the Long C and R1 & R2 Scales working together are the K Scale (hence Cubes and Cube Roots) – Note for example the L Scale can as I note be either the LL0+ or LL0- scale as well as the LL1+ and the LL1-. Imagine that

I can illustrate the power of the Fowler this way – say you wanted 17 to the 3.3 power. Simple Look up 17 on short C, read its L log value on the L scale now look up this value on the Long C and simply multiply it by 3.3 – remember to have a +1 in front of your L value since this is the power of 10 for 17. Now look up this number on the L Scale where the number before the decimal is the power of 10 involved and this is 4 here in this answer response. You find that this corresponds to 1150 on the short C scale and hence your estimated answer of 11, 500. The actual answer from a scientific calculator for 17 to the 3.3 power is 11, 494 – wow consider that less than 0.05 percent error and this tool acting as a LL0+ and LL1+ scales in the process – awesome sauce!

Here is a further illustration – say you want 0.9 to the 4th power. Here there actually two options read 9 and consider it as 0.9 from the Square Root Scale and read back to the short C Scale and find 0.81 and now look this up on the Square Root Scale and find its square on the Short C again and viola you have 0.656 – not bad considering the scientific calculator gives an answer of 0.6561. – What of the LL Capacity since here it is LL0-. Simple again you can read from the short C to the L scale but you must remember to add a -1 to your answer since you are looking not for 9 but for 0.9. But there is an easier way – use the C1 Scale and read from 9 since it is already 1/9 and find its L value – only bear in mind it is negative here. You find it to be -0.045 and now go to the Long C scale and multiply this by 4. You look up the answer on the L scale of 0.18 remembering it is negative meaning that you should use the C1 scale to find the answer and realize that you have to move one decimal place over so that the answer is 0.661 which is very close to the scientific calculator by-the-way – How Brilliant is That!

Short C, Short D, Short CF, Short DF – where CF and DF are found by setting the Mobile Cursor to Pi and the Permanent Cursor to 1 and hence all calculations are readily found, also note that you can readily get the Sine from the Short C, hence Cosine (90 – Theta) from the Short C scale as well,

How to Multiply and Divide with one Scale yet it is two –
Let's multiply
2.6 times 3.4. As with all Use of Slide Rules, it is best to estimate the answer 2 x 3 is 6 and 3 x 3 is 9 so the answer is between 6 and 9 since both of our values are near to 3.
Set the Permanent Cursor to the 1 on the Short C scale.
Spiral the Mobile Scale to 2.6.
Now Dial the Disk to 3.4 to the Permanent Cursor –
Viola the Answer is now readily read under the Mobile Cursor : 8.84 as read directly from the Short C Scale –
Try this on the R1 and R2 Scales too to
check the 3rd significant digit if you like as well going through the same steps

Note too this can be done on the extended or Long C Scale which is more thn 30 inches in length.

Consider that thought as you use this tool, which may be 100 years old approximately – in the palm and outstretched fingers of one hand you have 3 scales where one is more than a 10 inch slide rule, the Square Root Scales can be used as a 20 inch slide rule and the Long C scale can be a 30 inch slide rule – Brilliant and Cosmic are the only terms one can use here.

Other Multiplication and/or Division Considerations for Practice and Understanding the Fowler and its use

What if you have A x B x C? Let's use basic numbers to follow the steps but note that of course we need to keep track of powers of 10 as needed to find the final correct answer
For our A x B x C example let's use 2 x 4 x 6 = N
Set the Immobile Cursor to the Index and the Mobile Cursor to 2 (A) , next rotate the disk to place the 4 (B) under the Immobile Permanent Cursor now and you immediately find the answer to the first two multiplicands – answer 8 under the Mobile Cursor. – Return the Immobile Permanent Cursor to the Index (1) and place the Mobile Cursor on 8 the result of the first product and turn the Disk to the next value, C which we have as 6 and place it under the Immobile Cursor and viola the answer for the whole system is found under the Mobile cursor or 48.

What if you have A x B / (C x D) ? Let's use basic numbers to follow the steps but note that of course we need to keep track of powers of 10 as needed to find the final correct answer

For our A x B / (C x D) example let's use 2 x 4 / (6 x 8) = N

There are several options here – in case number one we do the work of the Numerator then follow by the Product of the Denominator and then Divide by the two Values for a Final Answer
In the Second and faster example consideration treat each ratio (A / C) then multiply by B and finally divide by D and along the way use the Inverse C scale for the Denominator Values to expedite the process I feel.

Case 1
Set the Immobile Cursor to the Index and the Mobile Cursor to 2 (A) , next rotate the disk to place the 4 (B) under the Immobile Permanent Cursor now and you immediately find the answer to the first two multiplicands – answer 8 under the Mobile Cursor. Keep this in mind or on a piece of scrap paper
Now Set the Immobile Cursor to the Index and the Mobile Cursor to 6 (C) , next rotate the disk to place the 8 (D) under the Immobile Permanent Cursor now and you immediately find the answer to the first two multiplicands – answer 48 under the Mobile Cursor. Keep this in mind or on a piece of scrap paper
In the final step place 8 on the Immobile Cursor and 48 on the Mobile Cursor and rotate the Disk Index to the site of the Mobile Cursor and under the Immobile Cursor read 166 which is 0.166 of course and guess what ! ? Wow! – below it on the Inverse C is the inverse value that is 1 / number that is the answer in this case which is 6 so it is 1 / 6 as the fractional Answer to the problem!

Case 2
Place the Immobile Cursor to the index (1) and the Mobile Cursor to A (2) and then rotate the disk to the Immobile Cursor for C (6) and find the answer in two forms – on the C1 Scale it is 1 / 3 and on the C Scale above it is 0.333.
Now return the Disk to place the Index (1) underneath the Immobile Cursor and Set the Mobile Cursor to our Prior answer of 0.333 (quickly found by using the C1 and placing the Cursor on 3) –
Rotate the Disk to B (4) and find the answer under the Mobile cursor of 1.33 which reads as 1 / 0.75 on the C1 scale (or 4 / 3 so far)
Place the Immobile Cursor once again on the Index and rotate the Mobile cursor to the prior answer 1.33 (over 0.75 on C1) and then rotate the Disk to the final value D (1 / 8) on the Immobile Cursor point on C1 to read the final answer on the Mobile Cursor as on C1 as 1 / 6 or 1.66 on C and viola these step are valid as well!

What of division –

Let's do 12 divided by 18
so place 12 of the Numerator on the Permanent Cursor and 18 on the Mobile Cursor and now move the 1 to the Denominator or Mobile Cursor turning only the Disk and read the final answer under the Permanent Cursor of 0.666

Realize this too – Dial the 1 to the Numerator and you have reversed the division to be 18 divided by 12 instead and you find the Answer where the Denominator or Mobile Cursor is at – Wow – either way and so easy peasy

Note that the Multiplication and Division directions here apply to all the scales in use for multiplying or dividing – for example the Short C, the R1 & R2 system or the Long C system

A second way of doing this is faster using the Short C and the C1 Scale. Place 1 on the Permanent Cursor from the Short C scale and Move the Mobile Cursor to 12 on the Short C Scale and now turn the Disk to 18 on the C1 Scale (since we are multiplying by the Inverse of 18 from the C1 Scale) and Read the Answer on the Mobile Cursor on the Short C Scale is 0.666 again

Realize the awesome Applications of These Basic Calculations I have outlined here – of Multiplication and Ratios and basic Division. – You can have any and all Ratios in everyday Life at your hands.

Say for example you wish Comparative Shopping – Set the Cost on the Immobile Cursor and the Mobile one at the Number of Items – such as Pounds – Ounces – et al. Turning the Dial to place the Index at the Immobile Cursor Point reveals the Cost per Unit Whatever at the Mobile Cursor.

Better Still you can compare to anything else since you have spanned the first value – now place the Cost of Item number 2 for comparison at the Immobile Cursor and see where the Mobile Cursor is at – does it match the Number of Unit Items at the point of the Mobile Cursor - meaning they both have the same Cost per Unit Item or do you see the expected number to the Left of the Immobile Cursor and this value matches the number you are purchasing in terms of unit items then this means that it Costs LESS Per Unit Item – Nice deal then or does the expected value from this calculation fall to the Right of the mark of the Immobile Cursor meaning that you are getting less per Unit Cost Per Item hence not a good deal. Realize your span of distance is the intended ratio for these cursors and how useful that may be!

So not only can you do Cost per Unit Item – Ounce – Pound and Number and such but also basic calculations of ratios – Pressure is Force per Unit Area, Density is Mass per Unit Volume, Mileage is Miles per Gallon, the Net Acceleration is the ratio of Net Force to the Mass of the Object (Newton's 2nd Law), the Number of Tiles needed on a Wall or Floor Project is the Ratio of the Square Area of the Wall or Floor in Question and the Area of the Tile in Question, the Average Speed of Your Trip – by walking – running – biking – or car is simply the Ratio of Total Distance divided by Total Time to do so. Then estimated Time is the Remaining Distance divided by the present Average Speed to estimate when you arrive at your destination!

How simple, How elegant, How Awesome and Brilliant – Simple Ratios as applied to the world. This then is the power of a Slide Rule, such as the Fowler Magnum and other types – your Knowledge, Imagination and journeying to Solve a Problem – be the girl seeking the solution – empowers your quest.

I address all of these ideas in my paper on Numeracy and Scientific Literacy and the Value of the Slide Rule in teaching. Take for example you cannot do a calculation such as with the 30 inch slide rule here with 6 rings without not first having an estimate in your mind – you are an active participant in the answer – no machine gives you a number you rely upon – you rely upon your knowledge, your measurements, your mathematical skills, your imagination, and your purposeful efforts. True compassionate rational reasoning employed. Also your purpose is often driven by your creative, imaginative, Artistic drive which is yet another article I have concerning such matters and their importance. Here then the Slide Rule is the nexus or crossroads of both worlds – the Science / Math Realms crossing roads with the Artistic / Imaginative / Creative roads inside us all.

Imagine that – How Brilliant, Cosmic are these notions.

Now let R1 be the inner Ring and R2 be the outer Ring and you have a > 20" functioning Slide Rule where the R's are now therefore W's since it acts as two Scales as a single scale with 2 cursors, plus when acting with the Short C Scale the R Scales are the Odd and Even system of the use of the Square Root Scales –

Here is an Example to Illustrate : Look up 4 on the Short C Scale and find the square root of the Odd number of digits, 1, on the R1 (inner) scale with a reading of 2 and now look to R2 (the outer Even Ring) where here is the Square Root of 40 (6.32).

This Rule Applies to Odd and Even Numbers of Zeroes 0 plus the first number greater than zero in front of the values past the decimal so 0.4 is Even since there are no Zeroes and zero Zeroes is the rule of an 'even' number of Zeroes so the R2 scale is read hence the answer is 0.632 whilst 0.04 has an Odd Number of Zeroes so the R1 scale (ie Inner) is read and the answer is 0.2 .

The Long C has as its compliment Long D of course and therefore like the Short C and D the Long C and D can be Long CF and Long DF as well.

Also realize that the Square Root Scales are 2 and the Long C is 6 hence the ratio is 3 which means that going from the Long C and reading to the R Scales you are Cubing and Reading from the R Scales to the Long C you are taking the Cube Roots – plus there are simple rules for these. For 1 Digit to Cube Read the Inner Two Long C Scales, for 2 Digits, Read the Middle Two Long C Scales Rings and For 3 Digits Read the Outer Two Long C Scales Rings

- And in all cases the Inner Ring of the Long C is for R1 (that is Long C Scale Rings,1, 3, 5) and the Outer R2 Scale is for the Outer of each set of Long C Scale Rings (ie 2, 4, 6) – here is a simple example place the cursor on 8 on the R2 Scale – this means you will read the inner 2nd Scale since it is on the Outer R Scale or R2 – If you were on R1 at 2.53 you would read the Inner Long C Scale Ring which reads 1.362, and this idea continues so that the next two rings say instead of 8 we call it 80, it has two digits so we read the middle Long C Scale Rings – the Outer one of course since we are on the Outer R2 Scale and we have the Cube Root of 80, what of 800 then We read the Outer Long C Rings

So this is how the K Scale System Works :
Read From the Long C to the R Scales and you are Cubing the Value
Read from the R Scales to the Long C Scale and you are Cube Rooting the given Value in question
What of the Digits? The Inner Long C Scale Rings (1 is for R1 (Inner) and 2 is for R2 (the Outer R Scale)) are for 1 Digit
The Middle Long C Scales are follow the same R pattern R1 is for Long C Scale Ring 3, whilst R2 is for the Long C Scale Ring 4 and
The Outer Long C Scales have the R1 for the Long C Scale in Ring 5, whilst R2 is for the Long C Scale Ring 6 and then
What of more digits?
The pattern is thus 1, 2, 3 for Inner two, Middle two, and Outer two then it is 4, 5, 6 and so on

Here is a great example that shows there is always more than meets the eye when it comes to the Fowler Magnum. Look up 2 on the Long C Scale and read its cube on the R2 Scale of 8 which is 2 to the 3rd power btw. Now read back from the R2 Scale to the short C Scale and Find that 8 is the square root of 64 here – wow. We have a 6th power scale in one direction and a 6th root scale in the opposite direction for any and all numbers in reality – super cool. My fictional writing btw is about the 6 sisters, the Galaxy Girls so personally I find this way cool and identify the Fowler Magnum being circular, mysterious, magical, mathematical and scientific as being a direct extension of each of them collectively and individually – Brilliant.

Notice that is it is your imagination and the curious questions you pose and explore that opens the doors for this wonderous piece of history nearly 100 years old.

Then there is the C1 scale which is another 10 inch plus Slide Rule in reverse and makes multiplying by the inverse as needed with numerators in the C Scale easy and convenient, plus the C1 functions as the Tan of all values of 45+ degrees to 90 as needed for the Trig Scale of the Tan Scale

The Power of the L Scale –
Clearly this is a base 10 Logarithm Scale so any and all Values read from the Short C Scale are found here and this is true for even when the number is not the one directly on the Scale – For Example the Log of 2 is 0.301 but what about 20 well that is 2 multiplied by 10 hence it is 1.301 where the 1 at the front is the power of 10 involved, so 200 is 2.301 and so on.

Can there be more with the L Scale – such as can we Change Basis or search for answer to other Bases as needed – Yes Most Definitely

The question to ask is, then, what is the next best Scale(s) to add to the Fowler Magnum? Just one is needed. The L Scale! Having a C Scale and an L Scale in combination allows not only for all of the basic calculations, but the ability to raise any number to a given power (2, 3, 4 et al), even if that power is not a whole number (2.4, 3.1, 5.5 et al). Also the inverse of this operation, taking roots, of any value, square roots $\sqrt{}$, cube roots $\sqrt[3]{}$, and even values like the 5.3^{rd} root is possible with these two scales (three if there is a C & D Scale with the L Scale – which are present on the Fowler Magnum since the short C is really a short D too).

<u>Roots & Powers :</u>
It is best to start with a simple example to illustrate the point. Let's try $\sqrt{3}$. This is $3^{1/2}$. When examined with a simple Slide Rule with only a few Scales, particularly the L Scale, the log(answer) = log($\sqrt{3}$). This means that we have to find the log of the value in question, multiply by its root or power using the C & D Scales. Take that number and look it up on the L Scale and find its antilog on the D Scale for our final answer.

This is from this property of Logarithms :

$$\text{Log} (A^N) = N*\log(A)$$

In this case it can then be seen as 0.5*log(3). Look up the log(3), it is 0.477. Take this value and on the C & D Scales divide 0.477 by 2 which appears between 0.238 and 0.239. With quick mental math it is easy to see it to be 0.2385. Now look up on the L Scale 0.2385 (as best that can be determined). Read from the L Scale to the Short C Scale and find the square root of 3, which turns out to be a bit more than 1.73 (check it on a calculator).

Try something more complex, like the 2.3rd root of 7, or $7^{(1/2.3)}$. The same process happens here. The log(7) is 0.845. 0.845/2.3 from the Short C Scale yields 0.367. Taking this from the L Scale back to the Short C Scale gives a final answer of 2.33 approximately. (A graphing calculator yields 2.3304***).

How about raising to a power? The same process is done for roots as it is with powers. With powers it is easier since typically they are whole numbers and one only has to take the number in question being raised to a power and look up its log value. Multiply this by the power in question and look up the decimal portion of the answer on the L Scale. The digits values in front of the decimal determine the whole number power of 10 to multiply by.

For example what is 4^6? The log of 4 is 0.602, multiplied by 6 is 3.612. The 3 is the power of 10 so we multiply our read out answer shortly from the Short C Scale by 1000.
Reading 0.612 on the L Scale and finding the corresponding value on the D scale it is just before 4.1 (this would be a good rounded off value here). So 4.1×10^3 = 4,100 is our approximate answer. The actual answer is 4,096. Like the prior example of non-whole number roots, the same can be done with non-whole number powers if desired.

Having this Scale combination (C, D, & L) allows for fast conversion of values into their respective log values. This can be useful for evaluating the product and division rules for the law of logarithms (Log(A*B) = Log(A)+Log(B)) (see Logarithmic Product Activity). Here it can readily be seen that the log of the product of any two values is the sum of the logs of the individual values. Also the construction of a logarithmic set of values for graphing purposes to evaluate the power relation between two variables (see Kepler Activity & Rate of Cooling Activity). Here plotting a log-log graph or a log-linear graph allows one to find the power relation between two variables in question. In these cases values are taken from the D Scale and the log of them are found on the L Scale as needed.

More than this, this simple Scale combination also allows for the evaluation of the Changing of Bases as well as other Laws of Logarithm considerations in Mathematics.

First consider the structure of the logarithm
(used in completing the solution) :

Log_{Base} (Number) = Exponent Value
$Base^{(Exponent\ Value)}$ = Number

<u>Formula for Changing Bases</u>
For any positive numbers, N, A, and B with A ≠ 1 and B ≠ 1,

$$\log_A N = \text{———}$$

Here is our first example :
Let's say we wanted to determine the natural log of 6, written as this $\log_{2.72}$ (6). The L Scale is set up in base 10, however. Using the formula above, we can evaluate the expression by doing this :

$$\frac{(\ )}{(\ .\ )}$$

So log of 6 is 0.778 and log 2.7 is 0.432 – we take the ratio of these two and can use the Short C or Long C and find 1.8 as the answer – whereas a calculator gives us when taking the natural log (2.71828...) which is what the goal of this question is

Let's say we wanted in a more advanced example this answer. To determine $\log_2$ (19). The L Scale is set up in base 10, however. Using the formula above, we can evaluate the expression by doing this :

$$\frac{(\ )}{(\ )}$$

The log(19) from the L Scale is 1.279. Note that there is a 1 in front of the decimal since it is the log(10*1.9) = 1og(10)+log(1.9). The log(2) from the L Scale is 0.301. With these two values in mind, we use the C & D Scales and divide them. Recognize that we can only estimate to at most 3 significant figures.

$\overline{\quad} \sim 4.25$

This means that the answer is $\log_2(19) \sim 4.25$, which is $2^{4.25} \sim 19$.
Awesome!
Now Let's extend the original question to this new one:

$$\log_2(19) = \log_3(X)$$

where we change bases and want to know what number equals the $\log_2(19)$ if it is the $\log_3$ of some unknown number?
Take our original answer $4.25 = \log_3(X)$
This means $X = 3^{4.25}$
To solve this with the C, D, L system, look up the log(3). This is 0.477. Multiply 4.25*0.477 on the C & D Scales. This turns out to be approximately 2.03. Look up 0.03 on the L Scale and find 1.07 on the D Scale. Since there is a 2 in the ones digits place, this means we multiply this value by 100 and arrive at an approximate answer of 107, which is X. (A graphing calculator yields 106.6) Our final answer : $\log_3(107) \sim \log_2(19)$

Not to Mention with the L Scale I argue that it acts as a sort of LL + / - Scale too since all you have to do is divide by the Root you are taking of the Log of the Value in Question and look it up on the L scale and find its connection on the Short C Scale for the Answer or if it is a Power then Multiply the Log of the Value by the Power and look up its value on the Log Scale (noting that the numbers before the decimal are the power of 10 involved in your answer) and find the Answer to the Question so easy peasy.
Here is an Example :

14 Rings, 8 Scales yet it can perform as a 20 plus scale Slide Rule smoothly, with little to no effort and awesome exactitude.

The next notion would and should be other than physical container ephemera what of the Operation Manual from that time – nearly a century ago when she was first fabricated and presented to the world stage and slide rule circular calculator circles and math sorceresses of those times? Well I have an entire reproduced manual with text modifications to make it useful here and one can readily use it – note not for commercial distribution and only for personal use and no monetary gains to be seen with this – all above board I say – I am a writer and an artist both and appreciate the works of the creators – be it nonfiction, fiction, pictures of any form, movies, music, animation and such.
Also keep in mind this piece is nearly 100 years old this decade (2020 – 2030 and made in the 1920s btw) hence can be seen as an Architectual piece, a Mathematical Tools, an Art Nuevo piece, an Art Deco and Industrial Piece, a Classic Office piece all of this just on looks and styling alone. But what of the Scales ? Just what are all these slides that the Amazing, Awesome and Brilliant Fowler has :

Due to the Two Cursors (Fixed and Mobile and the Rotating Disk) you have C, D, CF, DF, C1, D1, C1F, D1F, L scale, R1 & R2 Square Root Scales (which can be used with the short C scale as A and B Scales), the L Scale acts on its own but also extends itself to the realms of LL0+, LL1+, LL0-, and LL1- scales readily as shown in my examples. Less we forget we have the C, D, CF and Df Scales twice due to the Long C Scale too keep that in mind. Finally the Trig Scales : Then there is the S scale which implies it is a C scale reading the Cosine of (90-θ) from Sin (θ) readings as needed, plus consider the fact that after looking up Sine for any angle on the short C scale, look for it on the Square Root scale and then its square value on the short C again and subtract this from one and then read this value from short C to the Square Root scale and guess what – you have now a P Scale as well. Let's not forget the Long C and the Square Root Scales (R1 & R2) create the K Scale as well. Plus one might try and argue that the Trig scales are 6" only hence not as precise – true – know what I do? – I carry a book of tables of all sorts of values and a pamphlet I created of basic formulae and in this book there are of course the tables of Sine, Cosine and Tangent – realize that the goal of these scales is to read from them to have a value to use – a book of tables is best and now you can use the Long C Scale and perform calculations – say one of these is to find the Sine of a given angle – even with the 6" slide rule capacity you will find at least the range for the answer and your booklet precisely pins it down. How convenient and brilliant is this. So that gives us More Than 20 Scales and on the order of 24 approximately some with great precision. Also there is more to this wonderous, adventurous, mathematical yet magical brilliant piece. Not to mention there are two 20+ inch slide rules you can choose to use – this girl loves that part. All in the span of your hand in a classic piece that is nearly a century old – Wow – Brilliant!!!
Here is the entirety of that awesome manual with many insights and examples to explore with one's thoughts, imagination, and spirit of adventure – enjoy :

The next most important element is this about Circular versus Rectilinear Slide Rules (which by-the-way I am actually very fond of and have articles on my all time favorite, the fabulous Pickett B1 Bamboo Slide Rule (aka the B1 Galaxy Girl, the B1 Briana Andromeda, and the B1 Andromeda – she is my personal favorite linear one), the Faber Castell 2/83n slide rule which I call the Cosmic Wonder Girl, and the Pickett aluminum metal slide rules of the N3 (the Unique Galaxy Girl) and the N600 (the NASA Girl)(the one used by NASA Astronauts and taken to the Moon btw)) but the Fowler Magnum is the Supreme Galaxy Girl Diva of the Circular Slide Rules in my book and my life – timeless, elegant, beautiful in function and design, mathematical yet magical in its nature, simple yet cleverly complex like a mathematical maze of wonder, plus from an era of beauty – a time of the neoclassic architecture, the power of the mind imagining, creating, and crafting awesome structures that still are here today and part of everyday life – all worked on with Slide Rules – the Golden Gate Bridge, the Hoover Dam, the Empire State Building, the early Jet engines up to the F16 versions, the earliest Rockets including the Saturn V which took the Apollo Astronauts to the Moon and their awesome equipment, not to mention innumerable electric, early versions of electronic equipment, radios, TVs, electric appliances, plumbing, piping, wiring, hydraulic considerations and creations, as well as cars and the earliest planes and trains as well.

But I digress, what is the next key element of the Fowler Magnum (who is the one and only Galaxy Girl) and Circular Slide Rules as compared to their cousins in the Rectilinear Family lines? Turns out to be the simple fact that a span that is set up with the cursors on a Circular Slide Rule allows fast and convenient rapid successive calculations as needed in a given process – any and all ratios as well as Proportions are set up instantly and one merely spins the disk dial to set up the next answer quickly and effortlessly – no need as often is the case in a Rectilinear Slide Rule of flipping it over to read the CF and DF rules if available and if not, oh well, change the 1 from one end of the Slide Rule to the other one at the other end so as to find the answer you seek – no issues or hassles with a Circular one ever.

Notice I mention the CF and DF Scales and one might raise an eyebrow and say well Briana the Fowler does not have this – ah but it does. What ! ? You might exclaim – Merely set the Permanent Cursor to One and the Mobile Cursor to Pi – conveniently marked on the outer ring and viola – any and all calculations are now multiplied by Pi as needed and since you have a two cursor system just as in the C Scale Case where it is both the C and D Scales, you now have the CF and DF Scales here – all on one Scale – Imagine that! So you have a great tool for Circles, Pipes and such due to the Pi Connection. Not to mention say you set the Permanent Index of 1 for the R Scales and set the Mobile Cursor to Pi on that Scale and then rotate the disk to any and all needed values for your Radius to Multiply Pi by – realize that above on the Short C Scale you now will have the R value squared and can then turn to this hence have Area of the Circle – $A = \Pi * R^2$ – How sweet is that

Thus the Fowler Magnum or Galaxy as I refer to her is clearly an exceptionally artfully and mathematically awesome and powerful tool and you can be the Art Devo Girl on an Adventure through Space and Time as I noted earlier with such a Cosmic Treat and Delight.

Also consider I use and hold a Calculator – Slide Rule that spans the user's imagination and mine is cosmic – and is not only Art Deco, Art Devo, Utilitarian , Office Chic, it is 100 Years old! How Cosmic is that – So it is my preferred Circular Slide Rule (though I am the B1 linear slide rule girl too though).

To turn 14 Rings, 8 labelled Scales into the Power of a 20 Plus Scale Slide Rule with vast potential is Brilliant as I often say Brilliant when the moment calls for it – and here it does. You have on it only a basic set of Scales – but with thought – math – imagination – and foresight you have a 20 plus scale slide rule for any and all times and spaces – and this is 100 years old! It has a 20 inch and a 30 inch slide rule in the palm of your hand – both!! Wow – Awesome Sauce and Cosmic!

Now - Set forth on an adventure and enjoy both your thinking and feeling, your thoughts and imagination and have a grand journey.

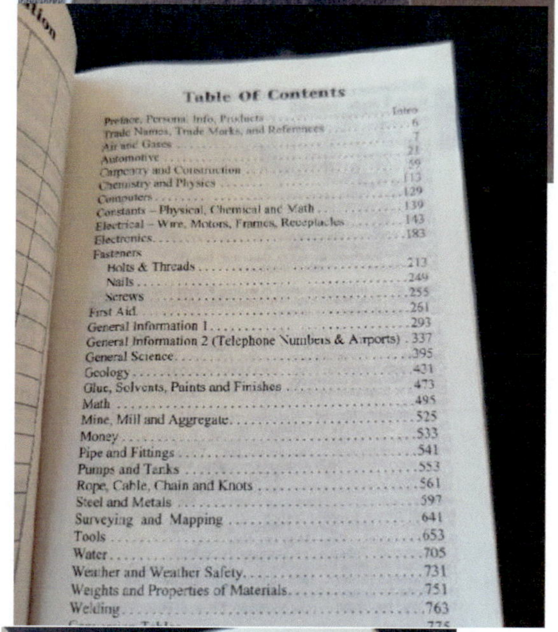

Table Of Contents

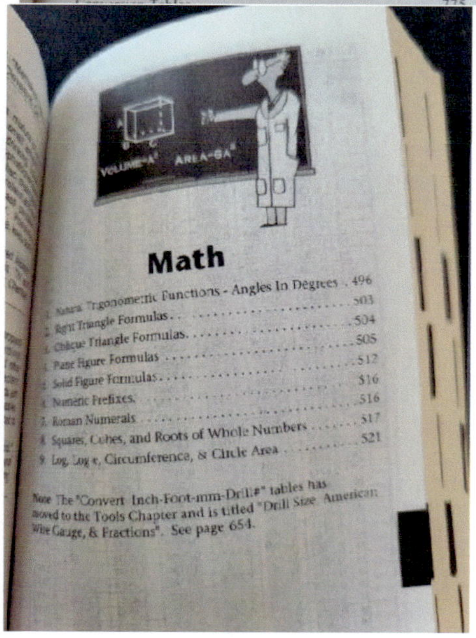

Math

Note: The "Convert Inch-Foot-mm-Drill#" tables has moved to the Tools Chapter and is titled "Drill Size, American Wire Gauge, & Fractions". See page 654.

SLIDE RULE COUPON from
Galaxy Girl's Inquisitive Pioneer Adventures

You receive :
1 set of paper templates for each of the 3 Slide Rules
2 Backing Boards
1 Plastic Sheet
Fastener Set for the Circular Slide Rule

Send $15 by Check or Money Order (preferable)
To
Briana Andromeda Purcell
7430 Plainfield
Dearborn Height 48127

Made in United States
Troutdale, OR
11/11/2026

57125320R00352